Earth Science:
Earth's Surface and History

Earth Science:

Earth's Surface and History

Cover Photo: Tree with Deep Roots © Southern Stock Corp/Corbis

The paper used in these volumes conforms to the American National Standard for Permanence of Paper for Printed Library Materials, Z39.48-1992 (R1997).

Library of Congress Cataloging-in-Publication Data

Earth science. Earth's surface and history / editor, David K. Elliott.
p. cm.
Includes bibliographical references and index.
ISBN 978-1-58765-989-8 (set) – ISBN 978-1-58765-977-5 (set (2 of 4)) – ISBN 978-1-58765-978-2 (v. 1) – ISBN 978-1-58765-979-9 (v. 2) 1. Geology. 2. Earth sciences. I. Elliott, David K., 1947- II. Title: Earth's surface and history.
QE26.3.E25 2012
550–dc23

2012005318

First Printing

PRINTED IN THE UNTED STATES OF AMERICA

CONTENTS

COMMON UNITS OF MEASURE

Notes: Common prefixes for metric units—which may apply in more cases than shown below—include giga- (1 billion times the unit), mega- (one million times), kilo- (1,000 times), hecto- (100 times), deka- (10 times), deci- (0.1 times, or one tenth), centi- (0.01, or one hundredth), milli- (0.001, or one thousandth), and micro- (0.0001, or one millionth).

UNIT	QUANTITY	SYMBOL	EQUIVALENTS
Acre	Area	ac	43,560 square feet 4,840 square yards 0.405 hectare
Ampere	Electric current	A *or* amp	1.00016502722949 international ampere 0.1 biot *or* abampere
Angstrom	Length	Å	0.1 nanometer 0.0000001 millimeter 0.000000004 inch
Astronomical unit	Length	AU	92,955,807 miles 149,597,871 kilometers (mean Earth-sun distance)
Barn	Area	b	10^{-28} meters squared (approx. cross-sectional area of 1 uranium nucleus)
Barrel (dry, for most produce)	Volume/capacity	bbl	7,056 cubic inches; 105 dry quarts; 3.281 bushels, struck measure
Barrel (liquid)	Volume/capacity	bbl	31 to 42 gallons
British thermal unit	Energy	Btu	1055.05585262 joule
Bushel (U.S., heaped)	Volume/capacity	bsh *or* bu	2,747.715 cubic inches 1.278 bushels, struck measure
Bushel (U.S., struck measure)	Volume/capacity	bsh *or* bu	2,150.42 cubic inches 35.238 liters
Candela	Luminous intensity	cd	1.09 hefner candle
Celsius	Temperature	C	1° centigrade
Centigram	Mass/weight	cg	0.15 grain
Centimeter	Length	cm	0.3937 inch
Centimeter, cubic	Volume/capacity	cm^3	0.061 cubic inch
Centimeter, square	Area	cm^2	0.155 square inch
Coulomb	Electric charge	C	1 ampere second

UNIT	QUANTITY	SYMBOL	EQUIVALENTS
Cup	Volume/capacity	C	250 milliliters 8 fluid ounces 0.5 liquid pint
Deciliter	Volume/capacity	dl	0.21 pint
Decimeter	Length	dm	3.937 inches
Decimeter, cubic	Volume/capacity	dm^3	61.024 cubic inches
Decimeter, square	Area	dm^2	15.5 square inches
Dekaliter	Volume/capacity	dal	2.642 gallons 1.135 pecks
Dekameter	Length	dam	32.808 feet
Dram	Mass/weight	dr *or* dr avdp	0.0625 ounce 27.344 grains 1.772 grams
Electron volt	Energy	eV	$1.5185847232839 \times 10^{-22}$ Btu $1.6021917 \times 10^{-19}$ joule
Fermi	Length	fm	1 femtometer 1.0×10^{-15} meter
Foot	Length	ft *or* '	12 inches 0.3048 meter 30.48 centimeters
Foot, cubic	Volume/capacity	ft^3	0.028 cubic meter 0.0370 cubic yard 1,728 cubic inches
Foot, square	Area	ft^2	929.030 square centimeters
Gallon (British Imperial)	Volume/capacity	gal	277.42 cubic inches 1.201 U.S. gallons 4.546 liters 160 British fluid ounces
Gallon (U.S.)	Volume/capacity	gal	231 cubic inches 3.785 liters 0.833 British gallon 128 U.S. fluid ounces
Giga-electron volt	Energy	GeV	$1.6021917 \times 10^{-10}$ joule
Gigahertz	Frequency	GHz	—
Gill	Volume/capacity	gi	7.219 cubic inches 4 fluid ounces 0.118 liter

UNIT	QUANTITY	SYMBOL	EQUIVALENTS
Grain	Mass/weight	gr	0.037 dram 0.002083 ounce 0.0648 gram
Gram	Mass/weight	g	15.432 grains 0.035 avoirdupois ounce
Hectare	Area	ha	2.471 acres
Hectoliter	Volume/capacity	hl	26.418 gallons 2.838 bushels
Hertz	Frequency	Hz	$1.08782775707767 \times 10^{-10}$ cesium atom frequency
Hour	Time	h	60 minutes 3,600 seconds
Inch	Length	in *or* "	2.54 centimeters
Inch, cubic	Volume/capacity	in^3	0.554 fluid ounce 4.433 fluid drams 16.387 cubic centimeters
Inch, square	Area	in^2	6.4516 square centimeters
Joule	Energy	J	$6.2414503832469 \times 10^{18}$ electron volt
Joule per kelvin	Heat capacity	J/K	$7.24311216248908 \times 10^{22}$ Boltzmann constant
Joule per second	Power	J/s	1 watt
Kelvin	Temperature	K	-272.15 Celsius
Kilo-electron volt	Energy	keV	$1.5185847232839 \times 10^{-19}$ joule
Kilogram	Mass/weight	kg	2.205 pounds
Kilogram per cubic meter	Mass/weight density	kg/m^3	$5.78036672001339 \times 10^{-4}$ ounces per cubic inch
Kilohertz	Frequency	kHz	—
Kiloliter	Volume/capacity	kl	—
Kilometer	Length	km	0.621 mile
Kilometer, square	Area	km^2	0.386 square mile 247.105 acres
Light-year (distance traveled by light in one Earth year)	Length/distance	lt-yr	5,878,499,814,275.88 miles 9.46 × 1012 kilometers
Liter	Volume/capacity	L	1.057 liquid quarts 0.908 dry quart 61.024 cubic inches

UNIT	QUANTITY	SYMBOL	EQUIVALENTS
Mega-electron volt	Energy	MeV	—
Megahertz	Frequency	MHz	—
Meter	Length	m	39.37 inches
Meter, cubic	Volume/capacity	m^3	1.308 cubic yards
Meter per second	Velocity	m/s	2.24 miles per hour 3.60 kilometers per hour
Meter per second per second	Acceleration	m/s^2	12,960.00 kilometers per hour per hour 8,052.97 miles per hour per hour
Meter, square	Area	m^2	1.196 square yards 10.764 square feet
Metric. See unit name			
Microgram	Mass/weight	mcg *or* µg	0.000001 gram
Microliter	Volume/capacity	µl	0.00027 fluid ounce
Micrometer	Length	µm	0.001 millimeter 0.00003937 inch
Mile (nautical international)	Length	mi	1.852 kilometers 1.151 statute miles 0.999 U.S. nautical mile
Mile (statute or land)	Length	mi	5,280 feet 1.609 kilometers
Mile, square	Area	mi^2	258.999 hectares
Milligram	Mass/weight	mg	0.015 grain
Milliliter	Volume/capacity	ml	0.271 fluid dram 16.231 minims 0.061 cubic inch
Millimeter	Length	mm	0.03937 inch
Millimeter, square	Area	mm^2	0.002 square inch
Minute	Time	m	60 seconds
Mole	Amount of substance	mol	6.02×10^{23} atoms or molecules of a given substance
Nanometer	Length	nm	1,000,000 fermis 10 angstroms 0.001 micrometer 0.00000003937 inch

Unit	Quantity	Symbol	Equivalents
Newton	Force	N	0.224808943099711 pound force 0.101971621297793 kilogram force 100,000 dynes
Newton-meter	Torque	N·m	0.7375621 foot-pound
Ounce (avoirdupois)	Mass/weight	oz	28.350 grams 437.5 grains 0.911 troy or apothecaries' ounce
Ounce (troy)	Mass/weight	oz	31.103 grams 480 grains 1.097 avoirdupois ounces
Ounce (U.S., fluid or liquid)	Mass/weight	oz	1.805 cubic inch 29.574 milliliters 1.041 British fluid ounces
Parsec	Length	pc	30,856,775,876,793 kilometers 19,173,511,615,163 miles
Peck	Volume/capacity	pk	8.810 liters
Pint (dry)	Volume/capacity	pt	33.600 cubic inches 0.551 liter
Pint (liquid)	Volume/capacity	pt	28.875 cubic inches 0.473 liter
Pound (avoirdupois)	Mass/weight	lb	7,000 grains 1.215 troy or apothecaries' pounds 453.59237 grams
Pound (troy)	Mass/weight	lb	5,760 grains 0.823 avoirdupois pound 373.242 grams
Quart (British)	Volume/capacity	qt	69.354 cubic inches 1.032 U.S. dry quarts 1.201 U.S. liquid quarts
Quart (U.S., dry)	Volume/capacity	qt	67.201 cubic inches 1.101 liters 0.969 British quart
Quart (U.S., liquid)	Volume/capacity	qt	57.75 cubic inches 0.946 liter 0.833 British quart
Rod	Length	rd	5.029 meters 5.50 yards

UNIT	QUANTITY	SYMBOL	EQUIVALENTS
Rod, square	Area	rd^2	25.293 square meters 30.25 square yards 0.00625 acre
Second	Time	s or sec	1/60 minute 1/3,600 hour
Tablespoon	Volume/capacity	T or tb	3 teaspoons 4 fluid drams
Teaspoon	Volume/capacity	t or tsp	0.33 tablespoon 1.33 fluid drams
Ton (gross or long)	Mass/weight	t	2,240 pounds 1.12 net tons 1.016 metric tons
Ton (metric)	Mass/weight	t	1,000 kilograms 2,204.62 pounds 0.984 gross ton 1.102 net tons
Ton (net or short)	Mass/weight	t	2,000 pounds 0.893 gross ton 0.907 metric ton
Volt	Electric potential	V	1 joule per coulomb
Watt	Power	W	1 joule per second 0.001 kilowatt $2.84345136093995 \times 10^{-4}$ ton of refrigeration
Yard	Length	yd	0.9144 meter
Yard, cubic	Volume/capacity	yd^3	0.765 cubic meter
Yard, square	Area	yd^2	0.836 square meter

COMPLETE LIST OF CONTENTS

Volume 1

Volume 2

CATEGORY LIST OF CONTENTS

Earth Science: Earth's Surface and History

Volume 2

Editor

David K. Elliot, Ph.D.

Northern Arizona University

Salem Press

A Division of EBSCO Publishing

Ipswich, Massachusetts   Hackensack, New Jersey

J

JOINTS

Joints form when rocks undergo brittle failure, usually during expansion. Their orientations, physical features, and patterns of occurrence can be used to help infer the physical conditions present at the time of failure. Because joints are important conduits for fluids, particularly oil and water, that move beneath the surface, understanding their formation and occurrence has economic benefits.

PRINCIPAL TERMS

- **chemical weathering:** changes in rocks produced by reactions with fluids near the surface of the earth
- **columnar jointing:** the formation of columns, often with hexagonal cross sections, as joints grow inward from the outer surfaces of cooling igneous rock bodies
- **conjugate shear sets:** two sets of joints that make angles with each other of close to 60 degrees and 120 degrees
- **exfoliation:** the splitting off of curving sheets from the outside of a body of rock; also called sheeting
- **extension:** expansion, or stretching apart, of rocks
- **fractography:** the study of fracture surfaces to determine the propagation history of the crack
- **joint:** a fracture in a rock across which there has been no substantial slip parallel to the fracture

Characteristics of Joints

Joints are the ubiquitous cracks found in nearly every outcrop. They are unquestionably the most common structure at the surface of the earth. They vary in size from microscopic fractures visible only within an individual grain in a rock to fractures kilometers in length, some of which are responsible for the magnificent scenery of Arches National Park in Utah. If one smashes a rock with a hammer, one produces joints. Joints also form as molten rocks solidify and cool, as weathering alters the volume of the outer layers of a rock, and even as erosion removes overlying layers, reducing the weight on the layers below and permitting them to expand and crack.

There is no appreciable slip across a joint. (Failure surfaces accommodating large amounts of slip are faults.) Often, the same mineral grain can be observed on both sides of a joint, neatly cut in two but otherwise not disturbed. This indicates failure in extension; such joints are sometimes called extension joints. The sides of the fracture moved away from each other but did not slide past each other at all. The forces producing such joints were literally pulling the rock apart. An engineer might call these tensile

Joints in granite cut by veins of quartz, feldspar, and biotite, Las Animas Canyon, San Juan County, Colorado. (U.S. Geological Survey)

forces. Geologists work with rocks that are nearly always in a state of compression, however, and true tension is uncommon. Therefore, geologists usually call such forces the forces of least compression. The direction of least compression is the direction in which extension occurs. That is, as the crack opens and the sides move away from each other, they will move in the direction of the least compressive force. Consequently, the plane of an extension joint will be perpendicular to the direction of the least compressive force.

Formation of Joints

Such extension may be produced in a variety of ways. Most obvious are mechanisms involving large-scale deformation of the rock. The folding of a unit of rock causes extension on the outside of the fold. Often, extension joints are found fanning around the "nose" of a fold. The injection of molten rock into cooler rock, and its subsequent cooling, can fracture the cooler rock, producing joints. Extension is also produced during weathering when, because of chemical reactions at the surface of a rock, the surface layer expands more than the interior does. This expanded surface pulls away from the rest of the rock much as an onion skin pulls off the outside of an onion. This process is called exfoliation or sheeting. One classic example of this is Half Dome in Yosemite National Park, California.

Contraction, too, can cause joints. The familiar cracks that form in dried mud puddles are an example: Moist mud contracts as it dries, and cracks in the surface result when the forces involved in this contraction overcome the cohesive strength of the mud. As rock cools, such as when a molten, igneous rock body solidifies and then cools further, it contracts. The cooling and contraction are greatest at the surface of such a body. The polygonal patterns of cracks that form on such surfaces are very similar to those seen in dried mud. As the hot rock continues to cool, these cracks extend into the interior of the body. This may result in spectacular columns, such as those seen at Devils Postpile National Monument in California. The process is called columnar jointing.

The most common way joints form, however, is probably when rocks that equilibrated at depth are brought to the surface, either by mountain-building forces or when erosion removes the overlying rocks. This means of forming joints was proposed by Neville Price in his 1966 book *Fault and Joint Development in Brittle and Semi-Brittle Rock*. Although the model cannot be allied directly at any particular location because the deformation history, local topography, and other factors vary too much from place to place, it is instructive to consider the process in general terms.

How can vertical uplift produce horizontal extension? Consider large suspension bridges such as the Verrazano Narrows in New York City or the Golden Gate Bridge in San Francisco. The vertical towers supporting these bridges diverge from one another by about one-hundredth of a degree because of the curvature of the earth. The tops of these towers are farther apart than their bases, so a rope that exactly reached between the bases would have to stretch a bit if it were raised to their tops. If it were unable to stretch, it would break.

Would uplift of 5 kilometers be sufficient to produce joints in a typical rock? If Earth's circumference is 40,074 kilometers, the circumference of a circle lying 5 kilometers beneath the surface would be 40,043 kilometers. If that circle were brought to the surface, it would have to be stretched by 31 kilometers, or by 0.078 percent. This amount of stretch might not seem like much, as a block of rock 100 meters long would need to extend only 7.8 centimeters. However, if one tried to stretch the block of rock by attaching a gigantic pulling apparatus on it, experimental data show that the block would break before stretching that much.

In addition to this geometric extension, the rock would generally cool as it came up to the surface, contracting and making more joints in the process. Because the state of compression at depth would likely be different from that at the surface, the changes that occur during uplift might encourage or inhibit joint formation, depending on local conditions. Still, if all the joints currently at the surface of the earth are considered, it appears as if the majority of extension joints may form by uplift. The fact that joints form during uplift and erosion does not mean, however, that they are necessarily unrelated to the structural history of the rocks in which they occur. Deformed rocks often contain stored-up energy, much like the energy stored in a spring, which was caused by the deformation. This energy, usually called residual stress, can influence the development of joints. Thus, joints that form hundreds of millions of years after a rock was initially deformed will often occur in patterns and orientations clearly related to that deformation.

Shear Joints

Some joints are not formed strictly by extension. These joints develop as a series of cracks, called shear joints, which break the rock into diamond-shaped pieces. These pieces slide slightly past one another, accommodating the deformation. Careful examination of these joints may show some slight offset of grains across the joint, but the displacement across any one joint is small. The cumulative effect across hundreds of joints, however, can be considerable. Often, these joints occur in two parallel sets, with angles of about 60 degrees and 120 degrees between them. Such sets are called conjugate shear sets.

Shear joints are not perpendicular to the direction of least compression, but it is possible to determine the direction of compression from the orientation of the conjugate shear sets. The direction of intermediate compression is indicated by the line of intersection of the joints. The direction of least compression bisects the obtuse angle between the sets, and the direction of maximum compression bisects the acute angle. In terms of a diamond-shaped piece, the direction of least compression is the short way across the diamond, and the direction of greatest compression is the long way across the diamond. It is not uncommon for conjugate shear sets to occur in conjunction with extension joints. In this case, the extension joints will be parallel to the long axes of the diamonds.

Consider the joints that might be associated with a fold that forms in a horizontal layer of rock not too far beneath the surface. The fold will form as the layer buckles in response to forces acting along it—in the north-south direction, for example—which is similar to the way a playing card flexes when one squeezes the edges between one's fingers. Early in the deformation process, least compression in the east-west direction results in extension joints running north-south and shear joints running northeast and northwest at a 30-degree angle. Eventually a buckle develops, folding the layer and producing extension fractures in the east-west direction. Much later, erosion and uplift may bring parts of this layer to the surface. The direction of the least compression at that time, which may have no relation at all to the forces that originally produced the fold, will control the vertical extension joints that develop because of this uplift. Finally, weathering and the vagaries of the topography at the time the rock is exposed to weathering control the exfoliation joints that will follow the shape of the exposed surface.

Exfoliation in granite near Shuteye Peak in the Sierra Nevada, Madera County, California. (U.S. Geological Survey)

Fractography

Fracture patterns—often called "decorations"—on the joint surfaces can yield useful information about the speed of fracture growth and the direction in which it grew. This field of study is called fractography and has been developed by ceramic engineers concerned with reconstructing the brittle failure of glass and ceramic objects in order to improve their design. It can be directly applied to the study of joint surfaces.

When a fracture begins to grow, it starts with a low velocity but accelerates quickly. While it is moving slowly, the front of the fracture is usually a smooth curve, and the decoration it leaves on the fracture surface may be perfectly smooth, called the mirror region, or slightly frosted in appearance, in which case it is called mist hackle. If the crack grows intermittently, arrest lines may result. These curves show where the crack front was at different times when it temporarily stopped growing. As it increases in speed, the fracture front divides into a number of fingerlike projections. These commonly move a bit beyond the initial plane of the fracture as the fracture continues to grow. The result is a pattern on the surface of the fracture that has long been called plumose structure by geologists but is known as twist hackle by fractographers. It looks very much like a feather. The directions in which the fracture grew are shown by the directions of each slightly offset, curving element. Many of these features can be seen on building stones, flagstones, and slate floor tiles.

Fluids and Joints

Joints provide conduits for the movement of fluids beneath the surface. Just as cracks in a pot permit water to leak through the pot, joints in the bedrock greatly enhance the rate at which water, oil, natural gas, and other fluids move through it. Near the surface, water is the fluid most likely to move through joints. As it does so, it is likely to attack the rock on both sides of the joint, chemically weathering it. This process enlarges the joint, increasing the flow of water through it, which in turn causes it to be weathered further, and the process continues. When the jointed rocks are limestone, the result may be elaborate systems of caverns, such as Carlsbad Caverns in New Mexico. Maps of such caves clearly demonstrate that joints controlled their development. In areas underlain by less soluble rock, joints may provide access to groundwater resources. By studying joint patterns displayed on geologic maps, aerial photographs, or satellite images, hydrologists are sometimes able to see where the natural underground flow of water may be greatest, and they can exploit this knowledge in their search for water.

Similarly, petroleum geologists seek conditions where joints may facilitate the movement of oil and gas toward potential well sites. Because the rocks of interest to them are often much deeper than those with useful water resources, petroleum geologists may be forced to guess the location of joints at depth. Although the surface traces of joints seen on maps can help, it is often necessary to apply an understanding of how and why joints form in order to predict where they may be at depth. In some cases, artificial joints are produced by pumping fluids under very high pressure into the rocks.

Otto H. Muller

Further Reading

Fossen, Haakon. *Structural Geology.* New York: Cambridge University Press, 2010. This text is well written and easy to understand. An excellent text for geology students or as a resource for geologists. Provides many links between structural geology theory and application. Photos and illustrations add great value to the text. Contains a glossary, references, an appendix of photo captions, and an index.

Grotzinger, John, and Tom Jordan. *Understanding Earth.* 6th ed. New York: W. H. Freeman, 2009. This comprehensive physical geology text covers the formation and development of the earth. Readable by high school students, as well as by general readers. Includes an index and a glossary of terms.

Hatcher, Robert D., Jr. *Structural Geology: Principles, Concepts, and Problems.* 2d ed. Englewood Cliffs, N.J.: Prentice-Hall, 1995. This undergraduate textbook covers joints and faults in six chapters. Intended for the more advanced reader.

Mandl, Georg. *Rock Joints: The Mechanical Genesis.* New York: Springer, 2005. A summary of joint types and processes, not to be confused with tectonic faults. Begins with fundamentals of fracturing, covers hydraulic fractures, tension joints, shear joints, faulting and folding, and echelon joints and veins. This book provides many diagrams to supplement the explanations in the text. Contains references and an index. For readers with some scientific background.

Marshak, Stephen, and Gautam Mitra, eds. *Basic Methods of Structural Geology.* Englewood Cliffs, N.J.: Prentice-Hall, 1988. Chapter 12, "Analysis of Fracture Array Geometry," by Arthur Goldstein and Stephen Marshak (18 pages), gives a good description of joint surface morphology, an overview of the field methods that have been used in studying them, and a summary of graphing and statistical techniques. The treatments are brief, but references are complete. The text seems to have been designed for the reader who already knows something about joints but wishes to learn more about how to study them. College level.

Michard, Andre, et al., eds. *Continental Evolution: The Geology of Morocco*. Berlin: Springer, 2010. This text is a summary of a decade's worth of research on the geology of Morocco. It focuses on the structure, stratigraphy, and tectonics of the three plates merging in that area.

Plummer, Charles C., Diane H. Carlson, and Lisa Hammersley. *Physical Geology*. 13th ed. Boston: McGraw-Hill, 2009. This is a straightforward, easy-to-read introduction to geology intended for those with little or no science background. Includes many excellent illustrations.

Price, Neville J. *Fault and Joint Development in Brittle and Semi-Brittle Rock*. London: Pergamon Press, 1966. This 176-page book is a classic in its field. Most of the work done on joints since its publication has been done by people who have read this book, so its influence is great. It suffers, however, from having been written before modern experimental methods for studying rock deformation had been fully developed and from its use of the British system of measurement. Contains only four photographs, of somewhat limited usefulness, and other figures that are schematic, though clear. Not overly technical, with a minimum of mathematical derivations, this book gives a fair summary of what was known about joints at the time it was written. Suitable for college-level students.

Ramsey, Jonathan M., and Frederick M. Chester. "Hybrid Fracture and the Transition from Extension Fracture to Shear Fracture." *Nature* 428 (2004): 63-66. This article addresses the two extremes of brittle fractures. Provides an experimental study showing a lack of evidence for hybrid or transitional types of fractures between these extremes.

Suppe, John, ed. *Principles of Structural Geology*. Englewood Cliffs, N.J.: Prentice-Hall, 1985. Chapter 6, "Joints," presents an excellent overview of the classification, appearance, and formation of joints. The sixteen photographs show all manner of joints at a variety of scales, and some of the fourteen line drawings provide good examples of how joint studies have been used in interpreting structures and the state of stress in the crust. At times, the discussion becomes a bit technical, and several references are made to Chapter 4, "Fracture and Brittle Behavior," which is more technical yet. A useful digression concerns the state of stress in the upper and lower crust and in the mantle. College-level reading.

See also: African Rift Valley System; Continental Rift Zones; Displaced Terranes; Fold Belts; Folds; Geoclines; Karst Topography; Ophiolites; Thrust Belts.

K

KARST TOPOGRAPHY

Karst topography is a landform produced by the dissolving action of surface water and groundwater on the underlying bedrock of a region. The landforms produced are unique because they represent internal or underground drainage, forming such features as sinking streams, sinkholes, caves, natural bridges, and springs.

PRINCIPAL TERMS

- **carbonic acid:** a weak acid formed by mixing water and carbon dioxide; it is important in the dissolving of the most common karst rock, limestone
- **cave:** an opening or hole in the ground enterable by people; in karst topography, caves have been dissolved out and act (or once acted) as underground conduit flow routes for water
- **limestone:** a common sedimentary rock containing the mineral calcite; the calcite originated from fossil shells of marine plants and animals
- **natural bridge:** a bridge over an abandoned or active watercourse; in karst topography, it may be a short cave or a remnant of an old, long cave
- **sinkhole:** a hole or depression in the landscape produced by dissolving bedrock; sinkholes can range in size from a few meters across and deep to kilometers wide and hundreds of meters deep
- **sinking stream:** a stream or river that loses part or all of its water to pathways dissolved underground in the bedrock
- **spring:** a place where groundwater reappears on the earth's surface; in karst topography, a spring represents the discharge point of a cave

DISSOLVING BEDROCK

Karst topography is a unique landscape produced by the dissolving of the bedrock of a region, with the consequent development of underground drainage. Most common bedrock materials, such as granite, sandstone, and shale, are resistant to the process of dissolving (also known as dissolution), and landscapes are carved into these bedrock materials by the mechanical action of water, wind, and ice. Some bedrock materials, such as limestone, dolomite, gypsum, and rock salt, dissolve relatively easily. Gypsum and rock salt are soluble in plain water and in most landscapes are chemically destroyed very quickly. Karst landscapes on these two rock types persist only in dryer climates, such as in the American Southwest. Gypsum and rock salt are also mechanically weak rocks and are destroyed rapidly by mechanical erosive activities. Limestone, and its close cousin dolomite, are mechanically strong rocks; therefore, they resist normal erosive activity. Yet calcite, the mineral of which limestone is composed, is especially vulnerable to dissolving by water that is slightly acidic.

Under normal conditions, the source of such acidity is carbonic acid, a natural combination of carbon dioxide and water. Rainwater absorbs carbon dioxide from the atmosphere and is naturally slightly acidic (modern pollution has accentuated this natural tendency to produce acid rain). In the soil, organic activity can increase the amount of carbon dioxide to levels well above those found in the atmosphere, and water moving through such soils can become very acidic. When rainwater or soil water charged with carbonic acid meets limestone bedrock, it will slowly dissolve the limestone. Limestone, mechanically resistant, supports the development of karst topography well. Dolomite reacts more slowly to acid waters, and generally the karst features found on dolomites are subdued and take longer to form than those on limestones. Discussion of karst topography is, therefore, in most situations a discussion of the dissolving of limestone.

The rate at which limestone dissolves depends on the amount of water in the environment and the amount of carbon dioxide available. Atmospheric levels of carbon dioxide are similar worldwide, but the amount of carbon dioxide in the soil varies greatly. Organic activity controls the amount of carbon dioxide available, and for that reason, the most acid groundwater tends to be found in the warm, wet zones of the tropics, where organic activity is high. Dry, cold climates have erosion rates as little as a few millimeters per one thousand years, whereas warm, wet climates can have rates up to 150 millimeters per one thousand years. These erosion rates are averages of the rate of surface lowering for a region. Over the course of hundreds of thousands of years, significant landforms can be developed.

The dissolving action of carbon dioxide-charged water produces a number of etching patterns on exposed limestone bedrock, from microscopic features to large trough structures more than 1 meter deep and many meters long. Karst topography develops when the water penetrates down pores, cracks, and openings into the limestone. Once inside the rock, the water is capable, over long periods of time, of dissolving voids and passageways in the rock. These openings in the rock become integrated into underground flow networks similar to stream patterns on the earth's surface. The result is a series of underground passageways called caves, which collect water from sinkholes and sinking streams on the earth's surface and transmit the water back to the earth's surface at springs. While the flow pattern of caves is often similar to the pattern of surface streams, caves are tubes in bedrock and may migrate up and down, as well as from side to side, in a manner in which surface streams cannot. As the cave system grows and matures, it can capture larger volumes of surface water and enlarge. Eventually, it will not only carry material in solution but also mechanically transport sediment underground. On the land surface, this underground or internal drainage produces sinking streams and sinkholes. In mature systems, the sinkholes may be kilometers across and hundreds of meters deep, containing many sinking streams.

Depending on the exact nature of the limestone, the climate, and the presence of mountains, a wide variety of karst landscapes may appear. The landscape may be as simple as rolling hills or a flat plain with mostly normal surface drainage and only a few scattered sinkholes, sinking streams, and caves. Other landscapes, however, may have no significant surface drainage and have sinkholes covering the land surface, producing a sinkhole plain. One of the best examples of a sinkhole plain is in the Mammoth Cave area of Kentucky. Under extreme conditions, usually associated with the tropics, the landscape is so altered by dissolution that the sinkholes deepen faster than they widen, producing a landscape of tall limestone towers and pinnacles standing above a flat plain. This landform is called tower karst and is best known from southern China, where thousands of towers several hundred meters high dot the landscape.

The downward erosion by groundwater can produce cave systems at lower elevations, causing earlier, higher cave systems to become abandoned. These abandoned caves may persist for hundreds of thousands of years, developing complex mineral displays of stalactites, stalagmites, and other deposits. Animals and people may use the cave for shelter, leaving behind important fossil and cultural material. Continuing erosion of the land surface will eventually breach into underlying cave systems

Under extreme conditions, the landscape is so altered by dissolution that the sinkholes deepen faster than they widen, producing a landscape of tall limestone towers and pinnacles standing above a flat plain. China's Shulin (Stone Forest) provides many examples of these so-called tower karst formations. (AP/Wide World Photos)

to produce new entrances and truncated cave fragments called natural bridges. Sinkholes become natural traps, collecting unwary animals as well as plant remains, soil, and pollen. The transport of material into the subsurface allows the material to be preserved from destruction in the earth's surface environment.

Developmental Stages

Like all landscapes, karst topography can go through a series of developmental stages. Karst development begins when the soluble rock is first exposed at the earth's surface. In some cases, groundwater reaches the soluble rock before it is ever exposed, and cave development can begin before the rock is present on the surface. In either case, the progressive development of a karst landscape by chemical and mechanical erosion will remove the layer of soluble rock responsible for the karst processes. In time, karst processes become less important in the landscape as the soluble bedrock disappears, and the landscape will begin to revert to a more typical, mechanically produced form.

A sinkhole 350 feet wide and 150 feet deep in central Alabama in December 1972. (U.S. Geological Survey)

In the United States, limestones form important areas of karst topography in the Virginias, Pennsylvania, Indiana, Kentucky, Tennessee, Alabama, Florida, Missouri, Arkansas, and Texas. Many other states have minor amounts of limestone karst, such as New York, Minnesota, Iowa, Colorado, and New Mexico. Some states have major cave systems located in minor areas of karst, such as Carlsbad Caverns in New Mexico and the Jewel and Wind Caves in South Dakota. Gypsum karst is locally important in Kansas, Oklahoma, and New Mexico, while dolomite karst is found in Missouri. Marble—limestone altered by heat and pressure—forms small karst areas in California, New England, and Oregon. Rock salt karst is very rare in the United States but can be found in Spain and the Middle East. Almost every state and every country has some form of karst development.

The unique features of karst topography can be mimicked by other landscapes in special cases. Such cases are called pseudokarst because they imitate karst topography but are produced by other phenomena. For example, volcanic areas produce eruptive craters and lava tubes that look like sinkholes and caves from karst areas. In deserts, streams can dry up and appear to be sinking, and winds can produce excavated hollows in the sand. Glaciers can produce depressions in the ground that look like sinkholes. In fine-grained clay deposits—which can be seen in the Badlands of South Dakota—caves, bridges, and sinkholes are produced by mechanical flushing of the clay particles, a process called suffusion. Wave activity on rocky coastlines can carve sea caves in a variety of rock materials.

Karst topography is unique because of the chemical manner in which the bedrock is destroyed and because of the internal or underground drainage that results. Most karst is developed in limestone because of this rock's abundance, slow chemical dissolution rates, and great mechanical strength. In certain areas of the world, karst topography is the dominant landscape for thousands of square kilometers. Cave systems in excess of 500 kilometers of surveyed passage exist, and caves have been followed as deep as 1,700 meters beneath the surface.

John E. Mylroie

Further Reading

Beck, B. F., and W. L. Wilson, eds. *Karst Hydrogeology: Engineering and Environmental Applications.* Boston: A. A. Balkema, 1987. This book is the proceedings volume of the Second Multidisciplinary Conference on Sinkholes and the Environmental Impacts of Karst and contains sixty-three papers on a variety of topics concerning human activity and karst, with an emphasis on the high rate of mistakes made as a result of ignorance of karst processes. Accessible to the nonspecialist.

Bloom, Arthur G. *Geomorphology: A Systematic Analysis of Late Cenozoic Landforms.* 3rd ed. Long Gove, Ill.: Waveland Press, 2004. This college-level text covers the basics of geomorphology. Includes a chapter on the development of karst topography. Index and bibliography.

Chernicoff, Stanley, and Donna Whitney. *Geology: An Introduction to Physical Geology.* 4th ed. Upper Saddle River, N.J.: Prentice Hall, 2006. This is a good overview of the scientific understanding of the geology of the earth and surface processes. Includes sections on the development of karst topography. Includes a link to a Web site that provides regular updates on geologic events around the globe.

Clary, Jordan. "Exploring Karst in Guilin, China." *Geotimes* 53 (2008): 34-37. This article discusses the karst topography found in Guilin, China. The cavern and sinkhole terrain is a top tourist attraction in the area. The author describes how these features are formed from groundwater or rainwater dissolving the limestone.

Gillieson, David. *Caves: Processes, Development, and Management.* Oxford, England: Blackwell, 1998. A description of how caves form, what can be learned from them, and how they can best be managed. The first one-third of the book deals with the development of karst and its relation to cave formation. Well illustrated and written in an engaging style that makes the material comprehensible to the general reader. Glossary, bibliography, index.

Jennings, J. N. *Karst Geomorphology.* 2d ed. New York: Basil Blackwell, 1987. A college-level text that emphasizes karst processes overall, as opposed to only their role in cave formation. The text requires some knowledge of the basic sciences. It is interesting to read, as it uses many Australian examples and contains the lifetime knowledge of one of the world's leaders in karst research.

Middleton, John, and Anthony C. Waltham. *The Underground Atlas: A Gazetteer of the World's Cave Regions.* 2d ed. Darby, Pa.: Diane Publishing, 1999. An easy-to-read guide to the famous long and deep caves of the world, this atlas covers the continents in a general review and then describes the cave resources of the world's countries, from Afghanistan to Zimbabwe. Simple maps are presented to portray the largest and deepest caves. The appendices supply a listing of record-holding caves in terms of length, depth, and volume, as well as data about caving organizations and sources of additional reading.

Sasowsky, I. D. *Cumulative Index for the National Speleological Society Bulletin, Volumes 1 Through 45, and Occasional Papers of the N.S.S., Numbers 1 Through 4.* Huntsville, Ala.: National Speleological Society, 1986. *The Bulletin of the National Speleological Society* is carried by many libraries and contains papers on the science of speleology and karst processes. This index to that journal allows quick access to published papers on specific topics, by topic, author, or geographical area.

Taminskas, J, et al. "Karst and Ecosystems." In *Geology and Ecosystems,* edited by Igor S. Zektser. New York: Springer, 2010. This chapter focuses specifically on karst topography and its role in the ecosystem. Discusses hydrology and processes that result in karst topography. Suited for scientists and other professionals with an interest in environmental protection.

Van Beynen, Philip E. *Karst Management.* New York: Springer, 2011. This text discusses the management of karst environments. Topics include public education, policy and regulation, and cave and groundwater management. Written for university students, professional geologists, and land management committees.

White, William B. *Geomorphology and Hydrology of Karst Terrains.* New York: Oxford University Press, 1988. An advanced college textbook written by one of the leaders in karst research in North America, this volume contains a well-written and comprehensive discussion of karst topography.

See also: Deserts and Dunes; Fluvial Frocesses; Fossilization and Taphonomy; Geomorphology and Human Activity; Geomorphology of Dry Climate Areas; Geomorphology of Wet Climate Areas; Glacial Landforms; Grand Canyon; Groundwater; Marine Terraces; Permafrost; River Valleys.

KRAKATAU

Residents of Sumatra and Java thought that the quiet volcano Krakatau was extinct until 1883, when it violently erupted, destroying three hundred villages and generating sea waves that killed more than 36,000 people. Perhaps the loudest noise in recorded history, the paroxysmal eruption was heard more than 4,600 kilometers away, and dust from the explosion reddened skies around the globe.

PRINCIPAL TERMS

- **basalt:** a type of volcanic rock that cooled from magma relatively low in silica, which enabled it to flow freely
- **caldera:** a large circular or oval depression formed by the explosion and collapse of a volcano
- **ejecta:** material, such as rock fragments and ash, thrown out by a volcano
- **ignimbrite:** volcanic rock, high in the element silica, formed by the cooling of layers of volcanic ash mixed with rock fragments
- **plate tectonics:** the theory of geology that relates volcanism, mountain building, earthquakes, and other phenomena to the movements of "floating" plates of the earth's crust
- **pumice:** a type of volcanic rock that is actually glass filled with air bubbles
- **pyroclastic:** formed by the accumulation of rock fragments and ash
- **tsunami:** a large, destructive wave caused by earthquakes or other short-duration disturbances of the sea floor; tsunamis are characterized by great speed and increased height when piled up in shallow water

Early Observations

Before 1883, the island Krakatau was noteworthy only as a landmark to the hundreds of ships sailing from the Indian Ocean through the high-traffic sea lane of Sunda Strait to the ports of Singapore and Jakarta in the China Sea. The 9-kilometer-long island, formed by the volcanoes Perboewatan, Danan, and Rakata, was the largest of four that marked the rim of a prehistoric volcanic crater. Java, to the north of Krakatau, contains forty-nine volcanoes, more than half of which have erupted in the past three hundred years or so. Sumatra and the smaller islands of Bali, Lombok, Timor, and Flores also are packed with volcanoes and accompanying geysers and hot springs. There is little argument that Indonesia has the highest concentration of volcanic activity in the world.

Yet the possibility that Krakatau was capable of eruption was ignored by the Dutch and other European colonists of the late nineteenth century. Unlike the neighboring island of Sebesi, Krakatau, a crater rim segment, was not an easily recognized cone-shaped volcano. Forgotten was the historical account of an eruption from two hundred years earlier: a roar like thunder, earthquakes, and a burned forest. Ignored were the legends of fifteen centuries earlier, the Javanese chronicles, that describe an enormous eruption and accompanying fires, shaking, and fearful roars from the mountain called Kapi. This same account explains the division of a single pre-eruption island into Sumatra to the north and Java to the south by the rising ocean that now ran between them as the Sunda Strait. The ancient chronicles are supported by contemporary geologic evidence. Volcanic deposits from the nearby Java Trench and the huge crater formed by Krakatau and neighboring islands imply the existence of an earlier gigantic volcano about one-half the size of Rhode Island.

Accounts of volcanic activity on Krakatau before the 1880's are sparse, but eyewitness accounts of the eruptions of August 26 and 27, 1883, and studies of the aftermath are extensive. Passing ships and local residents detail the deafening sounds, the darkness, falling pumice, choking ash, bolts of lightning, the eruption itself, and the subsequent destruction by 12- to 40-meter-tall tsunami waves. The tremendous noise, startling red sunsets, and other unusual phenomena were heard and seen around the world.

Krakatau's Reawakening

Krakatau began its return to life on May 20, 1883, when it ejected pumice, steam, and ash. During the next three months, residents became accustomed to the continual artillery-like booms that rattled doors and windows, the shaking from earthquakes, and the views of molten lava and columns of dust. Observers from an exploration party to the island witnessed new vents on the crater floor of Perboewatan, the smallest of the three volcanoes.

At 1:00 p.m. on Sunday, August 26, the first of a series of violent explosions shook the volcano. Within one hour, a 27-kilometer-high black cloud hid Krakatau from view. For the next twenty-three hours, eruptions burst forth every ten minutes and grew more ferocious through the night. Passing ships reported hellish conditions. From 65 kilometers away, the schooner *Sir Robert Sale* witnessed a great vapor cloud, lit by bursts of forked lightning, "like a large serpent rushing through the air." On the nightmare journey past Krakatau, Captain Watson of the *Charles Bal* kept a log that details falling hot pumice stones and the buildup of muddy ash that kept his crew busy shoveling for their lives. If a sailor paused in his duty, he became more aware of the hot, choking sulfurous fumes and the almost supernatural pink glow, called St. Elmo's fire, on the rigging above. On the horizon, sailors observed flashes of lightning jumping back and forth from Krakatau to the sky.

The volcano's eruption was not yet over. At 5:30 a.m. on Monday morning, Krakatau radically changed styles of eruption in a grand finale of four explosions. The third explosion, at 10:02 a.m., brought down the island. Where there had been several peaks, the highest at 4,267 meters, a gaping crater, 300 meters below sea level, appeared. That was not known, however, until twenty-four hours later, when the pitch-black darkness, produced by about 9 cubic kilometers of ash and rocks ejected an estimated 80 kilometers into the atmosphere, lifted.

Two-thirds of the island's disappearance was not, however, the main subject of concern. The explosions, equivalent to 1.5 million tons of dynamite, generated sea waves (tsunamis) that drowned more than 36,000 residents and air waves that produced what is considered the loudest sound in human history. The largest sea wave, cresting at 40 meters, or the height of a twelve-story building, swept away the coastal town of Merka in Sumatra one hour after the explosion. This first giant wave—and nine others—drowned about three hundred towns and villages on Java and Sumatra. Ships close to shore, such as the gunboat *Berouw*, were lifted out of the water and carried 3 kilometers inland to lay stranded 9 meters above sea level. Blocks of coral weighing 600 tons were carried 11 kilometers inland. The waves circled the globe twice, rocking ships in South African harbors 8,000 kilometers away as they passed.

The Anak Krakatau (Child of Krakatau) volcano sends up powerful clouds of hot gasses and rocks Wednesday November 7, 2007 in the Sunda Straits between Java and Sumatra, Indonesia. Sending a boom echoing across the bay, the volcano known as the "Krakatau's Child" unleashed another eruption. While impressive, this eruption was nothing compared to what took place in 1883 at this spot, when Anak Krakatau's predecessor blew apart in one of the most devastating eruptions in recorded history. (AP Photo)

THE LOUDEST NOISE EVER HEARD

Before and immediately after the colossal 10:02 a.m. explosion, continual thunder deafened Java and Sumatra residents. During the later eruptions, the nearby islands, thickly blanketed in ash, heard nothing, even though windows and walls were blown out. Yet on Rodriguez Island 4,800 kilometers to the west, in Australia 3,200 kilometers to the south, and in places in between, such as Celebes, Ceylon, and the Philippines, loud explosions were heard. People on more than one-thirteenth of the earth's surface, about 40 million square kilometers, heard the noise. At no other time in recorded history has a sound been heard at such a distance without technological amplification. If Krakatau had exploded in Denver, Colorado, residents from Alaska to Florida would have heard the noise. Pressure differences on barographs at a number of widespread locations recorded seven distinct airwaves bouncing back and forth up to five days later.

During nearly twenty-four hours of the eruptions on Sunday and Monday, ash completely darkened the area within 200 kilometers of Krakatau. Ships as distant as 6,000 kilometers to the northwest swept ash from their decks. Clouds of ash, an estimated 8.5 cubic kilometers in volume, spewed 50 kilometers into the stratosphere, where they were transported globally in the jet stream. This volcanic veil of dust caused spectacular atmospheric effects around the earth lasting for three years. Exceptionally colored and prolonged sunsets visible to three-quarters of the world population became common; the sun, and sometimes the moon or Venus, often appeared blue or green; and a white or bluish white corona, called Bishop's Ring, circled the sun. These finely ground ash particles created more than the gaudy optical displays. The ash became a filter to solar radiation and reduced heat reaching the earth by 1 percent. Global temperatures, averaging 0.05 degree Celsius lower in 1884, decreased the British growing season by one week. For the next four years, reduced solar radiation and resulting lower-than-average global temperature were recorded.

Aftermath

News of Krakatau's disastrous eruption spread quickly. Telegraph-cable carried the news to Singapore and then to the world. Only after the dark cloud of ash over Sumatra and Java had dispersed was the magnitude of the death toll known. Ships returned to Great Britain and the United States with blocks of pumice and ash samples; tidal gauges and barographs recorded the passage of sea and airwaves. Almost around the globe, every "man on the street" witnessed garish sunsets and unusual optical effects. Within three weeks after Krakatau's final blast, a group of scientists commenced detailed studies on the eruption, subsequent tsunamis, and global atmospheric effects.

Loudest, not Biggest

Unequivocally, Krakatau made the loudest noise of any volcano in history. Certainly it was one of the most violent. An eruption of Krakatau's energy and duration occurs about once or twice every one hundred years; yet in the context of global volcanology, other volcanoes have superseded Krakatau in the amount of ash and magma ejected and the size of caldera created. The 1815 eruption of Tambora, also in Indonesia, released about fifty times more energy and produced five times as much pyroclastics (mixed pulverized rock or ash and lava). In Indonesia, blasted by 17 percent of the world's volcanoes, Krakatau's eruption was not unusual.

Elementary plate tectonics, the theory that the earth's outer crust is composed of moving plates or blocks, provides an explanation for why Indonesia is the most volcanically active region in the world. Indonesia rides on the Indian Ocean plate, which plunges northeastward beneath the buoyant continental Eurasian plate at geologically rapid speeds of up to 6 centimeters per year. A 3,000-kilometer stretch of volcanoes in Sumatra, Java, the Lesser Sunda Islands, and northward toward the Philippines marks the collision zone, the Java Trench, between these two plates. Volcanoes that lie along this type of contact, called a subduction boundary, are more explosive and produce lavas distinct from volcanoes exploding within a plate or under water where two plates pull apart.

All Bottled Up

Krakatau resembles Mount Vesuvius in the Mediterranean and Mount St. Helens in the western United States in the properties of the lava emitted, the explosiveness and collapse into a empty magma chamber, and the enormous amount of ash and other ejecta shot into the atmosphere. Krakatau blew out pyroclastics that welded into a type of rock called ignimbrite, rich in silica. An estimated 18 to 21 cubic kilometers of ignimbrite and 8.5 cubic kilometers of ash poured out during the eruptions. Unlike the more easily flowing basaltic lavas of the Hawaiian volcanoes, high-silica lava is thick enough to plug up the vent behind itself and build up pressure in volcanic gases unable to escape.

At the time of the eruption, mining engineer R. D. M. Verbeek, sent to research the volcano, conjectured that the exceptional explosiveness of Krakatau was caused by seawater seeping into a collapsed magma chamber, where it was converted to steam. Later twentieth century studies argue that significant amounts of seawater did not enter the vent, which had not collapsed during the early stage of the eruptions. Scientists suggest that a partially solidified plug developed to block the vent. Pressure built up and eventually exploded into the Monday series of four eruptions, each producing a pyroclastic flow into the surrounding sea.

One significant remnant of such an explosion is the giant caldera formed by the collapse of the volcano into the drained magma chamber within. Krakatau collapsed under water into the Sunda Strait, where a 6-kilometer-diameter caldera edged by three small islands marked its resting place. The flooded caldera was not immediately identified. At first, scientists assumed that the island blew itself apart during the series of explosions. If this were so, then newly formed volcanic rocks should be of the same composition as the preeruptive Krakatau. Yet analysis of the ash and pumice rock indicated that 95 percent of this material formed from cooled lava rather than from pulverized rock. The missing island was, therefore, underwater in the caldera, not deposited as new strata or carried around the globe as ash.

Cory Samia

FURTHER READING

de Boer, Jella Zeilinga, and Donald Theodore Sanders. *Volcanoes in Human History: The Far-Reaching Effects of Major Eruptions.* Princeton, N.J.: Princeton University Press, 2002. Provides an overview of volcanism followed by specific volcanoes and eruptions. Includes the interaction of human history and volcanic activity. Covers the Hawaiian Islands, Crete, Vesuvius, Iceland, Tambora, Krakatau, Pelee, Tristan da Cunha, and Mount St. Helens.

Decker, Robert, and Barbara Decker. *Volcanoes.* 4th ed. New York: W. H. Freeman, 2005. This readable and well-illustrated text is geared toward the general public. The authors explain geologic phenomena in a clear and simple manner. Includes a good section on the explosion of Krakatau. The appendices list the world's 101 "most notorious" volcanoes, as well as Internet sources for volcano information.

Francis, Peter, and Clive Oppenheimer. *Volcanoes.* 2d ed. New York: Oxford University Press, 2004. This text covers volcanic activity ranging from lava flows to pyroclastic currents. Provides many examples from Hawaii, Italy, Ethiopia, Japan, Mount St. Helens, and Krakatau. Written for the layperson, but some knowledge of geology is helpful.

Guo, Jerry. "Volcanology: Fire and Life." *Nature* 454 (2008): 930-932. The author discusses aspects of volcanology in general terms. Examines the Krakatau volcano from geological and ecological perspectives. Mentions volcanic activity as a cause of succession processes.

Prothero, Donald R. *Catastrophes!: Earthquakes, Tsunamis, Tornadoes, and Other Earth-Shattering Disasters.* Baltimore: Johns Hopkins University Press, 2011. This text provides a detailed and clear explanation of the many natural and anthropogenic disasters facing our planet. Each chapter is devoted to a different catastrophe, including earthquakes, volcanoes, hurricanes, ice ages, and current climate changes.

Ritchie, David. *The Ring of Fire.* New York: Atheneum, 1981. The author relays many eyewitness accounts of destructive forces, such as volcanoes and earthquakes, that rim the Pacific Ocean. The section on Krakatau summarizes for the lay reader scientific studies of the eruption and aftermath.

Sigurdsson, Haraldur, ed. *Encyclopedia of Volcanoes.* San Diego, Calif.: Academic Press, 2000. This book contains a complete summary of the scientific knowledge of volcanoes. Includes eighty-two well-illustrated overview articles, each of which is accompanied by a glossary of key terms. Although this is a college-level text, it is written in a clear and comprehensive style that makes it generally accessible. Cross-references and index.

Simkin, Tom, Richard S. Fiske, and Sarah Melcher, eds. *Krakatau 1883: The Volcanic Eruption and Its Effects.* Washington, D.C.: Smithsonian Institution Press, 1983. The definitive work, between 1883 and 1983, on the volcano. The Smithsonian Institution and the U.S. Geological Survey collected and reinterpreted information on Krakatau for the one hundredth anniversary of the eruption. Most articles are condensed versions of the original studies.

Time-Life Books. *Volcano.* Alexandria, Va.: Author, 1982. Like other volumes in the *Planet Earth* series, geared to the general earth science reader, this text is extremely readable and enhanced by excellent graphics. The eruption of Krakatau is viewed within the framework of contemporary plate tectonics theory.

Waters, Tom. "Return to Krakatau." *Discover* 9 (October 1988): 64-67. A short article that discusses the contemporary work of ecologists studying the recolonization of Krakatau by plants and animals.

See also: Calderas; Eruptions; Forecasting Eruptions; Hot Spots, Island Chains, and Intraplate Volcanism; Island Arcs; Lava Flow; Mount Pelée; Mount Pinatubo; Mount St. Helens; Mount Vesuvius; Ocean Ridge System; Popocatépetl; Recent Eruptions; Shield Volcanoes; Stratovolcanoes; Volcanic Hazards.

L

LAND BRIDGES

The theory that the earth's continents were once connected by land bridges that accounted for the migration of flora and fauna was considered viable by many leading geologists during the last half of the nineteenth century and for the first three decades of the twentieth century.

PRINCIPAL TERMS

- **Cenozoic era:** the geologic era dating from the present to about 65 million years ago
- **continental drift:** the theory that landmasses separated and drifted apart in prehistoric times
- **Cretaceous era:** a geologic period ending some 65 million years ago, during which seas covered much of North America and the Rocky Mountains were formed
- **Eocene epoch:** part of the Cenozoic era, dating to about 37 million years ago
- **glaciation:** the process of being covered by an ice sheet or glacier
- **isthmian links:** chains of islands between substantial landmasses
- **land bridges:** narrow land formations that connect landmasses
- **Mesozoic era:** the geologic era spanning the period from about 245 million years ago to about 65 million years ago

Changing Earth

For geologists, the earth has long been a huge, complicated jigsaw puzzle waiting to be reconstructed. The difference between this jigsaw puzzle and the ones with which most people are familiar is that this one is constantly changing. The earth changes in small ways—and sometimes in larger ways, such as when a volcano erupts—on a daily basis. It changes more drastically over longer periods of time. It is now thought to be changing as global warming slowly melts the polar ice caps, with a resulting increase in the volume of ocean waters and a corresponding decrease in the shorelines adjacent to the rising oceans.

It has long been acknowledged that drastic climatic changes have occurred on the earth over millions of years. The Arctic and Antarctic, now solidly frozen, once had tropical and subtropical vegetation, as is evident from impressions of plants found in mineral deposits in these regions and from other paleontological evidence. Now-moderate regions were, during the Ice Age, much colder. Glaciation pushed debris unrelentingly from one region to another as the glaciers moved toward the equator.

Long baffled by the existence of similar flora and fauna in areas seemingly unrelated to each other climatically—Africa, Australia, and Antarctica, for example—geologists arrived at various conclusions regarding the phenomenon. It must be remembered that geologists talk in terms of hundreds of millions of years more often than they do in terms of a century or a millennium. Geological change often occurs so slowly as to be virtually unnoticeable within such geologically abbreviated time spans as thousands of years.

Land-Bridge Theories

Among the theories promulgated to explain the presence of similar flora and fauna in widely disparate areas was that of land bridges. Some land bridges obviously existed, such as the one across the Bering Strait that linked Siberia to what is now Alaska. It is apparent that two continents, Asia and North America, were at one time linked by this broad bridge. Similarly, some geological conjecture also has it that land bridges existed between present-day Ireland and Scotland and between present-day Gibraltar and North Africa.

A Viennese geologist, Edward Suess, proposed an extensive land-bridge theory in his influential book *The Face of the Earth* (1885). Suess coined the term "Gondwanaland," named after Gondwana in east-central India, an area with rocks that revealed unique fossil plants and showed indisputable signs of glaciation dating from the late Paleozoic to the early Mesozoic eras.

As Suess pieced together the discrete elements of his geological puzzle, he found that separate continents had the sorts of fossil plants and glacial rocks that he found in Gondwana. He attributed these occurrences in widely separated areas to land bridges rather than to continental drift. Suess and the scores of notable geologists who accepted his thesis, contended that the land bridges connecting the continents sank beneath the ocean at some time in prehistory.

Other geologists were trying to solve the same riddles with which Suess was confronted. In 1908, American geologist F. B. Taylor published a privately printed pamphlet in which he suggested that present mountain ranges occurred when enormous landslides advanced slowly and steadily from the polar regions toward the equator. He conceived of the world's great continents as originally consisting of huge sheets of rock that were torn apart by glaciation as gigantic fields of ice moved toward zero latitude.

Taylor advanced two fundamental ideas, both of which have been opposed by other geologists to the point that these two hypotheses have been discredited. First, he speculated that mountain ranges were products of thousands of miles of lateral movement; second, he believed that the moon became associated gravitationally with Earth during the Cretaceous period, about 100 million years ago. In this period, there was a considerable increase in the variety of flora and fauna found on the earth. Taylor suggested that in the earliest times, the moon created phenomenal tidal forces that slowed the rate at which Earth rotated and pulled the continents away from the poles. Both of these theories were subsequently disproved.

H. H. Baker, an American geologist, considered the supercontinent of the earth's earliest years to have been split along the Atlantic and Arctic Oceans as the Miocene period neared its end some 7 million to 8 million years ago. He asserted that great tidal action had torn a huge piece from Earth. The resulting void, according to Baker, became the Pacific Ocean. The portion that was catapulted into space, he speculated, became the moon.

In his landmark 1915 book *Die Entstehung der Kontinente und Ozeane* (*The Origin of Continents and Oceans*, 1924), Alfred L. Wegenerproposed that the earth had at one time consisted of a single land mass that he called "Pangaea," the Greek term for "all Earth." He concluded that during the Mesozoic era, the single landmass was torn apart, with the southern continents (Africa, Australia, South America, Antarctica) being pulled away first. Looking at the globe as though it were a jigsaw puzzle, one can imagine that the east coast of South America would fit easily into the declivities in the west coast of Africa.

Eduard Suess (1831-1914), Austrian geologist whose theories about the changing face of the earth foreshadowed the development of plate tectonics. He mainly studied mountains, volcanic island, and deep-sea trenches. Suess proposed that there had once been a single supercontinent that had split up to form the present continents. He named this supercontinent Gondwanaland after the Gonds, the supposed human inhabitants. His most important book was Das Antlitz der Erde (The Face of the Earth), *published during 1885-1909.* (SPL/Photo Researchers, Inc.)

Evidence for Land Bridges

As this investigation into the earth's origins was ongoing, the fossil record uncovered during the 1870's and 1880's led to the inevitable conclusion that land links, either land bridges or isthmian links, had existed between Africa, India, Australia, South America, and Antarctica. These areas of vastly differing climates in modern times (geologically speaking) appeared to share a fossil plant record that suggested a close affinity among them.

F. H. Knowlton, speaking in 1918 about Mesozoic floral relations, supported the notion that land bridges from Antarctica once linked most of the southern landmasses. Writing in 1947, however, after the land-bridge theory had been largely discredited, Theodor Just suggested that "the needs of animal and plant geographers . . . vary sufficiently with their respective interests and so do the land bridges assumed by them." Just discounted land links as deciding factors in the migration of flora.

During the Eocene epoch, the southeastern United States experienced an influx of tropical flora, suggesting to some that a land bridge or isthmian link had once existed between South America and what is now the United States, possibly between South America and southern Florida. Modern researchers suggest that this tropical foliage actually came to the southeastern United States from an intermediate region such as Central America or the Antilles.

During the Oligocene epoch, a considerable exchange of plant life appears to have taken place between present-day Panama and the Antilles. Petrified wood found on Antigua reflects several forms of flora found in both Panama and southeastern America. While the land-bridge or isthmian-link theory might explain this coincidence, other, more viable theories exist to explain it. Among these is the theory that certain flora simply drifted on the open sea, perhaps hitchhiking on debris that eventually washed ashore in a place where the climate could support its germination and growth.

During the Pleistocene epoch, North and South America were connected. The Isthmus of Panama did not exist, and the region of what is now the Antilles was considerably higher than it is today, possibly with a land bridge to South America, less than a hundred miles away, and certainly with isthmian links to that continent. Vegetation common to the rain forests of the Amazon and Orinoco basins occurred in southern Florida, but most geologists consider it improbable that a land bridge existed between that region and the Antilles or South America.

By comparison, those who dispute the existence of a land bridge between North and South America are on shakier ground than those who dispute a land bridge between Florida or the Antilles and South America. Vertebrate paleontologists have made a convincing case for animal migration (including human) in this area, as they have for animal migration between Asia and North America by way of the land bridge that is known to have existed between Siberia and present-day Alaska. In 1917, Edward W. Berry postulated that the Antilles were once a part of South America but that continental drift separated the region from that continent. Such a theory would explain much of the coincident flora and fauna in both places.

Numerous catastrophic theories regarding continental drift have been used to explain the separation of the earth's landmass into continents. Most of these theories have to do with the separation of the moon from Earth, accompanied by the violent tidal action that followed this separation. As late as 1932, Charles Schuchert supported the notion that land bridges connected parts of an enormous supercontinent and that these bridges subsequently sank beneath the deep waters of the Atlantic, Pacific, and Indian Oceans, as suggested several decades earlier by Edward Suess.

The major portions of the jigsaw puzzle on which geologists have worked for the past century are now firmly in place. Other portions do not yet fit into any reasonable pattern. The land-bridge theory, although it explains certain animal migrations in a limited geographical range, has been displaced by other theories relating to diverse flora and fauna in such widely separated areas as India, Antarctica, Africa, and South America. The major importance of the land-bridge hypothesis is that it demanded testing, and this testing led to valuable insights about the earth's earliest history.

R. Baird Shuman

Further Reading

Carey, S. Warren. "A Tectonic Approach to Continental Drift." In *Continental Drift: A Symposium. Being a Symposium on the Present Status of the Continental Drift Hypothesis, Held in the Geology Department of the University of Tasmania, in March, 1956.* Hobart: University of Tasmania, 1958; revised 1959. Carey presents one of the most balanced overviews of the continental-drift/land-bridge controversy in this somewhat specialized article. Important, but not easily accessible to the beginner.

Fuentes, Carlos. *The Buried Mirror: Reflections on Spain and the New World.* New York: Houghton-Mifflin, 1992. Celebrating the five hundredth anniversary of Christopher Columbus's first trip to the New World, Fuentes traces with considerable skill the migration patterns that accounted for the presence on the American continent of a large population of people whose ancient ancestors were of Asian origins and who probably arrived in America by way of the Bering Sea land bridge. A most engaging book.

Grotzinger, John, and Tom Jordan. *Understanding Earth.* 6th ed. New York: W. H. Freeman, 2009. This comprehensive physical geology text covers the formation and development of the earth. Discusses the idea of land bridges and why such a theory was not accepted. Readable by high school students, as well as by general readers. Includes an index and a glossary of terms.

Marshal, L. G. "Land Mammals and the Great American Interchange." *American Scientist* 76 (1988): 380-388. This article describes the various mammals that migrated north and south along the isthmus of Panama during the Great American Interchange. Discusses the diversity and effects on the ecology of the two continents.

O'Neill, Dan. *The Last Giant of Beringia: The Mystery of the Bering Land Bridge.* Cambridge, Mass.: Basic Books, 2005. Discusses the past existence of the land bridge Beringia, over which animals and humans crossed to what is now Alaska. The author recounts the studies of the many scientists who pieced together the current knowledge of Beringia.

Ross, Charles A., ed. *Paleobiogeography.* Stroudsburg, Pa.: Dowden, Hutchinson, and Ross, 1976. Ross has gathered in one compendious volume the salient works of most of the geologists and paleontologists who worked toward evolving the land-bridge/isthmian-link theory, as well as the work of those who questioned and eventually discredited the theory.

Shea, James H., ed. *Continental Drift.* New York: Van Nostrand Reinhold, 1985. The section entitled "Gondwana" and Alfred Wegener's chapter "The Origins of Continents" are particularly valuable to those interested in the land-bridge theory. The presentations are clear and informative.

Suess, Edward. *The Face of the Earth.* Oxford, England: Oxford University Press, 1909. This translation of Suess's *Das Anlitz der Erde* (1885) is essential reading for anyone seriously interested in this field. Suess was the major early proponent of the land-bridge/isthmian-link theory. He presents his salient arguments, since discredited, in this book.

Sullivan, Walter. *Continents in Motion: The New Earth Debate.* 2d ed. Melville, N.Y.: American Institute of Physics, 1993. Sullivan presents considerable information about land bridges and isthmian links, offering a comprehensive overview of the theory and its eventual loss of credibility. The presentation is well balanced and accessible to people not specifically trained in the field.

Webb, S. David. "The Great American Biotic Interchange: Patterns and Processes." *Annals of the Missouri Botanical Garden* 93 (2006): 245-257. The article addresses the rate of species radiation in North and South America following the interchange of the Panama land bridge. Discusses the new ecosystems created by the changing faunal composition. Written for the advanced undergraduate.

Wegener, Alfred. *The Origin of Continents and Oceans.* New York: Dover Books, 1966. This translation of Alfred Wegener's landmark book *Die Entstehung der Kontinente und Ozeane* (1915), in combination with Suess's *The Face of the Earth*, helps one to understand the considerable controversy that Suess's hypothesis evoked and some of the reactions to it. Fortunately, this relatively recent translation of the original work is available in many college and university libraries.

See also: Alpine Glaciers; Biostratigraphy; Cenozoic Era; Continental Glaciers; Cretaceous-Paleogene Boundary; Geologic Time Scale; Glacial Deposits; Glacial Landforms; Glacial Surges; Gondwanaland and Laurasia; Ice Ages; Mass Extinctions; Mesozoic Era; Paleoclimatology; Paleogeography and Paleocurrents; Paleoseismology; Paleosols; Paleozoic Era; Pangaea; Proterozoic Eon; Stratigraphic Correlation; Transgression and Regression; Unconformities; Uniformitarianism.

LAVA FLOW

The Hawaiian words "pahoehoe" and "aa" are used all over the world to designate two very common types of lava flow. The outward appearance of lava flows tells much about their eruption temperatures, chemical compositions, and viscosities. The surfaces of lava flows vary from smooth and glassy to rough and clinkery, depending on the viscosity of the lava and whether it is emplaced on land or under water.

PRINCIPAL TERMS

- **aa:** a Hawaiian term (pronounced "ah-ah") that has been adopted for lava flows with rough, clinkery surfaces
- **autoclastic breccia:** the clinkery or blocky rubble that forms on some lava flows
- **block lava:** lava flows whose surfaces are composed of large, angular blocks; these blocks are generally larger than those of aa flows and have smooth, not jagged, faces
- **breccia:** a general term for any deposit composed mainly of coarse volcanic rock fragments
- **pahoehoe:** a Hawaiian term (pronounced "pa-hoy-hoy") that is used in reference to lava flows with smooth, ropy surfaces
- **pillow lava:** a type of bulbous, glassy-skinned lava that forms only when basaltic lava flows erupt under water

Lava Flows

Aa and *pahoehoe* are Hawaiian words that describe the surface textures of lava flows. Aa is a rough, clinkery variety of lava, and pahoehoe has a smooth, ropy surface. Although aa and pahoehoe are the most common types of lava flow on land, other forms occur if lava is erupted under water or if it is extremely viscous. An explanation of why lava flows assume different forms requires a basic understanding of the physical properties of magma.

Magmas

Magma is molten rock that originates by partial melting of either the crust or the mantle of the earth. "Lava" is a general term for magma that has erupted onto the earth's surface. Lava flows are bodies of magma that have been emplaced onto the surface as coherent, flowing masses of lava. The viscosity of lava—its resistance to flowage—largely controls the external appearance of the lava. In turn, the viscosity of lava is controlled mainly by its temperature and chemical composition. Most magmas are dominantly composed of the elements silicon and oxygen, which form strong chemical bonds with each other in the magma. Together, these elements are referred to as silica. The silica content of magma exerts a major influence on magma viscosity. Similar to adding flour to batter, increased silica content means higher magma viscosity.

Although geologists recognize hundreds of kinds of volcanic rocks, there are three fundamental types of magma. Basalt magma is produced in the earth's upper mantle and is poorer in silica (about 50 percent by weight) and dissolved gases than other types of magma. It erupts at the highest temperatures of all lava, usually about 1,100 to 1,200 degrees Celsius. Because of its low silica content and high temperature, basalt lava has relatively low viscosity, or resistance to flowage. With the consistency of honey or peanut butter (although considerably denser, at around 2.7 grams per cubic centimeter), basalt lava is relatively fluid and usually erupts as thin, sheetlike lava flows.

Andesite magma erupts at slightly lower temperatures than does basalt (around 1,000 degrees Celsius) and is richer in dissolved gases and silica (about 60 percent by weight). Andesite magma has a viscosity much greater than that of basalt. Andesite lava viscosity is similar to that of cold putty or frozen caulking compound, although its density of about 2.6 grams per cubic centimeter is considerably higher. Because of their relatively high viscosity, andesite lava flows are thick and tonguelike in shape and do not travel as far from the vent as do fluid basalt lava flows.

Rhyolite, the third fundamental type of magma, is formed by melting of the continental crust. It is very rich in silica (greater than 70 percent by weight) and often contains high amounts of dissolved gases. Thus, it tends to erupt explosively and to form voluminous deposits of frothy pumice. The viscosity of molten rhyolite lava is extremely high. Even at its white-hot eruption temperature of around 900 degrees Celsius, rhyolite behaves more like solid material than a liquid. It fractures when struck with a hammer, and its viscosity is sometimes similar to that of glacial ice. Thus, rhyolite magma moves very slowly and oozes onto the earth's surface as thick, pasty lava flows or steep-sided volcanic domes. The latter are steep piles of lava that grow directly on top of volcanic vents because the lava is too viscous to flow away from its source. Because the crystallization of rhyolite is very sluggish, its lava flows and domes are often composed of obsidian, a type of silica-rich volcanic glass.

Pahoehoe

Pahoehoe seldom forms on lava flows other than basalt, which is the hottest and most fluid type of magma. The smooth surface texture on pahoehoe lava flows is formed during quenching of the flow surface against the earth's atmosphere. A several-centimeter-thick crust of brittle lava is formed, commonly with a thin skin of glass. The smooth, brittle crust forms a barrier to rising gas bubbles (vesicles), and there is often a frothy zone of round vesicles directly underneath the crust, which rides atop the hot, fluid interior of the lava flow. The lava crust is an insulating barrier that prevents the flow of heat from the interior of the lava flow, which can remain hot for weeks, months, or even years, depending on its thickness. During flowage of the molten interior, the smooth crust is commonly deformed into folds and ridges, as when a carpet is shoved against a wall. The folds and ridges assume many forms, as they are repeatedly stretched and refolded onto themselves. Several varieties of pahoehoe have been recognized. Entrail pahoehoe is commonly formed when lava tubes are breached, spilling their contents as entwined, elongate bulbs of glassy-skinned lava. Shelly pahoehoe forms only near volcanic vents, where centimeter-thick lava crusts are stacked atop one another and are separated by cavernous gas pockets.

Lava flows into the sea, Big Island, Hawaii Volcanoes National Park. (Art Wolfe/Photo Researchers, Inc.)

Pahoehoe lava flows are able to travel tens of kilometers from their vents because rocks are poor conductors of heat. As the top and sides of the lava flow solidify, an insulating barrier is formed around the interior of the flow, which remains hot and fluid. Like blood within a system of blood vessels, fluid lava moves within circular channels known as lava tubes, and the advancing nose of the lava flow is fed by this network. When eruption of lava ceases, some of the tubes drain out and their roofs may collapse in places, leaving accessible caves with flow marks on the walls and lava dripstones hanging from the ceilings.

Pillow Lava

Pillow lava is the subaqueous counterpart of pahoehoe and forms only when fluid basalt lava is erupted under water. Pillow lava has been observed forming around the Hawaiian Islands, and beneath its blanket of sediment, most of the sea floor is composed of this type of basalt lava. When basalt lava erupts under water, large quantities of steam are formed, sapping heat from the lava surface. As a result, a glassy skin quickly forms. Back-pressure from the still-fluid lava inside the lava flow eventually bursts the brittle skin, which is shattered into plates, chunks, and small shards of glass. New bulbs of lava ooze out of the cracks in a sort of budding process. Continuation of the process builds entwined masses of bulbous, glassy-skinned lava that look like toothpaste extrusions. The lava is mingled with layers of glassy fragments from the brittle skin of the advancing lava flow. Because it looks like a stack of pillows, the bulbous, glassy lava is appropriately called pillow lava.

Aa

Aa is a type of lava with a rough, clinkery surface. The term "clinkery' refers to hardened lava. It is typical of basalt and andesite lava flows that are somewhat fluid but are too viscous to have pahoehoe surfaces. Jagged, spiney blocks of lava are formed during crumbling of the viscous mass. The spiney fragments continually rub against one another, eventually forming a thick outer envelope of debris that grades into the hotter, more fluid interior of the lava flow. The clinkery rubble rides on the flow until it eventually tumbles down the nose of the moving flow front. Similar to an advancing tractor tread, the fluid interior of the lava flow overrides its own debris. Eventually, the interiors of aa flows may become so viscous that they can no longer flow as liquids and instead begin to shear along horizontal fractures (platy joints) that form near their bases. The shear planes are similar to those that develop near the bases of glaciers, and the process is much like spreading out a deck of cards across a tabletop.

Cooling of Lava

Single lava flows can change from pahoehoe into aa, but the reverse of this process has never been observed. When lava is first emitted, it is hottest and most fluid. As the lava cools, loses its gases, and slowly crystallizes, its viscosity is irreversibly increased. The originally smooth lava crust becomes thicker, and slabs of it begin to grind against one another. Eventually, smooth crust can no longer develop because the outer portion of the lava flow has become too brittle.

In addition to cooling and the resulting increase in viscosity, the pahoehoe-to-aa transition has been observed to occur when lava flows undergo high rates of internal shear. This can occur, for example, when a pahoehoe flow travels over steep terrain such as a cliff face, sometimes continuing as an aa flow at the base of the cliff. Under such circumstances, the lava must flow faster, thus increasing internal shearing. During the pahoehoe-to-aa transition, the brittle lava crust becomes thicker, and the interior of the flow becomes more sluggish, as shown by the generally slower rates of movement of active aa flows (meters to hundreds of meters per hour), as compared to pahoehoe flows (hundreds of meters to tens of kilometers per hour). Pahoehoe flows have been clocked at speeds up to about 60 kilometers per hour in open channels and in lava tubes.

Block Lavas

A third type of surface is formed on lava flows with extremely high viscosities. In some andesite and most rhyolite lava flows, fragmentation of the lava is very thorough because of the high viscosity of the mass. Great volumes of large, angular blocks are formed, each block having relatively smooth (not sharp, clinkery) faces and sharp edges between the faces. Called block lavas, the flows are thick, crumbling masses of fine-grained lava or obsidian that can barely move away from the vent. The noses of block-lava flows are steep (30- to 35-degree) embankments of coarse rubble, and the tops of the flows are pocked with irregular depressions, lava spines, and automobile-sized lava chunks. The jagged rubble of block and aa lava flows is called friction breccia, crumble breccia, or autoclastic ("self-fragmented") breccia. "Breccia" is a general size term that refers to any deposit that is composed of coarse volcanic rock fragments (greater than 64 millimeters).

When the output of block lava is relatively low and magma viscosity is high, steep-sided volcanic domes may form over the vent. Most domes grow from within, by internal expansion, rather than by the addition of surface flows of lava. As a result, the thick carapace (shield or shell) of blocky rubble and lava spines is continually shoved aside, and the debris tumbles down the margins of the growing dome. Volcanic domes are very unstable and tend to collapse unpredictably into piles of blocky rubble. During collapse, ground-hugging avalanches of hot debris are sometimes jetted outward from the dome. Hence, volcanic domes are among the most dangerous of volcanic phenomena. Lava domes have repeatedly formed in the crater of Mount St. Helens since its catastrophic eruption of 1980, and the explosion of a Mount Pelée lava dome destroyed the city of St. Pierre on the Caribbean island of Martinique in 1904.

STUDY OF LAVA FLOWS

Lava flows are a major component of most volcanoes. Together with chemical analyses of lava, the physical features of ancient lava flows have revealed much about how lava flows are emplaced. From laboratory experiments and theoretical calculations based on the chemistry of lava, the viscosity of virtually any type of magma can be derived.

Observations on active lava flows have revealed much about how lava moves. The temperatures of active lava flows can be estimated in several ways. The incandescent color of the lava gives a general indication of temperature, in much the same way that blacksmiths judge the temperature of forged steel: dull red is about 600 degrees Celsius, orange is about 900 degrees Celsius, and golden yellow is about 1,100 degrees Celsius.

A more precise method uses an optical pyrometer to measure the wavelength of visible and near-infrared radiation, which varies uniformly as a function of temperature. The instrument is essentially a telescope in which a wire filament is mounted. The filament is heated by increasing an electric current until its color temperature matches that of the object being viewed. The temperature of the filament and hence, the lava, can then be calculated.

Both the visual method and the optical pyrometer can give inaccurate results because of atmospheric effects and because only the surfaces of objects are measured (the interiors of lava flows are much hotter than their surfaces). Another method that is less subject to error but considerably more hazardous in use is the thermocouple, a pair of wires of different composition welded together at both ends. When one end of the circuit is immersed in hot material (a difficult procedure in viscous lava), an electrical current is generated, its strength depending on the temperature difference between both ends of the circuit. An ammeter near the cold end of the circuit reads the electrical current, from which the temperature at the hot end of the circuit can be calculated.

Viscosities of active lava flows can be estimated by the use of penetrometers. When a known amount of force is applied to a steel rod of known diameter, the penetration of the rod into the lava will depend on the viscosity of the lava. Again, this is a superficial measurement that may not give an accurate estimate of the viscosity of the fluid interior of the lava. The chemical compositions of lavas are now routinely analyzed, and theoretical viscosities can then be calculated from the chemical data. Similarly, if the chemical composition of a lava is known, its crystallization temperature can be accurately estimated, and such a value is usually close to the eruption temperature of the lava. Field measurements are often used together with theoretical calculations in order to arrive at the best understanding of the flow behavior and physical properties of lava.

William R. Hackett

FURTHER READING

Cas, Ray A. F., and J. V. Wright. *Volcanic Successions: Modern and Ancient.* Winchester, Mass.: Allen & Unwin, 1987. This 530-page text is written for geologists at the college level. Presents very thorough coverage of volcanic materials at an advanced level; much of the information on lava flows is readily digested by persons with some college science background. Contains many black-and-white photographs of volcanic features.

Chester, Roy. *Furnace of Creation, Cradle of Destruction.* New York: AMACOM Books, 2008. This text discusses the turbulent processes of the planet. It covers earthquakes, volcanoes, and tsunamis in reference to plate tectonics, natural disasters, and predicting and mitigating these events. Multiple chapters explore seafloor spreading. The author also discusses hydrothermal activity. This text takes on an immense range of content, but still explains concepts clearly and with detail.

Decker, Robert, and Barbara Decker. *Volcanoes.* 4th ed. New York: W. H. Freeman, 2005. This readable and well-illustrated text is geared toward the general public. The authors explain geologic phenomena in a clear and simple manner. Includes a good section on lava flows. The appendices list the world's 101 "most notorious" volcanoes, as well as Internet sources for volcano information.

Francis, Peter, and Clive Oppenheimer. *Volcanoes.* 2d ed. New York: Oxford University Press, 2004. This text covers volcanic activity ranging from lava flows to pyroclastic currents. Provides many examples from Hawaii, Italy, Ethiopia, Japan, Mount St. Helens, and Krakatau. Written for the layperson, but some knowledge of geology is helpful.

Jerram, Dougal, and Nick Petford. *The Field Description of Igneous Rocks.* 2d ed. Hoboken, N.J.: Wiley-Blackwell, 2011. Begins with a description of field skills and methodology. Contains chapters on lava flow and pyroclastic rocks. Designed for student and scientist use in the field.

King, Geoffrey, and Geoff Bailey. "Tectonics and Human Evolution." *Antiquity* 80 (2006): 265-286. Covers the geology and tectonics of the African Rift Valley. The authors relate this information to the evolution and dispersal of humans. Some background in evolution and/or geology is beneficial.

Lockwood, J. P., and Richard W. Hazlett. *Volcanoes: Global Perspectives.* Chichester, England: Wiley-Blackwell, 2010. An up-to-date and well-illustrated undergraduate-level text that is also accessible to the interested reader.

Sigurdsson, Haraldur, ed. *Encyclopedia of Volcanoes.* San Diego, Calif.: Academic Press, 2000. This book contains a complete summary of the scientific knowledge of volcanoes. It includes eighty-two well-illustrated overview articles (including one on lava flow), each of which is accompanied by a glossary of key terms. Although this is a college-level text, it is written in a clear and comprehensive style that makes it generally accessible. Cross-references and index.

Williams, Howell, and A. R. McBirney. *Volcanology.* San Francisco: Freeman Cooper, 1979. Geared to geologists, this 400-page college textbook nevertheless contains much descriptive information that can be understood by nonscientists. Chapters on the physical nature of magma and on lava flows are excellent, and the text is liberally illustrated with graphs, drawings, and black-and-white photographs. A subject index and references list are included.

See also: Calderas; Flood Basalts; Geysers and Hot Springs; Hawaiian Islands; Hot Spots, Island Chains, and Intraplate Volcanism; Island Arcs; Mount Pelée; Mount St. Helens; Oceanic Crust; Ocean Ridge System; Ophiolites; Ring of Fire; Shield Volcanoes; Spreading Centers; Yellowstone National Park.

LIFE'S ORIGINS

Life originated about 3.4 billion years ago, arising most likely from organic molecules that formed in the environment of the early earth. A number of theories exist to explain the origin of the first organic molecules and the development of the first protocells from organic components. Researchers have also attempted to discover the most likely location of life's origin, suggesting potential locations that include deep-sea vents, frozen prehistoric oceans, and extraterrestrial planets.

PRINCIPAL TERMS

- **abiogenesis:** a process involving the generation of living systems from nonliving components
- **enzyme:** a substance that increases the rate of a certain biochemical reaction
- **hydrothermal vent:** a structure on the ocean floor through which superheated water and gases spout from Earth's core
- **inorganic compound:** a chemical compound that lacks carbon
- **microfossil:** a fossil or fossil fragment that cannot be seen by the unaided human eye
- **nucleic acid:** a molecule, such as deoxyribonucleic acid (DNA) or ribonucleic acid (RNA), used to store and transmit information about the formation of proteins; essential to reproduction in living systems
- **organic compound:** a compound containing carbon that constitutes the basic building blocks of any living system
- **panspermia:** a theory that life can be carried between planets in the form of bacteria integrated into the mineral components of meteorites
- **primordial soup:** a combination of molecules and gases that were present in the seas at the origin of life and may have contributed to the formation of the first organic compounds
- **protocell:** a theoretical structure representing a midpoint between nonliving chemical systems and the earliest true cells

DEFINITION OF LIFE

Science has no concise definition of what constitutes a living organism, relying instead on a list of characteristics that, when taken together, differentiate life from nonlife. Life is therefore understood as a biological system capable of growth, metabolism, reproduction, and a variety of other behaviors. The cell is the smallest unit of life, and all life forms comprise one or more cells. The origin of life on Earth is that moment when the first cell came into being. The study of life's origins has therefore become the search for a theory of abiogenesis, or the generation of life from nonliving components.

CHRONOLOGY OF LIFE

Physicists and geologists estimate that the earth formed approximately 4.6 billion years ago from the dense cloud of gas that also gave rise to the sun and the other planets of the solar system. For hundreds of millions of years, Earth was an unstable, molten mass with no atmosphere; meteors and comets frequently impacted Earth's surface. About 3.8 billion years ago, the earth stabilized enough for clouds of gas to interact and form water vapor. Water began to accumulate into the first oceans, which helped to further stabilize the climate and form the first atmosphere. Scientists believe that Earth was then primed for the formation of the first organic molecules.

In 2011, physicists David Wacey and colleagues reported the discovery of microfossils from bacteria in samples of Australian sediment that were aged 3.4 billion years or more. These microfossils are the metabolic remains of microorganisms that lived by metabolizing sulfur; modern bacteria similarly thrive in extreme deep-sea environments. This discovery indicates that life may have taken 400 million years or less to develop from inorganic molecules into living cells.

EARLY THEORIES ON THE ORIGIN OF LIFE

In the 1860's, French chemist Louis Pasteur noted that food left to spoil apparently "generates" insects and other organisms from within, an observation that led him to speculate that life could spontaneously generate from nonliving matter. Pasteur considered this spontaneous generation as proof of the influence of a god. One of Pasteur's most prominent contemporary opponents was English naturalist Charles Darwin, whose theories on the origin of life were based on his published theory of evolution by natural selection. Darwin believed that all life developed from less-complex life forms in a gradual process that took millions of years. He rejected the idea that complex life could generate spontaneously from nonliving matter.

Experimental evidence verified Darwin's theories by showing that a sterile sample of organic material will not generate life unless it has been contaminated by living cells from the environment. Insects appear to grow within rotten meat or spoiled fruit because the sample has been contaminated with insect eggs or larvae. Most scientists, following in Darwin's vein, then began searching for a new hypothesis to a scientific understanding of abiogenesis that could be understood given the known scientific laws and an understanding of Darwin's evolutionary theory.

To understand the process of abiogenesis, scientists first must determine how collections of inorganic chemicals formed into the first organic molecules, such as proteins, nucleotides, and lipids. From here, scientists must determine how these molecules formed into the first cells.

Origin of Organic Molecules

The environment of early Earth was very different from its environment in the twenty-first century. The atmosphere had a different gaseous composition and the climate was affected by regular disturbances, including volcanic activity and electrical storms. The early oceans were relatively shallow and, because there were no living cells, they were filled with high concentrations of dissolved molecules. This early marine environment became known as the primordial soup. In 1929, British-born evolutionary biologist and geneticist John B. S. Haldane proposed that ultraviolet (UV) radiation from the sun provided the energy to drive chemical reactions in the seas, thereby giving rise to the first organic molecules. Haldane's critics argued that UV radiation does not provide sufficient continuous energy to account for the necessary chemical reactions.

In the 1950's, University of Chicago biologists Stanley Urey and Stanley Miller conducted an experiment in which they attempted to replicate the chemical composition of the primordial soup; they then subjected samples of their soup to electrical energy, reasoning that electrical storms might have energized the early ocean. The Miller-Urey experiment, as it came to be called, succeeded in producing some types of organic molecules, though not a sufficient variety of molecules to assume that cellular formation would have followed.

Carnegie Mellon University chemist Jim Cleaves revisited the Miller-Urey experiment in 2011 using different gas concentrations from the original experiment, based on contemporary refinements regarding the atmosphere of early Earth. Cleaves believed that the products of volcanic eruptions, including hydrogen, methane, and hydrogen sulfide, were critical components in the formation of organic molecules. The experiment produced twenty-three types of organic molecules, many of which are essential for the formation of living cells, giving credence to the theory that the early environment facilitated the onset of abiogenesis.

Formation of Precellular Life

Most biologists believe that it would have been impossible for living cells to form directly from inorganic materials and that there must have been one or more stages at which life existed in a precellular form. Biologists postulate that protocells had formed from the organic molecules produced in the atmosphere of early Earth. These protocells, they believe, then underwent further waves of evolutionary refinement to give rise to modern cells.

For a cell to form, molecules must be present and capable of facilitating chemical reactions, thereby constituting a basic metabolism that fuels maintenance, growth, and replication. Protein molecules serve as the engines of metabolism in cells, facilitating and directing chemical reactions. Second, a cell requires a method of replicating and storing information to be passed to the cell's offspring, a method fulfilled by a cell's nucleotides, such as DNA and RNA. Cells also require a membrane, which establishes the cell's boundaries with respect to the environment and governs the passage of material to and from the cell. Researchers who study abiogenesis are divided into those who believe that metabolism came first, those who believe that reproductive structures arose first, and those who believe that cellular "containers" were the first to evolve.

Scientists who prefer a metabolism-first approach argue that proteins formed first, creating a disembodied system of metabolic reactions that were later associated with reproductive molecules and cellular containers. Proteins have a variety of functions, including functions as enzymes, which are molecules that catalyze or increase the speed or likeliness of certain chemical reactions.

The containers-first approach focuses on certain fatty acids forming naturally into spherical globules that scientists call vesicles. These vesicles can form around proteins and other molecules that occur in solution, thereby providing a contained environment for chemical reactions. When associated with certain proteins, vesicles will develop structural abnormalities that make it more likely for the vesicle to split apart, giving rise to two vesicles, each containing some portion of the associated proteins and other molecules.

Other researchers have argued for a gene-first theory, in which rudimentary DNA or RNA molecules were the first to evolve, creating self-replicating nucleotide structures that were later associated with fatty acid vesicles and proteins to facilitate the movement of energy.

RNA molecules have dual functions in cells: They can transmit genetic information, like DNA molecules, but they can also catalyze chemical reactions, like proteins. In a 1986 article, American molecular biologist and Nobel laureate Walter Gilbert used the term "RNA world" in theorizing that the early earth may have been populated by RNA-based organisms in which RNA fueled both the metabolism and replication functions of early cells. Whereas many biologists support the RNA world hypothesis, researchers have been unable to synthetically produce RNA molecules in laboratory conditions. Biologists and chemists are therefore uncertain whether it would have been possible for RNA to form in the early environment. In addition, no theory can explain how the first RNA molecules functioned in forming the earliest protocells.

Environmental Theories of Abiogenesis

Some biochemists have suggested that the key to understanding abiogenesis lies in determining the exact microenvironment in which the first cells formed. Since research in abiogenesis began, microbiologists have discovered that living cells populate nearly every environment on Earth. Cells have evolved to survive in highly acidic and basic environments and to endure extremes of temperature and pressure that would be fatal for most cells.

One of the most promising environmental theories of abiogenesis involves deep-sea formations known as hydrothermal vents, where gases and heated water from the earth's core spout into the oceans to create superheated and chemically rich undersea environments. Chemist Nick Lane and his team from University College London believe that life began in a unique type of hydrothermal vent, one with a highly specific complement of basic minerals, gases, and other molecules, and with abundant, naturally occurring crevices and chambers on its surface. The combination of geochemical forces surrounding the vents provides a proton gradient, in which the electrical properties of the outer area are different from the electrical properties within the various channels on the surface of the vent. This difference in electrical potential provides the energy needed for chemical reactions and powers the production of energy-storing molecules that can be used to fuel additional reactions.

Lane and other scientists who support the hydrothermal vent theory propose that lipid vesicles formed inside the inorganic chambers around the vent and acted as containers for other organic molecules produced in this chemically unstable environment. Eventually, the early protocells housed in the chambers of the vent evolved an independent proton gradient similar to the one maintained at the vent's surface. Modern cells maintain a similar proton gradient with respect to the environment by utilizing a system of chemical reactions that transfer electrons between donor and recipient molecules.

In the late 1980's, German chemist Günter Wächtershäuser proposed another theory of abiogenesis associated with the hydrothermal vent environment. Wächtershäuser proposed that the earliest cells formed with naturally occurring structures composed of iron sulfide and associated metals around the surface of volcanically active vents in the deep ocean. Wächtershäuser theorized that these composite structures may have acted to filter chemical reactions in the superheated environment of the vents. The internal metals reacted with chemicals entering the system, forming a basic metabolic system that further facilitated the rate of chemical reactions. In time, organic molecules associated with the iron sulfide structures and eventually played a role in the chemical reactions inside the structures. Over millions of years, protocells formed within these structures, which gradually evolved the capacity to maintain their metabolic activity without the iron sulfide structure.

While hydrothermal vent hypotheses are considered to be among the most promising theories regarding life's origins, other researchers have proposed alternative locations for life's initial genesis. Biochemists Stanley Miller and Jeffrey Bada, from the Scripps Institution of Oceanography and the University of California, proposed a theory in the mid-1990's that life may have begun during a time when the world's oceans were frozen, a condition that paleontologists believe may have occurred when the earth first formed because the sun was less luminous.

The formation of organic molecules may therefore have occurred in an ocean shielded by a layer of ice that may have been hundreds of feet thick in some areas. Organic molecules that formed under this sheet of ice may have lasted longer, because of the effect of cold temperatures in slowing decay and the protection that the ice layer provided against ultraviolet radiation and other disturbances. This "frozen soup" may therefore have allowed more time for molecules to interact and, potentially, to form into protocells.

Both the hydrothermal vent and the frozen ocean hypotheses have interesting implications for the search for life on other planets. Among the planets known to science, many have extreme environments characterized by intense heat or cold. If life first arose in one of Earth's extreme environments, it might mean that simple forms of life could exist on planets where conditions make it impossible for more complex forms of life to evolve.

Extraterrestrial Origins

The theory of panspermia, which was first developed in the mid-nineteenth century, holds that life was carried to Earth, and that this life may have been carried to numerous other planets on meteors, comets, and asteroids originating from elsewhere in the galaxy. Panspermia argues that when these natural projectiles impacted a planet, bacteria were deposited and began evolving to change the environment. The reasoning behind the theory is that it seems unlikely that life would have been able to evolve from nonliving systems in the 400 million years that passed between the formation of the earth and the appearance of the first cells.

One version of the panspermia theory holds that life developed on a distant planet where the conditions were different from Earth in such a way that biological life was more likely to develop. This "origin" planet may later have been obliterated, seeding the galaxy with meteors that carried dormant bacterium or protocells. Another version of the theory, more extreme and generally less respected, holds that life was deposited on Earth by a more advanced extraterrestrial race with an interest in seeding the galaxy. Research has demonstrated that bacteria can survive in a dormant state for millions of years and that some types of bacteria can survive the type of radiation, heat, and cold that might kill most cellular life on Earth.

Ultimately, the panspermia hypothesis does little to solve the problem of how life originated from nonlife. If panspermia were proven, the question of how, why, and where the first life appeared would simply shift to an extraterrestrial stage. Scientists theorize that humans may eventually use bacteria to initiate the process needed to colonize other planets. If this comes to fruition, there may one day be planets for which panspermia was the avenue for the generation of life.

Micah L. Issitt

Further Reading

Deamer, David W., and Jack W. Szostack. *The Origins of Life: A Subject Collection from Cold Spring Harbor Perspectives in Biology.* Cold Spring Harbor, N.Y.: Cold Spring Harbor Laboratory Press, 2010. Essays relating to the origin of life, including detailed discussions of competing theories to explain the origin of the first organic compounds and the first cells. Also covers a variety of alternative theories for abiogenesis and a complex discussion of the RNA world hypothesis.

Dyson, Freeman, *Origins of Life.* New York: Cambridge University Press, 1999. Physicist Dyson discusses various theories on the origins of life, covering the early history of the earth and the development of the first cells. Contains reviews of many of the major theories in the study of early life and the environment.

Goodsell, David S. *The Machinery of Life.* New York: Springer, 2009. Overview of cellular form and function in the natural world. Provides sections on cellular metabolism, reproduction, and the origin of early cells. Also contains sections covering the activity of cells in the human body.

Luisi, Pier Luigi. *The Emergence of Life: From Chemical Origins to Synthetic Biology*. New York: Cambridge University Press, 2010. Covers the major theories in abiogenesis, including early research and the theory of spontaneous generation. The book's three main sections give theories based on the metabolism, replication, and boundary theories of early cell formation.

Margulis, Lynn. *Acquiring Genomes: A Theory of the Origin of Species*. New York: Basic Books, 2003. Evolutionary biologist Margulis lays out the evidence for an alternative theory of evolutionary development, focusing on the innovations of the earliest cells to evolve on the earth. Also covers some of the problems facing the common conceptions of cellular formation and early evolution.

Yarus, Michael. *Life from an RNA World: The Ancestor Within*. Cambridge, Mass.: Harvard University Press, 2010. Yarus provides detailed evidence supporting the RNA world hypothesis, but also touches on some of the other competing theories of abiogenesis.

See also: Archaebacteria; Cambrian Diversification of Life; Catastrophism; Colonization of the Land; Dinosaurs; Ediacaran Fossils; Eukaryotes; Evolution of Birds; Evolution of Flight; Evolution of Humans; Evolution of Life; Fossilization and Taphonomy; Fossil Plants; Fossils of the Earliest Life-Forms; Fossils of the Earliest Vertebrates; Geoarchaeology; Geologic Time Scale; Mammals; Mass Extinctions; Microfossils; Neanderthals; Paleobiogeography; Paleobotany; Paleosols; Petrified Wood; Prokaryotes; Stromatolites; Systematics and Classification.

M

MAMMALS

Mammals are four-legged backboned animals distinguished by a number of unique characteristics, including hair, constant warm body temperature, mammary glands, and specialized teeth that are replaced only once. Mammals first descended from a group of animals known as synapsids, or mammal-like reptiles, about 200 million years ago but remained small (mouse-sized) and unspecialized until the dinosaurs died off about 65 million years ago. They then underwent a huge evolutionary radiation—that is, they spread into different environments and diverged in structure.

PRINCIPAL TERMS

- **Cenozoic era:** the period of geologic time from about 65 million years ago to the present
- **genus (plural, genera):** a group of closely related species; for example, *Homo* is the genus of humans, and it includes the species *Homo sapiens* (modern humans) and *Homo erectus* (Peking Man, Java Man)
- **mammary glands:** the glands that female mammals use to nurse their young
- **marsupials:** mammals that give birth to a premature embryo that then finishes its development in a pouch
- **monotremes:** primitive mammals, such as the platypus and echidna, that lay eggs and have other archaic features
- **order:** a group of closely related genera; in mammals, orders are the well-recognized major groups, such as the rodents, bats, whales, and carnivores
- **placentals:** mammals that carry the embryo in the mother until it is born in a well-developed state; it is nourished in the womb by a membrane (the placenta)
- **synapsids:** the mammal-like reptiles that lived from about 300 to 200 million years ago and evolved into mammals

THE AGE OF MAMMALS

Mammals are a group of vertebrates (animals with backbones) that have been the dominant animals on land and sea since the dinosaurs died out about 65 million years ago. Indeed, the present period of geologic time, the Cenozoic era, is often called the age of mammals. About 4,170 species of mammals are alive currently, but at least five times that many are now extinct. About 1,010 genera of mammals are living, but, according to a 1945 tabulation, an additional 2,000 genera were extinct, and that number has greatly increased since 1945.

Mammals have been very successful in occupying a great variety of terrestrial and aquatic ecological niches. They include terrestrial meat-eaters and plant-eaters, tree dwellers, burrowing forms, and even aquatic forms. The largest terrestrial mammals today are elephants, but the extinct hornless rhinoceros *Paraceratherium* was much larger, reaching 6 meters at the shoulder and weighing about 20,000 kilograms. The largest mammals, however, are whales; they can weigh up to 150,000 kilograms in the case of the blue whale, which is larger than even the largest dinosaurs.

Living mammals are easily distinguished from all other vertebrates by a number of unique evolutionary specializations. Unlike other vertebrates, mammals have hair, are equipped with mammary glands to nurse their young, and bear live young (except for the most primitive egg-laying mammals, the platypus and echidna). Mammals maintain a constant, relatively high body temperature. They have a four-chambered heart and a very efficient digestive and respiratory system, which includes a diaphragm in the chest cavity to aid in breathing. Mammals develop rapidly as juveniles and then stop growing when they reach adult size, unlike other animals, which grow continuously through life.

Fossil mammals are known only from their fossilizable parts, which are mostly their teeth and bones. Their fossil record can be traced back to their ancestors, the synapsids, or mammal-like reptiles, which are known as far back as 300 million years ago. Fossil mammals are usually distinguished from their ancestors by a number of skeletal features. These features include specialized teeth that are replaced only once (comparable to our "baby teeth" and adult teeth), a jaw joint between the dentary and squamosal bones, and a middle ear that is composed of three bones (the hammer, anvil, and stirrup).

MAJOR MAMMAL GROUPS

As the large synapsids were being replaced by the dinosaurs about 210 million years ago, mammals evolved as tiny, mouse-sized animals that fed on insects. Mammals remained in this state for almost 150 million years, living in the nooks and crannies of the world of the dinosaurs, which dominated the planet. A number of different "experimental" groups of insectivorous (insect-eating) mammals lived during this time, but all became extinct. About 100 million years ago, the three major groups of living mammals evolved. These include the monotremes, the marsupials, and the placentals.

The monotremes are the egg-laying mammals, the platypus and the echidna of Australia and New Guinea. The more advanced mammals bear live young. One group, the pouched mammals, or marsupials, give birth to a premature, partially developed embryo. The embryo then climbs to the mother's pouch and fastens to a nipple, where it finishes its development. The most familiar marsupials are the kangaroo, koala, Tasmanian devil, and opossum, although there have been many other types of marsupials in the past, and many are still alive today in Australia and South America. Where marsupials lived in isolation with no competition from placental mammals, they have evolved into many different body forms, which converge on the body forms of their ecological equivalents in the placentals. In Australia presently, there are marsupial equivalents of cats, wolves, mice, flying squirrels, rabbits, moles, tapirs, and monkeys. Among extinct marsupials of South America were the equivalents of lions and of saber-toothed cats. As similar as these animals look to their placental equivalents in their external body form, they are not related to true cats, wolves, or the rest, as they are all pouched mammals.

Unlike marsupials, placentals must carry the embryo in the womb through its full development. To allow this, the embryo is nourished by an extra membrane surrounding it in the womb. This membrane, the placenta, is shed when the baby is born and is part of the "afterbirth." This mode of reproduction makes the placental baby less vulnerable than a marsupial baby, but it means that the mother is more vulnerable, as she must carry a larger embryo for a longer time. A female marsupial can also reproduce faster, as she can carry one baby in a pouch and be pregnant with another.

APPEARANCE OF MAMMALS

After the extinction of the dinosaurs about 65 million years ago, the planet was ready for a new group of large animals to evolve and take over the vacant ecological niches. Placental mammals underwent a tremendous diversification until they occupied many ecological niches, and some reached the size of sheep. Most, however, were no larger than a cat. The placentals split into the edentates (anteaters, sloths, armadillos, and their relatives) and the rest of the mammalian orders (groups of genera). Like marsupials, edentates had their greatest success in isolation in South America, although presently the armadillo is successfully spreading northward. The most successful placentals in the Northern Hemisphere

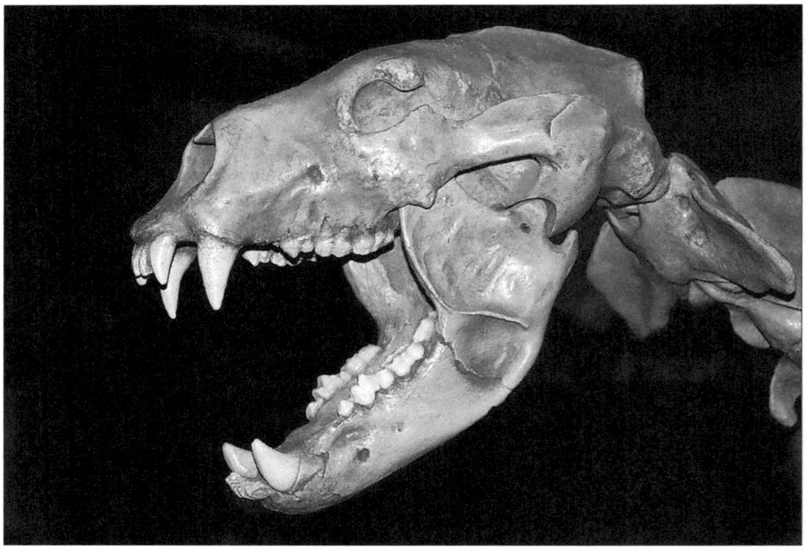

The fossilized skull of a cave bear at the Black Hills Institute of Geological Research in Hill City, South Dakota. (Larry Miller/Photo Researchers, Inc.)

in the early Cenozoic were the archaic ancestors of the insectivores (moles, shrews, hedgehogs, and their extinct relatives), primates (lemurs, monkeys, apes, humans, and their extinct relatives), the carnivores (meat-eating mammals), and a great number of extinct, archaic orders that have no living descendants.

About 54 million years ago, in the Eocene epoch, some of the modern orders of mammals began to appear. They were still small and unspecialized and would not be easy to recognize today. Some of these groups included the rodents (mice, rats, squirrels, guinea pigs); the even-toed hoofed mammals, or artiodactyls (pigs, camels, deer, cattle, antelopes, giraffes, and their relatives); and the odd-toed hoofed mammals, or perissodactyls (horses, rhinos, tapirs, and their extinct relatives). These groups lived along with many archaic groups that are now extinct, so the Eocene world had a strange mixture of modern and archaic mammals. During the Eocene, the first bats began to fly, and the first whales to swim.

About 38 million years ago, the world's climate got much cooler and more extreme. The tropical rain forests that had dominated the world in the Eocene were replaced by more mixed vegetation, including open grasslands. As a consequence, most of the archaic groups of mammals died out, and they were replaced by a great diversification of mammals from living groups. The archaic forms were mostly leaf eaters that could not survive the loss of the tropical forest or tree-dwelling forms, such as the ancient primates. At this time, the first true dogs, cats, elephants, rhinos, tapirs, camels, pocket gophers, rabbits, and shrews appeared. Although still very archaic representatives of their respective orders, the mammalian communities began to take on a more modern look. From this point on, environments (like the East African savanna today) were dominated by perissodactyls, artiodactyls, carnivores, rodents, rabbits, and insectivores.

Climate Changes

Since the end of the Eocene, the world's climate continued to get cooler and more extreme. More and more forests were replaced by open grasslands, and the mammals evolved in response to the changing environment. Most of the larger plant eaters (perissodactyls and artiodactyls) had to develop ever-growing molars to chew tougher grasses and long legs to escape their predators by running. This development can be seen not only in the evolution of the horse, but also in the rhinos, camels, and the many artiodactyls that chew their cud (including antelopes, deer, sheep, and cattle). Carnivores became more specialized into ambush hunters (cats), pack hunters (dogs), bone-crushing scavengers (hyenas), and a specialized type, the sabertooths, which evolved four different times independently. Rodents became the most common mammals of all, and they dominate the burrowing, ground-foraging, and tree-dwelling ecological niches.

During the last 5 million years, the earth's climate became increasingly severe as the ice ages developed. Many mammals became extinct, while others became adept at migrating away from ice sheets or developed thick coats of hair (such as the woolly mammoth and woolly rhino) in order to live in glaciated regions. The most severe extinction of large mammals happened only 10,000 years ago, when the ice sheets retreated and the present interglacial period began. Whether this extinction was attributable to the change in climate or to the severe overhunting by prehistoric humans is still disputed. Since that time, many mammals have been wiped out by overhunting, by the destruction of habitat, or by competition from mammalian "weeds" (such as rats, rabbits, and goats) that accompany human habitation. The extinction of mammals has reached its worst levels during the last two centuries, as human populations have exploded to more than 5 billion. Only a small remnant of the once-rich assemblage of mammals that dominated the Earth for 65 million years is left presently, and much of it is endangered in the wild.

Study of Mammal Remains

Because only the hard parts of the mammals commonly become fossils, mammalian paleontologists specialize in studying bones and teeth. Ideally, the paleontologist would study complete skeletons, but these are rarely found, as scavenging and stream erosion break up and scatter them. Most paleontologists thus make do with teeth, jaws, and skulls. Teeth, in particular, are valuable, as they are the hardest and most durable tissue in the body and thus are the most resistant to breakdown by erosion and stream abrasion. In addition, teeth are highly diagnostic of species and are influenced by the diet of the animal. Therefore, the paleontologist can not only identify a mammal by its teeth but also learn much about its ecology.

A paleontologist starts searching in sedimentary beds of the appropriate age and environment (river-channel sandstones and floodplain mudstones), particularly those that have produced fossil mammals in the past. The best results occur in areas with desert climates and badlands exposures, as there are few plants to cover the rocks, and the erosion is fast enough that new fossils are exposed each year. The most successful method is to prospect, or to walk head-down with the eyes "fastened" to the ground, for as many hours and days as it takes until bones or teeth are seen. When fossils are found, they may not be worth collecting if they are too fragmentary or are unidentifiable. If the fossil is worthwhile, the paleontologist collects not only the fossil itself but also exact data about where it occurred and at which stratigraphic level (so that it can be used for dating). The most common mistake made by amateurs is to collect a fossil without carefully noting this information.

Occasionally, mammal fossils are large enough or fragile enough that special methods are necessary. In this case, the paleontologist digs around the exposed fossil and clears the rock around it until it is almost free, resting on a thin pedestal of rock. Then it is encased in a "jacket" of burlap and plaster of Paris, much like the plaster cast used to set broken limbs. When the plaster cast is dry and hard, the fossil is removed from the pedestal, and more plaster and burlap seal it in the cast. When it reaches the lab, technicians can open the cast and carefully clean away the excess rock, treating the fossil with hardener and preservative as they expose it.

Fossils of small mammals (particularly rodents, rabbits, primates, and insectivores) are too tiny to see while prospecting. These fossils are more often collected by finding a rich, bone-producing level from which bags of fossil-rich material are then filled. This material is later placed in a wooden box with a window-screen bottom and allowed to soak in a stream. The water washes out all the silt and clay, leaving a concentrate of pebbles and fossils. This concentrate can then be dried and spread out on a table to be sorted through by hand.

Donald R. Prothero

FURTHER READING

Benton, Michael J. *The Rise of the Mammals.* New York: Crescent, 1991. A beautifully illustrated book, suitable for high school students, that explains when and how various mammal families developed.

__________. *Vertebrate Paleontology.* 3rd ed. Malden, Mass.: Blackwell Science, 2005. This comprehensive text provides a complete outline of the history and development of vertebrates. The text begins with background information on the evolution of vertebrates and the study of fossils in general, making this book accessible to the layperson. Chapters cover topics such as the vertebrate transition to land and the evolution of birds, mammals, and humans. Designed for college courses but also suitable for the informed enthusiast. The last one-third of the book deals with the rise and diversification of mammals.

Carroll, Robert L. *Vertebrate Paleontology and Evolution.* New York: W. H. Freeman, 1988. A complete, well-illustrated textbook on fossil vertebrates. Includes several chapters on fossil mammals. Although written at the college level, some parts can be understood by the general reader.

Feldhamer, George A., et al. *Mammalogy.* 3rd ed. Baltimore, Md.: Johns Hopkins University Press, 2007. Provides details on all topics of mammalogy, from population dynamics to biodiversity, from evolution to phylogeny. This text is easy to follow and filled with useful illustrations and diagrams. Well suited for undergraduate and graduate students, this is the go-to text for mammalogy courses.

Kemp, T. S. *The Origin and Evolution of Mammals.* New York: Oxford University Press, 2005. This text discusses the evolution of mammals, beginning with synapsid evolution and continuing beyond dinosaur extinction. Highlighting the fossil evidence and DNA sequencing, the author discusses comparative anatomy and geology, making this text best suited for advanced undergraduates, graduates and experts.

MacDonald, David W., ed. *The Encyclopedia of Mammals.* 2d ed. New York: Facts on File, 2006. An excellent, colorful review of the living mammals, with some very good sections on extinct mammals as well.

Savage, R. J., and M. R. Long. *Mammal Evolution: An Illustrated Guide.* New York: Facts on File, 1986. A beautifully illustrated account of mammal evolution for the general reader. It groups the mammals by their ecology, however, rather than by their normal classification, and contains some inaccuracies.

Wicander, Reed, and James S. Monroe. *Historical Geology*. Belmont Calif.: Brooks/Cole, Cengage Learning, 2010. This textbook is accessible to university students with or without a background in geology. It covers Earth's geological systems and plate tectonics, physical and biological history, geological time, and evolution. There are many diagrams, images, and charts along with a CD-ROM to supplement the text.

Woodburne, Michael O., ed. *Late Cretaceous and Cenozoic Mammals of North America: Biostratigraphy and Geochronology*. New York: Columbia University Press, 2004. A valuable review of the dating and correlation of strata by fossil mammals. Updated from the previous version temporally to include the Late Cretaceous and geographically to include Mexico. Intended for the advanced college student and the paleontologist.

See also: ArchaeB.C.E.ia; Cambrian Diversification of Life; Catastrophism; Colonization of the Land; Dinosaurs; Ediacaran Fossils; Eukaryotes; Evolution of Birds; Evolution of Flight; Evolution of Humans; Evolution of Life; Fossilization and Taphonomy; Fossil Plants; Fossil Record; Fossils of the Earliest Life-Forms; Fossils of the Earliest Vertebrates; Geoarchaeology; Life's Origins; Mass Extinctions; Microfossils; Neanderthals; Paleobiogeography; Paleobotany; Paleosols; Pangaea; Petrified Wood; Prokaryotes; Stromatolites; Systematics and Classification.

MARINE TERRACES

Marine terraces are ancient coastlines that are at elevations well above or well below present-day sea level. Global cycles of sea-level change and tectonic uplift and subsidence of coasts are responsible for creating multiple sets of parallel marine terraces.

PRINCIPAL TERMS

- **barrier island:** a long, low sand island parallel to the coast and separated from the mainland by a salt marsh and lagoon
- **bioabrasion:** physical and chemical erosion or removal of rock as a result of the activities of marine organisms
- **guyot:** an oceanic volcano, presently submerged far below sea level, with a top that has been beveled flat by wave erosion
- **isostatic readjustment:** rapid tectonic uplift or subsidence of continental areas in response to the addition or removal of the weight of overlying deposits of glacial ice or seawater
- **lithosphere:** the outer shell of the earth, where the rocks are less dense but more brittle and coherent than those in the underlying layer (asthenosphere)
- **notch or nip:** an erosional feature found at the base of a sea cliff, the result of undercutting by wave erosion, bioabrasion from marine organisms, and dissolution of rock by groundwater seepage
- **Pleistocene epoch:** the time of earth history, from about 2.58 million to 11,000 years ago, during which the earth experienced cycles of warming and cooling, resulting in cycles of glaciation and sea-level change
- **strandline:** the position or elevation of the portion of the shoreline between high and low tide (at sea level); usually synonymous with "beach" and "shoreline"
- **subsidence:** the sinking of a block of the earth's lithosphere because of a force pushing it down; coastal areas undergoing rapid subsidence tend to be submerged below sea level
- **tectonic:** pertaining to large-scale movements of the earth's lithosphere
- **uplift:** the rising of a block of the earth's lithosphere because of a force pushing it up; coastal areas undergoing uplift tend to emerge above sea level

Coastlines

The coastlines of continents and islands represent a fundamental boundary between the earth's solid landmass and the constructional and destructional energy of the sea. The landforms in coastal areas are the result of continuous dynamic interaction between these competing geological agents. Marine processes of erosion and sedimentation construct a shoreline profile on the edge of the landmass that defines the strandline, a narrow zone of wave- and tide-washed coast. The coastal strandline usually exhibits a gently sloping platform with its top at an elevation between high and low tides; it is often bounded by adjacent steeper slopes and is a reliable indicator of sea level. Familiar strandline features are beaches, coral reefs, and wave-cut platforms, strandline platforms cut into the bedrock by wave erosion.

If erosion and sedimentation were the only active geological conditions, most coasts would have a single type of strandline landform whose position remained constant through time. The volume of seawater in the world's oceans, however, fluctuates directly with changes in the volume of ice in continental glaciers, causing sea level to rise and fall in response. Glacial ice on the continents melts (sea-level rise) or accumulates (sea-level fall) in response to temperature changes in the earth's atmosphere. The temperature of the surface ocean reacts in a similar manner: Warming causes expansion of the seawater (sea-level rise), and cooling results in contraction (sea-level fall). It has been estimated from oxygen isotope evidence in deep-sea cores that worldwide sea level has been 130 to 160 meters lower than today several times during the last 2 to 3 million years of earth history. There is also abundant physical evidence for large-scale sea-level excursions. On the Atlantic coast of the United States, elephant teeth 25,000 years old have been dredged up by fishermen from more than forty locations on the continental shelf, some as far as 120 kilometers offshore. Three-thousand-year-old oak tree stumps and many cultural artifacts used by coastal American Indians have also been found below sea level in this region.

The landmasses are also surprisingly dynamic, rising or falling in elevation in response to a variety of tectonic factors. On some coasts, isostatic adjustment results in uplift and subsidence. The weight of increasing thicknesses of overlying glacial ice or seawater will cause the earth's lithosphere to be depressed downward; when the weight is removed, the landmass will rebound upward. Volcanic activity will cause expansion (uplift) when the earth's lithosphere is heated; contraction (subsidence) occurs as the lithosphere cools. An extreme example of this effect is the guyot, an inactive volcanic seamount that subsides far below sea level as the hot lithosphere on which it formed cools and contracts. Guyots have been beveled flat by wave erosion and rimmed with banks of dead coral reef during the period when their tops were in shallow water. Coastal landmasses also undergo uplift and subsidence in response to the forces of plate tectonic movement. Although usually a slow, incremental process, a sudden, rapid coastal uplift occurred in Prince William Sound, Alaska, during the 1964 Good Friday earthquake, which resulted in elevation of a wave-cut platform 400 meters wide.

Fossil Coasts

During the Pleistocene epoch, strandline features formed on coasts worldwide that were strongly influenced by these complex and rapidly changing geologic conditions. Most coasts are not tectonically stable (located at one elevation) for a very long period of geologic time, and processes of landscape erosion and decay often act much more slowly than the rates of sea-level and land-level change. Abandoned strandline features, or "fossil coasts," are common to many areas of the world. First recognized in nineteenth-century Europe, parallel sets of these distinctive landforms are present on cliffs and coastal hills at elevations high above present-day sea level and on the ocean floor far below present-day sea level. Many contain marine fossils—demonstrating their marine origin—and create a steplike topographic profile of the terrain, leading up the seaward-facing hills. These features are referred to as marine terraces.

Because their formation requires time, strandline landforms usually form when coastal uplift/subsidence and sea-level change are in balance and the position of the strandline is relatively stable for a period of thousands of years, a situation referred to as a sea-level stillstand. In many areas, sea level tends to vary in cycles separated by stillstand events, during which marine terraces can develop. Occasionally, the processes of terrace formation will destroy or obscure an older terrace, so most coasts exhibit only one or two terraces. An unbroken flight of marine terraces climbing inland indicates a coast that is being continuously uplifted. One of the most striking and well-developed sequences of marine terraces in the world is on the seaward slopes of the Palos Verdes Peninsula of Southern California. Here, thirteen distinct terrace levels, from 15 to 448 meters in elevation, form a complete record of coastal uplift and sea-level change spanning several hundred thousand years.

Submerged terraces are usually better preserved than elevated marine terraces because they are less affected by processes of erosion and landscape degradation. Because they are obscured from view and more difficult to study than terraces on land, less is known about them. The most distinctive submerged wave-cut platforms are present off the southwest coast of Great Britain, off northern Australia, and off the Southern California coast and its nearshore islands from Santa Barbara to San Diego. Study of marine terraces has been used to document sea-level variations and the tectonic uplift and subsidence of coastal landmasses through time and to estimate the magnitude of these changes.

Coastal Erosion

There are several types of marine terraces, differing in their mode of formation. Processes of coastal erosion form marine terraces on coasts where the shoreline is B.C.E. by a steep, irregular cliff, where the supply of sediment is too limited to build beaches, and where nearshore waters are so shallow that waves break directly on the sea cliff. The base of the cliff is continuously pounded by the full force of the waves, resulting in very large impact pressures. Beach cobbles and abrasive, sand-laden water are hurled at the base of the cliff. The force with which these stones attack the cliff is difficult to exaggerate; there are reports of beach cobbles thrown 45 meters above sea level.

Wave-cut platforms result from the rapid horizontal cutting away of rock at the base of a sea cliff. Wave erosion quarries a notch into the base of the sea cliff, which undercuts and oversteepens the cliff until landsliding causes the cliff to collapse; this loose rubble is rapidly carried away by the waves and is a source of new rocks to continue wave attack. In some places, notch formation is aided by the bioabrasion activities of marine organisms that live attached to and that feed on rocky shores, and that secrete chemicals to help dissolve the rock or abrade it away with rasping feeding appendages. Dissolution of the rock is sometimes aided by groundwater seeping out of the base of the sea cliff. The net effect of this process is the landward retreat of the sea cliff and formation of a platform that slopes gently seaward. The boundary between the "step" of a wave-cut terrace and the steeper "riser" of the sea cliff that terminates its landward edge is called the inner edge point. The elevation of this point is the position of highest sea level during formation of the feature, and it is used to calculate the height of ancient sea levels in marine terraces.

The platform is continually abraded by the waves transporting sand down the coast, and its width is determined by the amount of landward sea-cliff erosion. The rate of sea-cliff retreat can be surprisingly rapid. Average rates on the California coast exceed 15 centimeters per year for hard rock cliffs and up to 1 meter per year in cliffs composed of soft, unconsolidated sediment. The coast of East Anglia, Great Britain, experiences up to 4 meters of sea-cliff retreat annually. Continuous records are available for the Huntcliff coast of Yorkshire, Great Britain, at a former Roman signal station where the cliff has retreated 30 meters in eight hundred years.

Overlying many elevated wave-cut platforms is a mantle of unconsolidated sediment referred to as coverhead. This material, deposited by a combination of terrestrial and marine sedimentary processes, buries the abraded platform surface and may be as much as 30 meters thick. Typical coverhead deposits include a basal layer of well-sorted beach sand, rounded beach cobbles, and marine fossils—the shoreline sediments left on the platform as sea level retreated. These basal strandline sediments are often covered by a heterogeneous deposit composed of poorly sorted rock debris, soil, and stream gravel deposited by sediment washing or landsliding down from the sea-cliff slopes above and by coastal streams building their alluvial fans toward the coast.

Sediment Deposition

Marine terraces are also formed by sediment deposition, which usually occurs on relatively flat coasts with a wide continental shelf where the energy of approaching waves is dissipated by friction with the shallow sea bottom. In areas such as the Atlantic and Gulf coasts of North America, rivers transport large quantities of mud and sand into coastal waters. Waves and wind currents pile sand into a long, narrow strandline feature known as a barrier island, a common landform found fringing low-lying coasts worldwide. Once formed during a stillstand of several thousand years, a barrier island will migrate landward as sea level rises. Waves and tidal currents cause barrier sand to wash over the top of the island or around its ends, moving it grain-by-grain landward with the rising sea. When sea level begins to fall, the barrier does not migrate seaward but is left behind inland of the new strandline, where it remains as a record of the former high stand of sea level; this feature is usually referred to as a beach ridge.

Marine terrace at Natural Bridges Beach State Park near Santa Cruz, California. An emergent marine terrace appears in the distance to the west. (U.S. Geological Survey)

Subsequent to sea level falling and the strandline moving seaward, river and marsh sediment is often deposited on the seaward side of the abandoned barrier. When sea level again rises to its former position, a new barrier island migrates with it; this island is, in turn, abandoned on the coast at the highest position of sea level. If the new barrier migrates far enough inland to reach the older remnant barrier, it will be welded onto it, forming a wide composite barrier island composed of two or more barrier islands.

Sediment deposition has widened the Atlantic and Gulf coastal plains, and repeated cycles of sea-level rise and fall have formed concentric arcs of abandoned barrier islands stretching inland nearly 50 kilometers from the present-day coast. These beach ridge barrier islands are a type of marine terrace, each recording an ancient high stand of sea level. One of the most prominent is Trail Ridge in southern Georgia, a sand ridge nearly 60 kilometers long that encloses a low swampy area on its landward side known as the Okefenokee Swamp. Submarine beach ridges are less common than their inland counterparts but have been reported from western Brittany.

In tropical areas where the coastal waters are warm all year and clear of suspended mud, corals will flourish and often form massive reefs. Reefs that grow during a long stillstand will become large and well developed, with a gently sloping top that corresponds to sea level. These coastal depositional features are also found stranded above the shoreline when sea level falls or the landmass rises, resulting in reef terraces. This type of marine terrace is most common on island coasts but is also present on the coasts of the Mediterranean and Red Seas.

James L. Sadd

Further Reading

Grotzinger, John, and Tom Jordan. *Understanding Earth.* 6th ed. New York: W. H. Freeman, 2009. This comprehensive physical geology text covers the oceans and their mineral resources, as well as the formation and development of the continents. Readable by high school students and general readers. Includes an index and a glossary of terms.

Hearty, P. J., and P. Aharon. "Amino Acid Chronostratigraphy of Late Quaternary Coral Reefs: Huon Peninsula, New Guinea, and the Great Barrier Reef, Australia." *Geology* 16 (July 1988): 579-583. A technical article describing the application of the amino acid dating technique to the classic coral reef terraces of the Huon Peninsula and a comparison of the results with uranium-series ages. *Geology* is a monthly publication of the Geological Society of America.

Masselink, G., and M. G. Hughes. *Introduction to Coastal Processes and Geomorphology.* London: Hodder Arnold Publication, 2003. The book is an excellent resource for students of geomorphology. Contains many illustrations and diagrams.

Moore, W. S. "Late Pleistocene Sea-Level History." In *Uranium Series Disequilibrium: Applications to Environmental Problems,* edited by M. Ivanovich and R. S. Hannah. Oxford, England: Clarendon Press, 1982. Half of this chapter in an advanced textbook is a description of state-of-the-art terrace dating techniques written in a technical style. Included is a summary of all known dates for fossil-bearing marine terraces all over the world that is slightly easier to read.

Pethic, J. *An Introduction to Coastal Geomorphology.* Baltimore: Edward Arnold, 1984. Written primarily for the Western European region, this advanced text contains several good sections on the formation of marine terraces, their use in sea-level and tectonic uplift studies, and the correlation of terrace data with other studies.

Pirazzoli, P. A. "Marine Terrace." In *Encyclopedia of Coastal Science* , edited by M. Schwartz. Dordrecht: Springer, 2005. The article discusses basic properties of marine terraces and how they form. Describes the marine terrace of Sumba Island, Indonesia. This brief article is a good starting point for understanding marine terraces.

Plummer, Charles C., Diane H. Carlson, and Lisa Hammersley. *Physical Geology.* 13th ed. Boston: McGraw-Hill, 2009. This is a straightforward, easy-to-read introduction to geology intended for those with little or no science background. Includes many excellent illustrations.

Segar, Douglas. *An Introduction to Ocean Sciences.* 2d ed. New York: W. W. Norton & Company, 2007. Comprehensive coverage of all aspects of the oceans and the oceanic crust, including the development of coastal features. Readable and well illustrated. Suitable for high school students and above.

Trenhaile, A. S. "Modeling the Development of Marine Terraces on Tectonically Mobile Rock Coasts." *Marine Geology* 185 (2002): 341-361. The article contains mathematical models used to study terrace formation. Discusses the factors that positively and negatively influence terrace formation. Best suited for graduate students or professionals with a mathematics background.

See also: Coastal Processes; Geomorphology and Human Activity; Geomorphology of Dry Climate Areas; Geomorphology of Wet Climate Areas; Glacial Landforms; Grand Canyon; Karst Topography; Permafrost; River Valleys; Transgression and Regression.

MASS EXTINCTIONS

Mass extinctions demonstrate that the evolution and extinction of life are not smooth but rather are interrupted by mass dyings as a result of as yet poorly understood causes. Mass extinctions seem to be an integral part of the overall pattern of life processes on the earth.

PRINCIPAL TERMS

- **ecology:** the study of the relations between organisms and their environments
- **environment:** the conditions affecting the existence, growth, and overall status of an organism
- **era:** one of the major divisions of geologic time, including one or more periods
- **geologic record:** the history of the earth and its life as recorded in successive layers of sediment and the fossil specimens they contain
- **niche:** in an ecological environment, a position particularly suited for its inhabitant
- **period:** the fundamental unit of the geologic time scale
- **plate tectonics:** the theory and study of the formation and movement of the sections that make up the outer portion of the earth, which move independently over the interior
- **stratigraphic time scale:** the history of the evolution of life on the earth broken down into time periods based on changes in fossil life in the sequence of rock layers; the time periods were named for the localities in which they were studied or from their characteristics

Patterns of Extinction

The extinction of animal species appears to be as much a part of the pattern of life on the earth as is the evolution of new species. Indeed, the two processes go hand in hand, as new species fill the environments vacated by dying ones. Extinction is so common that more than 99 percent of the species that have ever lived are now extinct. At periods throughout the earth's history, extinctions of numerous and widely varied species have occurred within a relatively narrow range of time. Climatic changes, worldwide catastrophic events, and changing sea levels lead the list of possible causes of these mass extinctions. A species cannot be considered separately from the environment in which it lives. A change in the environment may occur faster than the organism can adapt and evolve along with it. This species may become extinct, while another species will develop to fill the environmental niche left behind. Thus, it appears that the ability of a species to survive may depend upon its ability to adapt quickly to changing conditions.

The pattern holds true for both plant and animal life. Mass extinctions of plant and animal life, however, do not seem to coincide. There have been three revolutions in principal plant forms. Each evolved quickly, and once a major group of plants was established, it continued for millions of years. Animal life is marked by frequent minor extinctions, and mass extinctions of animal life appear successively throughout the geologic record. In fact, it is these very extinctions that have, in many cases, provided the reference points that separate one geologic period from another. Catastrophic extinctions and revolutionary changes in animal life have occurred during at least six and perhaps as many as nine distinct junctures in the history of the earth. Hundreds of minor extinctions have occurred as well.

Periods of Mass Extinction

The earliest known mass extinction occurred some 650 million years ago toward the end of the Precambrian eon. Animal and plant life was primitive at this time, and it is possible that chemical changes in the atmosphere were occurring as the oxygen-producing plants developed. At the time of the mass extinction, more than 70 percent of the forms of algae disappeared. The end of the Cambrian period, about 500 million years ago, is generally seen as the first of the major extinctions of life. More than half of the existing species of animal life vanished at this time. At the end of the Devonian period, 360 million years ago, many types of fish became extinct, along with 30 percent of the animal life on the planet. The largest mass extinction in the earth's history happened some 248 million years ago at the end of the Permian period. This extinction also marked the end of the Paleozoic era. Half of all the families of animal life were exterminated, including 75 percent of all the amphibian families and 80 percent of the reptiles. Marine life, however, was hit hardest; it has been estimated that 95 percent of shallow marine organisms became extinct. It took 15 million to 20 million years for animal life to recover to anywhere near the previous variety. Another mass extinction took place at the end of the Triassic period some 213 million years ago. Up until this time, primitive amphibians and reptiles had dominated the land. By the end of the period, they had dropped out and were replaced by the early dinosaurs. It could appear that competition with the dinosaurs led to the demise of these creatures, but other extinctions took place at this time, particularly in the oceans, that point to a more general cause.

The most studied and best-known mass extinction occurred at the end of the Cretaceous period some 65 million years ago, at a point called the Cretaceous-Tertiary boundary. It is known mostly for the disappearance of the dinosaurs, but it also marks the extinction of many other forms of life, including many ocean reptiles, shelled creatures called ammonites, and forms of microscopic plankton. In all, 25 percent of all the known families and 75 percent of the known species of animals were eliminated. In this extinction, as is true with previous extinctions, there was a gradual decline in diversity among animal life for tens of millions of years before the final extinction at the end of the period.

The latter part of the Tertiary period, 10,000 years ago, saw the extinction of the giant land mammals. From the woolly mammoth and the mastodon to the saber-toothed cats, all perished. There is some evidence that implicates humankind in the extinction of these giant mammals. With organized hunting and the use of fire, it is conceivable that early humans played a role in the decimation of animal populations. Remnants at ancient kill sites show that thousands of animals were killed at a single hunt. Even in historical times, it has been seen that with the arrival of primitive people to a new area, extinctions of indigenous life-forms follow.

Although it has not been proved that humankind is responsible for the extinctions of the giant mammals toward the end of the Tertiary period, there is little doubt as to its role in the most recent mass extinction of life. The twentieth century witnessed the beginning of a widespread extinction of life-forms resulting from the explosive spread of the human species. While human demand for space increases, available habitats for other animals diminish. Remote areas that were once the havens of wildlife are now being penetrated by hunters, farmers, lumbermen, fishermen, and developers. Those areas devoid of humans are visited by the products of civilization—acid rain, air and water pollution, and destruction of the ozone layer. Thus, humankind is directly or indirectly responsible for the extinctions of more than 450 species of animals, and many more are endangered.

CAUSES OF MASS EXTINCTION

Current extinctions aside, why should a large percentage of life on the earth die out suddenly? The question can be misleading. For example, "suddenly" is a relative term when applied to geologic events. Changes happen over thousands of millions of years. An abrupt change in layers of sediment may indicate a change that took place over a period of a few days or a few million years. It is difficult to determine from the geologic evidence how sudden "suddenly" was.

Fossil showing part of the jaw and the teeth (about 180 mm long) of a mastodon, an elephant-like mammal. The teeth are designed for plant eating, and have fused crowns. Mastodon fossils are found in Tertiary period rocks worldwide. Human activity may have contributed to the extinction of mastodons and other large mammals in this period. (Biophoto Associates/Photo Researchers, Inc.)

Although it was long known that mass extinctions occurred from time to time throughout the earth's history, the causes were thought to be unknowable for lack of definitive evidence until high levels of the element iridium were found at the boundary marking the end of the Cretaceous period. With the possibility of testable evidence, the study of mass extinctions was opened again in earnest, and new relationships among available data began to be formed. As mass extinctions were documented exhaustively, what appeared to be a cyclic pattern emerged. According to one study, exterminations of life occurred about every 26,000 years. The periodicity remains in question, and theories range from an orbiting death star to the fact that random extinction events create an intrinsic periodicity.

Something in the earth's environment definitely changes at the time of a mass extinction. Many theories begin, then, by determining the changing conditions at the time of the extinction. Although valid, a weakness in this method is that the earth's environment is constantly in a state of flux. In whatever time period one examines, changes are bound to be found. It is therefore merely speculation whether any particular change or set of changes was sufficient to cause a mass dying. In addition, it seems that mass extinctions cannot be studied from the perspective of a single species or even a single extinction event. Only by studying them as a whole can scientists determine patterns that may explain them. In general, mass extinctions kill species on the land and in the oceans at the same time. They strike hardest at large animals on the land. Freshwater animals are generally left unscathed. Plants are not affected as severely as animals. Any theory of mass extinction must therefore address these issues.

Two Mass Extinction Theories

Two theories have emerged that seem to handle the criteria. Neither theory is new, but each has received renewed interest with the emerging evidence. One popular theory invokes a global catastrophe with a celestial origin. Impacts by meteorites or comets are proposed to be responsible for at least the mass extinction that killed the dinosaurs, if not others. A 1979 study by Walter Alvarez and Luis Alvarez showed that the presence of glass microspherules, shocked quartz, and the rare element iridium at the Cretaceous-Tertiary boundary provides evidence for the impact of a bolide (a meteorite or comet that explodes upon striking the earth). However, the effect of this on the extinction event has been hotly debated.

The other theory has to do with the changing sea levels of the earth, a theory from the 1920's that was revived during the 1960's and 1970's in the light of the theory of plate tectonics. As the sea level drops, warm, shallow seas recede. The habitat of some marine animals is eliminated, while chemical changes in the deeper ocean water cause other extinctions. On land, bridges appear between areas that were once separated by water, and the migration of animals follows. Competition for available habitats is generated among species that were previously separated, also leading to extinctions. The drifting together and breaking apart of continents would dramatically affect habitats. It has been shown that in general, the arrangement of continents affects the variability of climate, the variety of species on land, and the nature of habitats along the continental shelves in the ocean. Changes in these factors over long periods of time would lead to extinctions of animal species.

There is no single or simple answer to the cause of mass extinctions. The pivotal question seems to be whether they occurred instantaneously, requiring a single catastrophic event, or over an extended period. In a period of several million years, many changes will occur in the environment, some of them catastrophic. Volcanoes erupt, meteorites strike, climates change, and continents drift. There is no doubt that catastrophes occur, but how they affect life on the earth is as yet poorly understood. All things considered, it is most likely that mass extinctions occur as a result of a combination of many factors, including catastrophic ones that happen over long periods of time. Perhaps a cyclic pattern exists, or perhaps species survival is a matter of playing the odds on a changing planet. Yet, if the earth is considered to be a living organism, it is no wonder that as the planet ages and evolves, the life it supports also changes.

Study of the Geologic Record

The major clues to unravel the pattern of life on the earth come from the study of the geologic record. The sequential layering of rocks, along with their deformation and erosion, allows scientists to construct a history of the processes the rocks have undergone. By analyzing the chemical composition of layers of rock and sediment and studying the nature of the fossil remains in them, scientists can begin to understand which forms of life existed during which time periods and when (and sometimes how) they became extinct. It was discovered that the layers could be dated by the decay of radioactive elements and that, in this way, the ages and periods of geologic time could be dated accurately.

Of all the fossils in the geologic record, fossils of tiny marine animals (microfossils) give the clearest picture of mass extinctions. Produced prolifically in the world's oceans, they settled among the sediment on the ocean floor and fossilized in layers. These layers tell the story of how the creatures developed, diversified, and flourished, and finally when they ceased to exist. Found in diverse areas of the world, from desert to mountainous regions, the fossils also tell the story of the movement of continents. Certain of them form such a complete record that their forms and types can be used to date sediments all over the world. While fossil remains of land animals can also be dated and related to specific time periods, these larger fossils are much scarcer and therefore do not present as complete a picture.

By studying traces of magnetism preserved in rock along with fossil deposits, geologists can chart the drifting of the continents in the distant past. The emergence of the theory of plate tectonics has provided scientists with a framework for understanding why the earth's environment and all its features are in a constant state of change. This changing environment is a key clue to periodic mass extinctions.

Besides the rise and fall of fossil species, other characteristics of the geologic layers can be measured. The relative abundances of certain elements indicate the processes that may have been occurring at the time the deposits were laid down. An example is the iridium found at the site of the extinction of the dinosaurs. Iridium is uncommon on the earth but is found in certain types of meteorites, which has led many scientists to believe that the extinction of the dinosaurs was caused by the impact of a comet or meteorite. Particles of soot were also found in the layers from that time period, suggesting that a large part of the earth may have been burned. Shocked quartz, indicating violent activity, also supported the theory.

The story of life found in the geologic record leaves scientists with myriad clues but few answers. It is as if they had millions of minute pieces to an enormous puzzle; collecting the pieces does not solve the puzzle. The various clues must be interpreted, seen in the light of their historical significance, and put together in fresh ways. As patterns emerge in the study of mass extinctions, the clues are beginning to fall into place, and an understanding of the dynamics of life on the earth is emerging.

Divonna Ogier

Further Reading

Alvarez, W. *T. Rex and the Crater of Doom.* Princeton, N.J.: Princeton University Press, 2008. A history of the discovery of the data that have been put forward to support an impact-caused extinction of dinosaurs. For the general reader.

Archibald, J. D. *Dinosaur Extinction and the End of an Era.* New York: Columbia University Press, 1996. A thorough review of dinosaur extinction and advocacy of a prolonged event caused by falling sea levels culminated by the impact of an asteroid. Intended for general readers.

Bakker, Robert T. *The Dinosaur Heresies.* New York: William Morrow, 1986. An expert on current theories concerning the dinosaurs explains them in an easy-to-read and interesting way. Dinosaurs are studied in terms of their place in geologic history, and their extinction is explained as part of a larger pattern. The pattern is compared with other mass extinctions in the history of the earth.

Benton, Michael J. *When Life Nearly Died.* New York: Thames & Hudson, 2003. This text discusses the five major extinction events. The author covers concepts of uniformitarianism and catastrophism, and finishes the text with a discussion of current biodiversity and the possibility of a mass extinction event occurring at present.

Erwin, Douglas H. *The Great Paleozoic Crisis.* New York: Columbia University Press, 1993. Examines the events recorded at the Paleozoic-Mesozoic boundary, where the most important extinction event in the earth's history took place. Written for the specialist but accessible to students. Glossary and extensive references list.

Grotzinger, John, and Tom Jordan. *Understanding Earth.* 6th ed. New York: W. H. Freeman, 2009. This comprehensive physical geology text covers the formation and development of the earth. Contains extensive sections on plate tectonics and its effect on the distribution of continents through time and thus on extinction events. Readable by high school students, as well as by general readers. Includes an index and a glossary of terms.

Rampino, Michael. "Darwin's error? Patrick Matthew and the Catastrophic Nature of the Geological Record." *Historical Biology* 23 (2011): 227-230. The article discusses Matthew's view of evolution as cycles of stability followed by catastrophic events. Mass extinction is described in reference to the organization of the geological time scale. The author develops the argument that the theory behind the geological time scale ties in closely with catastrophism, more so than to the classic Darwinian theory of evolution via gradual change.

Wicander, Reed, and James S. Monroe. *Historical Geology*. Belmont Calif.: Brooks/Cole, Cengage Learning, 2010. This textbook is accessible to university students with or without a background in geology. It covers earth's geological systems and plate tectonics, physical and biological history, geological time, and evolution. There are many diagrams, images, and charts along with a CD-ROM to supplement the text.

Wilford, John Noble. *The Riddle of the Dinosaur*. New York: Alfred A. Knopf, 1985. A well-written, sparsely illustrated account of the life and death of the dinosaurs. More in-depth than most books about dinosaurs, the book deals with each area of the study of the dinosaurs, discusses the various and sometimes controversial ideas concerning them, and comes to a synthesis without espousing a particular conclusion.

See also: ArchaeB.C.E.ia; Biostratigraphy; Cambrian Diversification of Life; Catastrophism; Cenozoic Era; Colonization of the Land; Dinosaurs; Ediacaran Fossils; Eukaryotes; Evolution of Birds; Evolution of Flight; Evolution of Humans; Evolution of Life; Fossilization and Taphonomy; Fossil Plants; Fossil Record; Fossils of the Earliest Life-Forms; Fossils of the Earliest Vertebrates; Geoarchaeology; Ice Ages; Life's Origins; Mammals; Microfossils; Neanderthals; Paleobiogeography; Paleobotany; Paleoclimatology; Petrified Wood; Prokaryotes; Stromatolites; Systematics and Classification.

MASS WASTING

Mass wasting, also called mass movement, is the process through which the layers of soil and rock that sit atop the bedrock are moved in bulk downslope because of gravity and other factors, such as moisture level and soil composition. Such movement of significant amounts of matter can have a profound effect on humans and other species.

PRINCIPAL TERMS

- **avalanche:** a type of granular flow with a high velocity of large volumes of materials
- **creep:** a type of granular flow that exhibits a gradual and continual downslope movement of regolith
- **debris flow:** a type of slurry flow that often results from heavy rains and has velocities between 1 meter per year and 100 meters per hour
- **free fall:** a type of slope failure where individual rocks on a steep slope break loose and fall to the ground; also known as rock fall
- **granular flow:** a type of sediment flow that contains between 0 and 20 percent water
- **mudflow:** a slurry flow with soup-like consistency that moves at speeds greater than 1 kilometer per hour along valley floors
- **regolith:** loose rock, sediment, and soil that sits atop bedrock
- **sediment flow:** debris that flows downslope; mixed with air or water
- **slide:** a type of slope failure that occurs when rocks or debris slide down a preexisting surface; also known as translational slide
- **slope failure:** a sudden failure of a slope resulting in transport of debris downslope
- **slump:** a type of slope failure involving the downward rotation of rock or regolith along a concave upward curved surface; also known as rotational slide
- **slurry flow:** a type of sediment flow that contains between 20 and 40 percent water
- **solifluction:** a slurry flow measured at a rate of centimeters per year
- **talus:** an accumulation of fallen material

Introduction

Earth's landscape is in a constant state of flux because of the effects of gravity and a geological process known as mass wasting. Mass wasting occurs when rock and regolith at the earth's surface move downslope. The movement can occur gradually, as slowly as 1 centimeter per year, or it can occur rapidly, as quickly as 1 kilometer per hour or more.

In mass wasting the material that was at a higher elevation moves to a lower elevation. In wilderness areas this may or may not have a significant impact on the environment or on local animal populations. Mass wasting also affects humans, often resulting in the loss of property and human life.

Slope Failure and Sediment Flow

Mass wasting occurs as the result of two major processes: slope failure and sediment flow. In a slope failure, the failure is sudden and can be caused by a variety of means, but the result is that debris is transported downslope by sliding, rolling, falling, or slumping.

Slumping occurs with the downward motion of materials in a concave, upward-curved surface area. As a result of this underlying formation, the upper surface of each block, along with the material in the block, remains intact. (It looks as if the face of the slope moves downward and leaves the impression of steps in the hillside.) The material accumulates at the base of the slope. This type of mass wasting is often seen along coastlines or riverbanks, where water has eroded the base of the slope, transporting the materials that supported the slope to a place downstream. With nothing to serve as a footing, the material above that point is prone to sliding, causing the characteristic depressions. Any houses or other structures atop the moving material will slide with the material and will often be destroyed by the rotation that is part of this type of movement.

Rock falls occur when a piece of rock on a steep slope is dislodged. As it falls down the slope it may bounce against other material that breaks free or it may simply fall by itself. It will join the talus at the base of the slope. Debris also falls in this way. The debris falls often include vegetation and other materials. These types of falls are dangerous because they can occur without any warning and because they often occur along roads that have been built between steep areas of rock. Because the rocks fall, and even

Mud and debris slides, caused by heavy rains, stripped the upper surface of the taiga off in the Brooks Range, Alaska. (Ken M. Johns/Photo Researchers, Inc.)

"bounce," they can travel far and can rest on the road. As a precaution, wire mesh is often anchored at the top of these steep slopes. When wire meshing is not practical, tall fencing is used.

Slides can be composed of rock and other debris. Slides are not the same as slumping because they do not move along a concave, upward-curved surface. Instead, slides move down the bedding plane or frontal surfaces of the underlying structures. They literally slide down the slope and end up at the bottom. Anything that sits atop the bedding or joint planes slides along with the material but is not necessarily destroyed in the process because a slide lacks rotation.

Sediment Flows

Slurry Sediment Flows. Slurry sediment flows occur when the regolith and rocks are mixed with water and end up containing about 20 to 40 percent water. A sediment flow often occurs after heavy rains or flooding. These water-saturated flows include mudflows. Mudflows are highly fluid, fast-moving mixtures of sediment and water that tend to travel across valley floors. Often they are caused by stream flooding in valley floors. As mudflows travel, they accumulate significant amounts of loose material. Because they move so swiftly and are made of such dense material, they can be dangerous. They also can cause significant structural damage. The slowest form of slurry flow is solifluction, in which rates of flow are measured in centimeters per year. These flows occur in areas where the soil has been saturated for years. They leave behind distinctive impressions on a hillside.

Granular Sediment Flows. Granular flows involve materials that include little or no water. Usually these flows take on the characteristics of fluids because of the introduction of air. Creep is the slowest form of granular flow, taking place at rates of less than 1 centimeter per year and occurring as a continuous downslope movement. Because it occurs so slowly, creep is rarely the cause of significant damage, but the steeper the hill, the faster the rate of creep. Generally, creep occurs at a rate that is not noticeable to the human observer. One way to confirm creep in an area is to look for trees that are not growing at an angle to the hillside. In such cases, the tree root structure is instead adapting to the downward movement of the slope.

Earthflows occur during heavy rains and cause the saturated material to move rapidly downslope, leaving tongue-like marks to show their paths. Earthflows usually remain active for long periods of time. Grain flows are similar in nature but they involve dryer materials moving at a faster rate. Not much disturbance is needed to set off a grain flow; an animal or a hiker on a trail can easily serve as the catalyst.

Avalanches, another type of granular flow that can be made of snow or rock, move vast amounts of material at high rates of speed, causing destruction to all habitat forms. Snow avalanches, which can be precipitated by animals or humans moving across susceptible areas, are quite common in high mountain areas that have significant snowfall; they lead to scores of deaths around the world each year. Debris avalanches are often triggered by earthquakes and volcanic eruptions. The slope of the mountain does not have to be especially steep for a debris avalanche to take place.

Human Activity

Human activity can play a significant role in mass wasting events. When humans create steep slopes or build homes on bluffs along the coast, for example, they are creating opportunities for mass wasting. Situations that are prone to mass wasting are easy to anticipate. When a work crew blasts a landscape to create a pathway through a mountain or hill, creating steep sides in the process, they are setting the stage for mass wasting. Also, water and ice will form cracks and fissures in rock, weakening the face of the rock and making materials susceptible to the effects of gravity. Soon, the weight of the material and the force of gravity will move that material downward. Once it hits the ground, that material may join the talus at the slope bottom, but it is just as likely to continue bouncing and rolling for some distance from where it lands.

Just as predictable is the erosion that occurs at the foot of a bluff. Over time, coastal processes erode the sediment that makes up the beach between the bluff and the ocean. Large pieces of the beach can be eroded by significant weather events. Even if the erosion occurs more slowly, the base of the bluff can be weakened by wind, currents, and waves. As the bluff base is weakened significantly, so too is the soil at the top of the bluff weakened by heavy rains, seismic activity associated with road traffic, building projects, and tremors and earthquakes. When the force of gravity becomes too great, the soil beneath structures and other objects atop the bluff will slide downslope. This type of sliding is inevitable unless steps are taken to reinforce the integrity of the bluff.

Preserving the Environment

The mechanisms involved in mass wasting are well understood. As a result, it is possible for engineers to build structures and roadways in areas that are less likely to be lost to mass wasting. Also, relocating structures and roadways is possible because engineers can determine an area's susceptibility to a mass wasting event. Red flags for such events include the angle of the slope, the makeup and weight of the regolith and rock, and the amount of precipitating activity. Another red flag is heavy rain and snowfall.

Even with thorough study and planning, engineers cannot ensure that mass wasting will not occur in a given area. Gravity is an ever-present force, and the composition of regolith changes with the weather. The best that engineers can do is locate roads and structures in areas calculated to be the least influenced by mass wasting.

Gina Hagler

Further Reading

Fletcher, Charles. *Physical Geology: The Science of Earth.* Hoboken, N.J.: Wiley, 2011. An introductory look at physical geology that includes key concepts.

Grotzinger, John, and Thomas H. Jordan. *Understanding Earth.* New York: W. H. Freeman, 2010. Views the earth as a system with inputs from chemistry, physics, and geology. Includes discussions of plate tectonics, rock weathering, mass wasting, and coastal processes.

Locat, Jacques, and J. Mienert. *Submarines Mass Movement and Their Consequences.* Boston: Kluwer Academic, 2003. Gives a thorough analysis of a number of events associated with tsunamis and liquefied sediment flows.

Nash, J. Madeleine. *El Niño: Unlocking the Secrets of the Master Weather-maker.* New York: Warner Books, 2003. Discusses the extreme weather condition known as El Niño and its ability to cause changes to weather patterns worldwide. Important to an understanding of the role weather plays in mass wasting.

Petersen, James, Dorothy Irene Sack, and Robert Gabler. *Fundamentals of Physical Geography.* Belmont, Calif.: Brooks/Cole, 2011. Illustrates the ways the physical geography of Earth affects humans, and vice versa.

Pipkin, Bernard. *Geology and the Environment.* Belmont, Calif.: Brooks/Cole Cengage Learning, 2011. Covers humans, geology, and the environment as well as weathering, mass wasting, and environmental considerations.

Plummer, Charles, Diane H. Carlson, and Lisa Hammersley. *Physical Geology.* Dubuque, Iowa: McGraw-Hill, 2010. A thorough, lucid introduction to the concepts of physical geology. Includes material on mass wasting and other geological processes.

Wicander, Reed, and James S. Monroe. *The Changing Earth: Exploring Geology and Evolution.* 6th ed. Belmont, Calif.: Brooks/Cole, 2011. An often-updated introduction to geology with a focus on the core concepts of physical geology. Includes a chapter on mass wasting.

See also: Coastal Processes; Contintental Glaciers; Continental Shelf and Slope; Deserts and Dunes; Fluvial Processes; Geoclines; Weathering.

MESOZOIC ERA

The Mesozoic era was a major episode in earth history during which a primitive flora and fauna and physical environment changed and became progressively more familiar. During this era, continents and ocean basins nearly achieved their present configuration and dinosaurs and other large reptiles flourished.

PRINCIPAL TERMS

- **epicontinental sea:** a sea covering part of a continental block; such seas generally are less than 200 meters deep and are the depositional site of most exposed sedimentary rocks
- **Gondwanaland:** an ancient, large continent in the Southern Hemisphere that included Africa, South America, India, Australia, and Antarctica
- **Laurasia:** an ancient, large continent in the Northern Hemisphere that included North America and Eurasia
- **Pangaea:** the supercontinent containing all continental crust that existed at the beginning of the Mesozoic
- **period:** a unit of geological time forming part of an era and subdivided, in decreasing order, into epochs, ages, and chrons
- **rifting:** a process of faulting and basaltic intrusion occurring where crustal plates separate during continental drift; may cause mountains
- **subduction:** a process by which one crustal plate rides over another, which descends and melts, generating molten rock that then intrudes the deformed plate above; it causes mountains
- **system:** the rocks deposited during a period, which is defined by age of the rocks making up its system
- **Tethys:** a seaway embayed into Pangaea between the southeast corner of Asia, the western end of the Mediterranean, and the southeast end of Pangaea

Breakup of Pangaea

The Mesozoic era and system are, respectively, a major subdivision of geologic time and the rocks of that age. The Mesozoic began about 251 million years before the present and ended about 65 million years before the present. It was preceded by the Paleozoic era and followed by the Cenozoic era and is divided into three periods: Triassic (about 251 to 201 million years before the present), Jurassic (about 201 to 145 million years before the present), and Cretaceous (about 145 to 65 million years before the present).

All the continents were gathered in a single large landmass, the supercontinent Pangaea, at the beginning of the Triassic period. South America, Africa, Antarctica, Australia, New Zealand, Arabia, and Peninsular India previously formed the supercontinent Gondwanaland during the Carboniferous (about 359 to 299 million years before the present), while North America, Greenland, Eurasia (less Peninsular India), and Borneo formed Laurasia. Laurasia and Gondwanaland merged during the Permian (about 299 to 251 million years before the present) by welding northwest Africa and South America to the south and east margin of North America. At this time, the Atlantic Ocean and Gulf of Mexico did not exist, but a large wedge-shaped seaway, Tethys, separated the former Laurasia and Gondwanaland blocks between the Mediterranean and opened to the east. The equator transected Mexico, the Sahara Desert, and northern India, continuing eastward to divide the Tethys seaway. The South Pole was slightly offshore of the western base of the Palmer Peninsula, and the North Pole was in eastern Siberia.

During the Late Triassic, Pangaea began to break up as North America moved away from Europe and northwest Africa. The resultant narrow North Atlantic Ocean and Caribbean Gulf of Mexico widened in the Jurassic and Cretaceous, reaching something near the present size and shape by the end of the period. Starting in the Jurassic, North America pulled away from South America, leaving a Caribbean Gulf of Mexico seaway connecting the Atlantic and Pacific. The South Atlantic opened later in the Cretaceous, resulting in a relatively narrow strait between Africa and South America by the end of the Mesozoic. The Late Triassic also was the time during which India, Australia, and Antarctica began moving away from Africa as a single block. India separated from the Australian-Antarctica block as movement continued, and, in the Late Jurassic, Madagascar split from Africa. Cenozoic movement then divided Antarctica and Australia and fused Peninsular India to Asia, eventually resulting in the present configuration of lands surrounding the Indian Ocean.

By the end of the Cretaceous, the equator passed just north of South America, crossed Africa at the southern margin of the Sahara, clipped the southwest corner of Arabia, traversed India along a north-south line through western India, and crossed Sumatra and Java. The North Pole was in the Arctic Ocean off western Siberia, and the South Pole lay within Antarctica.

Movement of the continents and opening of ocean basins caused compressional deformation of the earth's crust accompanied by massive igneous intrusion. The Andes and the mountains of western North America resulted from North and South America moving westward over oceanic crust in the Pacific Basin. Thrust faulting, folding, and batholithic (igneous) intrusion began in the Late Jurassic from California and Nevada northward to British Columbia and Alaska. This process continued during the Cretaceous, spreading to South America, the Palmer Peninsula, and northwest Antarctica. Similar deformation associated with subduction also took place along the northern margin of the Tethys seaway from the Mediterranean through the Balkans, Caucasus, and into Indonesia. Rifting along the East Coast of North America from Nova Scotia to Florida caused fault-block mountains and major basaltic flows as the Atlantic Ocean began opening in the Triassic and early Jurassic. Similar rifting with mountain making and basaltic flows occurred in South Africa, India, and Australia as the eastern part of Gondwanaland fragmented during the Late Triassic through the Cretaceous.

Sedimentary Deposition

Triassic sedimentary deposits are characterized by terrestrial red beds laid down on the supercontinent, Pangaea, under arid climatic conditions. These conditions occur between paleolatitudes (lines of latitude shown on the present pattern of land and sea where those lines existed in the past) from 10 to 30 degrees north and south. Thus, red beds were deposited north of the equator, from the north coast of Africa through Spain and Germany to the Ural Mountains and northwestward into Britain and Scandinavia. Similar rocks in East Greenland and southwestern North America are part of the same belt, as well as red beds in the rift valley system between Nova Scotia and Florida. Red beds and evaporites (salt and associated minerals precipitated from evaporating water) also accumulated in a parallel southern belt in Brazil, South Africa, Madagascar, eastern India, Australia, and Antarctica. Triassic marine rocks are sparse because Pangaea stood well above sea level during most of the Triassic, but fairly complete sequences do occur between the Alps and southern China along the northern margin of the Tethys seaway. Similarly, marine Triassic sediments occur in the Alps and western Cordillera of North America, which then were on the continental side of a subduction zone. Epicontinental seas encroached on both the Americas and Eurasia from these bordering

At Alaska's Chulitna Terrane, Triassic and Jurassic rocks are exposed in an overturned (tilted) fold: Black and white banded rocks are limestone and pillow basalt overlain by red beds (Triassic age); above these bands are Jurassic sandstone and argillite with marine fossils. (U.S. Geological Survey)

areas, leaving marine rocks interbedded with the great red bed accumulations on their western and southern margins.

Growth and expansion of spreading centers as Pangaea began breaking up probably caused the advance of epicontinental seas in the Jurassic and through the Middle Cretaceous. Resultant marine sedimentary rocks are widespread in North America from the Rocky Mountains westward and in western Europe. In addition, growth of mountain ranges along continental margins, as a result of either subduction or collision, led to extensive alluvial (river or stream) deposition. Also, relatively deep sea deposition continued in subduction zones bordering the Pacific and the northern shore of Tethys. Epicontinental seas also encroached on the western Americas prior to the Late Jurassic and were responsible for extensive deposits in western Europe. As the North Atlantic opened, Late Jurassic deposits first filled the Nova Scotia to Florida rift zone and then overlapped the continent in the Middle Jurassic and into the Cretaceous. Also, a great evaporite basin in the Gulf of Mexico area accumulated Jurassic salt deposits. Rising mountains from California to western Canada shed extensive alluvial deposits eastward as far as central Kansas during the latest Jurassic. Widespread alluvial deposits associated with basalt flows in India, Australia, Antarctica, Madagascar, and New Zealand. Terrestrial, coal-bearing rocks also border the northern and western shores of marginal seas from Iran to Siberia.

Processes responsible for Jurassic depositional events accelerated in the Cretaceous, with maximum expansion of epicontinental seas in the middle of the period. Drying and filling of most of these seas followed as a result of expanded mountain making toward the end of the Cretaceous. Thus, deep-water, continental margin deposition continued around the Pacific perimeter and on the southern margin of Eurasia until these basins were engulfed by growing mountain chains. Epicontinental seas expanded over Europe west of the Urals, western and northern Africa, central Australia, western and northern South America, and on the Atlantic coast of North America. In addition, a new seaway opened between the Arctic Ocean and the Gulf of Mexico, separating the rising mountains of the western Cordillera from the eroded stumps of the Appalachians. Widespread chalk deposits characterize these seas in areas remote from actively rising mountains, including the notable chalk cliffs of Dover and Normandy. The rising western North and South American mountains, however, shed abundant detritus to their east in the form of alluvial plains, deltas, and muddy marine deposits, which very nearly filled the North American midcontinent seaway by the end of the Mesozoic. Most, but not all, of the other Cretaceous epicontinental seas either dried up or were greatly constricted at the end of the period or early in the Tertiary.

Plant Life and Dinosaurs

Early Mesozoic plants resembled those of the preceding Paleozoic. Primitive Carboniferous coal swamp plants persisted in restricted moist environments in the Northern Hemisphere but died out at the end of the Triassic. Cycadeoids, gingkoes, conifers, and ferns dominated Triassic through Early Cretaceous floras, with the cycads, primitive conifers—including the auracariaceans of the Petrified Forest—and gingkos first expanding in the Triassic. This plant life was joined by more modern conifers and early relatives of cypress in the warmer, moister Jurassic and Early Cretaceous. Abruptly, in the middle of the Cretaceous, modern broad-leaved trees appeared and quickly dominated the forests. Thus, Late Cretaceous forests closely resembled those of the present.

Large amphibians, along with an expanding contingent of reptiles, constituted the terrestrial vertebrate fauna at the beginning of the Triassic period. Phytosaurs (large crocodile-like reptiles) were the dominant reptiles, but the first dinosaurs appeared in the Late Triassic, along with the first ichthyosaurs (porpoiselike reptiles), nothosaurs (ancestors of plesiosaurs), and the first turtle. The first frogs also were Triassic. Mammal-like reptiles (reptiles ancestral to mammals), living mostly in the Southern Hemisphere, declined during the Triassic but probably gave rise to the first mammals late in the period.

Rapid dinosaur diversification in the Jurassic led to bipedal (two-footed) and quadrupedal (four-footed) forms as well as both herbivores and carnivores. *Brachiosaurus*, a quadrupedal herbivore, was probably the largest land animal of all time. Other, bizarre, nondinosaurian reptiles also appeared in the Jurassic. Ichthyosaurs and the first plesiosaurs (large animals with long necks and tails and four paddles for limbs) became abundant in the seas, and the first batlike pterosaursinvaded the skies. The only mammals were small, inconspicuous, and generalized. *Archaeopteryx*, from the Upper Jurassic of Germany, generally is considered to be the first bird.

Dinosaurs continued to dominate terrestrial Cretaceous faunas but began a slow decline culminating in extinction at the end of the period. Large sauropods (quadrupedal, herbivorous saurischians, saurischians being "reptile-hipped" dinosaurs) declined throughout and disappeared before the end of the period. Theropods (bipedal saurischians with short forelimbs), including *Tyrannosaurus*, the largest known terrestrial carnivore, also declined, but a few persisted to the end of the period. Ornithischians ("bird-hipped" dinosaurs) included a number of quadrupedal herbivores. One of these, *Triceratops*, with its distinctive three horns and fringed neck, persisted to the very end of the Cretaceous. Bipedal, ornithischians included the duckbilled dinosaurs. Pterosaurs also declined to extinction at the end of the period, but one of them, *Quetzalcoatlus*, was a giant, having a wingspan of 10 meters. Ichthyosaurs declined, but plesiosaurs were abundant, along with giant lizard-like mosasaurs in the marine fauna. All, however, became extinct at or before the end of the Cretaceous. The Cretaceous also saw the rise of turtles, snakes, lizards, and crocodilians—all currently prominent reptiles. Mammals of the time included both marsupials (mammals retaining the newborn in a pouch on the mother's abdomen) and placentals (mammals in which the young are not nurtured in a pouch). All of them were small, resembling modern shrews and insectivores, but included the ancestors of all modern mammals. Cretaceous birds were rare, but a diversity of toothed birds are known, including the flightless, marine, diving bird *Hesperornis*. More modern birds appeared in the very latest Cretaceous.

Marine and Insect Life

Chondrostian fish (ray-finned, but with largely cartilaginous skeletons) diversified greatly in the Early Triassic as sharks declined and more primitive fish characteristic of the Paleozoic persisted in low numbers and variety. Cartilaginous fish, including sharks, continued to be dominant in the Jurassic, but holostean fish gave rise to the first teleosts (bony fish) in the Late Jurassic. This group continued to expand and diversify in the Cretaceous and persists to the present.

Primitive insects occured sporadically in Triassic through Middle Cretaceous fossil faunas but were reduced to a minor role by an explosive appearance of very modern-appearing insects accompanying the appearance of broad-leaved forest trees in the Middle Cretaceous.

Marine invertebrates suffered extensive extinction in the Permo-Triassic transition. The Paleozoic corals, trilobites, and graptolites completely disappeared, and other groups, such as brachiopods, bryozoans, and crinoids, were decimated. By the Late Triassic, however, new organisms had occupied the environments inhabited by the extinguished animals. Modern corals, not at all closely related to Paleozoic corals, already were constructing reefs. The old bryozoans were replaced by new bryozoan types, and the reduction in brachiopods was matched by expansion among the pelecypods (clams, oysters, and scallops). Ammonites, descended from a single group of shelled cephalopods surviving into the Early Triassic, quickly evolved into many species, so much so that the Mesozoic is known as the age of ammonites as well as the age of dinosaurs. Other characteristic organisms appeared during the Jurassic and Cretaceous, including redistid clams (giant pelecypods that built reefs), a wide variety of echinoids (sea urchins and sand dollars), globigerinids and other planktonic foraminifera (single-celled animals secreting shells), and coccoliths (single-celled, planktonic algae secreting minute carbonate plates), along with crabs, shrimp, and lobsters. Globigerinids and coccoliths became sufficiently abundant in the early part of the Late Cretaceous to cause worldwide deposition of chalk and chalky limestone.

By the end of the Mesozoic, many organisms had become extinct, including dinosaurs and ammonites. These extinctions, though not as extensive as those at the end of the Paleozoic, are the reason for recognizing separate Mesozoic and Cenozoic eras. Not all the characteristic Mesozoic animals that failed to survive into the Cenozoic, however, became extinct at the very end of the period. In addition, very little change occurred in the flora, as the major change took place in the middle of the Cretaceous.

Close examination of the rocks at the Mesozoic-Cenozoic boundary as defined by occurrences of planktonic foraminifera, coccoliths, ammonites, and dinosaurs has uncovered the widespread occurrence of a thin layer containing anomalous concentrations of iridium and, more questionably, sooty material and minerals of the sort associated with meteoritic impacts. This evidence has been used to advance a theory that the end-of-Cretaceous extinctions were caused by meteoritic impact, perhaps through drastic cooling caused by vast quantities of dust and smoke thrown into the atmosphere. Physical evidence of an actual impact has not convinced all researchers, however, and many paleontologists do not believe that the gradual disappearance of the Mesozoic fauna can readily be explained by an instantaneous event, even though it might be synchronous with the final extinctions.

Mesozoic rocks are rich in mineral fuels. Coal is sparse in the Triassic, reflecting the aridity of the time. Much coal, however, occurs in the Jurassic of Asia, deposited on alluvial plains north of the Tethyan orogenic belt, and even more extensive deposits in the Cretaceous of North America and northern South America are in a similar position relating to the growing western American mountains. Jurassic and Cretaceous oil and gas are very important, especially around the Gulf of Mexico and in the Persian Gulf region. Triassic ore deposits are rare because of little mountain building and igneous intrusion, but Jurassic and Cretaceous metallic ores are very extensive. The gold of the California Mother Lode is in Jurassic rocks, and many of the precious and base metal reserves of the Rocky Mountains and the Andes are Cretaceous. Similarly, the mountains formed elsewhere on the Pacific Rim and on the northern margin of the Tethys are rich in metallic ores. Finally, the diamond pipes of southern Africa are Cretaceous.

Ralph L. Langenheim, Jr.

Further Reading

Arkell, W. J. *Jurassic Geology of the World*. Edinburgh: Oliver and Boyd, 1956. A summary description of the Jurassic rocks of the world, their location and age, as determined in detail by fossil content. A professional work with an extensive bibliography.

Brookfield, Michael E. *Principles of Stratigraphy*. Hoboken, N.J.: Wiley-Blackwell, 2004. Written for undergraduate students; this text provides an overview of the principles and applications of stratigraphy. Includes stratigraphic techniques and case studies. Organized into three sections, beginning with foundational material, followed by data collection and research topics, and completed with interpretations and analysis.

Dott, Robert H., Jr., and Donald R. Prothero. *Evolution of the Earth*. 8th ed. New York: McGraw-Hill, 2009. This basic textbook on historical geology is aimed at students of geology. However, it is very readable by anyone with a background in science. Presents an up-to-date account of the earth's history from the viewpoint of plate tectonics. Includes a glossary.

Grotzinger, John, and Tom Jordan. *Understanding Earth*. 6th ed. New York: W. H. Freeman, 2009. This comprehensive physical geology text covers the formation and development of the earth, including information on the Mesozoic. Readable by high school students, as well as by general readers. Includes an index and a glossary of terms.

Levin, Harold L. *The Earth Through Time*. 9th ed. Hoboken, N.J.: Wiley, 2009. A history of the earth, emphasizing North America. Especially good account of the fossil record. A beginning college text.

Ogg, James G., Gabi Ogg, and Felix M. Gradstein. *The Concise Geologic Time Scale*. New York: Cambridge University Press, 2008. This book is a complete overview of the geological time scale, including stratigraphy topics such as chronostratigraphy and magnetic stratigraphy. The book is organized by geological time periods. Also contains a reference appendix, the geological time scale table, and an index.

Prothero, Donald R. *Bringing Fossils to Life*. 2d ed. Boston: McGraw-Hill, 2004. This well-illustrated and entertaining text covers a broad range of paleontological topics, including biostratigraphy and the use of fossils in the study of Mesozoic rocks. Glossary, bibliography, and index.

Stanley, Steven M. *Earth System History*. 3rd ed. New York: Freeman, 2008. An introductory undergraduate level text integrating coverage of life and the environment. Good coverage of the Mesozoic.

Wicander, Reed, and James S. Monroe. *Historical Geology*. Belmont Calif.: Brooks/Cole, Cengage Learning, 2010. This textbook is accessible to university students with or without a background in geology. It covers Earth's geological systems and plate tectonics, physical and biological history, geological time, and evolution. There are many diagrams, images, and charts along with a CD-ROM to supplement the text.

See also: Biostratigraphy; Cenozoic Era; Continental Drift; Continental Glaciers; Continental Rift Zones; Continental Shelf and Slope; Continental Structures; Continents and Subcontinents; Cretaceous-Paleogene Boundary; Dinosaurs; Geologic Time Scale; Ice Ages; Land Bridges; Paleoclimatology; Paleogeography and Paleocurrents; Paleoseismology; Paleosols; Paleozoic Era; Proterozoic Eon; Stratigraphic Correlation; Transgression and Regression; Unconformities; Uniformitarianism.

MICROFOSSILS

Micropaleontology is the study of plant and animal fossils that are too small to be seen without magnification. These microscopic objects provide valuable information about the evolution of life on the earth and about changes that have occurred on the earth's surface through time. They also have great value as index fossils and as indicators of ancient environments, data useful in the search for oil and natural gas.

PRINCIPAL TERMS

- **foraminifera:** single-celled, amoeba-like animals
- **index fossils:** indicators of geologic age; they have short geologic (time) ranges and broad geographic distribution
- **isotopes:** atoms of an element that differ in weight because different numbers of neutrons are present in their atomic nuclei
- **morphologic evolution:** changes in the body or skeleton shape of organisms through time
- **parthenogenic:** organisms in which unfertilized females produce viable, fertile offspring without copulation with males of the species

MICROPALEONTOLOGY

When people think of fossils, they envision items such as the bones and teeth of dinosaurs, the petrified bones of Neanderthals, or the shells of oysters. Micropaleontologists study the preserved remains of organisms too small to be seen with the unaided eye. It is convenient to divide micropaleontology into two subfields: animal micropaleontology and plant micropaleontology. Animal micropaleontology encompasses the study of a wide variety of fossils. In most cases, the material studied is actually the shell constructed by the once-living animal. Shells are commonly composed of calcium carbonate (the common mineral calcite), calcium phosphate (the mineral apatite, a constituent of human teeth and bones), and opaline silica (quartz with water molecules in its crystal structure). Other microscopic animals build shells from mixes of sand grains, shell fragments, and volcanic ash, all glued together with organic or mineral cement. Thus, shells are referred to as calcareous, phosphatic, siliceous, or agglutinated.

Microfossils vary in size as well as in shell composition. As a rule, few are smaller than 0.05 millimeter in diameter; most are in the range of 0.75 to 2.0 millimeters, and a very few are as large as 5 centimeters. Among the larger microfossils are those called nummulites, which are major constituents of the limestone used in the Egyptian pyramids. In about 450 B.C.E., Herodotus described them, erroneously, as mummified lentils (the food of pyramid construction crews).

Systematic study of microfossils did not begin until the early years of the nineteenth century. Alcide Dessalines d'Orbigny published a paper describing some microfossils that he thought to be microscopic cephalopods related to the chambered nautilus. Subsequently, it was realized that d'Orbigny's microfossils were types of protozoans (single-celled animals) called foraminifera. Another pioneer was Christian Gottfried, who was the first to treat micropaleontology as a field of study. As is the case for larger fossils, microfossils can be used to determine the age of sedimentary rocks. European micropaleontologists used them for this purpose as early as 1874 and, by 1930, micropaleontology was aiding in petroleum exploration worldwide.

FORAMINIFERA AND RADIOLARIANS

Many types of animals can produce microscopic shells or other hard parts that may be preserved in sedimentary rocks. Among these are protozoans, gastropods (snails), worms, crustaceans (crabs, lobsters), sponges, echinoderms (starfish, sea urchins), and fish. Protozoans contribute significantly to the microfossil record. Two are particularly useful as indicators of geologic age and of ancient environments: the foraminifera and the radiolarians. Most living foraminifera are found in the ocean, where their distribution is controlled by water temperature, salinity, depth, turbulence, light intensity, bottom conditions, availability of food, predators, parasites, and other biologic factors. Foraminifera, or forams, usually have calcareous or agglutinated shells. The largest number of species are bottom dwellers (benthic), but one group, the globigerinids, evolved to live as passive floaters (or plankton) in the surface waters of the oceans. Benthic forams have a long geologic history extending back 500 to 600 million years. For the last 300 million years, their shells have been significant rock formers. The rock that forms the White Cliffs of Dover is composed largely of the shells of foraminifera. Planktonic forams evolved more recently, first appearing in rocks about 175 million years old. Since planktonic species live in surface waters, ocean currents may carry them thousands of miles. This wide geographic distribution coupled with their rapid morphologic evolution makes planktonic foraminifera especially valuable as index fossils.

Radiolarians are also planktonic protozoans, but they differ from foraminifera in their soft-body-part anatomy and in their shell construction. Radiolarian shells are composed of opaline silica, are usually in the 0.05 to 0.5 millimeter size range, and display a bewildering array of shapes. Most are variations on three shapes: spheres, cones, or disks. Some radiolaria secrete their shells as spongy masses, others build shells of perforated sheets of silica, and still others construct lattice-work shells of great delicacy and beauty. Radiolaria are found in rocks almost 600 million years old, giving them the longest geologic range of planktonic microfossils. Mesozoic and Cenozoic radiolarians are better known than are Paleozoic ones.

Ostracods

Microscopic crustaceans called ostracods (or ostracodes) have a 500-million-year-long fossil record and are abundant in many sedimentary rocks. Unlike the familiar macroscopic crustaceans, ostracods encase their minute, shrimplike bodies in a pair of tiny, calcareous, bean-shaped shells 1 to 5 millimeters long. Outer surfaces of these shells may be smooth, or they may bear spines, wartlike bumps, grooves, ridges, flanges, and pores. Hinge structures and muscle scars, found on the inner surfaces of the shells, are also useful features in distinguishing different ostracods. Although some species are planktonic, most ostracods are benthic. Living ostracods can be found from the deepest ocean floor to the shoreline and landward into lakes and streams. A few species have adapted to live on land in moist ground litter. Playa lakes in desert regions often are inhabited by ostracods. Species living in these harsh environments are often parthenogenic: Unfertilized females lay fertile eggs, which lie dormant in the muddy bottom of a dried-up lake; the next time the lake fills, the eggs hatch into a new generation of females, and the cycle continues. Dormant periods of a decade or more have been reported.

Two different kinds of toothlike structures are frequently found in assemblages of microfossils. Annelid worm jaws, called scolecodonts, are composed of a resistant organic material (chitin) and have been found in rocks up to 600 million years old. The second type are called conodonts ("cone-toothed" animals). These fossils are 0.2 to 6.0 millimeters in greatest dimension, are composed of calcium phosphate, and occur as isolated specimens or in clustered assemblages. The composition suggests that they were produced by an animal, and their distribution in sedimentary rocks suggests that these "conodont animals" (ordinal name Conodontophorida) were marine and planktonic. They first appeared in Cambrian time, flourished through the rest of the Paleozoic, and became extinct during the Triassic period. Specimens of conodont animals are known from the Carboniferous of Scotland and the Ordovician of southern Africa. They appear to have been eel-like animals in which the conodonts themselves formed a grasping structure in the throat. Although the affinities of this animal remain obscure, it is generally thought to have been a chordate of some kind.

Study of Microfossils

While microfossils can be recovered from many types of marine sedimentary rocks, they are most abundant in fine-grained rocks such as shale. Samples of the rocks can be obtained from surface outcrops or from subsurface boreholes in the form of cores or cuttings. In collecting samples for study, care must be taken to eliminate any contamination, and accurate records of sampling localities must be maintained.

In the laboratory, a variety of techniques are used to extract the microfossils from the host rock and to clean them. Composition of the fossils, composition of the host rock, and the kind of study to be done dictate the separation methods used. Some foraminifera, for example, can be studied profitably only in thin section. To prepare them, fossiliferous samples are glued to microscope slides and are ground and polished until the rock is paper thin and the internal features of the shell can be seen. Acids are used to dissolve sedimentary rocks effectively, in order to release insoluble siliceous or phosphatic microfossils. Acetic acid will dissolve limestone without damaging conodonts, and radiolarians can be freed best by dissolving the host rock in hydrochloric acid.

Light micrograph of Nummulitic foraminiferans in limestone from an Egyptian pyramid, magnification 4× at 35mm. (M. I. Walker/Photo Researchers, Inc.)

Calcareous-shelled ostracods and foraminifera are usually extracted from shales in the following manner. A clean, dry, crushed (to pea size) sample is soaked for thirty to sixty minutes in kerosene. The kerosene is poured off, filtered, and saved for reuse. The sample container is filled with water, and a wetting agent is added. This mixture is boiled for twenty to thirty minutes, and then it is poured through a 200-mesh screen. Fine clay particles pass through, leaving the microfossil residue on the screen. Usually it is necessary to repeat the boiling and screening process several times to yield a good, clean residue. After drying, the residue is ready for "picking." A one-grain-thick layer of residue is spread on a picking tray, which is then placed on the stage of a binocular microscope. Magnification of thirty to forty times is required. A picking brush is also needed. The micropaleontologist slides the picking tray back and forth so that all the residue is scanned. Once a microfossil is found, the picking brush and a steady hand come into play: The brush is moistened and guided to the fossil, which sticks to the brush while being transferred to a microscope slide for future study. Routinely, several hundred specimens will be picked from each sample; each specimen will be identified and tallied in the sample census. Specimens can be repicked for further study using the greater magnification of the scanning electron microscope (SEM). Thin coatings of a conductive metal are vacuum-plated on the specimens, which are then ready for SEM analysis. In a vacuum chamber, a beam of electrons is focused on the specimen; reflected electrons are collected and are converted electronically into an image of the specimen. Magnification of fifty thousand times or more is possible, and the electronic image has a three-dimensional appearance.

Biostratigraphy

Probably the single most important application of microfossils is in biostratigraphy, or the use of fossils to determine the age of rocks. Fundamental to biostratigraphy has been the establishment of reference sections. These are sequences of rock that have been precisely dated and whose fossils have been described in considerable detail. Fossils from a rock sequence of unknown age are compared with fossil sequences from reference sections and the best match is obtained. Because of their rapid evolution and their broad geographic distribution, fossils of planktonic organisms are particularly well suited for this work. For example, radiolarian assemblages from Japan, western Texas, and the Ural Mountains of Eurasia can be compared directly with one another.

Using a picked collection of foraminifera, a number of other studies can be carried out. If the assemblage contains both planktonic and benthonic species, the planktonics-benthonics ratio gives a measure of water depth at the time of deposition. Another depth indicator is the ratio of calcareous-shelled benthonics to agglutinate-shelled benthonics. The diversity of the assemblage may also provide useful information. In a tropical assemblage, many species are usually present, but only a few individuals of each species occur (high diversity). In a high-latitude assemblage, only a few species are present, but each is represented by a large number of individuals (low diversity). Diversity measures must be used carefully, as similar effects can be seen in relation to water depth—low diversity in very shallow or very deep water and higher diversity at intermediate depths.

Calcareous-shelled forams can also be analyzed to give a direct measure of temperature of the water in which they lived. Oxygen atoms in two forms (isotopes) occur in seawater. The ratio of oxygen-18 to oxygen-16 is a function of water temperature. When forams extract calcium carbonate from seawater and build their shells with it, the oxygen isotopes are incorporated as well. In the laboratory, the shells are converted to carbon dioxide, and the ratio of oxygen-18 to oxygen-16 is determined. The measured ratio is then plugged into a mathematical formula to calculate water temperature. A second temperature-measuring technique can be used with planktonic forams. As the animals grow, they add chambers to their shells in a spiraling pattern. In warm water, the spiral is clockwise (right-handed); in cold water, it is counterclockwise (left-handed). The coiling ratio for planktonic forams in a sample thus gives an indication of water temperature. Although this technique is less precise than isotopic measurements, it is faster and less expensive.

Studies of evolution can also be conducted using microfossils. Because planktonic animals seem to evolve faster than do benthonic ones, plankton are often preferred in these studies. Ideally, one needs closely spaced samples from a core in an area where insignificant environmental change occurred while deposition of the planktonic shells was continuous. Some lineages of Cenozoic radiolarians have been traced for several million years, during which time small morphologic changes accrued so that the descendant species are quite different from their ancestors in morphology. In some of these studies, hybridization between different but related species seems to have been recorded by the fossils.

William C. Cornell

Further Reading

Anderson, O. R. *Radiolaria.* New York: Springer-Verlag, 1983. This volume considers the biology of living radiolarians and includes discussion of morphology, ultrastructure, physiology, ecology, and evolution. A glossary and appendix are included, as is an index. College level.

Armstrong, Howard, and Martin Brazier. *Microfossils.* 2d ed. Malden, Mass.: Blackwell Publishing, 2005. Discusses taxonomy, phylogeny, and ecology of the major microfossil groups. This concise volume treats fifteen groups of microfossils, emphasizing their features as seen in a microscope. Illustrated with high-quality line drawings and photographs. Improves upon the original text by including recent advances in scientific study. Discusses isotopes and their application in stratigraphy and the study of evolution. Also covers the study of palaeoclimatology using microfossils. An appendix summarizes sampling and preparation techniques. The bibliographic entries are chiefly twentieth-century articles and books. Both a general index and a systematic index are included. College level.

Briggs, Derek E. G., and Peter R. Crowther, eds. *Paleobiology: A Synthesis.* Oxford: Blackwell Scientific Publications, 1990. This multiauthored text deals with all aspects of paleontology in a series of short articles by experts in each field. Intended for university students but also useful to high school students and interested readers.

__________________. *Palaeobiology II.* New York: Wiley-Blackwell, 2001. This text is a sequel to the acclaimed *Palaeobiology: A Synthesis.* A compilation of articles discussing fossil systematics; combining genetics, geochemistry and other disciplines.

Feldmann, Rodney M., Ralph E. Chapman, and Joseph T. Hannibal, eds. *Paleotechniques.* Knoxville, Tenn.: Paleontological Society, 1989. This extremely useful volume discusses specific techniques for collecting, preparing, and studying microfossils. Extensive bibliography.

Hag, Bilal, and Anne Boersma, eds. *Introduction to Marine Micropaleontology.* 2d ed. New York: Elsevier, 1998. Articles by specialists on fourteen groups of microfossils. Illustrations are outstanding; indexing is thorough. A helpful glossary of terms is included. Each chapter includes lists of general references and of cited references. College level.

Jenkins, J. M., ed. *Applied Micropaleontology.* New York: Springer, 2010. A very technical presentation of seven case studies covering the various topics in micropaleontology. This text discusses stratigraphy and sedimentology, foraminifera, and ostracods. Best suited for advanced students with a strong geology background.

Pokorný, Vladimír. *Principles of Zoological Micropaleontology.* Translated by K. A. Allen. 2 vols. Elmsford, N.Y.: Pergamon Press, 1965. Translated from the German, this 650-page book presents an excellent overview of animal micropaleontology. Includes discussions of micropaleontological methods and of the major groups of animal microfossils. Profusely illustrated with high-quality photomicrographs and line drawings. References are grouped by chapters and include good coverage of literature published outside of North America. Fossil and subject indexes are given. College level.

Prothero, Donald R. *Bringing Fossils to Life.* 2d ed. Boston: McGraw-Hill, 2004. This well-illustrated and entertaining text covers a broad range of paleontological topics, including micropaleontology. Glossary, bibliography, and index.

See also: ArchaeB.C.E.ia; Biostratigraphy; Cambrian Diversification of Life; Catastrophism; Colonization of the Land; Dinosaurs; Ediacaran Fossils; Eukaryotes; Evolution of Birds; Evolution of Flight; Evolution of Humans; Evolution of Life; Fossilization and Taphonomy; Fossil Plants; Fossil Record; Fossils of the Earliest Life-Forms; Fossils of the Earliest Vertebrates; Geoarchaeology; Geologic Time Scale; Geomicrobiology; Gondwanaland and Laurasia; Ice Ages; Life's Origins; Mammals; Mass Extinctions; Molecular Paleontology; Neanderthals; Paleobiogeography; Paleobotany; Paleoclimatology; Petrified Wood; Prokaryotes; Stromatolites; Systematics and Classification.

MID-OCEAN RIDGE BASALTS

Mid-ocean ridge basalts are slowly extruded and build up oceanic ridges so that they or their coarse-grained equivalent make up the bulk of the oceanic crust. The basalts form by the melting of peridotite in the upper mantle. Most of these basalts are characterized by lower potassium than basalts formed in other tectonic environments.

PRINCIPAL TERMS

- **basalt:** a dark-colored, fine-grained to porphyritic igneous rock
- **gabbro:** the coarser-grained equivalent of basalt
- **magma:** molten rock material or melt, mixed with crystals, that occurs below the earth's surface
- **peridotite:** a coarse-grained rock that consists of olivine, pyroxene, and garnet; it is likely the main rock in the upper mantle of the earth
- **porphyritic rock:** igneous rock with large mineral crystals embedded in much finer-grained minerals
- **tholeiitic basalt:** a type of basalt with calcium-rich plagioclase, monoclinic pyroxene, orthorhombic pyroxene, and olivine or quartz; it has low potassium concentrations compared to the more alkali-rich basalts
- **upper mantle:** the mantle occurs between the core and the crust of the earth; portions of the upper part of the mantle are believed to have a zone of partial melting that may produce melts that crystallize to basalts

COMPOSITION OF BASALTS

Plate tectonics is the theory that the earth's crust is divided into seven major plates as well as many minor plates and microplates that are being created at divergent boundaries on one edge and destroyed by subduction at another edge. Divergent boundaries are ones in which basaltic magma is being produced and added as new oceanic crust to the plate as it moves slowly away from this boundary. Large mountain chains called mid-ocean ridges, which encircle the earth mostly below the ocean's surface, are formed at the divergent boundaries. The plate gradually moves away from the divergent boundary and is eventually subducted or thrust under another plate.

Basalts are dark, fine-grained to porphyritic rocks that contain calcium-rich plagioclase, pyroxene, olivine, and minor spinel. Basalts or their coarser-grained equivalents, gabbros, are the main rock types produced at oceanic ridges. Basalts and gabbros, then, make up most of the oceanic plate that moves in conveyer-belt fashion out from the oceanic ridges. The basalts at oceanic ridges are reasonably homogeneous; for example, they mostly range from 48 to 52 weight percent silicon dioxide. Most of the basalts at oceanic ridges are in a subgroup of basalts called tholeiitic basalts. They have the same composition as other basalts except they contain orthorhombic pyroxene and generally have low potassium contents.

Tholeiitic basalts are also found in other tectonic environments with other rock types, such as on ocean floors away from plate boundaries (as in the Hawaiian Islands), subduction zones, and continental rifts. They can also occur as flood basalts on continents. The main difference between most oceanic ridge tholeiitic basalts and tholeiites formed in other tectonic environments is that the oceanic ridge tholeiites have lower amounts of potassium and certain trace elements (rubidium, barium, and the light rare-earth elements) than the others.

VARIATION IN COMPOSITION: FAST-MOVING VERSUS SLOW-MOVING RIDGES

The variation in composition of mid-ocean ridge basalts has not been studied in the same detail as basalts at the surface have been studied, since they usually occur several kilometers below the surface of the ocean. Detailed sampling in some places, however, suggests interesting variations in composition and style of eruption.

Oceanic ridges vary a great deal in the speed at which they separate. Fast-moving ridges, such as the East Pacific Rise, spread at rates of 8 to 16 centimeters per year, and there is no distinct valley produced at the highest portions of the ridge. Also, the width of the volcanic zone is narrow, suggesting a narrow but large magma chamber below the surface. In contrast, slow-moving ridges, such as the Mid-Atlantic Ridge, move at rates of only 1 to 4 centimeters per year, and there is a central valley about 8 to 20 kilometers wide and 1 to 2 kilometers deep with a wider zone of volcanic activity. This suggests that the magma chambers below the surface are wider than those at fast-moving ridges.

In the case of fast-moving oceanic ridges, sheets of lava are extruded fairly rapidly and continuously along a low-relief ridge. In contrast, slow-moving ridges have lavas slowly extruded at only certain points along the ridge, where the ridge is built to fairly high relief. In addition, fast-spreading ridges have frequent eruptions that are fairly homogeneous in composition, while slow-moving ridges have less frequent eruptions that are more heterogeneous in composition. The fast-moving ridges are believed to have narrowly focused magma chambers where the magma is injected along a narrow region all along the ridge crest, while slow-moving ridges are believed to have only a few, less focused magma chambers that erupt lavas only at different points along the ridge over a wider area.

Slow-moving ridges seem to exhibit regular variations in the composition of basalts from the center of the ridge out to the valley walls. Basalts in the center of the rift have a greater abundance of large olivine crystals relative to plagioclase and monoclinic pyroxene crystals than those closer to the valley walls. The volcanic glasses in the center of the ridge also contain less silica and potassium than those near the valley walls. Spreading of ridges at speeds intermediate between these two extremes has characteristics intermediate between those of the fast- and slow-moving ridges.

Variation in Composition: Eruption Styles

Volcanic eruption rates at undersea ridge crests are not continuous, and they are of low volume compared to eruptions in other tectonic environments. Nevertheless, the continued eruption of the tholeiitic basalts over long periods of time on the ridge crests results in the largest total volume of basaltic eruptions of anywhere in the world.

Basalts formed at mid-ocean ridges likely form by partial melting of rocks in the upper mantle of the earth. The melts formed may be modified in composition by the crystallization of minerals that settle out of the melt, thus changing their composition.

The main rock in the upper mantle is probably peridotite. Up to 30 percent melting of this rock produces tholeiitic basaltic melts similar in composition to many found at oceanic ridges. The peridotite must rise upward from some depth in the mantle as a blob or plume. The peridotite is close to its melting point, so it can begin to melt as the pressure is reduced on rising to shallow depths of about 80 kilometers below the surface. This melting produces a liquid-crystal mush that melts further as it rises to shallower depths. Eventually, a point is reached where most of the melt can leave the crystals behind as the liquid rises into the oceanic crust.

The large differences in composition among tholeiitic basalts formed in different regions may be explained by differences in the composition of the peridotite source and by differences in the degree of melting of the peridotite. For example, the relatively high potassium, rubidium, barium, and light rare-earth element concentrations along the Mid-Atlantic Ridge near Iceland can be explained by the melting of previously unmelted peridotite with high concentrations of these elements. Most peridotite in the upper mantle below oceanic rises, however, appears to have melted at least once before. Melting of peridotite causes it to be depleted in potassium, rubidium, barium, and light rare-earth elements as these elements move into the melt. Thus, melting of this peridotite a second time produces the abundant tholeiitic basalts depleted in these elements.

The smaller differences in concentrations of elements across the valley at the top of the mid-ocean ridge may be explained by the crystallization and settling out of minerals such as olivine, pyroxene, and plagioclase from the melt. These minerals have different compositions from those of the melt, so that as they settle out of the melt or float, the composition of the melt slowly changes. For example, as olivine crystallizes, it picks up less silica than does the melt, so the melt will gradually increase in silica as the olivine settles out. The melt and some suspended crystals may periodically squirt upward along fractures as crystallization takes place. Some melt and crystals may reach the surface of the ridge as a lava flow, or some may crystallize completely within the fractures below the surface.

Over long time periods, such processes result in the zonation of the oceanic crust. Depths of about 4 to 8 kilometers below sea level contain mostly coarse-grained gabbros with varied amounts of olivine, pyroxene, and plagioclase. Depths of 2 to 4 kilometers below sea level contain mostly fine-grained basalts produced by periodic extrusions of lava on the seafloor ridge. Between the gabbros and basalts, there are a lot of injections of intermediate-grain-size gabbros along fractures. Sometimes these rocks are so abundant that it is difficult to tell the composition of the original rock.

Interaction of Basalt with Seawater

Some basalts sampled on the sea floor appear to have had sodium added to them and calcium, iron, and silica removed from them. These basalts likely came into contact with hot seawater enriched in sodium. Hot springs have been observed to exist along the East Pacific Rise. The waters emitted at the hot springs have temperatures of up to 350 degrees Celsius. Some hot springs have a black or gray, smoky appearance. The dark color is the result of the precipitation of tiny sulfide and oxide minerals from the hot water as it interacts with cold seawater. As a result of this activity, large columns of sulfide minerals are built up, and a unique ecosystem of B.C.E.ia, giant worms up to 2 meters long, clams, and crabs occurs near the vents of hot water. They obtain their energy from the heat of the water.

These gray and black smokers appear to be the surface expression of an elaborate hot-water plumbing system in which seawater percolates down fractures in an area several kilometers wide along the ridge axis. The seawater likely percolates downward to depths of at least 2 to 3 kilometers and is gradually warmed to temperatures of 400 to 450 degrees Celsius. The seawater then runs out of fractures and eventually percolates back up toward the surface, and some is liberated at the ridge axis as smokers. During the course of its circulation through the fractures in the rocks, the hot water likely reacts with the basalts and forms new minerals.

Significance

Some geologists study the potential ore deposits in association with mid-ocean ridge basalts to learn about their origin. In some places, for example, copper sulfide deposits are formed as a precipitate around the vents of hot springs. The island of Cyprus has abundant copper sulfide deposits that now occur at the surface that were originally formed this way. These deposits have been mined since ancient Greek times.

Manganese is also given off by these hot waters, but not much of it is deposited around the vents. Instead, the manganese slowly precipitates out of seawater in manganese nodules on large portions of the ocean floor. Although not yet mined, these nodules are a potentially abundant source of manganese and other metals. Other ore deposits associated with mid-ocean ridge basalts are related to the precipitation of minerals out of the basaltic melts below the rise. In some cases, the mineral chromite (an ore of chromium) is one of the first minerals to crystallize out of the magma. Since this mineral is dense, it settles out of the melt as pods and layers. Some of the pods are small, but others may contain up to several million tons of chromium. One such deposit is now exposed at the surface in Oman.

Geologists also study the mid-ocean ridge basalts to build theories about how they form and evolve. Much evidence must be accumulated to develop such theories. For example, experiments in furnaces that approximate the conditions in the upper mantle suggest that the main rock in the upper mantle, peridotite, is the only rock that can melt to produce basaltic melts. Also, certain isotopic and trace element concentrations in the basaltic melts are consistent with their origin by melting of peridotite. Estimates of temperatures and pressures in the upper mantle suggest that peridotite is close to its melting point in many places. Experiments in furnaces suggest that the release of pressure on peridotites under these conditions as they rise toward the surface will cause them to begin to melt.

Robert L. Cullers

Further Reading

Blatt, H., R. J. Tracy, and Brent Owens. *Petrology: Igneous, Sedimentary, and Metamorphic.* 3rd ed. New York: W. H. Freeman, 2005. This introductory text about the formation of rocks is somewhat technical. The section on mid-ocean ridge basalts is very detailed. Contains a glossary and index.

Chester, Roy. *Furnace of Creation, Cradle of Destruction.* New York: AMACOM Books, 2008. This text discusses the turbulent processes of the planet. It covers earthquakes, volcanoes, and tsunamis in reference to plate tectonics, natural disasters, and prediction and mitigation of these events. Multiple chapters explore seafloor spreading. The author also discusses hydrothermal activity. This text takes on an immense range of content, but still explains concepts clearly and with detail.

Davidson, J. P., W. E. Reed, and P. M. Davis. *Exploring Earth.* 2d ed. Upper Saddle River, N.J.: Prentice Hall, 2001. This introductory geology text has a chapter on divergent plate margins. Includes a glossary and an index.

Garrison, Tom S. *Oceanography: An Invitation to Marine Science.* Belmont, Calif.: Brooks/Cole, Cengage Learning, 2010. Chapter 3 provides excellent coverage of plate tectonics, ocean formation, seafloor spreading, and island arc formation. Chapter 4 covers topography and bathymetry of deep-ocean basins. Abundant diagrams aid readers from the layperson to advanced undergraduates.

Haymon, Rachel M., et al. "Manifestations of Hydrothermal Discharge from Young Abyssal Hills on the Fast-Spreading East Pacific Rise Flank." *Geology* 33 (2005): 153-156. This article discusses oceanic hydrothermal heat released via black smokers. The authors comment on the research possibilities available due to fluid expulsion from the abyssal sea floor. This article is best suited for advanced undergraduates and graduate students. It provides a view of where research of the sea floor is heading.

Kennish, Michael J. *Practical Handbook of Marine Science.* 3rd ed. Boca Raton, Fla.: CRC Press, 2001. Chapter 4 covers marine geology topics such as seafloor topography, sediments, plate tectonics, mid-ocean ridges, and hydrothermal vents. Best suited for advanced undergraduates and graduate students. References are provided for each chapter.

Murck, B. W., and B. J. Skinner. *Geology Today.* New York: John Wiley and Sons, 1999. This readable introductory geology text includes a section on mid-ocean ridge basalts, as well as a glossary and index.

Nicolas, A. *The Mid-oceanic Ridges.* Berlin: Springer-Verlag, 1995. This nicely written account of the formation of oceanic ridges and associated ore deposits is suitable for the introductory reader. Includes a glossary.

Seibold, Eugen, and Wolfgang H. Berger. *The Sea Floor: An Introduction to Marine Geology.* 3rd ed. New York: Springer-Verlag, 2010. As the title suggests, this is an introduction to many topics in marine geology. This text covers geological structures from the continental shelf to deep-ocean trenches. Processes such as seafloor spreading, the sediment cycle, currents, and pelagic rain are discussed.

Singh, Satish C., et al. "Discovery of a Magma Chamber and Faults Beneath a Mid-Atlantic Ridge Hydrothermal Field." *Nature* 442 (2006): 1029-1032. This article provides an example of current seafloor exploration and research. It discusses the use of seismic analysis for developing models of mid-ocean ridges. Also provides some information on hydrothermal activity and submarine volcano structure. Understood by the layperson with careful reading.

See also: Abyssal Sea Floor; Eruptions; Ocean Basins; Ocean-Floor Exploration; Oceanic Crust; Ocean Ridge System; Ophiolites; Seafloor Spreading.

MINERALS

Minerals are naturally occurring solids made of inorganic matter arranged in crystalline structures. They are classified into major groups that depend upon their chemical composition, and they can be identified in several ways. Minerals are essential to human, animal, and plant health. They are major components of rocks and make up about 90 percent of the earth's surface. Also, minerals are used in industrial and commercial applications.

PRINCIPAL TERMS

- **anion:** a negatively charged ion
- **cation:** a positively charged ion
- **cleavage:** describes the way a rock or mineral breaks in a certain direction, or plane
- **crystal:** a solid substance with a regular form and symmetrical faces
- **element:** a substance that cannot be broken into more simple substances
- **fracture:** the distinctive way a mineral breaks
- **gemstone:** a mineral that is valued for its color, sparkle, rarity, and hardness
- **geode:** a small rock cavity lined with crystals or other mineral matter
- **habit:** the general shape of a mineral
- **ion:** an atom or molecule in which the total number of electrons is not equal to the total number of protons; has a net positive or negative charge
- **luster:** the way light reflects off the surface of a mineral
- **ore:** a rock or mineral from which a metal can be extracted
- **unit cell:** the basic building block of a mineral

Characteristics

Minerals are structures that form naturally through geologic processes when inorganic matter is arranged in crystalline structures. These structures are formed around what is known as a unit cell through crystallization, evaporation, and precipitation. Minerals form as crystalline structures or as true crystals. When minerals form as true crystals, their faces reflect their internal structure, or the shape of their unit cell. Whether crystalline or crystal in form, all minerals are made of atoms in highly ordered geometric structures.

Minerals are vital to the health of humans, animals, and plants. Also, they are important to scientists, who use minerals to learn the history of an area, as they reveal information about environmental conditions at the time of their formation.

Mineral Identification

Minerals can be identified in many ways, including by classifying them according to their chemical composition, a process first suggested by American mineralogist and geologist James Dwight Dana in the mid-nineteenth century. Dana's original system made use of seven major chemical groups. Today, there are nine classification groups, including the following:

Halides. Minerals in the halide group combine the halogen elements chlorine (Cl), bromine (Br), fluorine (F), and iodine (I) with one or more other metals to form minerals such as halite (NaCl), sylvite (KCl), and fluorite (CaF_2). Because of the type of attraction between their atoms, halide minerals are poor conductors of heat and electricity. Also, they melt at moderate to high temperatures and rank relatively low on the Mohs scale of mineral hardness. (The Mohs scale was developed by German mineralogist and geologist Friedrich Mohs in 1812.) Minerals in this class are most likely found in evaporitic settings such as salt lakes or landlocked seas, where they are left behind as the water in these bodies evaporates. Halide is salt, and it is harvested and used commercially as table salt and as a salt for covering icy or ice-prone roads in winter.

Sulfides. The sulfide group includes those minerals formed when metal cations with a +2 charge combine with sulfur (S). This group includes most of the ore minerals such as galena (PbS), pyrite (FeS_2), sphalerite (ZnS), and calaverite ($AuTe_2$). Minerals in this class are opaque, and they exhibit distinctive colors and colored streaks. Many of the sulfide minerals are important as metal ores and are therefore mined.

Carbonates. Minerals in the carbonate group are formed when carbon (C) atoms join with oxygen (O) to form the carbonate ion (CO_3), which then combines with metal cations. Calcite ($CaCO_3$), dolomite ($CaMg[CO_3]_2$), and malachite ($Cu_2CO_3[OH]_2$) are part of this mineral group. These minerals are found in many settings, including marine and evaporitic.

Sulfates. Sulfate minerals form when SO_4 groups (which contain the sulfate ion) combine with metals. Sulfates also are found in evaporitic settings and in hydrothermal vein systems. They are uncommon and include barite ($BaSO_4$), crocoite ($PbCrO_4$), and antlerite ($Cu_3[OH]_6[SO_4]_2$).

Silicates. The silicates make up more than 90 percent of the earth's crust. Because most of the silicates are igneous rock formations, they can be used by scientists to reveal environmental conditions that were present as they formed. The silicon-oxygen tetrahedron (SiO_4) is the fundamental unit for silicates. When silicon (Si) and the oxygen atoms combine, they form an anion with a -4 charge. Positively charged atoms then combine to form minerals. The most common positively charged elements that combine with the silica tetrahedral at the earth's crust are aluminum (Al), iron (Fe), calcium (Ca), sodium (Na), potassium (K), and magnesium (Mg). It is also possible for a silica tetrahedron to join with other silicon tetrahedra to form a variety of silicate structures.

Crystal geode. A paleontologist holding a geode found in a rock on a seashore. A geode is a concretion, an irregular or ball-shaped object created when minerals fill a cavity in rock. Geodes form in different ways in different rock types. Here the geode formed from a cavity left by a fossilized ball of algae. (Pascal Goetgheluck/Photo Researchers, Inc.)

Oxides. Oxides are formed when oxygen combines with one or more metals. This group includes hematite (Fe_2O_3), magnetite (Fe_3O_4), and manganite ($MnO[OH]$). These minerals are of extreme importance in mining because they form many of the ores from which it is possible to extract valuable metals. Oxides also are among the best recorders of information about changes in the earth's magnetic field.

Elements. Minerals in the elements class include native metals and intermetallic elements such as gold (Au), silver (Ag), and copper (Cu). Each of these minerals is composed of atoms of a single element that is called the native element. Also in this group are semimetals and nonmetals such as sulfur (S).

Phosphates. The phosphates include any mineral with the tetrahedral unit AO_4. In these minerals A can be phosphorus (P), antimony (Sb), arsenic (As), or vanadium (V).

Organics. The organic minerals include those minerals that have an organic chemical component in their formulas. This would seem to contradict the definition of minerals as "inorganic," but this is not the case because these minerals are not formed as a by-product of the organisms. Minerals in this class include amber ($C_{10}H_{16}O$).

PHYSICAL PROPERTIES OF MINERALS

Minerals also have distinct physical properties that are used in their identification. These properties include the following:

Hardness. Hardness is a measure of a mineral's ability to resist scratching; it is tested using the Mohs scale. This hardness scale is based on harder materials scratching softer ones. On the Mohs scale, diamond is assigned a hardness of 10. A fingernail is assigned a hardness of 2, whereas a copper penny is a 4 and glass is a 7. As a result, a person's fingernail will scratch gypsum or talc, both of which have a hardness of 2 or less, while a copper penny will scratch anything with a hardness of 4 or lower. Glass will scratch anything with a hardness of 7 or lower. A diamond will scratch anything, including another diamond. The hardness ranking is important to mineral identification because, when compared with attributes of known specimens, it assists in identifying new specimens.

Cleavage. Cleavage is the tendency of a mineral to break (because of its atomic structure) along a plane where weakness exists. The cleavage is always parallel to crystal faces and is said to be good or perfect depending on how predictable it is that a mineral will break as expected. Because of the weakness of the bond between the potassium and oxygen atoms in micas, a mica ($X_2Y_{4\text{-}6}Z_8O_{20}[OH,F]_4$), for example, is said to have highly perfect cleavage because it predictably cleaves in sheets.

Fracture. A fracture is a mineral break against a cleavage plane. The fracture can be conchoidal in nature and can exhibit a smooth curved fracture with concentric ridges. Fractures can also be hackly (jagged with sharp edges), earthy (similar in appearance to freshly broken soil), splintery (sharp, long points), or irregular.

Luster. Luster is a measure of how light is reflected from the surface of a mineral. Luster can be classified as metallic or nonmetallic. It is metallic when a mineral looks like metal. Nonmetallic luster can be defined in one of several ways, ranging from vitreous (glass-like) to dull.

Color. Color, which also can be used to identify a mineral, is unreliable in identification because several minerals each can have different colors. When this is the case, however, these minerals will share their shapes and other attributes.

Diaphaneity. Diaphaneity, or how well light passes through a mineral, describes the degree of transparency of a mineral. The transparency can range from opaque, with no light passing through the mineral, to transparent, with the ability to see right through the mineral.

Streak. Streak is the color of the powder left (as a streak plate) when the mineral is scraped across a piece of unglazed porcelain. The streak plate has the hardness of glass, so anything with a Mohs hardness above 7 will scratch the plate instead, leaving no useful powder.

Specific Gravity. Specific gravity (SG) also is used routinely to identify minerals. It measures the mass of a specimen when compared with the mass of a volume of water equal to the mass of the specimen. SG is expressed as the ratio between the weight of the mineral and the weight of the equal volume of water. Water has a specific gravity of 1. The SG of a mineral depends on the types of atoms that make up the mineral and on how tightly packed the atoms are. Sophisticated equipment is often used to make this type of measurement.

Other physical properties used to identify minerals are fluorescence (the measure of the mineral's response to ultraviolet light), magnetism (the ability of the mineral to attract metal particles), tenacity (the mineral's response to a mechanically induced change in shape or form), radioactivity (a measure of the rate at which ionizing particles are emitted from an unstable atom), the presence of striations, a salty taste, a rotten egg smell, and the presence of effervescence when a drop of concentrated lemon juice is place on the surface of a mineral.

Minerals and Human Health

Minerals play a vital role in ensuring good health in humans. They are not directly metabolized by the body but are involved in the functioning of the body's electrolytes. Also, minerals are not a source of energy, but they do aid in muscle contraction, nerve transmission, blood clotting, immunity, regulation of blood pressure, and growth and development.

Minerals make up about 5 percent of human body weight, with one-half of this percentage (2.5 percent) in the form of calcium. The calcium is essential to maintaining strong bones. The other 2.5 percent is made up of phosphorus and other minerals.

Minerals are divided into two groups. The major minerals (macrominerals) are calcium, phosphorus, magnesium (Mg), sulfur (S), sodium, chloride (Cl^-), and potassium. For good health, a human requires 100 milligrams or more per day of these minerals. The trace minerals (microminerals) are required at only about 15 milligrams per day. Among the trace minerals are iron (Fe), zinc (Zn), copper (Cu), and manganese (Mn).

Mineral Mining

Humans have mined for centuries, and there is evidence that mining has taken place in various forms since humans were first able to manipulate their environments for their own purposes. Minerals are integral to rocks and other land formations and do not exist in a form that is readily available. Therefore, the minerals that are valuable to humans must be mined. In mining, the ore containing the minerals is extracted and the minerals are extracted from the ore. Mining also is used to extract diamonds.

Today mining occurs as surface mining and subsurface mining. Surface mining is far more common, accounting for 85 percent of the minerals and 98 percent of the metallic ores obtained in the United States. In surface mining, to reach the ore deposits, the earth's surface is stripped of vegetation, dirt, and any layers of bedrock. Once the minerals are reached, they are cut from the earth and processed. Quarrying is another surface mining method in which material is gathered from an open pit. The pit may reach hundreds of feet into the ground. Mountaintop removal is common with coal mining.

In subsurface mining, tunnels are made deep inside the earth. The buried ore is extracted and brought to the surface for processing. The desired metals are then extracted from their ores. Once in their desired forms, they can be used commercially and ornamentally.

Minerals in Daily Life

Humans use minerals in many ways. They use them to improve their health by taking dietary supplements with minerals. They use minerals as ingredients in cosmetics. They polish gemstones and rocks to make jewelry, cutting them along predictable cleavage planes to create gems worth much more than the raw materials that produced them.

Headstones are created from marble, and the marble is used to build massive structures. Rocks with desired minerals are used ornamentally in buildings, walkways, and exhibitions. Diamonds are fashioned into drill tips and other cutting implements. The metals extracted from ore are used in construction and industry. Steel (an alloy of iron and carbon) is used in the construction of bridges and buildings. Calcite is used to make cement, and hematite (Fe_2O_3) is used as a component in such items as steel door hinges and handles. Chromite ($FeCr_2O_4$) is found in chrome plating, quartz (SiO_2) in mirrors, copper in wiring, and cassiterite (SnO_2) in solder. Minerals also are collected by hobbyists for personal use.

Gina Hagler

Further Reading

Klein, Cornelius, and Barbara Dutrow. *Manual of Mineral Science.* New York: John Wiley & Sons, 2008. This book is the latest incarnation of the original *Manual of Mineralogy* published by James Dwight Dana in 1848. It provides a comprehensive treatment of the topic.

McLaughlin, Eugene, and Levan Breaux. *Chemical Mineralogy, Smelting, and Metallization.* New York: Nova Science, 2009. Presents research on mineralogy, which includes discussions of the chemical mineralogy of ornamental rock and the waste of water treatment plants.

Nesse, William. *Introduction to Mineralogy.* New York: Oxford University Press, 2011. This book includes material on traditional and optical mineralogy and a thorough exploration of minerals and their formation, structure, and properties.

_______. *Introduction to Optical Mineralogy.* New York: Oxford University Press, 2004. Covers the nature of light, the interaction of light and matter, the use of the petrographic microscope, and other topics relevant to optical mineralogy.

Perkins, Dexter. *Mineralogy.* New York: Prentice Hall, 2010. Begins with an overview of the topic and drills down to the details, which are essential to an understanding of mineralogy.

Perkins, Dexter, and Kevin R. Henke. *Minerals in Thin Section.* New York: Prentice Hall, 2003. Includes practical aspects of optical mineralogy and the identification of minerals in thin section.

Warner, Terence. *Synthesis, Properties, and Mineralogy of Important Inorganic Materials.* New York: Wiley, 2011. A textbook on preparative solid-state chemistry.

Wenk, Hans-Rudolf, and Andrei Bulakh. *Minerals: Their Constitution and Origin.* New York: Cambridge University Press, 2004. This textbook covers mineralogy from its definition to its structures and uses.

See also: Abyssal Sea Floor; Andes; Biogeography; Biostratigraphy; Carbonate Rocks; Clastic Rocks; Continental Shelf and Slope; Groundwater; Ophiolites; Paleoecology; Petrified Wood; Radiometric Dating; Rocky Mountains; Trace Fossils; Transverse Ranges; Weathering.

MOLECULAR PALEONTOLOGY

Molecular paleontology involves the recovery of deoxyribonucleic acid, proteins, and other molecules from ancient animal, human, and plant remains for study and classification. Science had believed that an organism's fossilized remains retained little molecular information. However, molecular fragments are now often found in such remains, primarily because of advances in research practices and technologies.

PRINCIPAL TERMS

- **amino acid:** a simple compound that serves as the building block of protein and therefore the development of all organic life on Earth
- **collagen:** a protein that is a building block for connective tissue
- **DNA (deoxyribonucleic acid):** the nucleic acid that carries the genetic information of an organism
- **phylogenetics:** a field of study in which the genetic profiles of long-extinct organisms are determined and their respective similarities and differences from modern species are established
- **protein sequence:** the order of amino acids that form a polypeptide chain, which in turn forms part or all of a protein molecule

Basic Principles

Molecular paleontology involves the extraction and study of ancient human, plant, and animal remains. This practice includes the examination of deoxyribonucleic acid, (DNA, which carries the genetic information of an organism), lipids (fats), carbohydrates, and other molecular-level materials found in the fossilized remains of ancient humans, dinosaurs, and florae. This pursuit is made easier with the discovery of actual soft tissues, bones, and other components. It had always been presumed that such materials are destroyed in the fossilization process and that the materials, therefore, could not yield valuable information. However, improvements in technologies have enabled scientists to extract usable subcellular materials from crystallized bone from animal and early human remains.

Molecular paleontology is part of a broader attempt to understand how life on Earth has evolved. Paleontologists search through strata layers of the earth for the fossilized remains of plants and animals (including proto-humans). Molecular paleontologists use the remains to uncover the genetic composition of these organisms. Using the latent materials present in the tissue or fossilized remains, scientists may, for example, piece together the genetic profiles of long-extinct organisms and their respective similarities and differences from modern species (a field known as phylogenetics). Scientists also may use such information to study the origins and transmission of certain diseases.

Background and History

In any science, a theory cannot progress into fact without concrete evidence and, in particular, without evidence that may be independently verified. One of the greatest challenges facing paleontology (the study of life during prehistoric periods using fossilized remains) is that the evidence needed to support a theory likely disappeared millions of years ago. Paleontologists have learned a great deal about dinosaurs, prehistoric plants, and even the earliest humans by assembling the bones and fragments buried under the strata or by studying the fossilized remains of such life forms that have been etched into solid rock. Until the mid-1950's, however, it was presumed that DNA, amino acids, or other molecular-level genetic material could never have survived fossilization.

In 1956, however, renowned American geochemist, geophysicist, and molecular biologist Philip Abelson dissolved a piece of a fossilized stegosaurus (a dinosaur from the late Jurassic period) in hydrochloric acid. He reported that from that solution he was able to find small amounts of amino acids (simple compounds that serve as the building blocks of protein). Two decades later, several scientists used Abelson's principles to analyze the ability of certain prehistoric organisms to combat infections. In the 1980's, molecular paleontology was used to build on this immunological study, citing similarities between species' ability to create antibodies (which help an organism fight foreign cells and invading viruses). Based on this and similar studies, scientists uncovered many evolutionary connections between prehistoric and modern species that might otherwise have gone unconfirmed.

Tissue Analysis

There are two general scenarios in which molecular paleontology is made possible. The first is through the analysis of actual tissue. For example, during the Pleistocene period (between 1.8 million and 100,000 years ago) woolly mammoths thrived in the permafrost of the cold Eurasian regions. Eventually, the species became extinct, but scientists have found a number of woolly mammoths encased in ice of what is now Siberia (and elsewhere). A question that has vexed paleontologists for decades before these creatures were uncovered was how an animal that originated in much warmer climates (in locations now populated by elephants) was able to survive after migrating north. Experts theorized that the answers lie in the creature's blood, specifically that the mammoth's blood contained some special element that enabled oxygen to travel quickly through its large body in cold temperatures. However, although the mammoth's tissue might still exist, its blood would have disappeared long ago.

Still, when paleontologists unearthed frozen mammoth carcasses in northern Siberia, these scientists were able to extract amino acids and DNA from the tissue samples. Using a chemical compound known as a blood substitute, the scientists re-created the mammoth's hemoglobin (which carries oxygen through the bloodstream). After careful genetic analysis, the molecular paleontologists determined that the mammoths had undergone a significant mutation, allowing them to use different amino acids to cope with the dramatic changes in their environment. The implications of this find are significant: Scientists believe that, based on this discovery, it may be possible to create new artificial blood products for persons placed in a state of hypothermia for brain and heart surgeries.

Tissue samples taken from frozen woolly mammoth carcasses often provide invaluable information about the creatures, but it has been much more difficult to locate the tissue samples of dinosaurs. Indeed, the dinosaur bones and fossils unearthed by paleontologists are often hundreds of millions of years old. Although rare, there have been instances in which small amounts tissue have been found deep in the bones of dinosaurs.

In 2005, for example, paleontologists made such a find inside one of the bones derived from one of the most prized finds in paleontology, a *Tyrannosaurus rex* (or T. rex) unearthed in 2003 in Montana. A femur recovered at the site actually contained the remains of soft tissue and collagen (a protein that serves as the building block for connective tissue). Scientists were able to remove the collagen and analyze the protein sequences (the order of amino acids that form a polypeptide chain, which in turn forms part or all of a protein molecule). The researchers then compared the protein sequences of the T. rex with those of modern animals, confirming what paleontologists have for many years suspected but were never able to verify: that some dinosaurs bore more similarities to birds, particularly ostriches and chickens, than to modern reptiles. This project represented the first time that molecular data placed a nonavian (bird-like) species along the evolutionary scale for birds.

Crystallized Remains

The second area in which molecular paleontology is possible is much more difficult to employ, as it involves the extraction of crystallized fragments of amino acids from fossils instead of from soft tissue. DNA degrades quickly after an organism dies. When an organism dies and is ultimately buried under the soil, the DNA on the bone becomes both degraded and potentially contaminated by other DNA. However, scientists have discovered that crystallized cells may be found within the collagen in fossilized bones. Placing ground bone powder into a solution of sodium hypochlorite, scientists were able to extract strings of DNA from the clumps of crystals. These strings were found to be reproducible, enabling scientists to create a more complete genetic profile of the organisms in question.

The implications of such a find are significant. By analyzing the DNA of early organisms, scientists can research the organisms' migratory patterns, how they reacted to disease, and how they were genetically related to other organisms (including those living now). There is a risk involved with this new procedure, however. Reproducing DNA strands based on the fragments found within bone may be possible, but it may not yield reliable data. Still, many scientists believe that this approach does create results that are as authentic as possible.

Electron Microscopy

Electron microscopes can magnify images by as much as several hundred thousand times, with an image clarity and depth that can reveal not only molecular-level objects, but also their composition. An electron microscope focused on a sample can produce images that provide clues about that objects' origins and similarities to other organisms. For example, in 1997, scientists used an electron microscope to study fossilized spores from prehistoric mushrooms. The microscope revealed the spores' distinctive shapes and surface textures, which, when compared with other, similar spores, revealed how the mushrooms migrated during continental drive.

Electron microscopes are also quite useful for analyzing crystals buried within dinosaur bones. In one study of sauropods (a class of large, four-legged dinosaurs that includes brontosaurus and brachiosaurus), molecular paleontologists used an electron microscope to analyze the crystal patterns within the bones of juveniles and adults. The images the microscope produced showed a diverse grouping of crystals in the juvenile, confirming a theory that young sauropods had a wide range of crystals within their bones that enabled them to perform certain mechanical functions with greater ease.

Spectrometry

Spectrometry involves the use of a device (the spectrometer) to emit light, sound, or particles onto an object under study. Spectrometers then record the object's reaction, such as the quantity of light it absorbs or the energy it itself projects upon exposure. Spectral analysis has proved extremely useful in discovering the physiological makeup of certain organisms.

Some scientists, for example, believe that certain organic substances, such as blood and bone cells, can, under the right conditions, survive in fossilized remains. One paleontologist, who was studying a slice of bone from a T. rex as a graduate student in 1992, noticed what appeared to be ancient blood cells. In an effort to confirm her findings, she subjected the sample to a laser spectrometer to detect heme (a molecule that contains iron found in red blood cells). Heme is known to absorb a distinctive amount of light when exposed. Indeed, her theory was confirmed when the sample was exposed to the spectrometer. The object she had found was indeed a red blood cell.

Two types of spectrometers that are occasionally used in the field of molecular paleontology are the inductively coupled plasma atomic emission spectrometer (ICP-AES) and the electrothermal atomic absorption spectrometer (EAAS). The ICP-AES involves targeting energy emissions at an object (after it has been immersed and dissolved in acidic solution) in the hope that certain elements contained in the sample will react and emit energy along a certain wavelength. The EAAS focuses a beam of ultraviolet light through a flame, using a detector to capture the elemental response along a certain wavelength. Each type of spectrometer is primarily focused on the search for metal, such as iron and sodium. It is, therefore, useful in the pursuit of many of the fundamental elements found in blood and related tissue. However, both types of spectrometers are expensive to use in molecular paleontology, which has limited their application in this particular field.

Chemical Analysis

As Abelson demonstrated in his 1950's experiment, chemistry is an invaluable tool in the pursuit of DNA and other molecular remnants within fossilized organisms. Chemical compounds have proved highly effective in breaking down the various elements contained in fossilized bone and rock, making it easier to detect and extract important evidence. For example, in 2004, scientists in China analyzed the skeletal fossils of four dinosaurs found in China's Sichuan Province. Rather than employ ICP-AES or ETAAS, the molecular paleontologists immersed samples of the fossils in a bath of boiling water and a highly corrosive combination of hydrochloric and nitric acid known as aqua regia. After the fossil samples were digested, the scientists detected and recorded large quantities of arsenic molecules. The presence of this toxic substance in the creatures' bones provided some clues as to why dinosaurs died in Sichuan specifically, and why they died perhaps elsewhere.

Relevant Organizations and Institutions

There are two general areas in which molecular paleontology serves a significant purpose. First, the science attempts to answer many of the questions about the forms of life that existed on Earth in the prehistoric eras. Second, the science seeks to determine how those organisms ceased to exist or to evolve into the types of plants, animals, and mammals found on Earth today. Also, a number of groups and organizations play major roles in the field of molecular paleontology.

Because paleontologists and molecular paleontologists conduct scientific research and experimentation, they frequently work in the zoology, biology, or paleontology departments of major universities. It is here that paleontologists conduct experiments, study fossil contents, collate data, and produce scientific papers and hypotheses.

Museums play a major role in paleontology, largely because so many fossilized remains of dinosaurs, plants, and other prehistoric organisms are housed and preserved at such institutions. Most museums are supported by donors and by admission fees, which are used to contribute to research. Some museums are supported by universities; the Museum of the Rockies, for example, has the world's largest collection of dinosaur fossils, and it is affiliated with Montana State University.

Governments play an important role in paleontology and molecular paleontology. Primarily, government agencies such as the National Science Foundation provide grants and support to help scientists perform their respective research. Such support includes funding for equipment (including spectrometers) and funding for travel to remote sites. Because molecular paleontology is a relatively new field, the National Science Foundation is positioned to invest significantly in such high-risk, high-payoff research.

Implications and Future Prospects

According to some observers, paleontology before the 1950's was considered a branch of geology. In time, however, greater attention has focused on paleontology as the study of ancient life, not of the fossils into which ancient life evolved. Molecular paleontology may be seen as a continuation of this changing attitude. This approach adds a new dimension to the science, utilizing twenty-first century understandings of genetics and biodiversity and using the latest in technology.

Molecular paleontology uses a number of different techniques, some of which include state-of-the-art technology (such as spectrometers and electron microscopes). The use of such systems is sometimes cost-prohibitive, especially for nonprofit museums and university professionals. The high price of using the technological tools of molecular paleontology, as a result, could limit the field's growth.

Paleontology has implications for modern science. Specifically, some scientists, when attempting to predict how life in the modern world might react to global warming and other climate changes, are considering how organisms adapted genetically to climate changes during prehistoric times. As technologies and research techniques continue to evolve, it is likely that molecular paleontology will continue to be relevant to the study of life on Earth.

Michael P. Auerbach

Further Reading

Dyke, Gareth. "Winged Victory." *Scientific American* 303, no. 1 (2010): 70-75. Argues that, based on molecular studies of ancient bird fossils, some modern species of birds actually lived alongside dinosaurs. This finding is counter to a long-standing theory that such birds appeared after the dinosaurs became extinct 65 million years ago.

Horner, John R., Jack Horner, and James Gorman. *How to Build a Dinosaur: Extinction Doesn't Have to Be Forever.* New York: Penguin, 2011. The authors discuss the combined sciences of molecular biology and paleontology and explains how this combination may be used to reproduce prehistoric organisms using and new technologies and what is known about dinosaurs' genetic codes.

Quental, Tiago B., and Charles R. Marshall. "Extinction During Evolutionary Radiations: Reconciling the Fossil Record with Molecular Phylogenies." *Evolution* 63, no. 12 (2009): 3158-3167. This article discusses the rates of extinction that occurred while the planet changed during prehistoric eras. Analyzes the diversity of species during periods of change, applying computer models to study how those species adapted and evolved.

Sepkoski, David, and Michael Ruse, eds. *The Paleobiological Revolution: Essays on the Growth of Modern Paleontology.* Chicago: University of Chicago Press, 2009. Provides a chronology of the development of paleontology and explains how the biological sciences (including molecular biology) have been gradually introduced to this field of study, painting a more comprehensive illustration of prehistoric life.

Weinstock, Jaco, et al. "Evolution, Systematics, and Phylogeography of Pleistocene Horses in the New World: A Molecular Perspective." *PLoS Biology* 3, no. 8 (2005): 1373-1379. Discusses how molecular paleontology helps explain how various species of horses evolved from the prehistoric era to today. The data found in the authors' research have provided information about the origins of certain species and their migratory patterns during certain ancient periods.

Wiens, John J. "Paleontology, Genomics, and Combined-Data Phylogenetics: Can Molecular Data Improve Phylogeny Estimation for Fossil Taxa?" *Systematic Biology* 58, no. 1 (2009): 87-99. Examines the value of molecular paleontology for connecting modern species to extinct organisms. Argues that the molecular data uncovered in this approach lends accuracy to the classification of prehistoric organisms and, ultimately, modern mammals, plants, and animals.

Wiley, E. O., and Bruce S. Lieberman. *Phylogenetics: Theory and Practice of Phylogenetic Systematics.* Hoboken, N.J.: Wiley-Blackwell, 2011. This book discusses the evolutionary relationships among modern and prehistoric organisms and how the development of this field impacts the field of taxonomy.

See also: Biogeography; Dinosaurs; Eukaryotes; Evolution of Birds; Evolution of Humans; Evolution of Life; Fish-Amphibian Transition; Fossil Fish; Fossil Record; Fossil Reptiles and Amphibians; Fossils of the Earliest Vertebrates; Geomicrobiology; Life's Origins; Microfossils; Minerals; Molecular Paleontology; Paleobiogeography; Paleobotany; Paleoecology; Trace Fossils.

MOUNTAIN BELTS

Mountain belts are products of plate tectonics, produced by the convergence of crustal plates. Topographic mountains are only the surficial expression of processes that profoundly deform and modify the crust. Long after the mountains themselves have been worn away, their former existence is recognizable from the structures that mountain building forms within the rocks of the crust.

PRINCIPAL TERMS

- **continental shelf:** the submerged offshore portion of a continent, ending where water depths increase rapidly from a few hundred to thousands of meters
- **epeirogeny:** uplift or subsidence of the crust within a region, without the internal disturbances characteristic of orogeny
- **gabbro:** a silica-poor intrusive igneous rock consisting mostly of calcium-rich feldspar and iron and magnesium silicates; its volcanic equivalent is basalt
- **granite:** a silica-rich intrusive igneous rock consisting mostly of quartz, potassium- and sodium-bearing feldspar, and biotite or hornblende
- **igneous:** from the Latin *ignis* ("fire"), a term referring to rocks formed from the molten state or to processes that form such rocks
- **metamorphism:** the change in the mineral composition or texture of a rock because of heat, pressure, or the chemical action of fluids in the earth
- **orogeny:** the profound disturbance of the earth's crust, characterized by crustal compression, metamorphism, volcanism, intrusions, and mountain formation
- **plate tectonics:** the theory that the crust of the earth consists of large moving plates; orogeny occurs where plates converge and one plate overrides the other
- **sedimentary rocks:** rocks that form by surface transport and deposition of mineral grains or chemicals
- **subduction:** the sinking of a crustal plate into the interior of the earth; subduction occurs at subduction zones, where plates converge

OROGENY

Mountains have many origins. They can be volcanic, like Mount Vesuvius, or they can be formed by vertical movements along faults, as the Sierra Nevada or the Ruwenzori of central Africa were. Some mountains are the result of relatively gentle epeirogenic (deformational) uplift of the crust—for example, the Black Hills or the Adirondack Mountains. The causes of epeirogeny, however, are poorly understood. The great mountain chains of the earth, such as the Andes, the Rocky Mountains, or the Himalaya, however, formed not only from uplift but also from internal deformation of the crust, volcanic activity, metamorphism, and the intrusion of vast quantities of molten rock into the crust, especially granite and related rocks. These processes are collectively called orogeny, and mountain chains that form from such processes are called orogenic belts. Orogeny is one of the most important consequences of plate tectonics. It occurs when plates collide and one plate overrides the other—a process known as subduction—in response to compressional forces and heating generated by the plate collision.

The earth's crust consists of two types of plates, called continental and oceanic by geologists. The continents and the adjacent continental shelves are underlain by granitic crust, averaging about 40 kilometers thick. The ocean floors are made of gabbro and basalt, averaging about 5 kilometers thick. The true edge of a continent is not the shoreline, which is constantly changing, but the boundary between continental and oceanic crust. The edge of the continental shelf coincides closely with this boundary.

Many mountain belts form at subduction zones with a continental overriding plate and an oceanic descending plate. The downward bending of the plate creates a deep, narrow trench on the ocean floor, sometimes more than 10 kilometers below sea level. The descending oceanic plate sinks into the earth's interior, eventually to be reabsorbed. The overriding plate experiences orogeny. All orogenic belts differ in detail, but most have certain major features in common. A typical orogenic belt consists of parallel zones, which may be defined by distinctive rock type, type of metamorphism, level of igneous activity, or type of deformation of the rocks. The zones, generally parallel to the boundary where the

Mountain showing geological fault and uplifting. Northern British Columbia, Canada. (Mark Newman/Photo Researchers, Inc.)

two plates collide, are the result of different crustal conditions and processes at different distances from the plate boundary. It is useful to regard orogenic belts as having an "outer" side, adjacent to the plate boundary, and an "inner" side within the overriding plate.

The first zones recognized in orogenic belts were those defined by environment of deposition: an outer zone of thick, deep-water sedimentary rocks and volcanic rocks and an inner zone of thinner, shallow-water sedimentary rocks without abundant volcanic rocks. The nineteenth-century American geologist James Hall first described these zones, which he envisioned as parallel troughs formed by downward folding of the crust. Because these troughs were viewed as immense versions of ordinary downward folds, or synclines, James Dana later called the troughs "geosynclines." The outer trough was called the eugeosyncline, and the inner trough was called the miogeosyncline.

The original concept of the geosyncline disturbed many geologists, because it did not quite match the structure of active mountain belts. In 1964, Robert Dietz reexamined the geosyncline concept and showed that the rocks need not have accumulated in troughs. With this insight, it became clear that the rocks of miogeosynclines corresponded closely to those of the continental shelves, while eugeosynclines were a good match to the rocks of many volcanic island chains. It eventually became clear that the rocks of the eugeosyncline often formed separately from the miogeosyncline and were later juxtaposed by plate motions.

Because of these revisions of the original geosyncline concept, many geologists have abandoned the original terms and prefer the terms "geocline," "eugeocline," and "miogeocline" instead. The eugeocline is the outer belt of deep-water sedimentary rocks and volcanic rocks. The miogeocline is the inner belt of shallow-water sedimentary rocks. Beyond the miogeocline is the platform, where thin shallow-water or terrestrial rocks were deposited on the stable interior of the overriding plate. In addition, most orogenic belts have an inner belt of coarse sedimentary rocks deposited late in the history of the orogenic belt. This belt, called the molasse basin, consists of debris eroded from the mountains and deposited at their base. Molasse basins consist mostly of rocks deposited in shallow-water or land environments.

MOUNTAIN BELT STRUCTURE

Much of the structure of a mountain belt is related to processes in the descending plate. As the descending plate reaches a depth of about 100 kilometers, it begins to melt, and molten rock, or magma, invades the overriding plate. In general, volcanic rocks in orogenic belts become progressively richer in silica with increasing distance from the plate boundary, because the rising magma has more time to react with silica-rich continental crust. Also, volcanic and intrusive rocks in mountain belts tend to become more silica-rich over time. Most orogenic belts have a main axis of igneous activity, the igneous arc, where volcanic and intrusive activity are concentrated. The igneous arc is generally on the inner side of the eugeocline. In deeply eroded orogenic belts, intrusive rocks of the igneous arc, usually granitic in composition, are exposed as great masses known as batholiths.

Different thermal conditions in different parts of the overriding plate give rise to two distinct zones of metamorphism. Adjacent to the descending plate, rocks are carried downward to great depths, often 20 kilometers or more, but because the rocks are in contact with the still-cool descending plate, they remain unusually cool. Temperatures in this zone generally average 200 to 300 degrees Celsius, instead of the 500 to 600 degrees Celsius that might be expected at 20 kilometers depth. This low-temperature, high-pressure metamorphism is known as blueschist metamorphism because many of the minerals that form often impart a bluish color to the rocks.

Generally coinciding with the igneous arc is an inner belt of metamorphism where temperatures are high but pressures are moderate. Peak temperatures commonly exceed 600 degrees Celsius, with pressures typically reflecting depths of 5 to 10 kilometers. Such conditions are called amphibolite metamorphism. Adjacent to the region of highest temperature is a region of lower-temperature metamorphism, where temperatures of 400 to 500 degrees Celsius prevail. This type of metamorphism is called greenschist metamorphism, from the greenish color of many of the minerals formed. The outer zone of blueschist metamorphism generally occupies the outer part of the geocline. Amphibolite metamorphism generally coincides with the igneous arc and the inner part of the eugeocline. Greenschist metamorphism commonly extends into the miogeocline.

The deformation of rocks in the overriding plate depends on the nature of the rocks, stress, temperature, and confining pressure. Orogenic belts display several zones of distinctive structures. The most important of these belts are the accretionary prism, zone of basement mobilization, and the foreland fold-and-thrust belt. The eugeocline in general is a region of intense deformation; deformation in the miogeocline is less intense. The outermost edge of the orogenic belt is occupied by the accretionary prism, and it often forms much of the eugeocline. Where the colliding plates meet, sediment is scraped off the descending plate. Other sediment is eroded from the continent and pours into the trench. The sediment from the continent is deposited rapidly, with little weathering or sorting, to form impure sandstone called graywacke.

Submarine landslides and slumps are common in the unstable setting of the trench; the resulting complex of chaotically deposited greywacke is called flysch. A wedge of intensely deformed sediment accumulates on the edge of the continent, much the way a wedge of snow accumulates ahead of a snowplow. This wedge of sediment is the accretionary prism. Frequently, fragments of oceanic crust break off the descending plate and are incorporated into the accretionary prism. These slices of oceanic crust, called ophiolites, are of enormous geologic value. Not only do they mark the location of former subduction zones, but they also provide otherwise unobtainable cross sections of oceanic crust exposed on dry land. The actual contact between the two plates is marked by mélange, a chaotic mixture of broken rock with fragments ranging from microscopic to kilometers in size.

The high temperatures in the igneous arc and amphibolite zone of metamorphism can make the rocks of the crust plastic. That is, the rocks flow like stiff fluids, even though they do not melt. Because the hot rocks are less dense than the cooler crust around them, they rise upward. A mass of rock that flows upward in this manner is called a diapir. The process of heating deep crust (or "basement") so that it rises is called basement mobilization, and evidence of it occurs in many mountain belts. The mobilized crust appears as intensely deformed and highly metamorphosed rock called gneiss. Often the rising mass of gneiss appears to have shouldered the overlying rocks aside, so that the rocks are arched upward with a central core of gneiss. Such a structure is called a gneiss dome.

Compressional forces arising from plate convergence thicken the crust of the overriding plate in a number of ways. Within the accretionary prism, sheets of sedimentary rock are thrust downward beneath the overriding plate, resulting in a stack of faulted slices of rock. These slices may thicken internally by fracturing along small faults and stacking the resulting small slices one above the other, a process called duplexing. Within the igneous arc, crustal thickening occurs when magma invades the crust, increasing its volume. Magma can also be added to the base of the crust, a process called underplating. Heating of the crust within the igneous arc makes much of the lower crust plastic, permitting the plastic crust to be squeezed upward by compression. This process probably assists the upward movement of gneiss domes.

The thickening of the crust results in the uplift of the surface and the formation of topographic mountains. In the foreland, which basically coincides with the miogeocline, deformation is largely a response to events in the active core of the mountain belt. Some of the deformation in the foreland seems to be driven by rising masses of mobilized basement. Rocks of the miogeocline and some of the underlying crust fracture into sheets that are shoved over the rocks beneath. Fractures or faults where one mass of rock overrides another are called thrust faults. These rocks may also be buckled into folds by compressive forces. Rocks nearer the surface often slide off the rising mountain belt. The rocks may break into thin sheets called nappes that stack one atop the other or may crumple into folds. Often the folded rocks have detached from the rocks beneath, much like a carpet slides and folds when a piece of furniture is pushed over it. This process of detachment is called décollement. Because this deformation involves only the surface layers of rock and not the underlying basement rocks, it is called thin-skinned deformation.

Other Types of Plate Collision

Other kinds of plate collision result in different combinations of structures. Oceanic-oceanic subduction zones have somewhat simpler orogenic belts. When the descending plate begins to melt, magma rises and breaks through the surface to create a volcanic island arc such as those found in the Aleutian Islands or the Lesser Antilles. Since both plates are oceanic crust, made largely of basalt, the magma also is basalt. Erosion strips sediment off the volcanic islands and dumps much of it in the trench to form an accretionary prism. Over a very long time, the island arc may be built up into a continuous belt of intensely deformed volcanic rocks and sedimentary rocks derived from them. The Greater Antilles and the Isthmus of Panama probably formed this way. Such orogenic belts consist essentially of a eugeocline and igneous arc, with the associated metamorphic zones and deformation structures.

Continent-continent collisions start out as continent-ocean subduction zones, but eventually the convergence of plates brings two continents together, one of which is pushed beneath the other. Continent-continent collisions include the Himalaya, where India is being pushed beneath Tibet, and the Persian Gulf, where the Arabian Peninsula is being pushed beneath Iran. Because continental crust is thick and relatively light, the descending plate cannot be subducted. Instead, one continent rides onto the other, creating a double thickness of crust. Eventually, resistance to further movement may cause plate motions to change on a regional or even global scale.

Usually the overriding continent has had a long history of orogeny before the collision, whereas the other continent may have had none. Orogeny results in such a wide range of structures that it is usually immediately obvious which of the continents is or was the overriding plate. The collision boundary between the two continents, called a suture, may display relics of the former accretionary prism, including melange, fragments of ophiolites, or evidence of blueschist metamorphism.

Often a small block of crust, called a terrane, collides with a larger plate. The terrane may be a volcanic island chain or a small fragment of continental crust. The northern coast of New Guinea is an area where terranes (in this case volcanic island chains) are colliding with a continent. The addition of terranes to a larger plate is called accretion. Terranes are recognizable as distinct blocks of crust separated from adjacent rocks by major faults. In many cases, eugeoclines did not form near their corresponding miogeoclines but as separate terranes. Repeated accretion of terranes can add large areas to a continent. Roughly 1,000 kilometers of the western United States was accreted to North America in the last 500 million years.

After orogeny ceases, it is common for mountain belts to experience a period of crustal extension and faulting. Once the compressional forces that uplifted the mountains subside, many mountain ranges simply cannot support their own mass and begin to spread under their own weight. It requires about 20 million years for erosion to level a mountain range. Nevertheless, long after the topographic mountains are gone, the structures created by orogeny remain. The most conspicuous markers of ancient orogenies are usually the eugeocline and miogeocline, igneous arc, and molasse deposits.

If mountains are worn away within a few tens of millions of years, it follows that the present Urals and Appalachians, the products of continental collisions more than 200 million years ago, cannot be remains of the original mountains. In the case of the Appalachians, this point is clear because rivers such as the Potomac flow across the structures in the mountain belt. Clearly, the Appalachians were once level enough (or buried) that rivers could flow across them. The Appalachians, the Urals, and even the Alps are the result of recent epeirogenic uplift after erosion had largely, or entirely, leveled the original mountains. Why mountain belts sometimes experience renewed periods of uplift long after orogeny ceases is unknown.

Steven I. Dutch

Further Reading

Condie, Kent C. *Earth as an Evolving Planetary System.* 2d ed. Waltham, Mass.: Academic Press, 2011. Covers the earth's crust, mantle, core and other spheres. Also provides detailed discussion of tectonics, the supercontinent cycle, and planetary evolution. A good resource for undergraduates.

Cook, Frederick A., L. D. Brown, and J. E. Oliver. "The Southern Appalachians and the Growth of Continents." *Scientific American* 243 (October 1980): 156-168. Seismic probing has revealed the fault where ancient Africa rode onto North America to create the southern Appalachians. *Scientific American*, which is aimed at the scientifically informed but nonspecialist reader, roughly at college level, is probably the best source of information on recent advances in science for nonspecialists. Its coverage of advances in the earth sciences since 1970 has been especially thorough.

Frisch, Wolfgang, Martin Meschede, and Ronald C. Blakey. *Plate Tectonics: Continental Drift and Mountain Building.* New York: Springer, 2010. This textbook provides a basic overview of continental drift and plate tectonics focusing on the resulting changes in the earth's surface.

Grotzinger, John, and Tom Jordan. *Understanding Earth.* 6th ed. New York: W. H. Freeman, 2009. This comprehensive physical geology text covers the formation and development of the earth, including extensive sections on the development of mountain belts through time. Readable by high school students, as well as by general readers. Includes an index and a glossary of terms.

Howell, David G. "Terranes." *Scientific American* 253 (November 1985): 116-125. Terranes are small blocks of crust added to a mountain belt by plate collisions. This article describes types of terranes and what happens when they collide.

Joly, John. *The Surface-History of the Earth.* Boston: Adamant Media Corporation, 2005. A reprint of the classic text written in 1930. Covers many basic topics of the earth's surface. Well-drawn discussions provide background on orogeny, mountain building, geosynclines, the African Rift Valley and the geological time scale.

Jones, David L., A. Cox, P. Coney, and Myrl Beck. "The Growth of Western North America." *Scientific American* 247 (November 1982): 70-84. The westernmost 1,000 kilometers of North America is a mosaic of at least 200 small blocks added by plate collision in the last 500 million years. This article was written by some of the scientists who were most influential in discovering this process.

Molnar, Peter. "The Structure of Mountain Ranges." *Scientific American* 255 (July 1986): 70-79. A summary of discoveries about the internal structure of mountain ranges. The apparently solid crust of a mountain range is actually brittle and unable to support its own weight without external compressive forces.

Ollier, Cliff, and Colin Pain. *The Origin of Mountains*. London: Cambridge University Press, 2000. This text encompasses the geographic, geological, and geophysical concepts of mountains. Topics include fault-block mountains, fold belts and folding, gravity structures, rift valleys and river valleys, plateaus, and mountain erosion. Begins with a general overview of subduction and plate tectonics. Mountain ranges discussed in detail include the Appalachians, Andes, Cascades, Himalayas, Mount St. Helens, and the Rockies.

Plummer, Charles C., Diane H. Carlson, and Lisa Hammersley. *Physical Geology*. 13th ed. Boston: McGraw-Hill, 2009. This is a straightforward, easy-to-read introduction to geology intended for those with little or no science background. Discusses plate tectonics and its relation to the development of mountain belts. Includes many excellent illustrations.

Spencer, Edgar W. *Introduction to the Structure of the Earth*. 3rd ed. New York: McGraw-Hill, 1988. A college-level textbook on structural geology. Chapters 1-3 survey plate tectonics and its role in deforming the crust. Chapter 19 describes mountain ranges that form from continent-ocean plate collision, particularly the structure of western North America. Chapter 20 describes mountains that form by continent-continent collision, with emphasis on the Appalachians and Alps.

Stanley, Stephen M. *Earth System History*. 2d ed. New York: W. H. Freeman, 2005. A comprehensive overview of the development of the earth. Chapters 8 and 9 deal with continental tectonics and the development of mountain belts. Illustrations, glossary, bibliography, and CD-ROM. Readable by high school students.

Willett, Sean D., and Mark T. Brandon. "On Steady States in Mountain Belts." *Geology* 30 (2002): 175-178. Provides an overview of steady-state processes. Discusses the four types of steady state in orogenic systems. Accessible to academia.

See also: Alps; Andes; Appalachians; Basin and Range Province; Cascades; Continental Crust; Continental Growth; Continental Structures; Displaced Terranes; Fold Belts; Folds; Geoclines; Himalaya; Island Arcs; Rocky Mountains; Sierra Nevada; Thrust Belts; Transverse Ranges.

MOUNT PELÉE

Mount Pelée, located on the island of Martinique in the West Indies, gained its notoriety on May 8, 1902, when all but two of about thirty thousand persons in the town of St. Pierre were suddenly killed by the eruption of a hot cloud of ash, rock fragments, and gases called a nuée ardente—the first time such an eruption was recognized.

PRINCIPAL TERMS

- **andesite:** a volcanic rock of intermediate silica content that contains plagioclase (a calcium, sodium, and aluminum-silicate mineral) and often hornblende (a calcium, iron, magnesium, and aluminum-silicate mineral) and biotite (a potassium, iron, magnesium, and aluminum-silicate mineral)
- **dacite:** a volcanic rock of fairly silica-rich composition with plagioclase, alkali feldspar (a potassium, sodium, and aluminum-silicate mineral), and quartz (a silica mineral)
- **magma:** a naturally occurring liquid that is usually composed of silicate material and contains suspended minerals or rocks
- **nuée ardente:** a hot cloud of rock fragments, ash, and gases that suddenly and explosively erupts from some volcanoes and flows rapidly down the volcano's slope
- **subduction:** the theory of plate tectonics that assumes that the earth's surface is divided into a number of large plates that either are colliding or are being pulled apart
- **viscosity:** how readily liquids flow; silica-rich magmas flow less readily than do silica-poor magmas
- **volcano:** a vent at the earth's surface in which gases, rocks, and magma erupt at the surface and build a more or less cone-shaped mountain

Nuées Ardentes

Mount Pelée, on the island of Martinique, is one of many volcanoes located along the sinuous arc of islands of the Lesser Antilles in the West Indies. One oceanic plate is being thrust or subducted underneath another oceanic plate along the Lesser Antilles, resulting in the generation of silica-rich magma such as andesite and dacite. The silica-rich magma is very "stiff," or viscous. Therefore, it will not readily flow up the conduit of the volcano, which may result in a plug of magma in the volcano crater that prevents the escape of magma and gas. The gas pressure inside the volcano may thus build until it explodes at its weakest point.

The explosions of this type often blast horizontally out below the volcanic plug, resulting in an exceedingly hot cloud of glowing ash (fine dust), rock fragments, and gases called a nuée ardente that cascades rapidly downslope. Initial temperatures may be approximately 1,200 degrees Celsius, but the temperature may rapidly drop off as the nuée ardente moves downslope. The rapid expansion of gases after eruption apparently produces much of the cooling, but the nuée ardente of May 8, 1902, at Mount Pelée was still hot enough to soften glass (about 700 degrees Celsius) 8 kilometers from the volcano. The speeds of nuées ardentes average 100 kilometers per hour. Lava does not extrude during a nuée ardente eruption.

This type of eruption was first recognized on May 8, 1902, when Mount Pelée erupted, instantly killing all but two of nearly thirty thousand people in the town of St. Pierre. The nuée ardente eruption is therefore often called a Peléan type of eruption. The journalists and scientists who descended on the area shortly after the eruption left many fascinating eyewitness accounts.

Warning Signs

The many signs of unusual volcanic activity that occurred prior to May 8 are now recognized as warnings of an imminent eruption. The emission of steaming gases was first noticed on April 2 in the upper portions of a stream draining off of Mount Pelée. There were minor earthquakes and an odor of sulfur, and ash was noticed in St. Pierre on April 23. Explosions in the cone of the volcano threw ash and rocks into the air on April 25. People who climbed up to the crater on April 27 said that it contained a small cinder cone, a lake, and columns of steam that were not previously noticed. The fine particles of dust, called ash, thrown out of the volcano during loud explosions gradually became heavier and eventually blocked some roads and forced residents indoors. Many residents left St. Pierre in fear; many from the countryside, however, migrated to the city. As a result, the population of the city was much larger than normal by early May. Authorities issued assurances that residents were in no danger and that they should not panic.

Volcanic Weather

Often, the hot updraft around active volcanoes produces thunderstorms. This often results in a large amount of water suddenly mixing with the abundant volcanic ash. Heavy rains occurred on the southwest slope of Mount Pelée on May 5, which resulted in water that was ponded in one of the craters bursting through the crater walls and forming a scalding torrent that rushed down the river valley at speeds of up to 90 kilometers per hour. As the mudflow, carrying huge trees and boulders, hit the ocean, it produced a large wave that overturned boats and washed up on the low-lying areas of St. Pierre. Residents of the lowlands panicked and fled to higher areas, which caused more of an exodus from St. Pierre. Explosions on May 6 were quite loud, and the authorities had to block the roads to keep people from leaving the city. The newspaper and governor issued assurances that there was no cause for alarm. By May 7, there were continuous downpours and clouds of ash that produced muddy floods in the swollen streams.

Violent Blasts, Sudden Death

Then, on May 8, the volcano erupted with four loud blasts. One black cloud full of lightning rose upward from the crater. A nuée ardente moved rapidly down the volcano, over the town of St. Pierre, and out over the ocean, killing nearly everyone in its path within two minutes. People died suddenly either from burns or from breathing the heated gas and ash. Houses lost their roofs, and walls 1 meter thick were ripped apart. Most of the ships in the harbor were destroyed. The two ships that were not capsized had a few survivors who were in inner cabins with the doors closed. There were two survivors in St. Pierre, including a prisoner in a windowless cell. A small grate in the door had allowed enough heat into the cell to burn parts of his body. The second survivor was located in an inner room of a house in which all other occupants died. Much of the city had burned after the initial blast and the city itself resembled an ancient ruin. Foundations of houses were left standing mixed with mangled bits of metal and furnishings. Everything was covered with a mantle of ash nearly 1 meter thick. A few tree trunks remained, devoid of their bark and scorched by the heat.

Continuing Eruptions

Mount Pelée continued to erupt more or less continuously for several hundred days. Nuées ardentes were observed on May 20, June 6, July 9, August 30, and December 6. The nuée ardente of August 30 moved down a different slope and killed several thousand people. The amount of ash given off in these continuous ash eruptions was prodigious. One conservative estimate of the quantity of ash given off in twenty-four hours was that it was about equal to one and one-half times that of the sediment carried by the Mississippi River in one year. Brilliant afterglows and modified sunsets were observed for several years after the eruptions. Deflections of the earth's magnetic field related to a volcanic eruption were observed for the first time.

A giant obelisk, or spine, of stiff lava gradually projected upward from the crater of Mount Pelée. This spine first appeared in photos taken in June 1902,

Debris on Morne d'Orange, Martinique, after the 1902 eruption of Mount Pelée. (U.S. Geological Survey)

and it pushed upward out of the summit for more than one year before it broke up and shattered. The obelisk reached its maximum height of about 330 meters above the crater floor and about 160 meters across at the base in June and July of 1903. It continued to crack and break at the top as it continued to grow in irregular spurts at rates of up to 17 meters per day. The obelisk was reduced to a height of approximately 160 meters by August 1903. The continued eruptions from August to November reduced it to a pile of rubble.

Mount Pelée was quiet until September 1929. Then it erupted in a pattern similar to that of the earlier eruptions. Fortunately, the one thousand people living in St. Pierre were evacuated at the first signs of volcanic activity, so there was no loss of life. During this period of activity, there were a number of nuées ardentes during the early stages of eruption; none of them, however, was as violent as those that occurred in 1902. As the volcanic activity waned, a new spine grew in the crater near the old spine and gradually crumbled. The new spine never reached the height of the 1903 spine.

Similar Eruptions

The volcano La Soufrière exists on St. Vincent Island, 150 kilometers from Martinique. It also erupts a silica-rich magma and has a pattern of eruption similar to that of Mount Pelée. La Soufrière erupted in 1718, in 1812, only one day before Mount Pelée in 1902, and in 1971. The eruption of La Soufrière on May 7, 1902, had a much smaller and more vertical initial blast than that of Pelée because the blast from La Soufrière occurred through an open rather than a blocked vent. The resultant cloud, consisting of sand-sized particles, descended down the slope induced only by gravity; the speed of the flow (about 40 kilometers per hour) was not nearly as great as at Pelée. The initial destruction of buildings in this area was much less than that caused by Pelée. Fewer buildings caught on fire than at Pelée because the cloud was much cooler from its longer contact with the atmosphere. The area covered by the cloud was much larger at La Soufrière than at Pelée because of the vertical nature of the eruption. The people on one side of the mountain in the town of Soufrière noticed that the volcano was erupting, and most evacuated. Clouds covered the other side of the volcano; the people there did not evacuate in time. Approximately two thousand people lost their lives. This variation of a nuée ardente is sometimes called the Soufrière type of eruption.

Another nuée ardente occurred at the Merapi volcano in Java. Here, a dome or spine built up much like that at Pelée, except that the dome extended out beyond the rim of the crater. The spine would periodically break off, sending a nuée ardente down the side of the volcano. There nuées ardentes moved quickly, like those at Pelée. This variation of a nuée ardente is also known as the Merapi type of eruption.

Study of Mount Pelée

Information about the nature of volcanic eruptions such as Mount Pelée's comes from two types of observation. One is the visual observation of the eruption while it is occurring. The second is the examination of the deposits of erosional features that result from the eruption. These two types of observation were made during and after the eruption of Mount Pelée.

The stiff spine that protruded from the volcanic vent at Mount Pelée has been seen at other locations that also have silica-rich and viscous magma. It is believed that the viscous magma plugs the vent of the volcano in these types of eruption. The gas pressure builds until gas-rich magma is suddenly released at a low angle. During the sudden decompression, the magma froth—broken up into ash, fine dust, or large rock fragments—discharges like a cannon. Observers report deafening explosions during these eruptions.

In a 1930 eruption of Pelée, one trained observer, standing only 35 meters away from the eruption, confirmed that a lateral explosion did occur. In lateral explosions, the solid particles and magma are believed to release water vapor rapidly and continuously. In this way, the magma freezes, so little or no magma is left after the eruption's initial release of pressure. Much of the flow's initial speed is a result of this lateral blast. The lower portion of the gas-particle mixture takes on speed and moves rapidly and fluidly down the low portions of valleys on the cooler cushion of air beneath it.

Thus, nuées ardentes behave much like running water or avalanches, except that they move faster. The continual release of hot water vapor from the particles may cause rapid expansion and compression of the gas-particle fluid and add to its downslope speed. Some of the dust-sized particles rapidly escape upward from this moving gas-particle fluid and rise in a glowing, black cloud. These glowing avalanches erode gullies along the steep flanks of the volcano and scratch rocks parallel to their flow. Little material is deposited until the avalanches reach the lower slopes of the volcano and slow down. The deposits usually consist of a random mixture of various-sized particles—from fine ash to blocks several meters wide. Apparently, there is enough turbulence to keep the particles stirred up. Some deposits, however, have coarser material concentrated at the bottom, with the finer material near the top. A thin layer of dust sits atop the main deposit as this material slowly settles from the overlying ash cloud.

Geologists who study volcanoes observe the eruption processes and magma compositions as they are ejected at the surface. One volcanologist, Frank A. Perret, set up an observation shed on the slopes of Mount Pelée during the 1929 eruptions near the area where a number of nuées ardentes descended down the slopes. The edge of one of these flows hit his shed and nearly killed him. He survived to publish his observations and conclusions on the flow mechanisms of nuées ardentes. His theories provided the basis for the first real understanding of these processes.

Robert L. Cullers

Further Reading

Bullard, Fred M. *Volcanoes of the Earth.* 2d ed. Austin: University of Texas Press, 1984. A very readable and nontechnical discussion of the location and the origin of volcanoes of the world, suitable for the layperson. Includes a long section on nuées ardentes. Contains many pictures and illustrations. A glossary defines unfamiliar terms.

de Boer, Jella Zeilinga, and Donald Theodore Sanders. *Volcanoes in Human History: The Far-Reaching Effects of Major Eruptions.* Princeton, N.J.: Princeton University Press, 2002. Provides an overview of volcanism followed by specific volcanoes and eruptions. Includes the interaction of human history and volcanic activity. Covers the Hawaiian Islands, Crete, Vesuvius, Iceland, Tambora, Krakatau, Pelée, Tristan da Cunha, and Mount St. Helens.

Decker, Robert, and Barbara Decker. *Volcanoes.* 4th ed. New York: W. H. Freeman, 2005. This readable and well-illustrated text is geared toward the general public. The authors explain geologic phenomena in a clear and simple manner. Includes a good section on Mount Pelée. The appendices list the world's 101 "most notorious" volcanoes, as well as Internet sources for volcano information.

Ferguson, James. "The Tragedy of St. Pierre." *Geographical* 74 (2002): 14-19. The article discusses the impact of Mount Pelée's eruption in 1902 on the town of Saint Pierre. Mentions warning signs of the eruption, describes current knowledge of the event, and provides a comparison of Pelée to other eruptions in history.

Francis, Peter, and Clive Oppenheimer. *Volcanoes.* 2d ed. New York: Oxford University Press, 2004. This text covers volcanic activity ranging from lava flows to pyroclastic currents. Provides many examples from Hawaii, Italy, Ethiopia, Japan, Mount St. Helens, and Krakatau. Written for the layperson, but some knowledge of geology is helpful.

Heilprin, Angelo. *Mount Pelée and the Tragedy of Martinique.* Philadelphia: J. B. Lippincott, 1903. A readable account of the eruption of Mount Pelée by a trained observer who arrived shortly after its eruption of May 8, 1902. Many photographs of the destruction, the spine, and one of the subsequent nuées ardentes. Can be easily read by the layperson.

Hill, Robert C. "On the Volcanic Disturbance in the West Indies." *National Geographic* 13 (July 1902): 223-266. One of the best-illustrated summaries of the eruption and destruction at Pelée, written in lay terms.

Kennan, George. "The Tragedy of Pelée." *Outlook* 71 (June 28-August 16, 1902): 769-777. Another eyewitness account of the eruption and destruction at Pelée. Suitable for the layperson.

Perret, Frank A. *Eruption of Mount Pelée, 1929-1932.* Washington, D.C.: Carnegie Institution of Washington, 1935. An eyewitness account of the smaller eruptions of Pelée during 1929-1932 by an electrical engineer who was a volcanologist for many years. He explained the origin of nuées ardentes by observing them as close as 30 meters. Suitable for the layperson. No glossary.

Russell, Israel C. "The Recent Volcanic Eruptions in the West Indies." *National Geographic* 13 (July 1902): 267-284. An illustrated description of the eruptions on Martinique and the St. Vincent Islands. Suitable for the lay reader.

Sigurdsson, Haraldur, ed. *Encyclopedia of Volcanoes.* San Diego, Calif.: Academic Press, 2000. This book provides a complete summary of the scientific knowledge of volcanoes. It contains eighty-two well-illustrated overview articles, each of which is accompanied by a glossary of key terms. Although this is a college-level text, it is written in a clear and comprehensive style that makes it generally accessible. Cross-references and index.

See also: Eruptions; Forecasting Eruptions; Krakatau; Mount Pinatubo; Mount St. Helens; Mount Vesuvius; Popocatépetl; Recent Eruptions; Shield Volcanoes; Stratovolcanoes; Volcanic Hazards.

MOUNT PINATUBO

The eruption of Mount Pinatubo on June 15, 1991, was one of the most violent eruptions of the twentieth century. The use of modern techniques of eruption prediction and danger assessment allowed the safe evacuation of nearly 150,000 people from the area adversely affected by the eruption.

PRINCIPAL TERMS

- **ash:** small fragments of volcanic material less than 2 millimeters in diameter formed by explosive ejection from a vent
- **caldera:** a large circular depression around a summit vent that typically forms by collapse when large volumes of magma are rapidly ejected
- **dome:** a small, steep-sided mass of volcanic rock formed from nonexplosive, viscous lava that solidifies in or above a vent
- **fumarole:** a volcanic vent from which only gases are emitted
- **harmonic tremor:** a type of earthquake activity in which the ground undergoes continuous shaking in response to subsurface movement of magma
- **lahar:** a volcanic mudflow resulting from the mixing of erupted lava and ash with surface water, rain, or melted snow
- **pyroclastic flow:** a turbulent, dense mixture of volcanic gases, magma, ash, and rock that is ejected from a vent and flows rapidly down the flanks of a volcano
- **stratovolcano:** a large, steep-sided volcano consisting of alternating layers of coherent lava and explosively ejected fragmental material; also called a "composite volcano"

Dormant Since 1541

Mount Pinatubo is one of the nearly three hundred active volcanoes that rim the Pacific Ocean Basin in a narrow belt called the Ring of Fire. This stratovolcano had been dormant since the Philippines were first settled by the Spanish in 1541. The awakening of Mount Pinatubo was of considerable concern to the United States because Clark Air Base was located only 20 kilometers to the east of the summit, and the Subic Bay Naval Station was 40 kilometers to the southwest.

Pinatubo rumbled into activity on April 2, 1991, when a series of small explosions formed a row of craters 1.5 kilometers long just northwest of the volcano's summit. The explosions lasted several hours and deforested an area of several square kilometers. Villagers living on the volcano's flank reported hearing the explosions and awoke the morning of April 3 to find a row of fumaroles extending along the western side of Pinatubo. Civil officials evacuated a circular area in a 10-kilometer radius from the volcano's summit.

Seismic Wake-up

During all of April and the early part of May, Mount Pinatubo experienced dozens of small earthquakes each day as magma began to move beneath the volcano. Magma fractured the brittle rocks as it cleared a pathway upward to form a shallow magma body. The center of the earthquake activity was about 5 kilometers northwest of the summit and 4 kilometers below the surface.

During May, the earthquake activity increased, with more than 1,800 low-intensity quakes being recorded between May 7 and June 1. During the last two weeks of May, the fumaroles showed a tenfold increase in the emission of sulfur dioxide.

At the beginning of June, a second area began to experience a series of shallow earthquakes several kilometers closer to the summit. The magma was rising beneath the volcano, and it was forcing a conduit to form above the magma chamber. The rate of emission of gas reversed its May trend, decreasing from a production of 5,000 tons of sulfur dioxide on May 28 to a production of 1,800 tons on May 30. By June 5, the total had dropped to only 260 tons. Apparently, the rising magma had sealed the fractures that allowed the escaping gases to reach the fumaroles.

Harmonic Tremors

On June 3, a series of small explosions blew ash from the summit area, and the earthquakes moved into a type of activity called "harmonic tremor": a prolonged rhythmic shaking of the earth, in contrast to the single sharp jolts of earlier earthquakes. Harmonic tremor occurs when underground magma flows in a more continuous manner through established subsurface channels. On June 6, the summit region of the volcano began to bulge outward, with the upper flank showing a measurable increase in the slope of the land surface. As the tilting increased, there was an increase in earthquake activity, leading to a strong shock and explosion on June 7. This explosion sent a column of ash and steam to a height of between 7 and 8 kilometers. The explosion allowed the summit area to stop its inflation temporarily, and there was a brief period of reduced earthquake activity. Based upon these warning signs, the area of evacuation was extended to a radius of 15 kilometers.

Magma first appeared on the surface of the volcano the next morning, June 8. Observers reported that a small lava dome 150 meters in diameter was growing near one of the fumaroles just northwest of the summit. The vents associated with this dome emitted a series of weak ash clouds that rose only to the level of Pinatubo's summit over a three-day period ending on June 12. During this interval, harmonic tremors occurred almost continuously. An examination of the composition of ejected ash revealed that a new, very fluid magma had invaded an old magma chamber that contained residual magma from the last eruption nearly five hundred years ago. This intruding magma signaled the potential of a major eruption, and authorities extended the evacuation area to a radius of 20 kilometers from the volcano. On June 10, fourteen thousand U.S. servicemen left Clark Air Base along with their aircraft, never to return. They left behind three helicopters and a contingent of fifteen hundred security and maintenance personnel.

Aerial view toward the southern end of Mount Pinatubo, in the Philippines. White plumes rise after phreatic eruptions from fumaroles on the crater floor. (U.S. Geological Survey)

Multiple Eruptions

The first major explosion of Pinatubo occurred at 8:51 a.m. on June 12, signaling the final and most powerful phase of the 1991 eruption. This very violent phase lasted about ten days, with the most intense activity occurring on June 15 and 16. The June 12 eruption lasted thirty-five minutes and spewed a column of steam and ash to a height of 19 kilometers. A small pyroclastic flow (a turbulent, dense mixture of volcanic gases, magma, ash, and rock that is ejected from a vent and flows rapidly down the flanks of a volcano) traveled a short distance from the vent down the northern flank of the volcano into already evacuated villages along the Maraunot River. Clark Air Base evacuated six hundred maintenance personnel, and Filipino authorities extended the evacuation zone to a 30-kilometer radius. Ash from the eruption was so dense and spread so far that the airport in Manila, more than 50 kilometers away, was forced to close.

Similar explosions continued over the next few days, producing ash clouds that prevented visibility of the volcano. The arrival of Typhoon Yunya on June 13 meant that most of these explosions were completely unseen from the ground. The associated earthquake activity was recorded by seismic stations, and the presence of an eruption cloud was verified by military weather radar.

Beginning on June 14, the main, violent phase peaked in a three-day period with more than fifty short, violent explosions in which numerous vents erupted simultaneously. The intensity of the earthquakes generated by these eruptions was more than a hundred times greater than those recorded during May. These eruptions grew progressively more violent during this time, sending ash to a height of more than 40 kilometers. At 5:55 a.m. on June 15, the ejected ash switched from a vertical orientation to a more horizontal one. The onset of laterally directed explosions indicated that the summit region was beginning to collapse. Several pyroclastic flows erupted and traveled up to 13 kilometers from the summit. The remaining personnel at Clark Air Base were evacuated that morning.

At about 3:30 p.m. on June 15, the climax occurred. A series of strong earthquakes started, lasting all afternoon and throughout the night. The seismographic equipment on the volcano was destroyed during the night by a series of large pyroclastic flows. Pyroclastic deposits that exceeded 200 meters in thickness formed in many valleys around the volcano. A 2-kilometer caldera formed when the summit area collapsed, causing the volcano to drop by more than 300 meters in elevation.

Ashfall, Mudflow, and Poisonous Gas

About 0.8 cubic kilometer of ejected material covered the west-central portion of the island of Luzon. The falling ash blanketed villages and buried crops around the volcano. Many rooftops collapsed under the added weight of the ash, causing the majority of deaths attributed to the volcanic eruption. The total volume of ash ejected was close to 5 cubic kilometers, most of which fell into the South China Sea.

The water runoff from the heavy rains associated with the typhoon mixed with the recently fallen ash and caused disastrous volcanic mudflows known as "lahars." The lahars swept down twelve river valleys, including the Abacan and Sacobia River Valleys, where the town of Angeles City was completely destroyed. Seven other towns were damaged by these mudflows. The July monsoons caused secondary lahars, one of which inundated the city of Pabanlog along the Gumain River. All these lahars destroyed bridges and farmland along the floodplains and caused widespread economic and social disruption. Special early-alert systems detected the lahars and sharply reduced the loss of life. However, the region continued to experience secondary lahars caused by heavy rains, and this threat was expected to last well into the twenty-first century.

The cloud associated with the June 15 eruption released four billion pounds of chlorine gas and forty billion pounds of sulfur dioxide into the stratosphere. By June 25, satellite images showed that a 7,750-kilometer-long cloud of sulfur dioxide had spread across the tropical Northern Hemisphere.

Damages and Death Toll

After June 16, the eruption slowly grew less intense. There were occasional ash explosions into 1992 that sent ash columns upward of 10 kilometers in height. During July and August of 1991, a dome grew in the fuming caldera that was 300 meters across and nearly 100 meters high. With an average of more than twenty centimeters of ash covering the region around Clark Air Base, base personnel were unable to stop ash infiltration into the jet engines, and the base had to be abandoned.

More than 108,000 homes were partially or totally destroyed. The final death toll for the region was 722; of these, 281 died as a result of ejected material either as ashfalls or pyroclastic flows, 83 died from primary and secondary lahars, and 358 died from disease related to the social turmoil associated with the interruption of the country's infrastructure. The loss of life was undoubtedly compounded by the typhoon that hit the island during the climax of the eruption.

Global Impact

The world's largest volcanic eruptions have all produced global climate and atmospheric changes. Fine volcanic ash and gases are blasted into the high atmosphere and dispersed around the world. Upper-level winds can keep volcanic ash suspended in the atmosphere for many years. The suspended ash from Mount Pinatubo's eruption produced colorful sunsets worldwide in 1991 and 1992.

The gases and ash combined to produce an aerosol that blocked both incoming sunlight and infrared radiation emitted by the earth. The loss of solar radiation causes cooling, whereas the absorption of the earth's transmitted infrared radiation leads to global warming. The two effects did not balance equally, and by 1993, measurements from the National Aeronautics and Space Administration's Earth Radiation Budget Satellite provided the first conclusive evidence of a significant change in global energy as the result of a volcanic eruption. The net effect of a loss of solar radiation and a greater retention of infrared radiation resulted in a period of global cooling in which the average global temperature dropped by one-half degree Celsius (about one degree Fahrenheit). Approximately 2 to 3 percent of the sun's energy was blocked out, counteracting prevailing global-warming trends and temporarily setting the earth's climatological clock back to the 1950's.

The gases also altered the chemistry of the upper atmosphere. Three months after the eruption, there was 50 percent less ozone in the tropical stratosphere over an area roughly coincident with Mount Pinatubo's volcanic plume. The ozone layer over the United States was 10 percent thinner than normal, translating to a 20 to 30 percent increase in the amount of cancer-causing ultraviolet radiation reaching the earth's surface. There was fear that an ozone hole might open over the populated areas of the Northern Hemisphere; by 1996, however, measurements of the levels of ozone-depleting chemicals had dropped to lower-than-average values.

ASSESSING THE RESPONSE

The successful evacuation of nearly 150,000 people during the Mount Pinatubo eruption required both successful short-term forecasting of eruption events and a well-organized program of danger assessment. The activities that transpired during the ten weeks from the first steam explosion on April 2 to the culminating eruption of June 15 were a model of cooperation between scientific personnel, Filipino authorities, and the local citizenry. Immediately following the first gas explosion, volcanologists from the U.S. Geological Survey joined forces with other scientists from the Philippine Institute of Volcanology and Seismology to coordinate the scientific work of eruption prediction. They jointly established the Pinatubo Volcano Observatory (PVO).

Modern eruption prediction involves determining the nature of the volcano's past eruptions and monitoring the active volcano for any changes in its physical and chemical behavior. An array of scientific equipment must be deployed around the volcano and the active vents to measure earthquake activity, volumes and composition of emitted gases, changes in the slope of the land surface, changes in the horizontal distance across vents and fissures, and fluctuations in the temperature of the volcano.

Active volcanoes have sporadic eruptions that are separated by time intervals called "periods of repose." Most active volcanoes display a repetitive history in terms of their repose interval, their eruptive violence, and the kinds of materials that they expel. The nature of a future eruption can often be established by an examination of the geological record of previous eruptions.

The PVO team realized the value of knowing the previous eruptive history at Mount Pinatubo and did a rapid geological reconnaissance of the volcano during the month following the initial steam explosion. They found that three previous eruptions had occurred at about 500 years ago, 2,500 years ago, and 4,800 years ago. Each eruption had been dominated by highly explosive activity. Pyroclastic flows had swept down the volcano's flanks, leaving deposits more than 20 kilometers from the summit. Debris from lahars from these prehistoric eruptions were found in six river valleys leading away from the volcano for a distance of more than 40 kilometers.

On the basis of their geological study of past eruptions, the PVO team compiled a hazard map showing the regions that were likely to be affected by pyroclastic flows, ashfall, and lahars. The map was used by both civil defense officials and military commanders during the various stages of the eruption. A map of the areas affected by the actual 1991 eruption shows a remarkable correlation with the pre-eruption hazard map. A videotape depicting the various volcanic hazards was produced by filmmaker and geologist Maurice Krafft, and the film aired repeatedly on local television.

Danger assessment is a complex task that goes beyond the accurate forecasting of eruptions. For example, in 1985, the Nevado del Ruiz volcano erupted in a region of Colombia that was much less populated than the area around Mount Pinatubo. The eruption generated a mudflow that killed 25,000 people. Although volcanologists were successful in predicting the eruption and recommended evacuation of the region, the civil authorities did not view the impending eruption to be particularly dangerous, and no evacuation occurred.

The PVO team worked closely with civil and military personnel to assess the level of danger posed by the various stages of the Mount Pinatubo eruption. Using the hazard map and the video, the Filipino scientists developed a five-level alert system. Alert level 1 was to be declared when low-level seismic activity was coupled with fumarolic emission. The level 1 alert occurred on April 3, when fumaroles developed in a row along the northwest flank of Mount Pinatubo. Alert level 2 was to be invoked when moderate levels of seismic activity occurred and there was positive evidence for the existence of subsurface magma. Level 2 was reached in late May, when 1,800 earthquakes were recorded within a three-week interval.

Authorities were instructed to interpret alert level 3 to mean that the magma was intruding into the volcano and could eventually lead to a major eruption. Level 3 was to be issued when high levels of gas emission occurred with simultaneous ground deformation. This level of danger was attained on June 5. Level 3 meant that a major pyroclastic eruption could occur within two weeks; it actually occurred ten days later. Alert level 4 was to be issued when extensive harmonic tremors indicated the magma was moving more freely beneath the volcano. Level 4 was announced immediately following the June 7 explosion at Pinatubo that generated the 7-kilometer-high ash column. Civil authorities were told that level 4 meant that a major pyroclastic eruption was possible within twenty-four hours. The highest level of danger assessment was alert level 5, which meant that an eruption was in progress. This level was issued on June 9, after a pyroclastic flow engulfed several evacuated villages on the northern flank of Pinatubo. The day after the level 5 alert was issued, the U.S. military evacuated fourteen thousand personnel from Clark Air Base.

Lessons Learned

The violent eruption of Mount Pinatubo was classified as a magnitude 6. The largest recorded eruption has been classified as magnitude 7, and the largest geologically known eruption has been ranked as a magnitude 8. Such violent eruptions as these have historically caused high death tolls. However, as shown by Mount Pinatubo experience, the most dangerous volcanoes can cause far fewer deaths when proper surveillance is employed.

There are nearly three hundred active volcanoes that should be monitored for the eruption of pyroclastic flows and lahars, and many of these volcanoes are located in underdeveloped countries. Surveillance and risk-reduction programs are very expensive, requiring equipment, trained scientists, education programs for local populations, and development of cooperative military and civil planning. The expenses associated with such programs are beyond the capabilities of many less-developed countries. Consequently, the United States, Russia, Iceland, Italy, and Japan lead the world in this field of volcanology.

The United States has two facilities dedicated to monitoring active volcanoes, both staffed by geologists from the U.S. Geological Survey. The Hawaiian Volcano Observatory has been successfully predicting eruptions of Mount Kilauea for more than forty years. After the eruption of Mount St. Helens in 1980, the Cascades Volcano Observatory was established to monitor magma amounts and movements beneath the active volcanoes in Washington, Oregon, and northern California. Japan has twenty-one volcano observatories.

It is very likely that Mount Pinatubo will erupt again, although the volcano has entered a period of repose that may last as long as a few hundred years. The volcano is located on the margin of two converging plates that have produced a very active chain of volcanoes. For example, Mount Mayon lies 300 kilometers to the south of Pinatubo and has had more than forty historical eruptions, with the last as recent as 1993. A key factor indicating continued life for Mount Pinatubo is that a 1996 study found that between 40 and 100 cubic kilometers of magma remained in a reservoir below the volcano's summit. This volume is the largest quantity of magma (by almost a factor of three) ever detected beneath any volcano.

Dion C. Stewart

Further Reading

Bullard, F. M. *Volcanoes of the Earth.* 2d ed. Austin: University of Texas Press, 1984. A good scientific treatment of volcanoes and volcanic products. Well written; does not require a geological background. The author, an experienced volcanologist, relies heavily on his own experiences, photographs, and drawings.

Francis, Peter, and Clive Oppenheimer. *Volcanoes.* 2d ed. New York: Oxford University Press, 2004. This text covers volcanic activity ranging from lava flows to pyroclastic currents. Provides many examples from Hawaii, Italy, Ethiopia, Japan, Mount St. Helens, and Krakatau. Written for the layperson, but some knowledge of geology is helpful.

Marti, Joan, and Gerald G. J. Ernst, eds. *Volcanoes and the Environment.* New York: Cambridge University Press, 2005. This text provides information on volcanoes ranging from the physical mechanisms and behaviors to hazards and effects on humans and nature. Written for students and geologists studying environmental science and geology.

Oppenheimer, Clive. *Eruptions That Shook the World.* New York: Cambridge University Press, 2011. Describes the mechanics of volcanoes and types of eruptions. The author also details the effects of eruptions on the atmosphere, humans, and other organisms. Specific volcanoes are examined and an appendix of major eruptions is provided.

Robinson, Andrew. *Earth Shock, Hurricanes, Volcanoes, Earthquakes, Tornadoes, and Other Forces.* New York: Thames & Hudson, 1993. Chapter 4 is devoted to volcanoes and gives a humanistic accounting of many historic eruptions, interwoven with excellent scientific commentary.

Sigurdsson, Haraldur, ed. *Encyclopedia of Volcanoes.* San Diego, Calif.: Academic Press, 2000. This book provides a complete summary of the scientific knowledge of volcanoes. It contains eighty-two well-illustrated overview articles (including one on calderas), each of which is accompanied by a glossary of key terms. Includes an extensive section on eruption forecasting. Although this is a college-level text, it is written in a clear and comprehensive style that makes it generally accessible. Cross-references and index.

Stimac, J. A., et al. "The Crater Lake and Hydrothermal System of Mount Pinatubo, Philippines: Evolution in the Decade After Eruption." *Bulletin of Volcanology* 66(2005): 149-167. Provides a technical description of the eruption of Mount Pinatubo. Discusses thermal and chemical composition and measurements. The authors explain the formation of a crater and lake. This article contains detailed data for comparison to other volcanic events.

Wolfe, Edward W. "The 1991 Eruptions of Mount Pinatubo, Philippines." *Earthquakes and Volcanoes* 23, no. 1 (1992): 5-35. The public report of the successful emergency response of the United States to the geologic danger presented by the 1991 eruption of Mount Pinatubo.

See also: Eruptions; Forecasting Eruptions; Krakatau; Mass Wasting; Mount Pelée; Mount St. Helens; Mount Vesuvius; Popocatépetl; Recent Eruptions; Shield Volcanoes; Stratovolcanoes; Volcanic Hazards.

MOUNT ST. HELENS

The eruption of Mount St. Helens in 1980 and the series of geologic and volcanic events that followed provided a unique opportunity for geologists to study a classic volcanic disturbance in detail using modern technological tools.

PRINCIPAL TERMS

- **caldera:** the sunken area at the summit of a volcano caused by the internal collapse of magma
- **fissure:** an extensive crack or fracture; a linear volcanic vent
- **magma:** molten rock material formed deep in the earth's interior; when thrown out of a volcano, it is known as lava
- **pyroclastic material:** rocks formed from the debris of explosive volcanic eruptions and fragments from the walls of the vents
- **subduction:** a process that occurs when one tectonic plate dives beneath another tectonic plate; it may occur over thousands or millions of years
- **tectonics:** the study of the processes that formed the structural features of the earth's crust, especially the creation and movement of immense crustal plates
- **tunnel vent:** the central tube in a volcanic structure through which material from the earth's interior travels
- **vent:** a break or tear on the side of a mountain through which magma and pressure can escape

WARNING SIGNS

In the weeks leading up to May 18, 1980, Mount St. Helens exhibited warnings of an impending eruption, including internal rumbles and small bursts. A large bulge had developed over 98 meters on the north face of the mountain, indicating a highly pressurized buildup of old, trapped magma just below the mountain's surface. As days passed, the bulge grew larger while the tiny eruptions at the summit continued.

What followed, however, could not have been predicted. First, deep internal disturbances forced magma (molten rock) up to the surface of the earth through the volcano's central, or tunnel, vent. As the magma spread upward, it exerted a large amount of pressure on the sides of the central vent and, as the pressure increased, caused the bulge of old magma to push even farther outward. Then, when the pressure became stronger than the mountain's ability to contain it, the magma blasted through two enormous holes—one at the summit and one on the north flank. The release of energy resulted in a huge explosion.

A MOUNTAIN BEHEADS ITSELF

At 8:32 on the morning of May 18, 1980, an earthquake in the Mount St. Helens region of the Washington State Cascade Range started a scenario of startling events, beginning with the sudden and spectacular eruption of Mount St. Helens and culminating in the destruction of hundreds of square kilometers of valuable timberland, vacation resorts, and foraging area for wildlife. Considered by many scientists to be a classic series of volcanic activities, the eruption blasted away the top 396 meters (one-third) of the mountain in a cataclysmic explosion that exuded 360 billion kilograms of rock and ash. Forests more than 27 kilometers away were completely leveled; pyroclastic debris (steam, gas-filled rock, and wet ash), moving at the amazing speed of 321 kilometers per hour, flew horizontally across the expanse of the Cascade Range together with suffocating volcanic gases, eventually taking the lives of sixty-one people and an estimated 2 million birds, fish, and animals; twenty-six lakes and 241 kilometers of freshwater streams were destroyed. In the end, the scene in the Cascades was one of utter desolation for some 300 square kilometers.

The loosed fury of the initial blast, comparable to a 10-megaton bomb, immediately tore away one-third of the entire mountain and was followed by a huge landslide. Hot gases, steam, and molten rock shot laterally from the mountain's flank at nearly the velocity of a rifle bullet. Explosion after explosion shook the volcano as debris poured down the mountainside, carrying hot lava, timber, and wet ash. Volcanic ash in heavy, dense clouds spewed out of the two gigantic fissures. Driven by mighty pressures from the earth's depths, the clouds of ash funneled outward, reaching beyond 33 kilometers in less than two minutes and beginning to spread over the northwestern region of the United States. Mount St. Helens' snow-capped summit quickly turned to water and steam under the heat and pressure of the eruption, releasing 174

Mount St. Helens Explosion, 1980

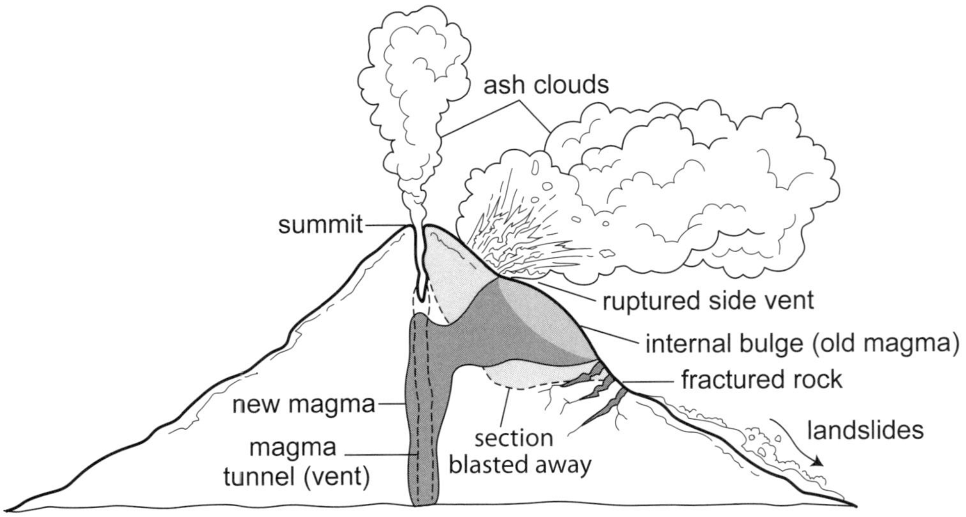

billion liters of water down the mountainside. The flood swept mud, solidified lava, tree splinters, pulverized rock, and other material in its path, creating an avalanche that clogged streams and bulldozed forestation as it continued its outward spread. Attacked by volcanic gases, fumes, ash, and intense heat from the bowels of Mount St. Helens, the atmosphere immediately surrounding the erupting volcano became a cauldron of earsplitting thunder and crackling lightning.

There was another eruption on May 25, 1980, spreading an additional ash deposit, up to 70 millimeters thick, for hundreds of square kilometers, mostly westward. Then, in October of the same year, Mount St. Helens erupted once more, this time sending yet more clouds of ash southward. The floor of the caldera grew to more than 260 meters in diameter, with a dome 49 meters high.

Ashes, Ashes

The picturesque valleys around the mountain were smothered with debris. In areas farther from the mountain, gray ash fell like a huge rainstorm, covering the ground with a blanket 1.5 to 2 meters thick. Even four days later, the ash was extremely hot, registering temperatures as high as 323 degrees Celsius. Rescuers had to move swiftly through the deep ash to keep their legs from burning. People in the town of Yakima reported removing 540 million kilograms of ash from streets, rooftops, and other structures. Elsewhere, the weight of the volcanic ash caused flattening of hayfields and similar agricultural problems, resulting in an estimated $300 million worth of damage to crops. Once fallen, the ash dried to a powder that was blown in all directions by new winds. Where the ash was thickest, people had to wear protective face masks to keep from choking. Not all the ice from the peak melted during the first several blasts; huge chunks—some larger than trucks—lay scattered along the valley floors, where they then melted.

The ecological damage from the eruption was widespread. Streams were choked with debris, killing fish and destroying watering holes; dense clouds of ash, composed of tiny glass particles driven like shrapnel by the foaming winds, sanded down the wings of birds and insects and clogged their breathing; the rolling landslides inundated timberland and covered wildlife with a shroud of ooze, ash, gases, and debris. Decades will be required for the forests and lakes to rebuild themselves.

Origins of the Blast

The Mount St. Helens eruption is among the most closely studied volcanic events in history. Scientists were able to record in detail the physical and chemical activities that occur during a cataclysmic volcanic eruption. Eyewitness accounts were available from people who had been caught in the midst of the eruption and narrowly escaped death. These accounts furnish important data about many different aspects of volcanism.

The Cascade Range in western Washington State was created by a subduction process beginning about 80 million years ago. Subduction is one of the main methods by which earthquakes are produced; it occurs where the edges of two tectonic plates meet. The stronger, or perhaps more rigid, plate "dives," or subducts, under the other plate. In the process of subducting, the plates scrape along each other and cause great jolts, which are experienced as earthquakes. In the Mount St. Helens scenario, the tiny Juan de Fuca plate subducted under the larger North American plate (the North American continent), a process that is still occurring. The Pacific plate has been moving closer to the North American continent as the Juan de Fuca plate disappears, and it is now abutted next to the San Andreas fault, which runs across western California. The Juan de Fuca plate is almost entirely subducted.

Study of Mount St. Helens

More scientific experience and technology were applied to the investigation of the Mount St. Helens eruption than to any other volcanic event in North America. One of the most valuable of all the data sources, however, was eyewitness accounts provided by both trained and untrained observers. Dozens of geologists and volcanologists worked in the immediate vicinity of St. Helens, some paying with their lives. In addition, hundreds of local citizens who witnessed the week-long scenario furnished information. The narratives, offered by citizens who had lived for many years in the region, enabled researchers to construct an almost minute-by-minute catalog of events, complete with time estimates, sizes, volumes, and intensities—much of which replaced the data that went unrecorded when scientific instruments were destroyed in the eruption.

The most frequently used instrument for studying volcanic activity is the seismometer, which measures the frequency, location, and intensity of tremors. Seismometers were placed in as many as twelve locations near Mount St. Helens, furnishing a constant monitoring system. In this instance, the instruments recorded dozens of fair-sized tremors up to the moment of the 5.0-magnitude earthquake that triggered the eruption.

Another valuable instrument is the geodimeter, which uses a laser beam to measure minute changes in ground swelling, such as the huge subsurface bulge on the mountain's north flank. By monitoring changes in the height of such a dome, geologists can determine its rate of growth or subsidence. A fast-growing bulge would signal an impending crisis, perhaps a vent-creating explosion.

Tiltmeters measure a mountain's angle of slope. They tell geologists about the rate of change of the mountain's growth and the altering angle of its sides. These measurements can help to predict an eruption because internal volcanic pressures and upwelling magma tend to push outward against the crust of the volcano, increasing the slope angle.

Mount St. Helens during its 1980 eruption, showing pyroclastic flow. (U.S. Geological Survey)

Regular cameras are used to record an eruption's progress. Cameras on tripods are wired to receive radio signals that automatically advance the film for the next picture, making it unnecessary for the cameras to be operated by humans. Cameras can therefore be placed in areas that would be life-threatening for researchers, perhaps because of intense heat or because they are in the path of a shock wave. Cameras in various positions can be loaded with color or black-and-white film. Color is extremely important in assessing various activities; variations in grayness in an ash cloud tell geologists about the density of different portions of the cloud. Cameras are also vital in recording the aftermath of an eruption. They help to determine the seriousness of such problems as stream choking, ash fallout, and forest destruction.

Geologists use a stream gauge to record water temperature, amounts of material suspended in the water, and the water's chemical content. By means of such measurements, scientists can determine the seriousness of the stream choking, its cause, and the rate at which the stream is being filled with alien material. It is important to know the temperature of the stream, too, because of the possible negative impact that sudden temperature changes might have on fish, fish eggs, and other aquatic life.

Finally, a vital instrument for the geologist is the gas sensor, which is used both at ground level and in airplanes. It is important to be able to assess the presence or absence of various gases in the region of an eruption or impending eruption, especially hydrogen, carbon dioxide, and sulfur dioxide. Although these gases are potentially life-threatening, they can reveal subsurface movement of magma.

Thomas W. Becker

Further Reading

Attenborough, David. *The Living Planet: A Portrait of the Earth.* Boston: Little, Brown, 1984. Prompted by the television series of the same name, this book explores the planet as an ecological and environmental entity. An easy-to-read text is one delightful aspect of the book; the excellent accompanying photography is another. The value of this volume lies in its attempt to integrate the earth's basic attributes. A good place for the interested layperson to begin.

de Boer, Jella Zeilinga, and Donald Theodore Sanders. *Volcanoes in Human History: The Far-Reaching Effects of Major Eruptions.* Princeton, N.J.: Princeton University Press, 2002. Provides an overview of volcanism followed by specific volcanoes and eruptions. Includes the interaction of human history and volcanic activity. Covers the Hawaiian Islands, Crete, Vesuvius, Iceland, Tambora, Krakatau, Pelée, Tristan da Cunha, and Mount St. Helens.

Decker, Robert, and Barbara Decker. *Volcanoes.* 4th ed. New York: W. H. Freeman, 2005. This readable and well-illustrated text is geared toward the general public. The authors explain geologic phenomena in a clear and simple manner. Includes a section on Mount St. Helens. The appendices list the world's 101 "most notorious" volcanoes, as well as Internet sources for volcano information.

Findley, Rowe. "St. Helens: Mountain with a Death Wish." *National Geographic* 159 (January 1981): 2-65. This series of well-written and spectacularly illustrated articles is perhaps the best narrative account of the Mount St. Helens eruption. Eyewitness accounts, scientific explanations, and interviews with survivors are woven together into an astonishing story. Some of the on-scene photographs carry great emotional impact. The best starting point for the reader interested in Mount St. Helens.

Francis, Peter, and Clive Oppenheimer. *Volcanoes.* 2d ed. New York: Oxford University Press, 2004. This text covers volcanic activity ranging from lava flows to pyroclastic currents. Provides many examples from Hawaii, Italy, Ethiopia, Japan, Mount St. Helens, and Krakatau. Written for the layperson, but some knowledge of geology is helpful.

Goodrich, Charles, Kathleen Dean Moore, and Frederick J. Swanson, eds. *In the Blast Zone: Catastrophe and Renewal on Mt. St. Helens.* Corvallis: Oregon State University Press, 2008. Provides ecological and geological views from 25 years following the event. A collaboration of nonscientific and scientific writers. Discusses the destruction and revival of the surrounding habitat.

Harris, Stephen L. *Fire Mountains of the West: The Cascade and Mono Lake Volcanoes.* 3rd ed. Missoula, Mont.: Mountain Press Publishing Company, 2005. Although this book focuses mostly on the Cascades, it includes information on Mount St. Helens, Mammoth Mountain, and Crater Lake.

Chapter 2 covers the Ring of Fire and hot spots. Written in a manner accessible to anyone with some interest in volcanoes.

Lockwood, J. P., and Richard W. Hazlett. *Volcanoes: Global Perspectives.* Chichester, England: Wiley-Blackwell, 2010. An up-to-date and well-illustrated undergraduate level text that is also accessible to the interested reader.

Ollier, Cliff, and Colin Pain. *The Origin of Mountains.* London: Cambridge University Press, 2000. This text encompasses the geographic, geological, and geophysical concepts of mountains. Topics include fault-block mountains, fold belts and folding, gravity structures, rift valleys and river valleys, plateaus, and mountain erosion. Begins with a general overview of subduction and plate tectonics. Mountain ranges discussed in detail include the Appalachians, Andes, Cascades, Himalayas, Mount St. Helens, and the Rockies.

Redfern, Ron. *The Making of a Continent.* New York: Times Books, 1986. Possibly no book on the subject of earth science can compare with this volume of excellent photographs, clear and concise diagrams, and compelling text. The author has created a record of the North American continent that lays bare all nature's secrets. All the photography was created by special camera techniques. A basic textbook for anyone interested in the earth sciences.

Sigurdsson, Haraldur, ed. *Encyclopedia of Volcanoes.* San Diego, Calif.: Academic Press, 2000. This book provides a complete summary of the scientific knowledge of volcanoes. It contains eighty-two well-illustrated overview articles, each of which is accompanied by a glossary of key terms. Although this is a college-level text, it is written in a clear and comprehensive style that makes it generally accessible. Cross-references and index.

See also: Calderas; Eruptions; Forecasting Eruptions; Island Arcs; Krakatau; Lava Flow; Mass Wasting; Mount Pelée; Mount Pinatubo; Mount Vesuvius; Popocatépetl; Recent Eruptions; Shield Volcanoes; Stratovolcanoes; Volcanic Hazards.

MOUNT VESUVIUS

Mount Vesuvius is one of the most famous, most studied, and most scientifically influential volcanoes in the world. The science of volcanology began with reports of its activity and several key terms and concepts originated from its example. The science of archaeology also began with Vesuvius because of the preservation of the ruins of Pompeii and Herculaneum as a result of its most famous eruption, in 79 C.E.

PRINCIPAL TERMS

- **cone:** the hill or mountain, more or less conical, surrounding a volcanic vent and created by its ejecta; it is normally surmounted by a crater
- **crater:** the circular depression atop a volcanic cone
- **eruption:** volcanic activity of such force as to propel significant amounts of magmatic products over the rim of the crater
- **lava:** molten rock (as opposed to ash) erupted by a volcano; the term originated with reference to Vesuvius
- **magma:** molten rock, normally subterranean
- **stratovolcano, or composite volcano:** a noticeably conical volcano composed of interbedded lava flows, ash, and cinders; also, the remnant of such a volcano
- **volcanology:** the study of volcanoes

Mediterranean Volcanoes

When most people think of a volcano, they think of the kind of volcano that Vesuvius is—a distinctly conical mountain that erupts spectacularly at intervals, blowing rocks and ash high into the air while sending fiery lava down its slopes. In actuality, however, there are several different kinds of volcanoes and several different kinds of eruptions. Vesuvius, the best-known example of what is called a stratovolcano, or composite volcano, features a prominent cone made up of lava, craters, and ash. Like most stratovolcanoes, it has erupted in more than one way. Two of the variously classified eruption types, the Plinian and the Vesuvian (or Vulcanian), were either first or best observed here.

Vesuvius is one of a series of volcanoes irregularly aligned throughout the Mediterranean region, from southern Greece down to central and southern Italy and then to Sicily and its adjacent islands. The best-known Greek volcano, now extinct, is the crater-remnant island of Santorini (or Thera), the eruption of which is sometimes credited with the destruction of Minoan Crete. Aside from Vesuvius, the best-known Italian volcanoes, all of them recently active, are Mount Etna in Sicily and the two Aeolian island volcanoes, Vulcano (for which all other volcanoes are named) and Stromboli, which has been in constant eruption for more than 2,500 years. Geologists explain this Mediterranean lineage of volcanoes through plate tectonics. According to this theory, the African plate is slowly moving northward, colliding with and sliding under the European plate. Its leading edge plunges downward beneath the European plate and eventually melts, creating large accumulations of liquid rock (magma) that later ascend to express themselves as volcanoes. This level of understanding, however, has only been achieved since about 1970, when plate tectonics theory was accepted.

Vesuvius, located 11 kilometers south of Naples and easily visible from there, is part of a larger volcanic complex surrounding that city. Other parts of the complex include the Phlegraean Fields, a group of nineteen small, closely grouped craters (regarded as the entrance to Hell in classical times); the so-called Temple of Jupiter Serapis at Puzzuoli, a Roman ruin that shows unmistakable evidence of abrupt vertical displacement within historical times (it was utilized by Charles Lyell as the frontispiece to all twelve editions of his *Principles of Geology*, 1830-1875); the Solfatara, a large, steaming crater whose last eruption (in 1198) probably displaced the Temple of Jupiter Serapis; and the buried cities of Pompeii and Herculaneum. All these phenomena are closely associated with Vesuvius and have been immensely influential in forming ideas about volcanoes and how they work.

Vesuvius is the most recently active (1944) member of a group of volcanoes—many of them now deeply eroded and scarcely recognizable—lying roughly parallel to the Apennine Mountains and the Tyrrhenian (west) coast of the Italian peninsula. In mid-Tertiary times (about 30 million years ago), through the agency of plate tectonics, the Apennine chain began to rise, and the Tyrrhenian Basin to subside. The

Bay of Naples is also essentially a downwarp, with the Triassic and Cretaceous limestone rocks of the Sorrento peninsula forming one of its uptilted edges; Vesuvius, the Solfatara, and the Phlegraean Fields are located roughly atop its center. Vesuvius probably began as a submarine volcano in the Bay of Naples, then emerged as an island and finally became part of the mainland through the constructive effects of its own eruptions.

Birth of Vesuvius

Vesuvius originated about 10,000 years ago; there were several cones before the present one. Archaeological evidence attesting human occupation of its slopes goes back to the seventh century B.C.E. The name "Vesuvius" (its origin is unknown) was used around 45 B.C.E. by Diodorus Siculus, a Sicilian writer who recognized the mountain as a former volcano. By around 30 B.C.E., however, the geographer Strabo regarded Vesuvius as a scene of local fires only. He described a flat-topped, barren summit and fertile slopes teeming with agriculture. Thus, no actual eruptions of Vesuvius had been recorded in historic times, and its identification as a volcano had been forgotten.

A destructive earthquake struck the Naples region on February 5, 63 C.E., causing extensive havoc at Pompeii and Herculaneum also. Subsequent tremors continued off and on for the next sixteen years, with one of the strongest occurring on the night of August 24, 79 C.E. It is apparent to modern scientists that such earthquakes are often caused by the upward movement of magma in the vent of an active volcano, but no such understanding prevailed at the time. The most famous volcanic eruption in history therefore took the population it affects almost completely by surprise.

Plinian Eruptions

The first reason that the Vesuvian eruption of 79 C.E. is so well known is that there is an eyewitness description of it written some years after the event by Pliny the Younger, in two letters to the historian Tacitus regarding the death of Pliny the Elder, the writer's unclear who was a distinguished naturalist and public figure. This two-part epistolary account, often reprinted in translation, is the first volcanological field report and more exciting than geological papers tend to be.

According to the younger Pliny, about one o'clock in the afternoon on August 24, a strange pine-tree-shaped cloud, unusually large and rapid, began to ascend over Vesuvius. The Italian pine tree, unlike North American pines, is narrow at the base and becomes increasingly broad as its grows upward. Pliny's description being so apt, an eruption cloud of the same shape today is still called a *pino* (Italian for pine tree); eruptions of this type are called Plinian. Realizing that a natural disaster was at hand, Pliny the Elder went by water to the home of a friend, where he slept for a time, until tremors and falling ejecta from the volcano awakened everyone. A corpulent man, he then died suddenly the next morning, from asphyxiation or possibly a heart attack—the first "scientist" ever to be killed by a volcano. Other portions of his nephew's account describe, for the first time, phenomena now frequently associated with volcanic eruptions: preternatural darkness, the retreat of the sea from its established shores, and repeated bursts of lightning within the eruption cloud.

Destruction of Pompeii and Herculaneum

The other reason that the Vesuvian eruption of 79 C.E. is so famous is that the nearby cities of Pompeii and Herculaneum were destroyed (and their remains preserved) by it. Pliny the Younger does not mention either city, though Tacitus may have in a portion of his history no longer extant. There is not another written account before about 230 C.E., when Dio Cassius erroneously claimed that the Pompeiians died while sitting in their theater, a scene later dramatized in Edward Bulwer-Lytton's novel *The Last Days of Pompeii* (1834) and numerous films. Archaeological evidence, however, has proved that Dio Cassius's account is false.

Except for some immediately posteruption attempts at salvage, Pompeii and Herculaneum lay essentially untouched beneath their respective volcanic coverings until well into the eighteenth century, when organized attempts at treasure hunting began. Skeletons of the deceased were then discovered rather frequently, especially at Pompeii, where the modern science of archaeology evolved. Destruction there had taken place so rapidly that the city's life ended with food still on tables and bread baking in ovens. Pompeii is one of the most often visited monuments in Europe because the evidence of its everyday life and the drama of its abrupt termination are apparent to all.

An aerial view of the crater of the Mount Vesuvius volcano, as seen near the Italian city of Naples on March 31, 2009. Nearly 200 years after the destruction of the famed town of Pompei, Mount Vesuvius is again threatening the surrounding Italian region. (AFP/Getty Images)

Pompeii and Herculaneum were destroyed in different ways. The greater number of bodies—some of them skeletons, others cast in plaster—are from Pompeii, and they often appear to indicate death by asphyxiation. No lava reached the city, though it was bombarded with ash to a depth of about 1 meter. It is believed, however, that even such an ashfall could not have killed the more than two thousand persons known to have perished there. Current theorizing holds that at least one phase of Vesuvius's eruption (which lasted two days in all) included a nuée ardente, a sudden basal surge of incandescent, heavier-than-air gas that poured out of the crater and ran down its slope like an avalanche instantly searing and suffocating whatever animate life lay in its path. Pliny the Elder may have died because of it.

Herculaneum, probably a slightly later victim, was inundated not by lava or ash but by mud. The torrential rains that sometimes follow a volcanic eruption (which ejects huge volumes of water vapor) can gather newly deposited loose particles of ash upon the volcano's slopes into a precipitous mudslide (lahar) that will then flow rapidly and in great volume for kilometers. Such a mudslide buried the city of Herculaneum, which remains less excavated than Pompeii because its eruption cover is much harder to dig through and because the modern city of Resina sits atop most of it. Both Pompeii and Herculaneum have fundamentally influenced archaeology, history, and the fine arts.

Plinian eruptions, such as that of 79 C.E., are noted for their explosive violence. The cones of stratovolcanoes, moreover, are surprisingly fragile and will often change significantly from one eruption to the next. In 79 C.E., this fragility was more than usually apparent, as the entire upper cone of Vesuvius was completely destroyed, leaving behind only an incompletely circular ring now known as Mount Somma. From it, the more general term "somma," which applies to similar structures everywhere, is derived.

Although some disagreement exists, for most volcanologists the mountain that erupted in 79 C.E. is the present-day Mount Somma and the cone within it now called Vesuvius did not arise until the quieter eruption of 172 C.E.

LATER ERUPTIONS

Since then, there have been dozens of further Vesuvian eruptions, most of them from the central vent but a few from the flanks as well. Various literary sources tell of eruptions in 203, 472, 512, 685, 993, 1036, 1049, 1138, and 1139; there was then nothing of importance for almost five hundred years, though the overnight eruption in the Phlegraean Fields that produced Monte Nuovo in 1538 is attested by no fewer than four eyewitness accounts. This extended period of quiescence ended on December 16, 1631, when, after an explosive beginning, seven streams of lava poured down the slopes of Vesuvius, destroying a dozen villages and some eighteen thousand people. Several accounts of this disastrous eruption were written, some of them book-length. It is the first volcanic eruption in history for which there is a published literature.

Vesuvius has been under almost constant observation ever since, with many subsequent outbursts. From 1631 until 1944, however, activity remained unusually frequent. During most of this time (including major eruptions in 1676, 1694, 1707, 1737, 1751, and 1760), Vesuvius was unquestionably the premier geological attraction in the world; it soon became a necessary destination for anyone who aspired to authority regarding the natural history of the earth. This prominence intensified as a result of the work of Sir William Hamilton, the British representative at Naples, who published highly influential accounts of the Vesuvian eruptions of 1766, 1767, 1770, and 1779. He is arguably the first person who might properly be described as a modern volcanologist.

Other major eruptions followed in 1793-1794, 1822, 1834, 1855, 1867-1868, 1872, 1906, and 1944. During this 150-year period, the nature of observations changed gradually from those of the intrepid amateur (such as Hamilton) to fully sophisticated, rigorous science. One may cite in particular the establishment in 1844 of the Royal Vesuvian Observatory, a then unique (but since copied) institution located on the very slopes of the volcano. Under Macedonio Melloni initially, the observatory made ongoing monitoring of day-to-day volcanic activity possible and became the training ground for a distinguished series of internationally known volcanologists, who naturally derived many of their concepts from Vesuvius itself. Among others, Luigi Palmieri, R. V. Matteucci, Giuseppe Mercalli, Frank Perret, and Giuseppe Imbo have all made important contributions.

Dennis R. Dean

FURTHER READING

Chester, Roy. *Furnace of Creation, Cradle of Destruction.* New York: AMACOM Books, 2008. This text discusses the turbulent processes of the planet. It covers earthquakes, volcanoes, and tsunamis in reference to plate tectonics, natural disasters, predicting and mitigating. Multiple chapters explore seafloor spreading. The author also discusses hydrothermal activity. This text takes on an immense range of content, but still explains concepts clearly and with detail.

de Boer, Jella Zeilinga, and Donald Theodore Sanders. *Volcanoes in Human History: The Far-Reaching Effects of Major Eruptions.* Princeton, N.J.: Princeton University Press, 2002. Provides an overview of volcanism followed by specific volcanoes and eruptions. Includes the interaction of human history and volcanic activity. Covers the Hawaiian Islands, Crete, Vesuvius, Iceland, Tambora, Krakatau, Pelée, Tristan da Cunha, and Mount St. Helens.

Decker, Robert, and Barbara DeC.E.. *Volcanoes.* 4th ed. New York: W. H. Freeman, 2005. This readable and well-illustrated text is geared toward the general public. The authors explain geologic phenomena in a clear and simple manner. Includes a section on Mount Vesuvius. The appendices list the world's 101 "most notorious" volcanoes, as well as Internet sources for volcano information.

Dobran, Flavio. *Vesuvius: Education, Security and Prosperity.* Oxford: Elsevier, 2006. Provides an all-encompassing account of Vesuvius, both historically and presently. This document represents a project to understand the volcanism of Vesuvius for the purpose of protection against and prevention of future disasters. Addresses geology, economics, historical progression of volcano knowledge, geophysics and ballistics, and computational analysis.

Francis, Peter, and Clive Oppenheimer. *Volcanoes.* 2d ed. New York: Oxford University Press, 2004. This text covers volcanic activity ranging from lava flows to pyroclastic currents. Provides many examples from Hawaii, Italy, Ethiopia, Japan, Mount St. Helens, and Krakatau. Written for the layperson, but some knowledge of geology is helpful.

Grant, Michael. *Cities of Vesuvius: Pompeii and Herculaneum.* New York: Spring Books, 1974. A reliable text about the cities destroyed by Vesuvius. Includes fine illustrations.

Hoffer, William. *Volcano: The Search for Vesuvius.* New York: Summit, 1982. A travel book centering on Vesuvius; short, accessible, and entertainingly (but responsibly) written. Recommended for high school students.

Leppmann, Wolfgang. *Pompeii in Fact and Fiction.* London: Elek, 1966. This source emphasizes cultural history—that is, the impact of Pompeii on later writers and artists.

Lobley, J. Logan. *Mount Vesuvius.* London: E. Stanford, 1868. Of the several nineteenth-century books devoted exclusively to Vesuvius, this one is the best. Useful primarily for its extensive history of Vesuvian eruptions; however, the science is dated.

Lockwood, J. P., and Richard W. Hazlett. *Volcanoes: Global Perspectives.* Chic.E.ter, England: Wiley-Blackwell, 2010. An up-to-date and well-illustrated undergraduate level text that is also accessible to the interested reader.

Sigurdsson, Haraldur, ed. *Encyclopedia of Volcanoes.* San Diego, Calif.: Academic Press, 2000. This book provides a complete summary of the sc.E.tific knowledge of volcanoes. It contains eighty-two well-illustrated overview articles, each of which is accompanied by a glossary of key terms. Although this is a college-level text, it is written in a c.E.r and comprehensive style that makes it generally ac.E.sible. Cross-references and index.

Sigurdsson, Haraldur, et al. "The Eruption of Vesuvius in A.D. 79." *National Geographic Research* 1 (1985): 332-387. This ariticle is an authoritative but technical account of the most widely known eruption of Vesuvius.

See also: Eruptions; Folds; Forecasting Eruptions; Krakatau; Mount Pelée; Mount Pinatubo; Mount St. Helens; Popocatépetl; Recent Eruptions; Shield Volcanoes; Stratovolcanoes; Volcanic Hazards.

N

NEANDERTHALS

Neanderthals are the best-known extinct members of the human lineage. Although it is generally agreed that they were close relatives of modern humans, the nature of this relationship is vigorously debated. They have been assumed to be a direct ancestor, a diseased member of the species, or an extinct side branch of the family tree.

PRINCIPAL TERMS

- **deoxyribonucleic acid (DNA):** the chemical that carries the instructions for all living things; closely related organisms have very similar DNA
- **genus:** the first part of the scientific name of an organism; members of the same genus but different species are closely related, but cannot mate and produce fertile offspring
- **mitochondria:** subcellular structures containing DNA used to estimate the relationships between groups of organisms; the more similar the DNA, the more closely related the groups
- **species:** the second part of the scientific name of an organism; members of the same species can mate and produce fertile offspring
- **subspecies:** the third part of a scientific trinomial, assigned to one of two groups that can mate and produce fertile offspring, but that have some strikingly different characteristics
- **taxonomy:** the science of classifying and naming living and fossil organisms, or the classification and scientific name of a living or fossil group

History and Characteristics

The fossil that gave the Neanderthals their name was found in a cave being quarried for limestone in Germany's Neander Valley in 1856. At least two Neanderthal fossils were discovered before the Neander Valley individual. However, neither was recognized as a member of an extinct human group until after the name "Neanderthal" was assigned. Many similar fossils have been found in scattered locations all over Europe and the Middle East since the Neander Valley discovery. Dates assigned to the various fossils indicate that the Neanderthals originated late in the Ice Age and became extinct a few thousand years before the last glacial retreat (from about 200,000 years ago to about 30,000 years ago). Thus the Neander Valley specimen lent its name to a fossil relative of modern humans that occupied Europe and the Middle East late in the Ice Age.

Though the Neanderthals were very similar to modern humans, they had several distinctive characteristics. Neanderthals were short and exceptionally stout-bodied with broad, supportive bones and joints. This body form suggests a life filled with intense physical effort. Perhaps the compact body also helped them cope with cold stress under Ice Age conditions. Their brains were somewhat larger than modern human brains. That size may have compensated for the more massive total body size of the Neanderthals, since large-bodied organisms generally have larger brains. Their foreheads sloped up from their exceptionally heavy eyebrow ridges, their jaws extended forward beyond the plane of the face, and their chins were weakly developed. These and several other characteristics are used to define a fossil find as a Neanderthal.

Structure and Behavior

Rudolph Virchow's initial interpretation of the Neander Valley fossil as a diseased human was popular for a time. Virchow held that the fossil was a modern human whose unique features were the result of disease. However, as more fossils with the same characteristics were discovered all around Europe and the Middle East, this explanation became untenable. Later, misinterpretation of the characteristics of Neanderthal fossils led Marcellin Boule and others to interpret Neanderthals as stooped, bent-kneed, apelike subhumans with an animal nature to match.

Additional fossil discoveries, including evidence for toolmaking and burials, sometimes with flowers placed in the grave, caused anthropologists to rethink the presumed animal nature of the

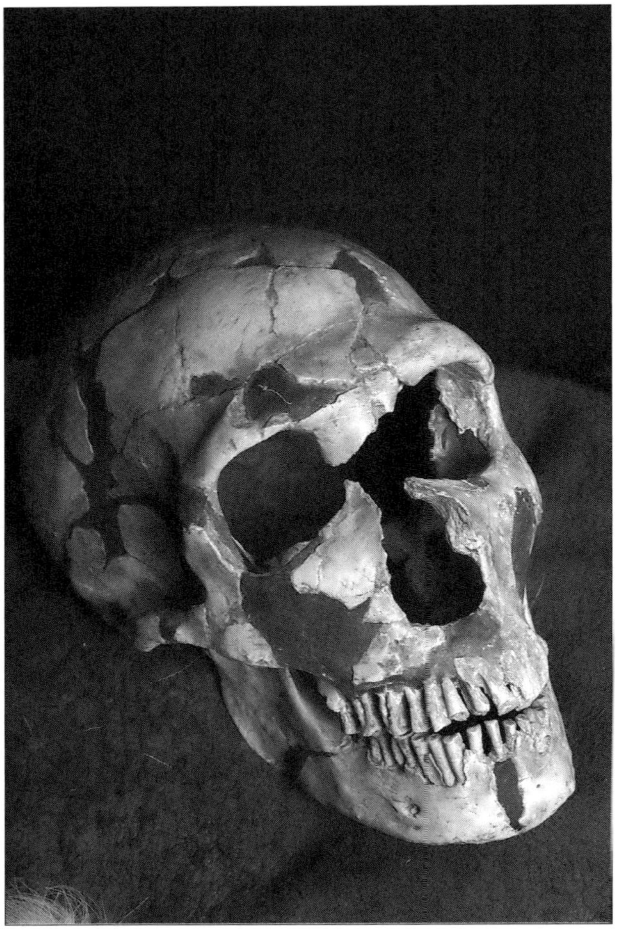

Neanderthal skull (Homo neanderthalensis). (Philippe Plailly/Photo Researchers, Inc.)

Neanderthals. Although the evidence for flowers has been challenged, the evidence for burials, presumably accompanied by mourning, is accepted by many anthropologists. In addition, fossils showed that some Neanderthals lived much of their lives with deformed limbs and other disabilities, which would have made it difficult or impossible for them to fend for themselves. Yet they apparently lived many years in that condition, suggesting the support of other members of a social group. Such behavior was not in keeping with Boule's picture of the Neanderthals as nonhuman animals.

Reinterpretation of the anatomic evidence also suggested that, instead of a bent-kneed, stooped posture, the Neanderthals walked on two rather straight legs and had hands capable of manipulating materials and making tools, much as modern humans do. All this indicated that the Neanderthals were more like modern humans than Boule's interpretation, and they came to be thought of in that light.

Taxonomic Relationship to Modern Humans

Neanderthals have always been recognized as close relatives of modern humans, but the specific taxonomy of the relationship is still a point of contention. They are placed in the same genus (*Homo*) as modern humans by almost all anthropologists, but researchers debate whether they were members of our species, *Homo sapiens*, or belonged in their own species, *Homo neanderthalensis*.

The discussion of the structure and behavior of Neanderthals bears directly on this question. If the Neanderthal characteristics were the result of disfigurement caused by disease, Neanderthals were simply aberrant humans and not especially interesting from the perspective of human evolution. However, if they were stooped, bent-kneed, and animal-like in behavior, they were probably a separate species, perhaps ancestral to modern humans, and therefore more interesting from the evolutionary perspective. In contrast, if they were upright in stature, were skilled toolmakers, were supportive of their handicapped and elderly, and buried their dead with mementos such as flowers, they might earn the designation *Homo sapiens* and take on an even greater interest to the more modern members of that species.

Such arguments are part of the practical taxonomy of the Neanderthals, but the real key to species identification and species separation is (at least theoretically) interbreeding. If the members of two groups can mate with each other and produce fertile offspring, and if these offspring can produce fertile offspring, the two groups are generally considered to be members of the same species. Therefore, the real taxonomic question becomes: Could Neanderthals and early-modern humans interbreed?

Because it is very difficult to determine whether fossil groups interbred with one another, Neanderthal taxonomy has been primarily determined by anatomic and presumed behavioral characteristics, such as those already discussed. That taxonomy has vacillated with changing interpretations of those characteristics. Neanderthals have been placed in their own species (*Homo neanderthalensis*) for much of their history, but they have been identified as a human subspecies (*Homo sapiens neanderthalensis*) at other times. The latter designation implies that the Neanderthals and modern humans (*Homo sapiens sapiens*) were members of the same species and, therefore, could interbreed.

Determination of the Neanderthals' taxonomic position is an integral part of arguments over the mechanism of the origin of modern humans. There are two main hypotheses for that origin: the replacement hypothesis of Christopher Stringer and the multiregional hypothesis vigorously supported by Milford Wolpoff. The replacement hypothesis is also designated "out-of-Africa" because it assumes that a population of African origin expanded throughout Africa, Europe, and Asia and rapidly replaced the more primitive humanlike species living there, including the Neanderthals. Whether this replacement was by competition or by more direct and violent means is undetermined. The multiregional hypothesis suggests that the widespread, more primitive humanlike populations evolved into modern humans rather than being replaced by new immigrants. Both hypotheses hold that the more primitive populations also originated in Africa and spread to Europe and Asia at a much earlier date.

Because the Neanderthals are the best-known and best-understood early human group, an understanding of the Neanderthal relationship is critical to an understanding of the evolutionary history of humanity. A Neanderthal contribution to modern human ancestry would support the multiregional hypothesis, and the lack of such a contribution would be consistent with the replacement hypothesis.

Advances of the 1990's

By the 1990's, the Neanderthals were well established as a group related to modern humans, but the questions remained: How close was the relationship? Did the two groups interbreed? Were Neanderthals a part of the evolutionary heritage of modern humans? During the 1990's, improved techniques and additional fossil discoveries led to greater understanding of the Neanderthals. A few examples will illustrate the situation.

In a 1996 study, Jean-Jacques Hublin and several coworkers determined that Neanderthals found at an archeological site in France made bone tools and wore decorative emblems on their bodies—behaviors not uncovered with older Neanderthal fossils. They concluded that the Neanderthals were influenced by early-modern humans who lived in the same area at the same time and that a reasonably elaborate cultural exchange must have occurred between the two groups. However, based on the strikingly different anatomy of the two groups' inner ears, they also concluded that the Neanderthals and modern humans did not interbreed. The investigators reasoned that if interbreeding had occurred, the two groups would have shared a common ear structure.

In 1997, Matthias Krings, Svante Paabo, and their colleagues isolated and engineered deoxyribonucleic acid (DNA) from the mitochondria of Neanderthal bones and compared it to DNA from modern human mitochondria. They found the Neanderthal DNA to be quite different from that of modern humans and concluded not only that the two groups were different species but also that Neanderthals were not ancestral to modern humans.

In 1998, Daniel Lieberman proposed that a reduction in the length of the sphenoid bone during embryology can explain most differences between the two groups' skulls. The sphenoid is a bone in the skull of both Neanderthals and modern humans, and Lieberman showed it to be shortened in modern humans but not in Neanderthals. He hypothesized that the impact of shortening the sphenoid resulted in the modern human skull characteristics, while the longer sphenoid resulted in the Neanderthal skull. Based on the fundamental nature of the change, he concluded that Neanderthals do not belong to the same species as modern humans and were probably not ancestral to modern humans.

In 1999, Cidàlia Duarte, Erik Trinkaus, and several colleagues discovered the buried remains of a four-year-old child in southern Spain. The skeleton was estimated to be about 24,500 years old, and they interpreted its anatomy to be a mixture of modern human and Neanderthal characteristics. Most anthropologists agree that southern Spain supported Neanderthal populations longer than other parts of the world, perhaps as late as 27,000 years ago, and that modern humans and Neanderthals coexisted in the region. Duarte, Trinkaus, and their group suggested that the skeleton they found demonstrated that the two groups did interbreed and that Neanderthals were part of the ancestry of *Homo sapiens.* This was supported in 2010 by genetic evidence indicating that interbreeding with *Homo sapiens* took place in the Middle East between 50,000 and 80,000 years ago.

Significance

Consideration of this short list of studies demonstrates the state of knowledge about the Neanderthals' place in human evolution. Viewed alone, each study seems to clinch the position of its authors. In fact, the first three reinforce one another so well that Neanderthals would seem to be eliminated from direct participation in the evolution of modern humans. However, Duarte and Trinkaus's study would seem to validate the opposite position, that Neanderthals were direct participants in the evolution of modern humans. This situation symbolizes the absence of consensus in the field. There are also established scientists with alternative viewpoints for each of these studies. Lieberman himself is a coauthor of a letter that criticizes his own conclusions about the sphenoid and points to the need for a better understanding of the development of primate skulls to help clarify the situation.

A number of anthropologists have pointed out that Krings and Paabo's conclusions are extrapolated from a single, short segment of the mitochondrial DNA and that more extensive studies, including studies of DNA from the nucleus, are necessary before definitive conclusions can be drawn. In fact, nuclear DNA studies of modern humans have suggested that modern human DNA comes from a number of sources rather than a single African source as in the out-of-Africa hypothesis. Clearly, extensive DNA comparisons would be helpful to settle this issue. However, DNA from fossils is difficult to find and work with, so an extensive collection of such studies is not likely to accumulate.

Ian Tattersall, who rejects the Neanderthals as direct contributors to modern human evolution, has criticized Duarte and Trinkaus's data and their interpretation of the data. The verbal exchange has been bitter, not an unusual circumstance for disagreements in this field.

Although anthropologists have learned an enormous amount about the Neanderthals, their relationship to modern humans continues to escape consensus. This is, without question, a result of the difficulty of the problem and the tentative nature of the evidence. Most agree that the Neanderthals were a successful group closely related to modern humans. Everyone's hope is that more fossils, improved technology, and fresh insight will clarify the question because understanding the Neanderthals is likely to contribute to an understanding of humanity.

Carl W. Hoagstrom

Further Reading

Akazawa, Takeru, Kenichi Aoki, and Ofer Bar-Yosef, eds. *Neanderthals and Modern Humans in Western Asia.* New York: Plenum, 1998. A scholarly but understandable group of papers with contributions from many of the major workers in Neanderthal evolution and biology.

Ciochon, Russell L., and John G. Fleagle, eds. *The Human Evolution Source Book.* 2d ed. Englewood Cliffs, N.J.: Prentice Hall, 2005. The broad spectrum of human evolution is covered in this book, but a large section titled "The Neanderthal Question and the Emergence of Modern Humans" covers Neanderthals. Several of the most fundamental questions are dealt with in a clear and interesting fashion.

Clark, G. A. "Neandertal Archaeology: Implications for Our Origins." *American Anthropologist* 104 (2002): 50-67. This article discusses the similarities and differences between Neanderthals and *Homo sapiens.* The author highlights the patterns of tool and material use, artwork, and other cultural aspects of each species. Lacking the most recent data, this article provides background information upon which current research builds.

Fox, Richard G. "Agonistic Science and the Neanderthal Problem." *Current Anthropology,* supp. 39 (June 1998). All articles in the supplement concern "The Neanderthal Problem and the Evolution of Human Behavior," the supplement's title. The interaction between Neanderthals and modern humans in Europe and in the Middle East and an archaeological consideration of the out-of-Africa model are the subjects of three other articles.

Green, Richard E., et al. "Analysis of One Million Base Pairs Neanderthal DNA." *Nature* 444 (2006): 330-336. This article discusses the DNA analysis of a Neanderthal fossil. Provides an estimate of the time period during which Neanderthals diverged from the human lineage. Also addresses the future of Neanderthal research.

Olson, Steve. "Neanderthal Man." *Smithsonian* 37 (2006): 76-82. An overview of the current knowledge (as of publication) of the Neanderthal, written in a manner accessible to the layperson. The author discusses the DNA sequence and evolutionary history of the species.

Shreeve, James. *The Neanderthal Enigma.* New York: William Morrow, 1995. Written by a science writer rather than a scientist, this is an interesting account of the history of Neanderthals and Neanderthal studies.

Stringer, Christopher, and Robin McKie. *African Exodus.* New York: Henry Holt, 1996. A small, interesting book on the out-of-Africa hypothesis written by an anthropologist intimately involved with the development of the hypothesis (Stringer) and a science writer.

Tattersall, Ian. *The Last Neanderthal.* Rev. ed. New York: Basic Books, 1999. An extensively illustrated account of what is known about Neanderthals and how it was learned. The prologue outlines two interesting, if imaginative, characterizations of the "last Neanderthal."

Tattersall, Ian, and Jeffrey H. Schwartz. "Hominids and Hybrids: The Place of Neanderthals in Human Evolution." *Proceedings of the National Academy of Science* 96 (June 1999): 7. This commentary is a reply to the paper by Duarte, Trinkaus, and their colleagues (pages 7604-7609 of this same issue) in which they report the description of a 24,500-year-old fossil with characteristics of both Neanderthals and early-modern humans. The two positions are clearly laid out in the article and the commentary.

Trinkaus, Erik, and Pat Shipman. *The Neanderthals: Changing the Image of Mankind.* New York: Alfred A. Knopf, 1992. An interesting and well-written account of the Neanderthals and the scientists who study them.

See also: Archaebacteria; Biostratigraphy; Cambrian Diversification of Life; Catastrophism; Cenozoic Era; Colonization of the Land; Dinosaurs; Ediacaran Fossils; Eukaryotes; Evolution of Birds; Evolution of Flight; Evolution of Humans; Evolution of Life; Fossilization and Taphonomy; Fossil Plants; Fossil Record; Fossils of the Earliest Life-Forms; Fossils of the Earliest Vertebrates; Geoarchaeology; Geologic Time Scale; Life's Origins; Mammals; Mass Extinctions; Mesozoic Era; Microfossils; Paleobiogeography; Paleobotany; Paleozoic Era; Petrified Wood; Prokaryotes; Stratigraphic Correlation; Stromatolites; Systematics and Classification.

O

OCEAN BASINS

Ocean basins contain basaltic crust produced by seafloor spreading at mid-ocean ridges, which may be covered with a thin layer of oceanic sediments. Seafloor sediments and rocks in the oceans may contain a record of the history of the development of ocean basins. Ocean basin deposits have provided evidence supporting the theories of seafloor spreading and plate tectonics.

PRINCIPAL TERMS

- **basalt:** a dark-colored, fine-grained rock erupted by volcanoes, which tends to be the basement rock underneath sediments in the ocean basins
- **biogenic sediments:** the sediment particles formed from skeletons or shells of microscopic plants and animals living in seawater
- **deposition:** the process by which loose sediment grains fall out of seawater to accumulate as layers of sediment on the seafloor
- **lithosphere:** the outermost layers (the crust and outer mantle) of the earth, which are arranged in distinct rigid plates that may be moved across the earth's surface by seafloor spreading
- **magnetic anomalies:** linear areas of ocean crust that have unusually high or low magnetic field strength; magnetic anomalies are parallel to the crest of the mid-ocean ridges
- **mid-ocean ridge:** a continuous mountain range of underwater volcanoes, located along the center of most ocean basins; volcanic eruptions along these ridges drive seafloor spreading
- **rifting:** the splitting of continents into separate blocks, which move away from one another across the earth's surface
- **seafloor spreading:** a theory that the continents of the earth move apart from one another by rifting of continental blocks, driven by the eruption of new ocean crust in the rift
- **seismic reflection:** study of the layered sediments in ocean basins by bouncing sound waves sent into the sea floor off the different rock layers
- **seismic refraction:** examination of the deep structure of the ocean crust using powerful sound waves that are bent into the crustal layers rather than being immediately reflected back to the ocean surface

Seafloor Spreading

Ocean basins make up one-third of the earth's surface, and the rocks and sediments in these basins may preserve an important record of the past history of the oceans. Earth materials in the ocean basins consist of a layer of volcanic basalts produced by seafloor volcanic eruptions at the mid-ocean ridges, which may be covered by layers of marine sediments and sedimentary rocks. The shape of individual ocean basins may be changed as a result of the interactions of lithospheric plates, such as plate collisions, plate accretion, and plate destruction. Ocean basin rocks and sediments may contain valuable deposits of metals and other economic minerals, which may represent important natural resources that could be extracted by humans at some time in the future.

Eruption of volcanic basalts in ocean basins makes up an important part of seafloor spreading. The creation of new oceanic crust by volcanic eruptions along the mid-ocean ridges provides the driving force to move blocks of continental lithosphere across the earth's surface. For example, the separation of South America from Africa during the past 200 million years has been driven by the creation of the South Atlantic Ocean by seafloor spreading along the Mid-Atlantic Ridge between these two continents.

Ocean basin shapes may be altered by plate interactions. As lithospheric plates are rifted and move apart from one another, new ocean basins are created between the continental landmasses. In contrast, lithospheric plates may run into each other, and plate collisions cause the ocean basin between continents to be destroyed by subduction, in which crustal slabs are forced downward into the mantle and are remelted. An example is seen in southern Europe, where the collision of the northward-moving African plate with the Eurasian plate has caused the Mediterranean Sea to become shallower and narrower at the same time that crumpling of the edges of the continents has caused mountain building of the Alps in Europe and the Atlas Mountains in northern Africa.

Preservation of Magnetic and Volcanic Records

The volcanic basement rocks in ocean basins may preserve a record of the earth's magnetic field during the past, through the record of oriented magnetic minerals contained within basalts. When basalt that erupted at mid-ocean ridges cools as a result of exposure to cold seawater, magnetic minerals within the igneous rock are aligned with the earth's magnetic field and are "locked" into position by the crystallization of adjacent mineral grains. Thus, the alignment of magnetic minerals within seafloor basalts in the ocean basins acts as an enormous magnetic tape recorder, which preserves a record of the alternating reversals of the magnetic field of the earth. Oceanographers investigating the magnetism of the sea floor during the 1950's discovered the existence of long, straight areas of ocean crust with unusual magnetic properties. These linear magnetic anomalies were parallel to the mid-ocean ridges but were offset from the ridge crests. The anomalies are symmetrical around the mid-ocean ridges. Anomaly records on both sides form identical "mirror images" of each other. This finding supports the theory of seafloor spreading, which predicted that the creation of new oceanic crust at mid-ocean ridges would cause rifting and separation of previously cooled basalts to either side of the ridge.

Seafloor spreading theory further predicted that the older the anomaly, the farther it has been pushed away from the ridge crest. This prediction was proved by deep-ocean drilling, which drilled into and determined the age of sediments immediately atop specific magnetic anomalies in the ocean crust. Estimates of the rate of creation of new oceanic crust along the ridges may be calculated from the distance between the crest of the mid-ocean ridge and specific magnetic anomalies whose age has been determined. These calculations have proved that creation of new oceanic crust does not occur at a constant rate through time and that there have been episodes of rapid seafloor spreading and of slower spreading during the past.

Examination of the chemistry and mineral composition of the volcanic rocks of an ocean basin may provide a record of the history of volcanic eruptions at the mid-ocean ridges and help geologists to determine the chemistry and type of igneous rocks being erupted at any point in the past. Understanding the mineral content of seafloor crust erupted at specific times in the past allows geochemists to make predictions about the nature of the deeper portions of the earth's crust. Chemical changes in seafloor basalts may reflect similar changes occurring in the lower crust or upper mantle of the planet.

Sediment Deposition

In addition to preserving historical information in the harder igneous rock basement, ocean basins provide a record of sediment deposition during the past. These regions of the ocean floor are among the flattest areas on the surface of the planet and have minimal relief. Most ocean basins are smooth and nearly flat, with less than 1 meter of vertical altitude change in 1 kilometer of horizontal distance. Their smoothness is the result of the burial of the blocky, irregularly faulted volcanic basement rocks beneath layers of slowly accumulating mixtures of biogenic sediments, turbidites, and other sediment particles derived from continental sources. Newly erupted basement rocks are gradually covered by oceanic sediments, so there is an overall correlation between crustal age and total sediment thickness within an ocean basin. Verification of this relationship by deep-ocean drilling provided further support for seafloor spreading.

Sedimentary rock layers provide information on the history of deposition in the ocean basins by their structure and by the fossils preserved in sediment layers. Marine geologists examine the types, sizes, and sorting of the individual grains that compose seafloor sediments. Geologists attempt to determine the sources of sediment particles deposited in ocean basins and to analyze both the changes in sediment particles as they fall through the water column and changes occurring on the sea floor after the sediments are deposited. Fine-grained particles derived from continental sources may be carried far out to sea by the winds to be deposited in the deep ocean basins. Also, biogenic particles either may dissolve as they sink through the oceans or may be dissolved on the sea floor by deep-ocean water masses.

Paleontologists study the fossils buried within layers of sedimentary rock. Marine sediments deposited in water depths shallower than the carbonate compensation depth have abundant microscopic fossil remains of ancient one-celled plants and animals (plankton) that lived in the shallow water of the oceans during past geologic time. As these organisms died, their remains sank to the sea floor to become an important part of the sedimentary rock layers. Examining the record of fossils preserved in seafloor sediments is like reading the pages of a book containing the history of the ocean basin: Changes in the type and number of ancient fossil organisms and fossil assemblages may be preserved in the microfossils contained in seafloor sediment layers.

Perhaps one of the most fascinating aspects of the study of seafloor sediments in the deep-ocean basins is that these materials contain significant amounts of micrometeorites and extraterrestrial material. Micrometeorites may fall on the continents, but their scarcity and small size make them difficult to identify. In deep-ocean basins far from the continents, however, sediment accumulates at a much slower rate as a result of a combination of the distance from continental sediment sources and the dissolution of biogenic sediment particles by corrosive bottom waters. As a result, deep-ocean sediments tend to be fine-grained red clays, which have few to no fossils and which may be deposited at rates as slow as 1 millimeter per million years. In these red clays, extraterrestrial materials may make up a significant portion of the sediment particles because of the extremely slow sediment deposition rates.

Study of Ocean Basins

Because the ocean basins contain a variety of geologic materials, including igneous, sedimentary, and metamorphic rocks, a number of different techniques are used in the study of ocean basins, depending upon the specific feature of interest. Methods that are suitable for the study of one aspect of the ocean basins may be completely useless for obtaining information about other features of the basin. Some of the methods used include acoustic profiling, seismic reflection and refraction studies, dredging, sediment coring, and deep-ocean drilling. In addition, information on ocean basins may be derived from ancient seafloor deposits that have been uplifted and are presently found above sea level on the continents.

The overall shape of the ocean basin and the water depths of individual parts of the basin may be studied by acoustic profiling, or echo sounding. This technique uses an acoustic transponder (a sound source) mounted on the hull of an oceanographic vessel to emit sound waves, which travel down through the water until they are reflected back by the sea floor up to a shipboard recorder, which measures the total time between emission of the sound pulse and its return. The water depth is equal to one-half the total time (sound must go down, then up again), multiplied by the speed of sound in seawater. Profiles of the ocean basin's shape are obtained by continuously running the echo sounder while the ship is sailing across the basin.

While acoustic profiling gives the water depth of the ocean basin, the energies of the sound waves are insufficient to provide information about the buried structure of the ocean floor. The shape and thickness of the basement rocks and the sediment cover on the floor of an ocean basin may be examined by seismic reflection profiling, which is somewhat similar to echo sounding. In seismic reflection studies, a large energy source (such as the explosion of a dynamite charge) is released in seawater to create high-energy sound waves, which move down through the ocean with sufficient energy to penetrate the sediment layers of the sea floor before they are reflected back up to the vessel by the different sub-bottom layers. Reflection profiling is also made by continuously producing these high-energy sound waves while sailing across a basin to obtain a record of the thickness and geometry of the sediment layers and the harder basement rocks of the sea floor.

The deep structure of basement rocks is investigated by seismic refraction and magnetic studies, which may be made by one oceanographic vessel and a stationary floating recorder (sonobuoy) or by two vessels. In seismic refraction studies, extremely large energy sources are released in the ocean to create powerful sound waves, which have the ability to penetrate through seafloor sediment layers into the deeper layers of igneous basement rocks. Sound waves penetrating layers in the sea floor are bent (refracted) into the layer and travel through it for a certain distance before they are refracted back up to the ocean surface. An acoustic recorder at the surface measures the depth of sound penetration below the sea floor and the time elapsed since the explosion. Refraction profiles may be done by exploding charges off the stern of a moving vessel, using a stationary sonobuoy as the recording device, or refraction profiles may be made in a "two-ship" experiment, where one vessel acts as the "shooter" and the second vessel is the stationary recorder. By alternately "leapfrogging" past each other, two ships may make a much longer continuous reflection profile than is possible with only one vessel and a sonobuoy. Magnetic studies measure tiny changes in gravity or magnetic field strength using instruments towed behind a research vessel. This can provide information on the presence of volcanic rock overlain by less dense sediment.

Study of Sediment Deposition

The history of sediment deposition preserved by marine sediments in ocean basins may be studied by obtaining long cores of seafloor sediments, either by sediment coring or by ocean-floor drilling programs. Once a long sediment core is obtained from the ocean floor, the sediment particles and fossils within the sediments are studied layer by layer in order to examine the sedimentation history of the basin. Younger sediments are placed atop older layers; thus, by beginning with the uppermost layers of the sediment core and continuing into deeper layers, the geologist can examine the record of progressively older deposits in the ocean basin.

Direct examination of the basement rocks of the ocean basins may be made by dredging rocks from the mid-ocean ridges, by drilling through the sediment cover to take cores of the volcanic basalts, or by studying portions of the ocean floor that have been uplifted above sea level by tectonic activity. Dredging uses a wire mesh bag attached to a rigid iron frame, which is towed on a long cable behind a vessel to obtain rock samples from the sea floor. As the ship moves across the surface, it drags the dredge along the bottom, and seafloor rocks are broken off by the frame and caught in the wire mesh bag attached to the rear of the dredge.

Deep-ocean drilling programs have provided long basalt cores that have been drilled from the sea floor in different ocean basins. The deepest seafloor borehole drilled in the oceans by the Deep Sea Drilling Project, site 504B, located near the Galápagos Islands in the eastern equatorial Pacific, has been extended to a depth of 1,350 meters below the surface of the sea floor. Drilling at this location recovered 275 meters of seafloor sediments, and more than 1 kilometer of seafloor basalts has been penetrated.

Information about the ocean basins has also been provided by uplifted sections of sea floor, located in areas of plate collisions. These ancient seafloor deposits, or ophiolite sequences, are found on the island of Newfoundland in eastern Canada, on the island of Cyprus in the eastern Mediterranean Sea, and on the island of Oman in the Persian Gulf, among other locations. Rocks in the ophiolite sequences are an important natural resource, because they contain copper and many other valuable metals interspersed between basalts and igneous rocks. These ancient ocean-floor deposits, which have been uplifted above sea level by the collision of two lithospheric plates, may contain enormous reserves of rare metallic minerals. The metallic deposits in ophiolites were originally deposited as vein minerals between the volcanic basalts in the deeper portions of ocean crust. By understanding the factors controlling the formation of ophiolites, scientists may be able to predict other locations where these rocks may be found, and humans will be better able to utilize these valuable minerals in the future.

Dean A. Dunn

Further Reading

Anderson, Roger N. *Marine Geology: A Planet Earth Perspective.* Rev. ed. New York: John Wiley & Sons, 1989. A textbook discussing various aspects of oceanography, whose content is aimed at readers with minimal scientific background.

Condie, Kent C. *Earth as an Evolving Planetary System.* 2d ed. Waltham, Mass.: Academic Press, 2011. Covers the earth's crust, mantle, core and other spheres. Also provides detailed discussion of tectonics, the supercontinent cycle, and planetary evolution. A good resource for undergraduates.

Dolgoff, Anatole. *Physical Geology.* Boston: Houghton Mifflin Harcourt, 1999. This is a comprehensive guide to the study of the earth. It is extremely well illustrated and includes a glossary and an index. Although this is an introductory text for college students, it is written in a style that makes it understandable to the interested layperson. Contains a section on the development of the ocean basins.

Garrison, Tom S. *Oceanography: An Invitation to Marine Science.* Belmont, Calif.: Brooks/Cole, Cengage Learning, 2010. Chapter 3 provides excellent coverage of plate tectonics, ocean formation, seafloor spreading, and island arc formation. Chapter 4 covers topography and bathymetry of deep-ocean basins. Abundant diagrams aid readers from the layperson to the advanced undergraduate.

Kennish, Michael J. *Practical Handbook of Marine Science.* 3rd ed. Boca Raton, Fla.: CRC Press, 2001. Chapter 4 covers marine geology topics such as seafloor topography, sediments, plate tectonics, mid-ocean ridges, and hydrothermal vents. Best suited for advanced undergraduates and graduate students. References are provided for each chapter.

LeGrand, H. E. *Drifting Continents and Shifting Theories.* New York: Cambridge University Press, 1988. A review of the history of the "modern revolution in geology," which culminated in the development of the theory of global plate tectonics.

Plummer, Charles C., Diane H. Carlson, and Lisa Hammersley. *Physical Geology.* 13th ed. Boston: McGraw-Hill, 2009. This is a straightforward, easy-to-read introduction to geology intended for those with little or no science background. Discusses the development of the deep ocean basins from the point of view of plate tectonics. Includes many excellent illustrations.

Prothero, D. R., and Robert H. Dott. *Evolution of the Earth.* 8th ed. New York: McGraw-Hill, 2009. This basic textbook on historical geology is aimed at students of geology. However, it is very readable by anyone with a background in science. Presents an up-to-date account of the earth's history from the viewpoint of plate tectonics. Includes a glossary.

Rao, T. Ramamohana. "Geological Evolution of the Caribbean Plate: Some Critical Aspects in the Two Divergent Models." *Current Science* 95 (2008): 736-742. This article discusses many geologic principles in reference to the Caribbean tectonic plate. Two major models of the plate's genesis are evaluated. The author examines oceanic basins, flood basalts, plumes, and hot spots. This is a good example of many geological terms in context.

Segar, Douglas. *An Introduction to Ocean Sciences.* 2d ed. New York: W. W. Norton & Company, 2007. Comprehensive coverage of all aspects of the oceans and the oceanic crust. Readable and well illustrated. Suitable for high school students and above.

Seibold, Eugen, and Wolfgang H. Berger. *The Sea Floor: A Introduction to Marine Geology.* 3rd ed. New York: Springer-Verlag, 2010. As the title suggests, this is an introduction to many topics in marine geology. This text covers geological structures from the continental shelf to deep-ocean trenches. Processes such as seafloor spreading, the sediment cycle, currents, and pelagic rain are discussed.

Van Andel, Tjeerd H. *New Views on an Old Planet: Continental Drift and the History of Earth.* 2d ed. New York: Cambridge University Press, 1994. A survey of the theories of continental drift and plate tectonics, written for an educated lay audience.

See also: Abyssal Sea Floor; Carbonate Rocks; Mid-Ocean Ridge Basalts; Ocean-Floor Exploration; Oceanic Crust; Ocean Ridge System; Ophiolites; Seafloor Spreading.

OCEAN-FLOOR EXPLORATION

The exploration of the ocean floor by probes, submersibles, and remotely operated vehicles is a relatively recent human enterprise. Much of what is known about the floor of the oceans from its geography, the details of its geology, and its unique biology has been discovered since 1960.

PRINCIPAL TERMS

- **acoustic echo sounding:** a method of determining the depth of the ocean floor that measures the time of a reflected sound wave and relates that to distance
- **cartography:** the science of mapmaking; maps or charts of the ocean floor are developed by linking the individual points of ocean-floor depths
- **chemosynthesis:** the synthesis of organic substances by living organisms through the energy of chemical reactions
- **coring devices:** devices that drill into ocean-floor sediments to provide scientists with information on the composition of the seabed
- **fracture zones:** areas that define the edges between the continents on the sea floor
- **geomorphic domain:** major underwater features that define the appearance of a seafloor area
- **hydrothermal vents:** areas where very hot water is expelled from volcanically active vents on the sea floor
- **remotely operated vehicle (ROV):** a submersible operated from a remote location; for example, a robot that explores the sea floor while operated by tether from a surface ship
- **sounding:** the measurement of depth; a sounding line is a line used for the measurement of depth
- **telepresence:** the ability of a human to explore an area remotely by live television

EARLY EXPLORATION

Scientists know more about the surface of the moon than about the floor of the earth's oceans. The reason for this surprising lack of knowledge is twofold. First, more than half of the planet lies at depths of more than 3 kilometers, greater than humans can explore at first hand. Second, engineering devices (crewed or uncrewed) that can travel to such great depths under positive control became technologically possible only relatively recently. Yet, from these remotely operated devices, scientists are discovering an often alien environment in a vast and unexplored landscape that covers three-fourths of the earth.

The first explorers of the ocean floor charted harbor basins by the topography they could actually see or from soundings taken with weighted lines, which represented only a tiny fraction of the earth's ocean bottom. The first recorded mid-ocean sounding was accomplished by the Spanish explorer Ferdinand Magellan in 1521. He spliced two sounding lines together and lowered them over the side of his ship in the Tuamotu Archipelago until they ran out. Although the 365 meters of line did not reach bottom, Magellan immodestly declared that he had discovered the deepest part of the ocean. The first modern sounding was taken in 1840 in the South Atlantic, measuring a depth of 4,434 meters. The first map of the ocean floor was constructed using approximately 7,000 soundings in 1895. All these soundings were taken with line and weight, which was the only method available at that time to explore the ocean floor.

The first comprehensive ocean-floor soundings were made possible with the advent of acoustic echo sounding in 1920, but the method was not widely used until the 1940's. Considering every recorded sounding prior to World War II, there was, on the average, only one sounding per 2,500 square kilometers of ocean bottom. The use of submarines for warfare became widespread during the 1940's. They required the ability to determine the exact ocean-floor topography. With the subsequent vast improvement of echo sounding and sonar technologies, a new age of exploration was engendered. By the early 1970's, a combination of satellite navigation (which enabled accurate geographical positioning of soundings) and precise echo sounders enabled a rapid charting of the ocean floor using a science called cartography. Sunlight does not penetrate to depths greater than about 100 meters under the very best of circumstances. Hence, for scientists to chart the geography of the sea floor, they must combine individual soundings to form seafloor charts, ultimately covering vast areas.

A picture has developed of the sea floor that has revealed a widely variable seascape, at least as mutable as the continents but quite unlike above-water, continent-born landscapes. The underwater topography is unaffected by the powerful erosional forces found on the surface; the submarine world is shaped by forces that are unique to the oceans. The major submarine features (called the geomorphic domain) are deep-sea trenches, rifts (deep valleys with steep sides), flat-topped undersea volcanic cones, and fracture zones of very long linear cracks and fissures.

Oceanographic research ships have prowled the oceans for the past two centuries. Such ships as *Challenger*, *Discovery*, and *Endeavor* used lines and weights to determine depths. With a slight modification, they used the same lines and weights to carry sampling devices to the ocean floor to obtain specimens for later examination on the surface. Such remote-sensing devices were called coring devices, and some were complex enough to carry several types of sampling equipment, from water to temperature probes and even sea-life traps of various kinds. Nevertheless, all these sampling devices were not usually a part of a single probe and, more often than not, were sent down independently.

The use of remotely operated submersibles, such as Alvin*—shown here in 1971 with its original support ship* Lulu*—has made it possible to examine phenomena at great ocean depths.* (National Oceanic and Atmospheric Administration)

Crewed Submarines and ROVs

Humans have also studied the ocean floor personally, first as free divers, then using the breathing apparatus invented in the 1940's. Oxygen and nitrogen are toxic and dangerous even at 100 meters, however, thus limiting a diver with a self-contained breathing apparatus to above this level. Thus, it became necessary to invent manned submarines that could dive deeper. There submarines are a form of remote-sensing device in that they shield the people within them from the tremendous ocean pressures by thick hulls.

Such a submarine is the *Alvin*, which is capable of dives as deep as 6,100 meters. Like most submersibles of this type, *Alvin* has a robot hand that extends from the body to enable the operators inside to perform work outside the submarine. This famous submersible has also allowed investigators to observe life-forms around submerged volcanic vents on the ocean floor. One of *Alvin*'s most famous dives was made in 1986 to investigate the sunken hull of the *Titanic* in 4,000 meters of water in the North Atlantic. It was on this dive, however, that the limitations of such a manned submersible became apparent. The trip from the surface to the ocean floor alone required more than two hours and the same time to return. That limited the amount of time *Alvin* had to work on the ocean floor to merely a few hours.

Oceanographers and other scientists were thus led to invent devices called remotely operated vehicles (ROVs). These devices perform work on the ocean floor with as much dexterity as a human inside a submersible. ROVs carry cameras with capabilities from still photographs to live television. Some maneuver independently on tethers and have remotely operated arms. One inventor of such devices, and the discoverer of the *Titanic*'s final resting place, is Robert Ballardof Woods Hole Oceanographic Institution in Massachusetts. Ballard, with funding from the U.S. Navy and the National Geographic Society, invented three different types of remote-sensing ROVs for exploring the ocean floor. Ballard's ROVs include the *Argo*, a camera sled equipped with still and live cameras mounted on a platform towed behind a surface ship. Others are the *Jason* and *Jason Jr. Jason Jr.* rides to the ocean floor on *Jason*, then detaches and is capable of executing a wide range of movements. According to Ballard, such devices free humans from the inherent dangers of piloting crewed submersibles to great depths, eliminate the excessive travel time from and to the surface, obviate the need for a complex manned vehicle, and give surface operators a "telepresence" on the ocean floor.

Ocean-Floor Characteristics

Undersea remote-sensing devices have revealed that the ocean floors are complex, distinctive, active landscapes with vast mountain ranges, plains, highlands, valleys, and active volcanic vents. Far from being a static, quiescent place, the ocean floor is a dynamic, constantly active geologic area of considerable interest to geologists.

The earth's continents drift across the planet in a slow, never-ending flux measured in billions of years; vital information describing this process may be found at the fracture zones that mark the boundaries between the earth's continental plates. Most of the earth's earthquakes are centered on such fracture zones. In addition, most of the planet's active volcanoes lie on the ocean floor. Only by extensive on-site study of these regions can scientists hope to fully understand what they seek to know about the planet's earthquakes and volcanoes. Because the very character of the earth's crust is different at these continental boundaries (as opposed to the upper, higher regions of crust on the continents' surface levels), scientists need to study the nature of the active regions far below sea level.

At the fracture zones, magma from deep inside the earth wells up to near the surface, heating the cold seawater to very high temperatures (hot enough, in some cases, to melt lead) and spewing forth mineral-rich deposits from hydrothermal vents into the water. At these places, strange life-forms have been discovered that do not rely on sunlight (photosynthesis) for their survival. They are entirely dependent on the minerals from the vents, surviving in a metabolic state called chemosynthesis. One of these creatures, called a tube worm, was discovered by Ballard in 1977 onboard the *Alvin* at a depth of 6,100 meters near hydrothermal vents off the Galápagos Islands.

Such astonishing discoveries give rise to questions that challenge fundamental biological assumptions. For example, did the life-forms evolve independently from surface, photosynthetically supported plants, and if so, what does that portend for the possibility of chemosynthetically evolved life throughout the universe? If such development is possible, then there may be vast reservoirs of life in the universe that have evolved in the absence of a neighbor star, long considered the most basic requirement for the development of life-forms.

Exploration of the ocean floor has applications for the economies and energy needs of modern civilization. On the vast ocean-floor regions lie metal ores (such as manganese nodules) that may one day supply a significant percentage of industrial requirements for this resource. Also, a blanket of very cold water (near or slightly below freezing) lies on much of the ocean floor, which may one day be exploited for its energy transfer capacity in ocean thermal energy power plants.

Significance

As spectacular as is the emergent image of the ocean floor derived from the millions of soundings taken thus far, it remains merely a coarse outline of the total picture of the underwater domain. As remote-sensing devices improve, five fundamental goals in addition to basic soundings will be accomplished: a more complete understanding of the extent and details of available resources on the ocean floor; an expansion of humankind's telepresence on the deep frontiers of the ocean floor effected from remote locations; continued exploitation of the available resources on the ocean floor, such as petroleum, mineral, and food resources; collection and categorization of a vast amount of basic scientific information on the geology, available energy, and biology of the sea floor; and ongoing exploration to discover and investigate life-forms previously unknown or entirely unsuspected.

These discoveries will both enrich science's base of knowledge about the earth and make the ocean's resources available to the world's peoples. The technology developed to explore the ocean floor will also be employed in other scientific endeavors. For example, such remote-sensing capabilities will probably be adapted for use in space exploration.

Dennis Chamberland

Further Reading

Ballard, Robert D. *The Discovery of the Titanic.* New York: Warner Books, 1995. Ballard, as the world's leading expert on seafloor ROVs and telepresence, describes the expedition that discovered the *Titanic.* The remote sensing of the ocean floor is discussed in detail, along with its future potential. A beautiful and important photographic work.

Bar-Cohen, Yoseph, and Kris Zacny, eds. *Drilling in Extreme Environments: Penetration and Sampling on Earth and Other Planets.* Weinheim, Germany: Wiley-VCH, 2009. A comprehensive presentation of current drilling techniques and technology. This book covers everything from ocean to extraterrestrial drilling and coring. Written by academics, researcher scientists, industrial engineers, geologists, and astronauts, in a manner accessible to the layperson, but also useful for scientists and engineers.

Dolgoff, Anatole. *Physical Geology.* Boston: Houghton Mifflin Harcourt, 1999. This is a comprehensive guide to the study of the earth. It is extremely well illustrated and includes a glossary and an index. Although this is an introductory text for college students, it is written in a style that makes it understandable to the interested layperson. Contains a section on the development of the ocean basins and the methods used to explore them.

Exon, Neville. "Scientific Drilling beneath the Oceans Solves Earthly Problems." *Australian Journal of Maritime and Ocean Affairs* 2 (2010): 37-47. Presents an overview of the purpose and future expeditions of the Integrated Ocean Drilling Program (IODP). Discusses the importance of deep-ocean drilling to understanding past and present earth processes.

Garrison, Tom S. *Oceanography: An Invitation to Marine Science.* Belmont, Calif.: Brooks/Cole, Cengage Learning, 2010. Chapter 3 provides excellent coverage of plate tectonics, ocean formation, seafloor spreading, and island arc formation. Chapter 4 covers topography and bathymetry of deep-ocean basins. Abundant diagrams aid readers from the layperson to advanced undergraduates.

Haymon, Rachel M., et al. "Manifestations of Hydrothermal Discharge from Young Abyssal Hills on the Fast-Spreading East Pacific Rise Flank." *Geology* 33 (2005): 153-156. This article discusses oceanic hydrothermal heat released via black smokers. The authors comment on the research possibilities available due to fluid expulsion from the abyssal sea floor. This article is best suited for advanced undergraduates and graduate students. It provides a view of where research of the sea floor is heading.

Kennish, Michael J. *Practical Handbook of Marine Science.* 3rd ed. Boca Raton, Fla.: CRC Press, 2001. Chapter 4 covers marine geology topics such as seafloor topography, sediments, plate tectonics, mid-ocean ridges, and hydrothermal vents. Best suited for advanced undergraduates and graduate students. References are provided for each chapter.

Koblick, Ian G., and James W. Miller. *Living and Working in the Sea.* 2d ed. Flagstaff, Ariz.: Best Publishing, 1995. Although this work is primarily directed to seafloor habitats, it offers a valuable discussion of the rigors of ocean floor exploration. Also discusses a wide range of ocean-floor exploration techniques and equipment. Of textbook quality, well indexed and illustrated.

Prothero, D. R., and Robert H. Dott. *Evolution of the Earth.* 8th ed. New York: McGraw-Hill, 2009. This basic textbook on historical geology is aimed at students of geology, but it is readable by anyone with a background in science. Presents an up-to-date account of the earth's history from the viewpoint of plate tectonics. Includes a glossary.

Seibold, Eugen, and Wolfgang H. Berger. *The Sea Floor: A Introduction to Marine Geology.* 3rd ed. New York: Springer-Verlag, 2010. As the title suggests, this is an introduction to many topics in marine geology. This text covers geological structures from the continental shelf to deep-ocean trenches. Processes such as seafloor spreading, the sediment cycle, currents, and pelagic rain are discussed.

Segar, Douglas. *An Introduction to Ocean Sciences.* 2d ed. New York: W. W. Norton & Company, 2007. Comprehensive coverage of all aspects of the oceans and the oceanic crust, including seafloor exploration. Readable and well illustrated. Suitable for high school students and above.

Singh, Satish C., et al. "Discovery of a Magma Chamber and Faults Beneath a Mid-Atlantic Ridge Hydrothermal Field." *Nature* 442 (2006): 1029-1032. This article provides an example of current seafloor exploration and research. It discusses the use of seismic analysis for developing models of mid-ocean ridges. Also provides some information on hydrothermal activity and submarine volcano structure. Understood by the layperson with careful reading.

Weiner, Jonathan Weiner. *Planet Earth.* New York: Bantam Books, 1986. This work gives an excellent accounting of historical seafloor exploration, from the earliest voyages of *Challenger* and the first soundings that revealed the vast seascape below. A beautifully illustrated and well-indexed volume for all readers interested in the human exploration of planet Earth.

See also: Abyssal Sea Floor; Geomicrobiology; Hot Spots, Island Chains, and Intraplate Volcanism; Hydrothermal Vent Faunas; Life's Origins; Mid-Ocean Ridge Basalts; Ocean Basins; Oceanic Crust; Ocean Ridge System; Seafloor Spreading.

OCEANIC CRUST

The oceanic crust is that portion of the outer layer of material forming the earth that underlies the world oceans. This crust is a dynamic layer, primarily composed of basalt, where new, submarine mountain ranges are continuously formed and old ocean floors are destroyed.

PRINCIPAL TERMS

- **fracture zones:** large, linear zones of the sea floor characterized by steep cliffs, irregular topography, and faults; such zones commonly cross and displace oceanic ridges by faulting
- **hydrothermal vents:** seafloor outlets for high-temperature, mineralized springs that are associated with seafloor spreading centers and that are often the site of deep-sea, chemosynthetic biological communities
- **oceanic ridges:** long, narrow elevations of the sea floor, some 2 to 3 kilometers higher than the surrounding ocean basins, that are associated with the creation of new seafloor material
- **ocean trenches:** long, deep (greater than 6,000 meters deep), and narrow depressions in the seafloor with relatively steep sides; these features mark the boundaries between ocean crust and continental crust and are associated with the subduction of oceanic crust
- **plate tectonics:** the theory of mobility within the earth's crust that accounts for mountain building at ocean ridges, spreading of the sea floor, and subduction at ocean trenches by dividing the crust into a series of plates that interact by colliding, rifting, or sliding past one another
- **seafloor spreading:** the process whereby crustal plates move away from mid-ocean ridges, creating new crustal material as molten rock moves upward through rifts at the ridge crests
- **seamounts:** isolated elevations on the sea floor, usually rising to higher than 1,000 meters, that are commonly the shape of an inverted cone reflecting their volcanic origin; a flat-topped seamount is known as a guyot
- **seismic activity:** a disturbance of the crust caused by earthquakes or earth movements, often associated with zones of seafloor subduction and ocean-ridge formation
- **subduction:** the process whereby old sea floor that was produced millions of years earlier at the ocean ridges is forced under continental crust in the vicinity of trenches

EARTH'S STRUCTURE

The earth is not homogeneous from its center to its surface, it is composed of three concentric layers: the core, the mantle, and the crust. The core of the earth is made up of a dense mixture of nickel and iron, with a solid inner portion and a liquid outer portion. The core is extremely hot, ranging from more than 6,600 degrees Celsius at the center to 4,500 degrees Celsius in the outer core. The core extends from a depth of about 2,900 to 6,378 kilometers (the center of the earth) and accounts for 31.5 percent of the earth's mass. The next layer, the mantle, consists of less dense rock and constitutes 68.1 percent of the mass.

The material of the mantle has the properties of iron-magnesium silicate rock rich in olivine and pyroxene minerals. The mantle is about 2,870 kilometers thick and is cooler than the core (1,500 to 3,000 degrees Celsius). The mantle also has two zones. The lower portion is presumed to be essentially rigid, but the upper mantle, or asthenosphere, is more plastic and flows when stressed. The asthenosphere extends to a depth of 700 kilometers and is likely the site of molten magma formation. The outermost layer, the crust, is the less-dense outer shell of the earth. Also known as the lithosphere, this layer consists of granitic continental crust and basaltic oceanic crust. The crust is underlain and likely fused with a layer of heavier mantle rock. The boundary between the crust and mantle is known as the Mohorovičić discontinuity(or Moho); this boundary occurs under continents at depths of 10 to 70 kilometers but only 5 to 10 kilometers under the oceans.

Above the Moho, both the oceanic and continental crusts have properties that resemble basalt. Basalt, a common volcanic rock found extensively on the earth's surface, is composed of silicates of calcium, magnesium, and iron. These rocks have an average density of 3 grams per cubic centimeter. Under the continents, but not the oceans, the basalt is overlain by a rock layer with properties similar to those of granite. Granite is a common igneous rock composed of silicates of aluminum and potassium. Such rocks are lighter in color and weight than basalt, with an average density of 2.8 grams per cubic centimeter. Thus, the continental crust "floats" as massive blocks on a layer of basalt. Because the densities of the continental and oceanic crusts are not greatly different, approximately 93 percent of the continental blocks are submerged in the underlying basaltic crust. Continental blocks are analogous to floating icebergs of various heights in that the Moho is pushed deeper under continental mountain ranges than it is under flat coastal plains. The Moho assumes a shape that reflects the surface of the continent but that is exaggerated to nine times greater. The bottoms of the continental blocks must rise as material is eroded to the sea, thus keeping the exposed-to-submerged ratio constant. This flotation phenomenon is known as isostasy, and the rising process is called isostatic adjustment.

Characteristics of the Ocean Crust

One of the most remarkable characteristics of the oceanic crust is its structural uniformity. Essentially, marine sediments overlie igneous rock, which forms three distinct layers. The sediments vary considerably in thickness and in composition. Shell material and debris from marine plants and animals form dominant sediments around the equator and near the polar seas; detritus from the land and glacial deposits are common near the continents; and chemical precipitates (oozes) are found elsewhere. The oceanic ridge crests are generally free of sediments. From the flanks of the ridges to the continents, the sediments generally increase in thickness to more than 3 kilometers at the continental margins.

The three igneous layers are each relatively uniform in composition and thickness. The upper layer has been penetrated by deep ocean drilling and is known to be composed of basaltic lavas, 1 to 2.5 kilometers thick. The basal layer, directly overlying the mantle, is thin (0.5 kilometer) and presumably formed of layered peridotite. Peridotite is a dense ultrabasic igneous rock consisting mainly of olivine minerals. Similar rocks are thought to be the principal constituent of the mantle. The main (middle) layer is 5 kilometers thick and has properties consistent with a gabbroic composition. Gabbro is a coarse-grained igneous rock consisting mainly of plagioclase feldspar, pyroxene, and olivine minerals. It is the deep-seated equivalent of the overlying, fine-grained basalt. In certain areas, metamorphism and hydrothermal processes have formed other rock and mineral types including amphibolite, greenschists, zeolites, and serpentine.

The surface features of the oceanic crust include such interesting and interrelated topographic features as oceanic ridges, fracture zones, seamounts, abyssal plains, deep-sea trenches, and island arcs. The existence of each of these features can be explained by the concepts of plate tectonics and seafloor spreading.

Oceanic Ridge System

The oceanic ridge system is the major topographic feature of the ocean basins, extending 80,000 kilometers as a continuous range throughout all the oceans. Ocean ridges generally rise 2 to 3 kilometers higher than do the bordering ocean basins. The ridge system in the Atlantic and Indian oceans lies equidistant between the adjacent continents, whereas in the Pacific Ocean, the system is highly asymmetric with respect to the continents. Passing out of the Indian Ocean between Antarctica and Australia, the ridge system continues eastward across the southern Pacific, then arcs northward toward the South American continent. Here known as the East Pacific Rise, it eventually passes into the Gulf of California and presumably under the Basin and Range Province of the western United States to reemerge in the Pacific Ocean as the Juan de Fuca Ridge off British Columbia. Although almost entirely submarine, the ridges do rise above sea level at a few places in the Atlantic and Pacific oceans where recently active volcanoes have formed islands (for example, Iceland, Tristan da Cunha, and the Galápagos Islands). The ridges are also seismically active, with frequent tensional earthquakes of intermediate strength. Such earthquakes are generally restricted to the oceanic crust within a few kilometers of the ridge crest. The crests of the ridges in the Atlantic and Indian Oceans are characterized by a central rift valley that is commonly 2 to 3 kilometers deep and 20 to 30 kilometers wide. Volcanism occurs along the centerline of the rift valley, which is also the site of most of the seismic activity. In the Pacific Ocean, earthquakes are confined to a similar narrow zone, but volcanism appears to have been much greater. Here, lava flows have filled the central rift valley to a large part so that the ridge crest appears smooth.

The ocean ridge system is clearly a continuous feature on a global scale, but when viewed in detail, the ridge crests are frequently offset by fracture zones. For example, the Mid-Atlantic Ridge has no fewer than forty such zones. These fracture zones are steeply cliffed features that vary in width from a few to 50 kilometers. They are mainly confined to the oceanic crust and only rarely approach the continental margins. Fracture zones are only seismically active along that portion of the fault line between the offset ridge crests; the segments extending toward the continents are seismically quiet. Earthquakes between the crests are associated with transverse motion, indicating that the fracture zones are the result of faults in which each side moves horizontally but in opposite directions. These displacements of oceanic crust are known as transform faults.

Seafloor Irregularities

Throughout the world's oceans, beyond the flanks of the ridge systems, numerous irregularities rise from the sea floor. Small volcanic extrusions that rise less than 1 kilometer from the ocean floor are known as abyssal hills. Larger volcanic features that reach 1 kilometer or more are called seamounts. Seamounts that have flat tops are known as guyots. The flattening is thought to have resulted when seamounts were near sea level and subjected to wave attack and erosion. The composition of seamounts is closely related to their proximity to oceanic or continental crust. For example, near the center of the Pacific Ocean, the seamounts (including those that broach the sea surface to become islands) are composed of basaltic-type rock characteristic of oceanic crust, whereas along the margin of the Pacific Ocean the islands are composed of the granitic-type rocks that are found on the continents. The boundary between these two regions has been named the Andesite Line for the type of rocks found in the volcanic mountains of South America. In tropical areas, seamounts often support coralline reefs in the form of atolls. Atolls are generally circular in plan, consisting of a central lagoon surrounded by a narrow carbonate reef dotted with elongated islands. Presumably, atolls form as volcanic islands subside at a rate that is matched by the upward growth of the encircling reef.

Abyssal Plains and Ocean Trenches

The relatively flat surfaces of the ocean floor, which extend from the mid-ocean ridges to either the marginal trenches or the continental slopes, are known as abyssal plains. Excluding the trenches, these plains are the deepest portion of the ocean. Abyssal plains account for nearly 30 percent of the earth's surface, accounting for 75 percent of the Pacific Ocean basin and 33 percent of the Atlantic and Indian Ocean basins. Oceanic rises are areas of the ocean floor that are elevated above the abyssal plain, distinctly separated from a continental mass, and that are of greater areal extent than are typical seamounts or abyssal hills. General oceanic rises lie at least 300 meters above the surrounding ocean floor. Rises are not seismically active and are thought to result from the uplifting of oceanic crust associated with volcanic hot spots (source areas for magma in the upper mantle). Examples are the Bermuda Rise in the North Atlantic Ocean and the Chatham Rise in the southwestern Pacific Ocean.

The oceanic trenches are the deepest parts of the oceans. They are elongate, narrow, and commonly arcuate in shape with the convex side facing to sea. With few exceptions, they occur at the margins of ocean basins. By convention, an ocean deep must be at least 6,000 meters below sea level to be considered a trench. Trenches are found in the Atlantic and Indian Oceans but are most common in the Pacific. The deepest are found in the western Pacific, where the Mariana Trench plunges to a depth of 11,033 meters. The largest, however, is the Peru-Chile Trench adjacent to South America; it is 5,900 kilometers long, averages 100 kilometers wide, and extends to depths below 8,000 meters. Most trenches are associated with island arc systems or with volcanic ranges adjacent to continents. Examples include the Guam and Saipan islands west of the Mariana Trench and the Andes east of the Peru-Chile Trench. Island arcs are volcanic belts that parallel the trench on the continental side. The profiles of the trenches are asymmetrical, with steep sides toward the island arcs. Trenches are also areas of high earthquake activity, low gravitational pull, and low heat flow from the earth.

Plate Tectonics and Hydrothermal Vents

The concepts of plate tectonics and seafloor spreading provide the mechanisms necessary for creating the features of the ocean floor. Plate tectonics proposes that the earth's lithosphere is composed of several plates of differing shapes and areas that glide over the plastic asthenosphere. Convection cells caused by radioactive decay of isotopes in the molten rocks of the mantle are the driving mechanism for this motion. These cells circulate the heat upward, causing upwelling in the mantle. The movement of the plates results in areas of separation where magma flows to the surface, creating the volcanic mountains of the mid-ocean ridges. Thus, the ocean floor spreads outward from the ridge crest expanding the dimensions of the ocean. This process occurs in the Atlantic Ocean at the expense of the Pacific Ocean. Where plates collide, such as off the coast of Southeast Asia and South America, deep trenches are formed as oceanic crust is forced under (or subducts) the lighter continental crust. The process is associated with volcanism as the oceanic crust is remelted and island arcs are formed by the accompanying submarine eruptions. The rate of spreading affects the form of the ridge system. Rapid spreading (up to 5 centimeters per year) produces a broad, relatively low ridge without a deep central valley, such as that found in the East Pacific Rise, west of South America. Slow spreading (1 to 3 centimeters per year), in contrast, results in a high-relief ridge with a deep central rift valley such as that of the Mid-Atlantic Ridge.

Hot waters are discharged by hydrothermal vents at active mid-ocean ridges. The chemical composition of ocean water and deep-ocean sediments is influenced by seawater circulating through hot oceanic crust, formed by volcanic eruptions. Some seawater enters the oceanic crust through faults and eventually reaches the vicinity of the magma chambers below the spreading center where molten rock collects before eruption. Reactions with hot basalt charge the seawater with metallic sulfides and remove magnesium and other elements. The hot water then flows into the ocean through irregular, chimney-like vents up to 10 meters high. The vent mounds are made of silica, native sulfur, and metallic sulfide minerals. The bright-colored chimneys and their surrounding deposits resemble valuable ore deposits of copper, zinc, and other metals that are found on the continents. This phenomenon may also be an important process in regulating the chemical composition of seawater, as well as providing a chemical base for a deep-sea biological community that uses chemical energy rather than sunlight to produce organic compounds (chemosynthesis instead of photosynthesis).

Seismic Refraction and Reflection

The oceanic crust covers about 70 percent of the earth's surface, yet it has received relatively little attention. For example, deep-sea drilling and sampling of the crust have been completed at only one site for every 500,000 square kilometers of ocean floor. The great depth of the oceans, the tremendous logistical problems of working on and in the sea, and the high cost of oceanic research have all acted to limit the amount of scientific information that is available. Technological advances in the second half of the twentieth century have permitted researchers to explore the oceanic crust with remote sensing techniques as well as through diving excursions to the ocean floor.

The structure of the earth and particularly the oceanic crust have been investigated indirectly by seismic refraction and reflection methods. Studies in the early 1950's showed that the crust was composed of several layers on the basis of the velocity of sound within each layer. Ocean sediment transmits sound at 2 kilometers per second, basalt transmits at 5.1, gabbro transmits at 6.7, and peridotite transmits at 8.1. Precise measurements of the length of time required for a seismic shock wave to penetrate these layers have permitted the thickness of each layer to be calculated.

Accurate and detailed maps of the ocean floor (bathymetric charts) have been compiled from enormous collections of sounding data. By the early twentieth century, electronic devices called precision depth recorders (PDRs) became available for oceanographic surveys. These early surveys gave the first realistic view of the ocean's major surface features. Side-scanning sonar, a later development, has provided three-dimensional illustrations of small-scale features of the seafloor. This type of bathymetric data can be recorded continuously as the research ship is underway, and locations can be determined precisely from navigational satellites.

Technological Advances

Ocean crustal rock is an average of 7 kilometers thick. This great thickness, plus the hundreds of meters of sediment and thousands of meters of seawater overlying the crust, have made the direct sampling of this rock very difficult. Deep-ocean drilling has, however, permitted samples to be collected from considerable depths below the ocean floor. Beginning in 1968, the Deep Sea Drilling Project (DSDP) has extensively explored the oceanic crust from the ship *Glomar Challenger,* operated by Global Marine. Sponsored by the National Science Foundation and the Office of Naval Research, this project has drilled a total of 160 kilometers of cores. The deepest penetration into the ocean floor was 1.7 kilometers; the deepest water in which drilling took place was 7,000 meters. More than 840 sites were drilled in all parts of the world oceans. Scientists from all over the world participated in this project, which confirmed much of the theory of how the earth's crust moves. DSDP also provided significant data on the age of the ocean basins and the rates of seafloor spreading.

Other important deep-sea data have been gathered by research submersibles and airborne remote sensors. Starting in the 1960's, submersibles such as Woods Hole Oceanographic Institution's *Alvin* have made some remarkable oceanographic discoveries, especially along the mid-ocean ridges. Much of the knowledge of hydrothermal vents has been obtained through submersible observations. A submersible possesses numerous advantages over surface vessels, including direct observation and sampling. Such vessels are, however, dependent on surface ships for support and transport to the dive sites. Aircraft magnetometer surveys have yielded valuable information on paleomagnetism and the earth's gravitational field. These data showed that the polarity of the earth's magnetic field is recorded in the crustal rocks as the ocean floor is formed. Thus, a record was revealed that demonstrated seafloor spreading through mirror-images of polar reversal patterns on the east and west sides of the Mid-Atlantic Ridge.

Satellites have also played a part in the exploration of the oceanic crust. The short-lived Seasat satellite carried a sophisticated altimeter that could measure the precise distances between the satellite and the ocean surface. Slight differences in sea level were observed that correspond to ocean deeps and density anomalies (such as accumulations of dense rock). For example, the sea stands higher over mid-ocean ridges and lower over trenches. Satellite images are now being used to map previously unknown topographic features of the ocean floor.

Charles E. Herdendorf

Further Reading

Chernicoff, Stanley, and Donna Whitney. *Geology: An Introduction to Physical Geology.* Upper Saddle River, N.J.: Prentice Hall, 2006. This is a good overview of the scientific understanding of the geology of the earth and surface processes. Includes a chapter on the oceanic crust. Includes a link to a Web site that provides regular updates on geologic events around the globe.

Davies, Geoffrey F. *Mantle Convection for Geologists.* New York: Cambridge University Press, 2011. Begins with strong foundational material upon which to build convection concepts. Although the title implies technical writing, the author intended the text to be for anyone studying geological processes.

Dolgoff, Anatole. *Physical Geology.* Boston: Houghton Mifflin Harcourt, 1999. This is a comprehensive guide to the study of the earth. It is extremely well illustrated and includes a glossary and an index. Although this is an introductory text for college students, it is written in a style that makes it understandable to the interested layperson. Contains a chapter on the oceanic crust.

Edmond, John M., and Karen Von Damm. "Hot Springs on the Ocean Floor." *Scientific American* 248 (April 1983): 78-84. A discussion of the role of deep-ocean hydrothermal springs in depositing metallic ores and sustaining life in the absence of sunlight.

Gross, M. G. "Deep-Sea Hot Springs and Cold Seeps." *Oceanus* 27 (Fall 1984). An introduction to a special issue devoted to the exploration, geochemistry, chemosynthesis, and biology associated with water discharge on the ocean floor.

Harmon, Russell, and Andrew Parker. *Frontiers in Geochemistry: Contribution of Geochemistry to the Study of the Earth.* West Sussex: Wiley-Blackwell, 2011. Covers geochemistry in two parts: the contributions to earth studies and use of the analytical tools in the future. Chapter 4 discusses the geochemistry of oceanic crusts. Chapter 12 discusses archaeological and anthropological applications of isotopic geochemistry. This textbook is best suited for advanced undergraduates or graduate students.

Kennett, James P. *Marine Geology.* Englewood Cliffs, N.J.: Prentice-Hall, 1982. A comprehensive treatment of the geology of the sea floor, including ocean morphology, geophysics, plate tectonics, oceanic crust, and marine sediments.

Kennish, Michael J. *Practical Handbook of Marine Science.* 3rd ed. Boca Raton, Fla.: CRC Press, 2001. Chapter 4 covers marine geology topics such as seafloor topography, sediments, plate tectonics, mid-ocean ridges, oceanic crust formation, and hydrothermal vents. Best suited for advanced undergraduates and graduate students. References are provided for each chapter.

Ogawa, Yujiro, Ryo Anma, and Yildirim Dilek. *Accretionary Prisms and Convergent Margin Tectonics in the Northwest Pacific Basin.* New York: Springer Science+Business Media, 2011. Discusses new techniques in plate tectonic studies. One volume of the series Modern Approaches in Solid Earth Sciences. Covers accretionary prisms, tectonics, and Pacific Ocean events.

Plummer, Charles C., Diane H. Carlson, and Lisa Hammersley. *Physical Geology.* 13th ed. Boston: McGraw-Hill, 2009. This is a straightforward, easy-to-read introduction to geology intended for those with little or no science background. Discusses the sea floor and the oceanic crust. Includes many excellent illustrations.

Seibold, Eugen, and Wolfgang H. Berger. *The Sea Floor: A Introduction to Marine Geology.* 3rd ed. New York: Springer-Verlag, 2010. As the title suggests, this is an introduction to many topics in marine geology. This text covers geological structures from the continental shelf to deep-ocean trenches. Processes such as seafloor spreading, the sediment cycle, currents, and pelagic rain are discussed.

See also: Abyssal Sea Floor; Mid-Ocean Ridge Basalts; Ocean Basins; Ocean-Floor Exploration; Ocean Ridge System; Spreading Centers.

OCEAN RIDGE SYSTEM

The ocean ridge system is a complex chain of undersea volcanic mountains that are found in all the oceans. These mountains contain rift valleys along their axes, which are believed to be spreading centers from which continental motion takes place. All existing evidence, such as volcanic activity, the flow of heat from within the earth, and various types of faulting and rifting, supports modern theories about seafloor spreading, tectonic plates, and continental drift.

PRINCIPAL TERMS

- **asthenosphere:** a zone of rock within the mantle that has plastic flow properties attributable to intense heat
- **basalt:** a heavy, dark-colored volcanic rock
- **converging plates:** a tectonic plate boundary where two plates are pushing toward each other
- **divergent plates:** a tectonic plate boundary where two plates are moving apart
- **metallogenesis:** the process by which metallic ores are formed
- **sediments:** the solid fragments of rock that have been eroded from other rocks and then transported by wind or water and deposited

SEAFLOOR SPREADING

The ocean ridge system is a complex chain of mountains about 80,000 kilometers in length that winds through the ocean basins. These mountain ranges vary from a few hundred to a few thousand kilometers in width and have an average relief of 0.6 kilometer. Studies of the sea floor indicate that ocean ridges are found in every major ocean basin. They are composed of basalt and covered by various types of sediments. Many of the ridges have narrow depressions that extend thousands of kilometers along their axes. Heat probes lowered into these rifts indicate much higher temperatures than on the flanks of the ridges. Another very significant finding has been that the rocks that make up the sea floor are much younger than those that make up the continents. This finding countered the pre-1960's belief that the rocks of the ocean basin were more ancient than those of the continents.

In the 1960's, the theory of seafloor spreading suggested that new sea floor is constantly being added by volcanic activity at the ocean ridges. The theory of plate tectonics proposes that the earth's crust is divided into several major plates. These plates extend down into the earth's mantle to a hot, semimolten zone known as the asthenosphere. Since the rock that composes the tectonic plates is less dense than the rock that forms the mantle, the plates may be considered to be floating on the asthenosphere. Because the interior of the earth is much hotter than the surface, a flow of heat toward the surface is a constant process. This method of heat transfer, known as convection, is a density type of current. Material such as molten rock or hot air or water is less dense than the same material in a cooler state. As a result, it flows upward, with cooler material filling in the space below. When the hot material reaches a higher level and releases its heat, it, too, returns to the depths to be reheated. It is this process, along with the relatively low density of the tectonic plates, that causes the plates to drift apart.

The location and symmetry of the mid-ocean ridges, especially in the Atlantic and Indian Oceans, suggests the configuration of the continents before they began to drift apart. The modern theories of plate tectonics and seafloor spreading are in basic agreement with the theory of continental drift as proposed by German geophysicist Alfred Wegener in 1915. However, modern ideas regarding the mechanics of drift differ. The basic concept of seafloor spreading at the oceanic ridges proposes that tension cracks form in the crust at these spreading centers. Molten rock from the mantle then flows upward through these fissures, both forming the volcanic ridges and creating new sea floor. As the fissures widen, new crustal material moves away on both sides, and additional new sea floor is created. It is in this way that the mid-ocean ridges such as the Mid-Atlantic Ridge, the East Pacific Rise, the Antarctic Rise, and the Carlsberg Ridge of the Indian Ocean were formed. The spreading away from the oceanic ridges takes place at a rate of about 2.5 centimeters per year in the Atlantic and about 8 to 12 centimeters per year in the Pacific.

As new sea floor is being created at the ridges, old sea floor is being destroyed at continental margins. Here, old sea floor is subducted beneath continental plates. At these converging plate boundaries, old sea floor is forced downward into the mantle, where it undergoes remelting. This molten rock may then find its way back to the surface through cracks or fissures in the overlying rock. Volcanic action is the result of this material reaching the earth's surface. Because old sea floor is destroyed in this manner, it is now understood why rocks that make up the ocean floor are relatively young. Deep-seafloor drilling projects in the Atlantic and Indian Oceans have failed to find rock or fossil samples that are older than the Jurassic period of earth history, which ended about 145.5 million years ago.

Rift Zones

The present ocean ridges show large offsets in some areas. This phenomenon results from transform faulting. As spreading of the plates took place at divergent boundaries, fracture zones developed at right angles to the axes of the ridges. Displacements of the ridges along these faults produced the observed structure.

The narrow rift valleys are believed to have been caused by downfaulting along divergent plate boundaries. Some rift zones, such as the one associated with the Carlsberg Ridgeof the northern Indian Ocean, link up with continental rift zones. This rift has been shown to be connected with the African rift zone. Studies conducted during the 1970's indicate that extensive amounts of volcanism and seismic activity exist along rift zones.

Hot Springs and Smokers

Also found along oceanic ridges are hot-water springs. The existence of these springs, or smokers, had been predicted, but direct evidence for them was not gathered until the early 1960's, when metal-bearing sediments were discovered on the East Pacific Rise. The first actual observation of a smoker took place in 1980 by the crew of the deep-diving research submarine *Alvin*. The *Alvin* was part of an underwater research program being conducted near the Galápagos spreading center in the eastern Pacific. Researchers were surprised to find a rather extensive plant and animal community living near the smoker at a depth too great for photosynthesis to be a factor. Clams as long as 30 centimeters were found, as well as white crabs and tube worms some 3 meters long.

Smokers have been found to emit great quantities of sulfur-enriched waters. Dissolved within the acidic water are various types of metals. These metals are dissolved from the rocks as the superheated water moves toward the surface. Metals such as copper, zinc, iron, silver, and gold are extracted and concentrated into a supersaturated fluid. When this fluid is discharged from a smoker at the sea floor, it will precipitate to form an ore body if cooling takes place rapidly. These

The Mid-Atlantic Ridge

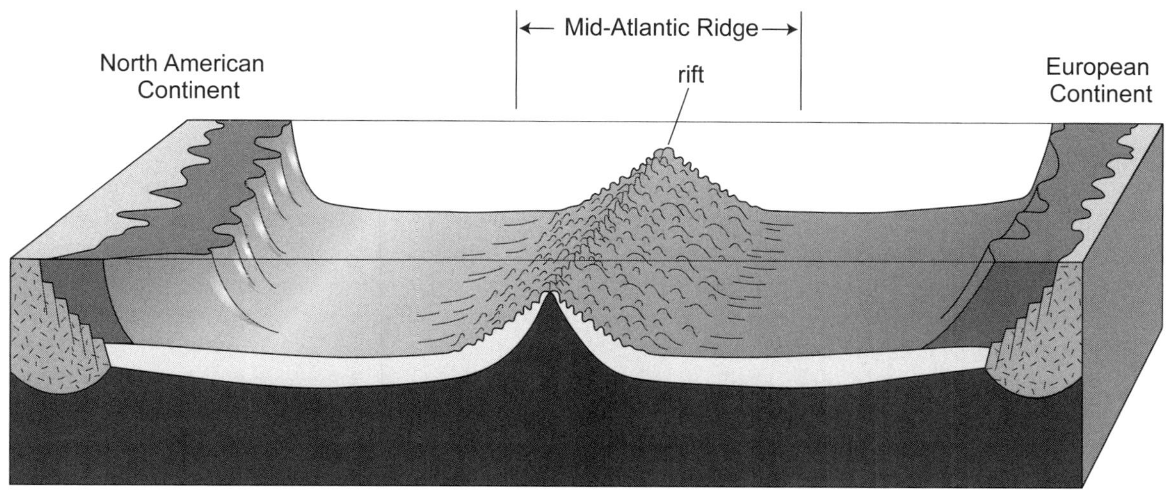

massive sulfide deposits are sometimes made permanent when volcanic eruptions cover them with basalt flows.

Mid-Atlantic Ridge

Although most of the knowledge regarding the nature of the sea bottom has been gained since the 1960's, the Mid-Atlantic Ridge system has been known to exist since echo-sounding studies were done after World War I. By 1960, it had been determined that the Mid-Atlantic Ridge was continuous with other oceanic ridge systems around the world. The same cannot be said of the rift system that was discovered in the Mid-Atlantic Ridge. Profiles taken across ridge systems do not always indicate such rift depressions. This anomaly can probably be explained by considering the possibility that these rifts have been filled with volcanic material and, therefore, may go undetected.

A rather extensive study was made along the Mid-Atlantic Ridge about 200 miles south of the Azores in 1974. It was found that the rift valley is bordered by a series of steep slopes that appear to be high-angle faults. Many open vertical fissures were observed, some of them as much as 8 meters across. Although no active volcanism was observed during this study, mounds of pillow lava along the fissures indicated that volcanism had taken place. The positioning of the faults and fissures indicated that the Mid-Atlantic Ridge is spreading outward east and west from its central axis.

East Pacific Rise

The East Pacific Rise, which is part of the system of mid-ocean ridges, stands some 2 to 3 kilometers above the ocean floor and is thousands of kilometers wide. The slopes in this area are not as great as those of the Mid-Atlantic system, but the mechanism of ridge formation is the same. It has been suggested that part of the East Pacific Rise extends under the North American continent and that the San Andreas fault may be part of this system. Evidence supporting this possibility is that the rate of displacement along the fault is comparable to the rate of ridge spreading on the East Pacific Rise near the Gulf of California.

Study of the Ocean Floor

The ocean floor has been investigated by the use of sonar since the early twentieth century. Using this technique, sound waves from a device are sent into the sea in all directions. As these waves strike objects, they are reflected back to the source. By analyzing the reflections, scientists are able to determine the nature of the sea floor. This technique was refined during World War II for the purpose of locating enemy submarines. By the early 1950's, new maps and charts of the sea floor had been made using sonar.

In the 1960's, a plan to drill a deep hole through the crust of the earth to the mantle was proposed. Since this deep hole was intended to intersect the Mohorovičić discontinuity (named after Andrija Mohorovičić, the Croatian geologist who discovered it), the project was called Mohole. Given that the crust of the earth is much thinner under the ocean basins than it is under the continents, the ocean was the most logical location in which to drill. Although the 6-kilometer-deep hole was never drilled, several preliminary holes were. Samples taken from this drilling revealed the presence of gray, claylike sediments overlying dark, heavy basalts.

Project Mohole was later replaced by a plan to drill many holes at various locations into the rock beneath the ocean depths. This project, under the direction of various American oceanographic institutes, became known as the Deep Sea Drilling Project. Much of the drilling was done from the deck of the now-retired *Glomar Challenger* of the company Global Marine. More than six hundred holes were drilled in the Atlantic, Pacific, and Indian Oceans and the Mediterranean Sea, revealing a wealth of data on the nature and evolution of the ocean basins.

In the past, ocean researchers had to gather data from the decks of ships by lowering various tools to the bottom to gather samples—there was a need for direct observation by the oceanographer. In 1948, the first minisubmarine, or bathyscaphe, made an unmanned dive to 759 meters. In 1954, a manned vessel was taken to a depth of 4,050 meters. During the 1960's, interest in these submersibles grew quickly in the United States. Since that time, submersibles such as the *Aluminaut* and *Alvin* have been used in various deep-sea research projects. Submersibles have been used for on-site studies of the rift zone of the Mid-Atlantic Ridge and of the hot-spring smokers of the East Pacific Rise.

Another technique, known as marine seismology, has been employed to study the ocean ridges. Oil companies have used seismic studies to explore for oil deposits on continents, and it was only a matter of time before this technology was adapted for study of the ocean depths. The process involves making a sound explosion in the sea, which is accomplished by the use of an air gun. The sound waves are produced by the rapid release of compressed air. The waves then travel through the water to the bottom, where they are reflected to seismometers located on the ocean floor. The data are collected, and a seismogram is generated by a computer. This technique has been used to locate active magma chambers in rift zone areas.

Economic Resources

Studies of submarine rift areas that have revealed the hot-water springs and smokers have shown these areas to offer potential economic riches. Smokers discharge water that contains sulfur in solution as well as various types of metals such as copper, iron, zinc, silver, and gold. These metals precipitate to the sea bottom and, if they are cooled rapidly enough, form deposits. If the ore bodies are covered with volcanic material shortly after their deposition, they become protected from the erosive processes of seawater. In time, these deposits drift from oceanic spreading centers to become parts of continents. Many mineral deposits formed in just this manner have been mined since ancient times.

In the late 1950's, metal-rich sediments were first found along mid-ocean ridges. By the late 1960's, metalliferous muds were found at the bottom of the Red Sea, the Gulf of California, and the East Pacific Rise near the Galápagos Islands and south of Baja, California. Beginning in 1977, marine geologists were able to observe the formation of metal deposits while observing smokers from the research submarine *Alvin*. Interest in ridge deposition of metals centers on the processes of metallogenesis. Mining of such deposits, however, even with advanced technology, is not profitable. It will no doubt be considered in the future, however, as the copper deposits in the area of the Galápagos spreading center alone are estimated to be worth in excess of $2 billion.

David W. Maguire

Further Reading

Anderson, Roger N. *Marine Geology: A Planet Earth Perspective*. Rev. ed. New York: John Wiley & Sons, 1989. A well-written, somewhat technical volume on the geology of the oceans. Contains excellent chapters on metallogenesis and mineral deposits. This volume is suitable for the college student of geology or the informed layperson.

Buck, W. Roger, Luc L. Lavier, and Alexei N. B. Poliakov. "Modes of Faulting at Mid-ocean Ridges." *Nature* 434 (2005): 719-723. The article presents new information on plate separation at ridges. Understandable to the layperson; the article has few, yet well explained, equations. It provides useful information for identifying processes that produce particular characteristics in faults.

Chester, Roy. *Furnace of Creation, Cradle of Destruction*. New York: AMACOM Books, 2008. This text discusses the turbulent processes of the planet. It covers earthquakes, volcanoes, and tsunamis in reference to plate tectonics, natural disasters, and predicting and mitigating these events. Multiple chapters explore seafloor spreading. The author also discusses hydrothermal activity. This text takes on an immense range of content, but still explains concepts clearly and with detail.

Dolgoff, Anatole. *Physical Geology*. Boston: Houghton Mifflin Harcourt, 1999. This is a comprehensive guide to the study of the earth. It is extremely well illustrated and includes a glossary and an index. Although this is an introductory text for college students, it is written in a style that makes it understandable to the interested layperson. Contains a section on the development of the oceanic ridge system.

Garrison, Tom S. *Oceanography: An Invitation to Marine Science*. Belmont, Calif.: Brooks/Cole, Cengage Learning, 2010. Chapter 3 provides excellent coverage of plate tectonics; ocean formation, seafloor spreading, and island arc formation. Chapter 4 covers topography and bathymetry of deep-ocean basins. Abundant diagrams aid readers from the layperson to advanced undergraduates.

Kennish, Michael J. *Practical Handbook of Marine Science*. 3rd ed. Boca Raton, Fla.: CRC Press, 2001. Chapter 4 covers marine geology topics such as seafloor topography, sediments, plate tectonics, mid-ocean ridges, and hydrothermal vents. Best suited for advanced undergraduates and graduate students. References are provided for each chapter.

Prothero, D. R., and Robert H. Dott. *Evolution of the Earth*. 8th ed. New York: McGraw-Hill, 2009. This basic textbook on historical geology is aimed at students of geology. However, it is very readable by anyone with a background in science. Presents an up-to-date account of the earth's history from the viewpoint of plate tectonics. Includes a glossary.

Segar, Douglas. *An Introduction to Ocean Sciences.* New York: W. W. Norton & Company, 2007. Comprehensive coverage of all aspects of the oceans and the oceanic crust. Readable and well illustrated. Suitable for high school students and above.

Seibold, Eugen, and Wolfgang H. Berger. *The Sea Floor: A Introduction to Marine Geology.* 3rd ed. New York: Springer-Verlag, 2010. As the title suggests, this is an introduction to many topics in marine geology. This text covers geological structures from the continental shelf to deep-ocean trenches. Processes such as seafloor spreading, the sediment cycle, currents, and pelagic rain are discussed.

Singh, Satish C., et al. "Discovery of a Magma Chamber and Faults Beneath a Mid-Atlantic Ridge Hydrothermal Field." *Nature* 442 (2006): 1029-1032. This article provides an example of current seafloor exploration and research. It discusses the use of seismic analysis for developing models of mid-ocean ridges. It also provides some information on hydrothermal activity and submarine volcano structure. Understood by the layperson with careful reading.

Skinner, B. J., et al. *Dynamic Earth: An Introduction to Physical Geology.* 5th ed. New York: John Wiley & Sons, 2006. A well-written, well-illustrated volume dealing with general topics in physical geology such as rocks, minerals, erosion, earth resources, sedimentation, and tectonics. Ideal for the college-level introductory physical geology course.

Wicander, Reed, and James S. Monroe. *Historical Geology: Evolution of Earth and Life through Time.* 6th ed. Belmont, Calif.: Brooks/Cole, Cengage Learning, 2010. A very well-illustrated volume dealing with such subjects as geologic time, origin and interpretation of sedimentary rocks, and a detailed account of the historical geology of the earth through various time periods. Excellent for a first-year college course in historical geology.

See also: Abyssal Sea Floor; Continental Rift Zones; Faults; Geoclines; Iceland; Mid-Ocean Ridge Basalts; Ocean Basins; Ocean-Floor Exploration; Oceanic Crust; Spreading Centers.

OPHIOLITES

Ophiolites are a unique assemblage of rocks found in many mountain belts throughout the world. They were formed in the oceans and subsequently transported to land during mountain-building processes. Ophiolites are useful to geologists as indicators of the location of ancient oceans. They are important to society because they are major sources of asbestos, nickel, and copper minerals.

PRINCIPAL TERMS

- **igneous rock:** a rock formed when magma cools and forms minerals; it can form on the surface of the earth when volcanoes erupt, or it can form at depth, without reaching the earth's surface
- **lithosphere:** the upper, rigid 100 kilometers of the earth that forms the moving plates; beneath oceans it is called oceanic lithosphere, and beneath continents it is called continental lithosphere
- **mafic:** a rock rich in iron and magnesium minerals; ultramafic rocks are very rich in these minerals
- **olivine:** a mineral consisting of magnesium, iron, silicon, and oxygen
- **peridotite:** an igneous rock consisting of minerals rich in magnesium; serpentinized peridotite has been heated such that the mineral olivine is partially converted into serpentine
- **pillow lava:** the lava that is formed when molten rock erupts into water and cools in the shape of a pillow
- **plate tectonics:** the theory that the upper part of the earth consists of a number of "plates," or rigid parts, that move relative to one another across the surface
- **serpentine:** a mineral consisting of magnesium, iron, silicon, oxygen, and water

Characteristics of Ophiolites

Ophiolites are unique assemblages of rocks that have fascinated geologists for centuries. The word "ophiolite" is derived from the Greek *ophis,* meaning snake or serpent. The term first appeared in the geological literature in the 1820's, when Alexandre Brongniartof France used it to describe rocks called serpentinite, which are made entirely of the mineral serpentine. The term "ophiolite" is appropriate because serpentinite, like some snakes, has a mottled green appearance.

The usage of the term "ophiolite" changed early in the twentieth century when the close association of serpentinite or serpentinized peridotite with deep-water sediments and with pillow lavas was noted. This assemblage of three rock types became known as the Steinmann Trinity. In the late 1960's, with the advent of plate tectonics as a unifying theory in geology, the definition of ophiolite again changed and was used to describe the Steinmann Trinity plus other rocks arranged in a particular sequence. At the same time, the interpretation that ophiolites formed in the oceans became the cornerstone for reconstructing the history of mountains. As defined for modern usage by a Geological Society of America conference in 1972, "ophiolite" refers to a distinctive assemblage of mafic and ultramafic rocks. A completely developed ophiolite is 10 to 12 kilometers thick and covers an area of hundreds of square kilometers.

Rocks in ophiolites occur in layers in the following order (starting from the bottom and working up): peridotite, gabbro, sheeted dike complex, pillow lavas, and chert. Peridotite is an ultramafic igneous rock formed below the earth's surface, dark in color, consisting of large grains (greater than 5 millimeters) of the minerals olivine (an olive-green mineral rich in magnesium, iron, silicon, and oxygen) and pyroxene (a black mineral rich in iron, magnesium, silicon, and oxygen). In many ophiolites, peridotite has been subjected to heat and hot water since its formation, resulting in some of the olivine minerals changing to serpentine, some of which is asbestos.

Gabbro is a mafic igneous rock formed below the earth's surface that has a salt-and-pepper appearance. It consists of large grains (greater than 3 millimeters) of the minerals pyroxene and feldspar (a white mineral rich in calcium, sodium, aluminum, and oxygen). Sheeted dike complex is made up of mafic igneous rocks formed just below the earth's surface, similar to gabbro but with smaller mineral grains (less than 3 millimeters). Pillow lavas are mafic igneous rocks formed on the earth's surface as lava erupts into water, similar to gabbro but with much smaller grains (less than 1 millimeter). Chert is a sedimentary rock formed in deep water consisting of very small grains (some too small to be seen with a microscope) of various minerals and organic remains.

Not all ophiolites exhibit the complete sequence of rocks; some were formed with certain layers missing, and others were dismembered subsequent to their formation. A very important observation is that beneath all ophiolites there is a fault (fracture in the earth) separating them from underlying rocks. Furthermore, there is evidence that ophiolites have probably moved hundreds of kilometers along those faults.

Formation of Ophiolites

Ophiolites can form in two different plate tectonic environments, both of which are in oceans. In one environment, such as in the middle of the Atlantic Ocean, two plates are moving away from each other. As they do so, magma from the interior of the earth moves up to the bottom of the oceans and cools to form long mountain chains called oceanic ridges (for example, the Mid-Atlantic Ridge). The solidified rock becomes part of an outer layer of the earth called the oceanic lithosphere. The newly formed oceanic lithosphere then moves away from the ridge at which it formed, resulting in widening of the ocean. (For example, the Atlantic Ocean is growing wider as North America moves westward, away from Europe.) Ophiolites are believed to be remnants of the top part of oceanic lithosphere that was created at ancient oceanic ridges. All the igneous rocks in ophiolites formed by the cooling of magma produced deep in the earth and spewed forth at an oceanic ridge. Pillow lavas form when the magma erupts into seawater; sheeted dikes, gabbros, and peridotites form when that magma cools at various depths below the surface. Chert forms when fine particles, including mineral grains, animal parts, and plants, fall to the bottom of the newly created ocean floor to form sedimentary deposits.

A second environment where ophiolites may form is the region where two plates are moving toward each other. In this environment, oceanic lithosphere that formed at oceanic ridges disappears down into the interior of the earth, resulting in melting at depth, the production of highly explosive volcanoes, and the generation of dangerous earthquakes. This general area is known as a subduction zone. When oceanic lithosphere has just begun to subduct (descend into the earth's interior) beneath another oceanic lithospheric plate, ophiolites may form in the region above the subducting plate. The processes involved in formation of these ophiolites are the same as at an oceanic ridge, though the environments are different.

Movement of Ophiolites

Another important part of the geological history of ophiolites is how they are moved from their place of formation in the oceans to their current location on continents. The key lies in understanding how mountains are built. Ophiolites have been described from all the world's major mountain belts: the Appalachian Mountains of eastern North America, the Rockies in western North America, the Alps in Europe, the Himalaya in Asia, the Ural Mountains, the Andes in South America, and the mountains of Papua, New Guinea. They range in age from about 600 million to 15 million years old. The movement of ophiolites from the ocean floor to mountains is related to plate tectonics. Subduction zones, where plates are moving toward each other, provide the mechanism for emplacement of ophiolites.

Gabbro, an igneous rock rich in feldspars and pyroxene minerals. (Biophoto Associates/Photo Researchers, Inc.)

To get a better idea of how ophiolites may have been put into place in the past, scientists consider what might happen to the Pacific Ocean in the future. The Pacific plate (that is, the oceanic lithosphere beneath the Pacific Ocean) is moving westward toward, and subducting beneath, the Asian plate along the Japanese islands, resulting in volcanoes and earthquakes along the zone. The Pacific plate is also subducting eastward beneath western North America. Therefore, the Pacific Ocean will become smaller, the North American continent will move closer to the Asian continent, and they will eventually collide. During this time, probably 99.99 percent of the Pacific oceanic lithosphere (potential ophiolites) will be lost to the interior of the earth at the subduction zones. At the time of the collision of North America and Asia, however, a small portion (less than 0.01 percent) of the oceanic lithosphere may not slip back into the interior but instead may be pushed up onto the continents, because when the two continents collide, mountains are being built, and some of the Pacific oceanic lithosphere, sandwiched in the zone of collision, may be pushed up with the mountains. These portions of the Pacific oceanic lithosphere will then be called ophiolites. In a similar fashion, the ophiolites that are now visible in mountain belts are tiny remnants of ancient oceanic lithosphere pushed onto continents during ancient mountain building.

The occurrence of ophiolites in mountains has helped geologists understand how mountains are built. Ophiolites lie in narrow bands that continue for thousands of kilometers along a mountain belt. This narrow band represents the zone where two continents have come together to build the mountains. Thus, by locating ophiolites, geologists can identify and then characterize the two (or more) continents that existed separately before the mountains were built. For example, in the Appalachian Mountains of eastern North America, a narrow band of ophiolites extends from western Newfoundland, Canada, through Quebec, Canada, into Vermont and as far south as Alabama in the United States. This zone may represent the boundary between the ancient North American continent to the west and the ancient African continent to the east; the two ancient continents were plastered together during the building of the Appalachians over a span of time between 500 million and 350 million years ago.

Study of Ophiolites

Geologists have used a variety of techniques to study ophiolites, including field mapping, seismic studies, electron microprobe techniques, and X-ray techniques. Field mapping of ophiolites involves identifying rock types, noting their locations on maps, and ascertaining the relationship of the ophiolites with neighboring rocks; it is the basic tool used by the geologist to analyze the origin of ophiolites, and it precedes any laboratory techniques.

Seismic studies have been employed to compare ophiolites to the modern oceanic lithosphere. Seismic waves travel through the earth after earthquakes; they can also be created by human-made explosions or by hitting the ground hard with a sledgehammer. Seismic waves travel at different speeds through different rock types. Thus, if scientists can identify the velocities of seismic waves as they travel through different types of rock, the type of rock can be identified even though it cannot be seen. In the oceans, seismic velocity measurements through rocks beneath the ocean floor have shown that the modern oceanic lithosphere is made up of layers that correspond exactly in rock type and in thickness to what is observed in ophiolites, which is very good evidence to support the interpretation that ophiolites formed as portions of oceanic lithosphere.

X-ray techniques have been used to determine the chemical composition of rocks in ophiolites. In a technique called X-ray fluorescence, X rays bombard a powdered rock sample and interact with electrons of the chemical elements to produce secondary X rays. The secondary X rays generated by specific elements can be isolated and counted to determine the amounts or percentages of specific elements present in the rock. Results of such studies on ophiolites have shown that pillow lava and sheeted dike rocks are similar to lavas erupted in modern oceans. In particular, low percentages of potassium oxide and of light rare-earth elements are common to both ophiolitic rocks and modern oceanic rocks.

The electron microprobe is also used extensively to help interpret the origin of ophiolites. In the electron microprobe, electrons hit a rock sample at high speed and interact with electrons in chemical elements in the rock to produce X rays. In the same way as in X-ray techniques, amounts of the chemical elements can be determined. In electron microprobe work, however, the rock sample, rather than being a crushed or powdered specimen, is a thin section (0.03 millimeter thick) of an intact rock that can be viewed through a microscope. Thus, minerals in the rock can be seen, selected, and analyzed for their chemical constituents.

Chemical analyses of minerals are most useful in the peridotite and gabbro layers of ophiolites. For example, chemical analyses of olivine, pyroxene, and feldspar from gabbro layers and upper parts of peridotite layers have shown that these layers precipitated from a magma of the same composition as the pillow lavas and sheeted dikes. Therefore, nearly all the igneous rocks in ophiolites can be related to one magma, part of which cooled at shallow depths (about 5 kilometers below the surface) to produce gabbro and peridotite and some of which was injected into higher levels in the earth to produce the sheeted dikes and pillow lavas. The lower parts of the peridotite layer, however, must have originated in a different way. The amounts of aluminum in the mineral pyroxene reveal that the lower peridotite layer formed under high pressure at considerable depth, as much as 30 kilometers below the surface of the earth—good evidence for the suggestion that ophiolites that now exist on the surface of the earth must have been moved since the time of their formation.

Raymond A. Coish

Further Reading

Ceuleneer, Georges. "Earth Science: Mantle Mapped in the Desert." *Nature* 432 (2004): 156-157. Provides a short overview of the processes that produced the Oman ophiolite. Discusses the supercontinents of Gondwanaland and Laurasia.

Chernicoff, Stanley, and Donna Whitney. *Geology: An Introduction to Physical Geology*. Upper Saddle River, N.J.: Prentice Hall, 2006. This is a good overview of the scientific understanding of the geology of the earth and surface processes. Contains information on the formation of ophiolites. Includes a link to a Web site that provides regular updates on geologic events around the globe.

Coleman, R. C. "The Diversity of Ophiolites." *Geologie en Mijnbouw* 63 (1984): 141. A very good review of the state of knowledge on the origin of ophiolites. Suitable for more advanced students of geology.

__________. *Ophiolites: Ancient Oceanic Lithosphere?* New York: Springer-Verlag, 1977. The most detailed treatment of ophiolites available. Includes sections on field descriptions, chemistry of the rocks, and models to explain the origin of ophiolites. Also includes detailed descriptions of four major ophiolites in the world. Suitable for those with previous exposure to geology.

Dilek, Y., and P. T. Robinson, eds. *Ophiolites in Earth History*. London: Geological Society Special Publication, 2003. A compilation of articles covering all aspects of ophiolites. Discusses Tethyan ophiolites and the creation and distribution of ophiolites. Articles are highly technical, best suited for graduate students or professional geologists.

Dolgoff, Anatole. *Physical Geology*. Boston: Houghton Mifflin Harcourt, 1999. This is a comprehensive guide to the study of the earth. It is extremely well illustrated and includes a glossary and an index. Although this is an introductory text for college students, it is written in a style that makes it understandable to the interested layperson. Contains a section on the formation of ophiolites.

Francheteau, Jean. "The Oceanic Crust." *Scientific American* 249 (September 1983): 130. Summarizes information on how modern oceanic crust is formed. Relevant to ophiolites because they are thought to have formed as ancient oceanic crust. Suitable for advanced high school and college students.

Gass, Ian G. "Ophiolites." *Scientific American* 247 (August 1982): 122. A comprehensive treatment of the evolution of ideas on the origin of ophiolites. Well illustrated with colored diagrams and photographs. Suitable for advanced high school students, college students, and interested adults.

Hamblin, William Kenneth, and Eric H. Christiansen. *Earth's Dynamic Systems*. 10th ed. Upper Saddle River, N.J.: Prentice Hall, Pearson Education, 2003. A very good general text to introduce readers to the field of geology. Particularly informative chapters on plate tectonics, ocean-floor processes, and mountain building. The text is well organized into four easily accessible parts. Suitable for anyone interested in geology.

Kusky, T. M., ed. *Precambrian Ophiolies and Related Rocks*. San Diego: Elsevier, 2004. The book is divided into four sections covering Proterozoic ophiolites, Archean ophiolites, oceanic crust models, and Precambrian analogs. The introduction alone provides an excellent summary of the collective knowledge of ophiolites. Background in geology is necessary when reading this text, as this is a compilation of scientific studies rather than a general textbook.

Prothero, D. R., and Robert H. Dott. *Evolution of the Earth.* 8th ed. New York: McGraw-Hill, 2009. This basic textbook on historical geology is aimed at students of geology. However, it is very readable by anyone with a background in science. Presents an up-to-date account of the earth's history from the viewpoint of plate tectonics. Includes a glossary.

Shervais, J. W. "Birth, Death, and Resurrection: The Life Cycle of Suprasubduction Zone Ophiolites." *Geochemistry, Geophysics, Geosystems* 2 (January 2001): 1-45. Provides detailed characteristics and processes of the life cycle of ophiolites along with examples from the coast of California and the Oman Ophiolite. The author uses many technical terms requiring background knowledge of geology. However the numerous diagrams and models supplement the text.

See also: African Rift Valley System; Carbonate Rocks; Displaced Terranes; Fold Belts; Folds; Geoclines; Joints; Minerals; Mountain Belts; Oceanic Crust; Seafloor Spreading; Spreading Centers; Thrust Belts.

P

PALEOBIOGEOGRAPHY

Paleobiogeography concerns the study of the geographic distribution of past life-forms. Paleobiogeographic studies, coupled with reconstructions of ancient plate configurations, serve as tools for reconstructing a detailed history of the earth's changing geography and environments through time and the concurrent evolution of the organisms living in these environments.

PRINCIPAL TERMS

- **barrier:** any sort of physical, chemical, or biological obstacle that prevents the migration of a species into or out of a given geographic area
- **biogeographic province:** a geographic region distinguished by a unique set of endemic organisms that live in the region
- **convergent organisms:** evolutionarily unrelated organisms that come to resemble one another through adaptation to the same life habit
- **cosmopolitan:** having a broad, essentially worldwide distribution
- **diversity:** the variety of life, usually described in terms of the number of species present
- **endemic:** restricted to a specific, limited geographic region
- **epicontinental seas:** shallow seas that periodically covered large portions of the continents in the geologic past
- **limiting factors:** the physical, chemical, and biological factors that govern the distribution of a given organism
- **paleobiogeography:** the study of the geographic distribution of past life-forms
- **plate tectonics theory:** the concept that the earth's exterior is divided into an interlocking mosaic of moving plates

Distribution of Life

Life is distributed on the earth's surface in complex patterns, governed by local environmental conditions and distributed in specific geographic regions. Very few living species of plants or animals are cosmopolitan in their distribution. *Homo sapiens* (modern humans) is one of the few cosmopolitan species. Most organisms evolve, exist, and eventually become extinct all in the same limited geographic region. These organisms are said to be endemic in their distribution. Endemic species may be restricted to a continent (the African elephant in sub-Saharan Africa), a portion of a continent (the snow leopard in the Himalaya), an island (the orangutan in Borneo), or even smaller geographic areas (the giant sequoia redwood in Northern California).

Most organisms are adapted to a specific set of narrowly defined environmental requirements that must be met if the organism is to exist in a given area. If only one of these environmental requirements or limiting factors is not present in the right proportion, the species will be unable to exist in a particular environment. These limiting factors include all the physical, chemical, and biological properties that define a given environment. For all organisms, an adequate food supply or source of nutrients is basic for continued existence. Factors controlling the distribution of organisms in terrestrial (land) environments include the availability of water, the temperature, and the altitude above sea level; humidity, precipitation, soil type, and sunlight intensity are limiting factors for plants, and types of vegetation are limiting factors for animals. For aquatic organisms, factors such as water depth, salinity, temperature, dissolved oxygen content, current flow, clarity, and bottom conditions become important. These various environmental factors in a particular geographic area are critical to determining the kinds of animal and plant life that can exist there.

Each environment within a given geographic region supports flora and fauna that are adapted to the specific conditions of that environment and are known collectively as biota. These geographically restricted biotas can be used to define biogeographic provinces, specific geographic regions distinguished by their endemic floras and faunas. The boundaries between biogeographic provinces are marked by the development of barriers to the free movement of species from one area to another. These barriers might be physical, chemical, or biological in nature. For example, land areas would be barriers to the dispersal of marine organisms, open oceans would be barriers to the dispersal of terrestrial organisms, and deep oceanic waters might be barriers to the dispersal of shallow-water, nearshore organisms. Many barriers differentiating biogeographic provinces today are climatic in nature, marking zones of changing atmospheric or water temperature.

Major Biogeographic Provinces

At the present, the major continental land areas can be divided into seven major biogeographic provinces, based primarily on the occurrence of distinctive groups of large mammals. These provinces are delineated by climate and vegetation, which, in turn, control the distribution of these animals. These land areas and the terrestrial organisms they support are isolated by large oceans. The greater the isolation of these provinces (that is, the more effective the barriers between provinces), the more distinctive the flora and fauna characteristic of each province. Similarities between faunas in North America and Eurasia result from similar climatic conditions in these land areas and the development of a "corridor," the Bering Strait land bridge, within the past 2 million years; this corridor allowed the unobstructed interchange of land animals between the two continents. Australia, by comparison, until the nineteenth and twentieth centuries maintained its geographic isolation and, therefore, developed distinctive marsupial mammalian fauna. The fossil record shows that a similar marsupial fauna existed in South America until the Pliocene epoch (about 5 million years ago), when the Central American isthmus was uplifted, allowing for the invasion of South America by more advanced placental North American mammalian forms. This endemic South American fauna is now largely extinct; survivors include the opossum and armadillo.

The great diversity of animal and plant life in the world at present is largely the result of the present-day configuration of the continents. The continents are spread all over the globe in rough north-south alignments that cross the entire spectrum of latitudinal climatic belts, from the Arctic Circle to the equator and from the equator to Antarctica. These climatic belts, as already noted, delineate many of the presently defined biogeographic provinces. The north-south arrangement of continents separated by large ocean basins allows cold-water polar currents to flow nearly unimpeded toward the equator, generating cooler climatic conditions on a global scale. In addition, large, deep oceanic basins separate the continents, forming formidable barriers to the dispersal not only of land animals and plants but also of shallow-water marine life.

The development of these biogeographic barriers is an important evolutionary mechanism, promoting increases in species diversity through the creation of new species by geographic isolation. The more effective these barriers are in preventing the interchange of species between provinces, the more genetically isolated faunas and floras in these provinces become, promoting greater differences between organisms living in these areas. As a result, similar environments in different geographic regions will support totally separate but often convergent faunas and floras. If these barriers did not exist, the same organisms would occupy the same environments in all geographic regions.

At present, nearly all the large continents are emergent, or above sea level. Their emergence resulted from a draw-down of worldwide sea level caused by relatively recent continental glaciation in the Northern Hemisphere (Pleistocene epoch, about 2 million to 12,000 years ago). The formation of this enormous mass of ice tied up much of the available surface water, which would otherwise have flowed into the ocean basins. Thus, sea level was depressed, and shallow seas were drained off most of the continents. Shallow epicontinental seas had previously covered portions of the continents and had had a climatically ameliorating effect, for bodies of water tend to hold heat longer than exposed land areas. Their disappearance contributed to the development of extreme climatic conditions. The development of emergent continents and intervening deep-water ocean basins also isolated shallow-marine faunas living in the narrow continental shelf regions ringing the continents. In this way, numerous marine biogeographic provinces have developed in these coastal areas, and the diversity of endemic shallow-marine faunas in these areas has increased.

Plate Tectonics and Species Diversity

The lateral motions of the continents, because of plate tectonic activity, have been a powerful force in reshaping the distribution of biogeographic boundaries in time and space. According to the plate tectonics theory, the earth's exterior is divided into an interlocking mosaic of moving plates composed of low-density crustal rocks and underlying, denser mantle rocks. These plates are created at spreading centers, where dense volcanic rocks well up from below and flow away laterally. Continental crust at these spreading centers is rifted apart, allowing for the formation of a new ocean basin dividing what was once a continuous continent. Such is the case in the Near East, with the formation of the Red Sea between the rifting African continent and the Arabian plate.

Plates can also move together, colliding with each other. At the point of impact, the crust shortens, thickens, and is deformed, leading to the formation of a new mountain belt that welds the colliding continents together into one larger continent. An example of this continent-to-continent collision is the collision between India and Eurasia, upthrusting the Himalaya. Plates also slide past each other along enormous linear cracks in the crust, termed strike-slip or transform faults. The San Andreas fault in California is such a fault, with California west of the fault moving north relative to the rest of North America. Such fault systems serve to displace continental crust horizontally along plate boundaries. All these plate tectonic activities cause the continents to change their shape and geographic position with time as well as result in the rise and fall of topographic features on the earth's surface.

Plate tectonic reconstructions for the major continents indicate major geographic realignments of these continents over the past 600 million years. These reconstructions show a pattern of cyclic periods of plate divergence (plates moving away from each other) followed by plate convergence (plates converging and eventually colliding). Plate divergence causes the formation of large ocean basins separating the continents. Convergent cycles culminate in the development of supercontinental masses composed of multiple, sutured continental plates. These "supercontinents" exist for several tens of millions of years, only to rift apart with the initiation of the next divergent cycle, which would fractionate the continents and disperse them over the globe. Detailed studies of the fossil organisms preserved in the sedimentary rocks on these continents indicate that the large-scale evolution of life over the past 600 million years parallels these plate tectonic events, with the changing geographic patterns having a profound effect on the evolution and extinction of organisms.

An example of this relationship is the Early Paleozoic interval of geologic time (Cambrian and Ordovician periods, about 600 to 450 million years ago). This was a time of plate divergence, with the continents aligned into rough east-west belts. The various continents that now make up the Northern Hemisphere were situated as isolated masses along the equator, in tropical latitudes. The present-day Southern Hemisphere continents were sutured into one large supercontinent, which scientists call Gondwanaland, located at higher, cool-temperate latitudes south of the equator. The continents were eroded down to flattened platforms and were flooded by shallow, epicontinental seas (perhaps more than 95 percent of their surface area was covered). Sea level was comparatively high because of the absence of continental collisions (which would elevate the continents through mountain building) and a lack of glaciation through much of the period. World climate was uniformly mild. The continents were separated by deep ocean basins and, as a result of the flooding of the continents by these epicontinental seas, shallow-marine environments were abundant.

This Cambro-Ordovician plate tectonics scenario coincides with a seemingly explosive evolution of shallow-marine animal life, especially the major groups of shell-secreting invertebrates (corals, brachiopods, snails, clams, nautiloids, trilobites, and echinoderms). These shallow marine faunas were highly provincial for much of this time interval because of the divergent nature of the major continental masses, with the intervening deep-water oceanic basins forming formidable barriers to the dispersal of these shallow-water marine animals. The highly endemic nature of these faunas and the mild climatic conditions promoted high levels of diversity. Terrestrial floras and faunas are unknown for this time interval, with the earliest land plants and animals being known from the succeeding Silurian period (about 425 million years ago).

Robert C. Frey

Further Reading

Briggs, Derek E. G., and Peter R. Crowther, eds. *Palaeobiology: A Synthesis II.* Oxford: Blackwell Scientific Publications, 2003. This multiauthored text deals with all aspects of paleontology in a series of short articles by experts in each field. Intended for university students but also useful to high school students and interested readers.

Crame, J. A., and Owen, A. W., eds. *Palaeobiogeography and Biodiversity Change: The Ordovician and Mesozoic-Cenozoic Radiations.* London: Geological Society, 2002. A compilation of articles discussing the ancient biogeography and biodiversity data. Focuses on two major radiation events, discussing the plate tectonic movements and related biodiversity patterns.

Fedonkin, Mikhail A., et al. *The Rise of Animals: Evolution and Diversification of the Kingdom Animalia.* Baltimore: Johns Hopkins University Press, 2007. Discusses the study of Ediacaran biota from initial identification through recent studies of the ecology and evolution of these species. Although the title suggests otherwise, this text focuses on Ediacarans, their diversity, and distribution.

Hallam, A., ed. *Atlas of Palaeobiogeography.* New York: Elsevier, 1973. One of the first volumes to be devoted exclusively to paleobiogeography and to incorporate plate tectonics information. Includes chapters by individual experts detailing the paleobiogeography of individual groups of fossil organisms for specific geologic periods. An important source of paleobiogeographic information. Most articles are well written and illustrated. Suitable for the informed layperson and college-level students with a basic knowledge of paleontology.

Grotzinger, John, and Tom Jordan. *Understanding Earth.* 6th ed. New York: W. H. Freeman, 2009. This comprehensive physical geology text covers the formation and development of the earth. Readable by high school students, as well as by general readers. Includes an index and a glossary of terms.

Prothero, Donald R. *Bringing Fossils to Life.* 2d ed. Boston: McGraw-Hill, 2004. This well-illustrated and entertaining text covers a broad range of paleontological topics, including biogeography. Includes a glossary, bibliography, and index.

Prothero, Donald R., and Robert H. Dott, Jr. *Evolution of the Earth.* 8th ed. New York: McGraw-Hill, 2009. This basic textbook on historical geology is aimed at students of geology. However, it is very readable by anyone with a background in science. Presents an up-to-date account of the earth's history from the viewpoint of plate tectonics. Includes a glossary.

Renema, Willem, ed. *Biogeography, Time and Place: Distributions, Barriers and Islands.* New York: Springer, 2010. The text is an interdisciplinary approach to the patterns of biodiversity on earth from the past and extant life.

Stanley, Steven N. *Earth and Life Through Time.* 2d ed. San Francisco: W. H. Freeman, 1989. A college-level historical geology textbook, this volume contains information on many aspects of plate tectonics, paleogeography, and evolution. Plate arrangements and the distribution of fossil organisms and environments are well illustrated for each geologic period.

See also: Archaebacteria; Biostratigraphy; Cambrian Diversification of Life; Catastrophism; Cenozoic Era; Colonization of the Land; Dinosaurs; Ediacaran Fossils; Eukaryotes; Evolution of Birds; Evolution of Flight; Evolution of Humans; Evolution of Life; Fossilization and Taphonomy; Fossil Plants; Fossil Record; Fossils of the Earliest Life-Forms; Fossils of the Earliest Vertebrates; Geoarchaeology; Life's Origins; Mammals; Mass Extinctions; Mesozoic Era; Microfossils; Neanderthals; Paleobotany; Paleozoic Era; Petrified Wood; Prokaryotes; Stromatolites; Systematics and Classification.

PALEOBOTANY

Only a small percentage of the plants that ever lived left a record of their existence. Others survive as fossils: mineralized wood, flowers in amber, leaf imprints in coal, or other indicators of life in an earlier era. Paleobotanists document this fossil record and use it to interpret the past evolution of plants.

PRINCIPAL TERMS

- **amber:** hardened resin from prehistoric trees; some amber contains intact, preserved organisms
- **fossilization:** the gradual process by which the remains of a once-living organism are naturally preserved, usually in sedimentary rock
- **microorganism:** a very small, usually one-celled, organism such as algae, bacteria, and protozoans
- **organism:** any particular living thing, whether a microorganism, plant, or animal
- **sedimentary rock:** a rock that forms from sediment deposited on the bottom of lakes and the ocean
- **taxonomy:** the branch of biology concerned with naming and classifying organisms
- **vascular plant:** a plant with vascular conducting tissues

IMPORTANCE OF PLANT FOSSILS

Paleontology (or paleobiology) is the science concerned with fossils, the physical evidence of prehistoric life—including plants, animals, and microorganisms—on the earth. Paleobotany focuses on plant fossils, including algae, fungi, and related organisms, as well as mosses, ferns, and seed plants. As most organisms decompose rapidly after death, their preservation in nature is a rare event. Most individuals are not represented in the fossil record, and even many species that must have existed have vanished without a trace.

As a branch of botany, paleobotany is of importance primarily because the record of fossil plants helps scientists understand the long process of plant evolution. Especially since the 1940's, fossil evidence has helped to explain the origin of major groups of lower plants, such as algae and fungi. Researchers now also have evidence for the origin of the earliest vascular plants and the formation of reproductive structures, such as cones of gymnosperms (evergreen trees and relatives) and flowers of angiosperms (flowering plants).

The location of fossils, including both their temporal (age) and their spatial (geographical) arrangement, is used to determine past climates. The climates of the world have changed continuously as continents have shifted over the earth's surface. For example, the location of coal deposits (which are the remains of giant tree ferns) in what is now Pennsylvania indicate the warmer climate that must have existed there in the past.

Although perhaps most of the contributions to paleobotany have been made by scientists with a background in geology, botany, and related sciences, amateurs have also made significant discoveries. Many valuable specimens of university and museum collections were made by people interested in paleobotany as a hobby.

HOW FOSSILS FORM

As already pointed out, the formation of a fossil is an exceptional event, one that requires a special combination of favorable environmental conditions. The most common scenario is a fossilization process in which the plant becomes covered by a soft sediment, which then hardens to form a sedimentary rock. This type of rock forms gradually over long periods of time as particles produced by erosion are compacted on the bottom of the body of water.

The large-scale process by which plant parts become impregnated with minerals produces what has traditionally been called "petrified wood." The modern term for this process is "permineralization." Soluble carbonates, silicates, and other compounds infiltrate plant cells and the spaces between them. Eventually, the mineral deposits may completely replace the naturally occurring organic matter, preserving the details of the plant's microscopic architecture. Well known are the petrified forests of western United States, many of which are protected within national parks, such as Petrified Forest National Park in Arizona.

Being trapped in a sedimentary rock does not automatically guarantee that the organism will be preserved. The environment must be an anaerobic one—that is, one in which oxygen is excluded—thus preventing the decay that would otherwise result.

The process may be interrupted by the action of waves or other erosive forces that re-expose the developing fossil before the process of fossilization is completed. Even after the process is completed, the well-preserved specimen may become distorted or altered in appearance because of the combined effects of time, pressure, and high temperatures that convert sedimentary rocks into metamorphic ones.

As one would expect, the harder cells and tissues of plants are more likely to be preserved as fossils than are softer ones. For example, the thick-walled cells of wood and bark (called xylem) are more often preserved than are those of the pith (center of a stem) or cortex (found in stems and roots beneath the bark or outer covering). Other cells that are often fossilized are pollen grains and spores, both of which have outer shells that are highly resistant to decay.

Limestone and dolomite are among the most common types of rocks that form in such a way that they trap plants and form fossils. Coal, a combustible sedimentary rock, is formed in much the same way as others but is distinctive because the sediment involved is of plant rather than mineral origin. Within this matrix of plant-derived material is often embedded a variety of plant parts.

Special Types of Fossils

Two special kinds of rock that may contain plant and animal fossils are diatomite and amber. Diatomite is a rock that forms from the silica cell walls of a group of unicellular algae known as diatoms. Since silica is the same material that sand and quartz are composed of, it is unusually permanent. Diatoms are found in both fresh and salt water in great numbers and diversity. When they die, their cell walls accumulate on the bottom of the water and become compacted over time into diatomite. The rock, itself formed by fossilization, may have fossil remains of various kinds of plants and animals preserved within it.

Amber, considered a semiprecious stone by gemologists and valued because of its beauty and distinctive appearance, is also of interest to paleobotanists. Amber is basically the fossilized resin produced by ancient cone-bearing evergreen trees. Sticky resins oozed from trees in response to injuries. Before such resins harden, various small animals, floral parts, pollen grains, fungal spores, and other plant parts may become trapped and be preserved intact. Deposits of amber valued for their use as jewelry and as fossils are recovered from two world areas: the Baltic region of northern Europe and the Dominican Republic in the Caribbean Sea.

Paleobotanists are sometimes challenged by puzzling specimens. Outright fakes are sometimes presented by pranksters, but more common are various mineral structures that bear a superficial resemblance to a plant. Such specimens are called pseudofossils. Mineral deposits called dendrites found in rock crevices bear a resemblance to fern leaves. A coprolite (fossilized feces) from the upper Cretaceous of Alabama was initially mistaken for the cone of a conifer (cone-bearing evergreen tree); these specimens may be referred to as pseudo-plant fossils, since they are true fossils of animals. During the formation of flint, bands are sometimes formed that suggest fossil mollusks or coral. Suspicious specimens require careful analysis by

Impressions of "seed ferns" Alethopteris *(center, with smaller leaves) and* Neuropteris *(surrounding* Alephopteris*) with individual leaves of* Cyclopteris *(base leaves of* Neuropteris*) at lower right. All of these plants were common in the Pennsylvanian epoch, Carboniferous period, some 300 million years ago. Though their foliage resembled that of true ferns, prehistoric "seed ferns" were among the first seed-bearing plants (gymnosperms) and the precursors of most plants living today. Specimen courtesy of Pennsylvania State Museum, Klose Coal Fossil Collection.* (Theodore Clutter/Photo Researchers, Inc.)

a specialist. In general, plants and animals, and therefore their remains, possess details and a characteristic regularity of form absent in pseudofossils.

Naming and Classifying Fossils

In order to prevent confusion, fossils, like living species, need to be named and classified in a consistent, systematic fashion recognized by paleobiologists throughout the world. The branch of biology devoted to the naming and classification of organisms is called taxonomy. Fortunately, the same system is applied to both living and fossilized plants.

According to binomial nomenclature (naming), each species is given a scientific name consisting of two parts: the genus name followed by the trivial, or species, name. The former is capitalized, whereas the latter is written in lowercase; both are italicized. Often a name follows that belongs to the person who assigned that name. As example, the scientific name of an extinct redwood tree is *Sequoia dakotensis* Brown, while that of a stemless palm is *Nipa burtinii* Brongniart. The ginkgo tree, *Ginkgo biloba L.*, is considered a "living fossil." Known from the fossil record, it persists as a commonly planted shade tree. The initial following its scientific name is that of Carolus Linnaeus, the Swedish botanist who established this binomial system of nomenclature in 1753.

Species, living or dead, are classified using a hierarchical system (also by Linnaeus), which reflects degrees of similarity or dissimilarity to other species. All three trees already mentioned, because they are (or were) vascular plants, are assigned to the division Tracheophyta. Within that division, the redwood and ginkgo, because they produce uncovered seeds, are placed into the class Gymnospermopsida (naked seeded plants), whereas the palm, a flowering plant, is assigned to the class Angiospermophytina.

The plant fossil record is often used to establish natural relationships among various extant plant species and other taxa (taxonomic categories) at higher levels. This is especially true of the vascular plants. In fact, the division Tracheophyta was established by A. J. Eames in 1936 to show the natural relationship between seed plants and ferns. The basis for this new category was the discovery, earlier in the twentieth century, of Devonian fossils of a group of primitive vascular plants known as "psilophytes." They were recognized as ancestral to both ferns and seed plants. Previously, ferns and seed plants had been assigned to a separate division of the plant kingdom.

As the plant fossil record becomes more complete, further revision of the classification system becomes necessary to allow the system to more nearly reflect the true or natural relationships among the various categories. This is, at least, the common goal of both paleobotanists and those who study modern plants.

Thomas E. Hemmerly

Further Reading

Bengtson, Stefan, ed. *Early Life on Earth.* New York: Columbia University Press, 1994. A collection of serious articles that relate early biological evolution to conditions on the primitive earth. Perhaps of most interest is the article written by Stephen Jay Gould, "Introduction: The Coherence of History."

Cleal, C. J., and B. A. Thomas. "Botanical Nomenclature and Plant Fossils." *Taxon* 59 (2010): 261-268. This article addresses the changes made to the Vienna Code for naming fossilized plant taxa. Discusses the concept of morphotaxa.

Cvancara, Alan M. *Sleuthing Fossils: The Art of Investigating Past Life.* New York: John Wiley & Sons, 1990. This interesting introduction to fossils is written in the first person for the amateur. A good place for beginners to start.

Golding, Roland. *Fossils in the Field.* New York: John Wiley & Sons, 1991. In lab manual format, this book is invaluable for giving specific directions for field and laboratory methods. Written for the more advanced amateur. Numerous interesting and instructive drawings.

Kenrick, Paul, and Paul Davies. *Fossil Plants.* Washington, D.C. Smithsonian Books, 2004. This text provides foundational information of fossilization, then focuses on the ecology and biology of ancient plants. Discusses plant-animal interactions and the evolution of flowering plants. Written for the layperson, yet still contains valuable information.

Roberts, D. C., and W. G. Hodson. *A Field Guide to Geology: Eastern North America.* Boston: Houghton Mifflin, 2001. Like other Peterson Field Guides, this is a small volume that is useful both in the field and as a reference.

Stewart, W. N., and G. W. Rothwell. *Paleobotany and the Evolution of Plants.* 2d ed. Cambridge: Cambridge University Press, 2010. Probably the most comprehensive and authoritative textbook on the subject. Of special interest to the beginner are Chapters 1 through 3, which give the scientific foundation for paleobotany.

Taylor, Thomas N., Edith L. Taylor, and Michael Krings. *Paleobotany.* 2d ed. Boston: Academic Press, 2008. Discusses fossil plants, fungi, and algae. Provides foundational material on geological time, fossilization, and the process of studying ancient organisms. Discusses analyses using phylogenetics and fossil discoveries occurring since the first edition in 1993. A full glossary, bibliography, and index.

Walker, Cyril, and David J. Ward. *Fossils: Smithsonian Handbook.* New York: DK Publishing, 2002. The nearly three hundred pages of full-color illustrations are the most striking features of this very useful book. The introduction contains an excellent orientation to fossil study.

Willis, K. J., and J. C. McElwain. *The Evolution of Plants.* New York: Oxford University Press, 2002. This text is written for the undergraduate student. Follows the history of plants from the earliest forms to extant species. Puts plant evolution in a context of environmental and geological changes throughout earth's history.

Wilson, Stewart, and Gar Rothwell. *Paleobotany and the Evolution of Plants.* 2d ed. New York: Cambridge University Press, 2009. Explains in a clear way the origin and evolution of plants. Well illustrated and written at the level of undergraduate students.

See also: Archaebacteria; Biogeography; Cambrian Diversification of Life; Catastrophism; Colonization of the Land; Dinosaurs; Ediacaran Fossils; Eukaryotes; Evolution of Birds; Evolution of Flight; Evolution of Humans; Evolution of Life; Fossilization and Taphonomy; Fossil Plants; Fossil Record; Fossils of the Earliest Life-Forms; Fossils of the Earliest Vertebrates; Geoarchaeology; Geologic Time Scale; Ice Ages; Life's Origins; Mammals; Mass Extinctions; Microfossils; Neanderthals; Paleobiogeography; Petrified Wood; Prokaryotes; Stromatolites; Systematics and Classification.

PALEOCLIMATOLOGY

Climate is the average of weather elements over long periods of time and over large areas. The study of ancient climates, termed paleoclimatology, utilizes sedimentologic, paleontologic, and geochemical data to reconstruct ancient temperature, wind patterns, precipitation, and evaporation.

PRINCIPAL TERMS

- **atmospheric circulation:** the movement of air as a result of regional pressure differentials
- **climate:** the accumulative effects of weather; its basic elements include radiation, temperature, atmospheric moisture and precipitation, evaporation, and wind
- **insolation:** incoming solar radiation; differences in global insolation at various places on the earth's surface create weather and climate patterns
- **weather:** the condition of the atmosphere at a given moment and at a given place

Climatic Processes

The basic elements of climate include radiation, temperature, atmospheric moisture and precipitation, evaporation, and wind. The features of ancient climates that have been most studied are temperature, wind patterns, amount of precipitation, and evaporation. In order to understand ancient climates, it is necessary to understand the fundamentals of modern climatic processes.

Solar radiation is the electromagnetic radiation emitted by the sun's surface; this radiation is primarily within the infrared, visible, and ultraviolet ranges. The incoming solar radiation is termed global insolation. The amount of energy received by the earth at any one time is essentially constant, although paleoclimatic evidence indicates that it has not always been so. The amount of global insolation received depends on the output energy of the sun, the distance between the earth and the sun, the angle at which the sun's rays strike the earth's surface, the duration of daylight, and atmospheric composition. Therefore, global insolation depends primarily on the latitude and the seasons: Regions within the higher latitudes receive the least amount of insolation, primarily because of the angle at which the sun's rays strike the earth, and they have the greatest seasonal variation. The result is an energy deficit poleward of 40 degrees north and 40 degrees south latitude, respectively; the equator has the least variation, so that between these degrees of latitude there is an energy surplus. Therefore, a continuous horizontal exchange of energy between these regions occurs, which is the cause of atmospheric circulation and weather patterns.

The unequal heating of the earth's surface creates pressure differences. Since gases move from areas of high pressure to areas of low pressure, this differential heating and pressure variation causes the wind to blow. If the earth were not rotating, its general circulation pattern would be one of ascending air at the equator and descending air at the poles, with continuous horizontal flow between. The Coriolis effect, however, changes this ideal pattern, and free-moving objects are deflected from a straight-line path in response to the earth's rotation: In the Northern Hemisphere, the deflection is to the right, and in the Southern Hemisphere, it is to the left. Ocean surface circulation patterns are therefore typically clockwise in the Northern Hemisphere and counterclockwise in the Southern, yet atmospheric patterns are in fact more complex because air parcels are ascending and descending as well as moving horizontally. General atmospheric circulation patterns from the equator poleward include the doldrums, the trade winds, the horse latitudes, the westerlies, and the easterlies. In the geologic record, general atmospheric circulation patterns are difficult to determine as a result of the scarcity of data points necessary for accurate calculation of such large-scale phenomena. Wind patterns, however, may be calculated for some areas on the basis of sedimentary structures, and general atmospheric circulation patterns may be inferred from models based on modern atmospheric circulation and the distribution of the earth's land and water areas through time.

In order for precipitation to occur, there must be atmospheric instability. An air parcel will rise until it reaches an altitude where the surrounding air is of equal temperature; this process may be enhanced where intense solar heating creates lower pressures, where the air mass is heated by a warm surface below it, where sloping terrain such as mountains forces air to ascend, or where cool air acts as a barrier over which warmer, lighter air rises. As the unstable air parcel moves vertically, it cools as a result of the expansion of gases (the adiabatic process). Condensation will occur when the dew point temperature is reached, which varies according to the initial relative humidity of the air parcel. Provided that there is sufficient moisture and that nuclei are present around which the moisture may accumulate, precipitation may occur. The major types of precipitation include rain, freezing rain, sleet, snow, and hail. The type of precipitation that occurred in ancient times is difficult to determine, although occasional examples of raindrop prints are found as casts in sediments, and ancient glacial deposits indicate the accumulation of frozen precipitation. In some situations, the amount of precipitation may be inferred, typically through the utilization of fossil plants. In other cases, the amount of precipitation versus evaporation may be indicated, primarily through the utilization of sedimentary mineral types such as evaporites.

Climate Changes

The result of the interaction of these climatic processes is the formation of regional climates. On the modern earth, a wet equatorial belt lies within about 20 degrees of the equator. The adiabatic cooling of very moist and warm air results in heavy precipitation within these regions, yielding a wet tropical climate. These air parcels then begin descending, they adiabatically heat up through compression, and the now-dry air encounters the earth's surface. The result is the formation of tropical deserts, centered at approximately 25 to 30 degrees north and south latitude. At the middle latitudes (approximately 35 to 65 degrees), the interaction of air masses between the polar and tropical regions and the variation in global insolation throughout the year result in large variations in temperature and seasonality. Polar regions are characterized by cold and dry conditions, because the sun's rays are typically at low angles and cold air parcels in those regions are unable to accumulate large amounts of moisture.

One phenomenon that seems to affect climate, at least for relatively short cycles of earth history, is sunspot activity. Although the data are incomplete, the increase and decrease in these solar disturbances seem to suggest that they create variations in global insolation and subsequent warming and cooling trends. The cooling of the earth's climate during the early-to-middle portion of the second millennium has been linked to variations in sunspot activity. Another explanation of causes of earth-climate cyclicity was put forth in the twentieth century by Milutin Milankovitch. Milankovitch suggested that the earth's orbit around the sun may vary from more elliptical (with more pronounced differences in global insolation during the summer versus the winter) to more circular (with less seasonality). Such a cycle would take about 100,000 years to complete. Other Milankovitch cycles proposed include changes in the tilt of the earth's axis in relationship to the sun (a 40,000-year cycle) and the wobble of the earth upon its axis (a cycle of 21,000 years). Studies on ocean-floor sediments for

Alluvial plain. These stones were deposited by flowing water during a more humid climatic phase. Photographed near Tiblist, Saharan Desert, Algeria. (David Parker/Photo Researchers, Inc.)

the Pleistocene (within the past 2 million years of earth history) and in lake basin sediments of the Triassic and Jurassic (approximately 200 to 180 million years before the present) of the East Coast of the United States indicate that such cycles may indeed exist.

Another possible cause of short-term changes in climate may be asteroid impacts. Scientists have theorized that such impacts would increase the particulate levels in the atmosphere to such an extent that a phenomenon similar to "nuclear winter" would occur. Overall global insolation would decrease dramatically, and temperatures would plummet. Suggestions have been made that this phenomenon could account for the periodic extinctions that take place on the earth.

Another possible cause of changes in the earth's atmosphere involves changes in carbon dioxide levels: a heating trend caused by increasing levels (the greenhouse effect) or a cooling trend caused by decreasing levels. Such changes may be caused by changes in the overall metabolism of organisms or by decreases and increases in volcanic activity. Increases in volcanic activity could also cause cooling trends because of a decrease in global insolation as a result of volcanic particles suspended in the atmosphere.

Ocean current patterns tremendously influence weather patterns and regional climates, and it has been suggested that the atmosphere and oceans can be viewed as a single, interacting unit. Changes in oceanic circulation would certainly modify climates, although the geologic evidence of a definite link between the two is equivocal. Stronger evidence for causes of variation may be related to plate tectonics. The movement of continental plates over polar areas may certainly account for the buildup of ice during certain periods of earth history, such as occurred on the southern continents during the Permian period (about 290 to 240 million years before the present). In part, such shifting of plates may also account for the last ice ages in Europe and North America, although the presence of several periods of glaciation and interglacial episodes indicates that each episode of glaciation during the Pleistocene cannot be explained this way.

Phillip A. Murry

Further Reading

Boggs, Sam, Jr. *Principles of Sedimentology and Stratigraphy.* 5th ed. Upper Saddle River, N.J.: Prentice Hall, 2011. As many of the paleoclimatological methodologies rely on sedimentological features, any detailed study of paleoclimates will require some knowledge of sedimentology. This text is one of the best books on the subject in terms of readability and accuracy. The chapters on sedimentary textures, sedimentary structures, and sedimentary environments are especially useful for understanding the sedimentological results of paleoclimatological processes. Designed for junior- or senior-level geology students.

Bolius, David. *Paleoclimate Reconstructions Based on Ice Cores: Results from the Andes and the Alps.* Saarbrücken: SVH-Verlag, 2010. Presents a study of climate change using samples taken from ice cores in the Andes and Alps. Includes sampling methodology, chemical analysis, and documents the history of anthropogenic air pollution. Best suited for graduate students and professional geologists and paleoclimatologists.

Cox, C. B., and P. D. Moore. *Biogeography: An Ecological and Evolutionary Approach.* 8th ed. Hoboken, N.J.: John Wiley & Sons, 2010. This book contains a wealth of information on many aspects of biogeography, but certain sections may be too detailed for the general college-level reader. Discussions of climate and other atmospheric limits on biogeography and coverage of ecology and plate tectonics serve as a useful overview for analysis of paleoclimate conditions and influences through time.

Dodd, Robert J., and Robert J. Stanton, Jr. *Paleoecology: Concepts and Applications.* 2d ed. New York: Wiley-Interscience, 1990. A comprehensive treatment of many aspects of paleoecology. Provides detailed coverage of various plant and animal groups as concerns their utilization in analysis of ancient ecologies and climates as well as detailed analysis of the use of isotope studies in paleoecology and paleoclimatology. Also included are discussions of ecosystems and communities in paleoecology and paleobiogeography. Written at the introductory college level.

Harmon, Russell, and Andrew Parker. *Frontiers in Geochemistry: Contribution of Geochemistry to the Study of the Earth.* West Sussex: Wiley-Blackwell, 2011. Covers geochemistry in two parts: the contributions to earth studies and use of the analytical tools in the future. Chapter 4 discusses the geochemistry of oceanic crusts. Chapter 12 discusses archaeological and anthropological applications of isotopic geochemistry. This textbook is best suited for advanced undergraduates or graduate students.

Pierrehumbert, Raymond T. *Principles of Planetary Climate.* New York: Cambridge University Press, 2011. This text is written for the undergraduate or graduate student, as it guides the reader toward original research and primary source material. Covers the fundamental principles of climate, climates around the solar system, and climates of past and present. Provides examples of current research practices on planetary climate, beyond the study of paleoclimatology.

Prothero, Donald R., and Robert H. Dott. *Evolution of the Earth.* 8th ed. New York: McGraw-Hill, 2009. This basic textbook on historical geology is aimed at students of geology. However, it is very readable by anyone with a background in science. Presents an up-to-date account of the earth's history from the viewpoint of plate tectonics. Includes a glossary.

Ruddiman, William F. *Earth's Climate: Past and Future.* 2d ed. New York: W. H. Freeman, 2008. Covers what has been learned of climate change from a study of earth history and provides predictions for the future. Written for undergraduate students but accessible to the interested reader.

Sharpton, Virgil L., and Peter D. Ward, eds. *Global Catastrophes in Earth History.* Boulder, Colo.: Geological Society of America, 1990. A compilation of fifty-eight papers on various aspects of catastrophic extinction and its causes. A large number of the essays deal with climatic changes associated with the boundary event. Although written for specialists, it is accessible to readers with some scientific background. Bibliography and index.

See also: Biostratigraphy; Cambrian Diversification of Life; Cenozoic Era; Colonization of the Land; Cretaceous-Paleogene Boundary; Deserts and Dunes; Fossil Plants; Fossil Record; Fossils of the Earliest Life-Forms; Geologic Time Scale; Geomicrobiology; Gondwanaland and Laurasia; Ice Ages; Land Bridges; Mesozoic Era; Microfossils; Paleobiogeography; Paleoecology; Paleogeography and Paleocurrents; Paleoseismology; Paleosols; Paleozoic Era; Proterozoic Eon; Stratigraphic Correlation; Transgression and Regression; Unconformities; Uniformitarianism; Weathering.

PALEOECOLOGY

Paleoecology is a branch of ecology and paleontology focused on studying the interrelationships between extinct organisms and the paleoenvironments in which they lived. The study of paleoecology developed when geologists and paleontologists began using fossils and geological data to examine environmental conditions in ancient ecosystems. Paleoecology has helped scientists understand the nature of environmental changes occurring on Earth and has enabled scientists to better predict how conditions might continue to change if global warming continues.

PRINCIPAL TERMS

- **actuopaleontology:** the linguistic act of applying organismic features found on modern living animals to extinct organisms
- **brachiopod:** any of a phylum of shelled marine animals that have formed valved shells superficially similar to such mollusk species as clams and oysters
- **foraminiferan:** a single-celled aquatic animal with a shell composed of compacted minerals or grains of sand from surrounding sediment
- **paleoclimatology:** the study of the climate of an extinct ecosystem as determined through geological evidence and the distribution of fossils
- **paleoecology:** the study of the interrelationships between extinct animals and the paleoenvironments in which they lived
- **paleoenvironment:** the extinct environment of a certain area during a chosen point in the history of the region
- **parsimony:** the theory that, given unexplained phenomena, one should assume the simplest possible explanation for the observed data
- **substantive uniformitarianism:** the theory that extinct animals would have been similar in both structure and behavior to living organisms, because the processes, rates of change, and the terrestrial materials have been relatively stable throughout the history of the earth
- **temperate zone:** those intermediate latitude zones of the earth marked by seasonal changes in climate

Development of Paleoecology

Ecology is a branch of biological and physical science that studies the relationships among individual organisms, populations, and the environments in which they live. Paleontologists developed a branch of research known as paleoecology, which focuses on the interrelationships between extinct organisms and the paleoenvironments in which they lived.

Early paleontologists often made inferences about ancient environments from the types of fossil organisms found in various sediment samples. Paleontologists uncovering marine fossils from sediment in a modern desert, for instance, would likely infer that the sediment in the area was once part of a submerged ocean environment. Similarly, an abundance of fossils representing herbivorous organisms might indicate that there was once abundant vegetation in a certain fossil site.

American geologist Kirk Bryan (1888-1950) is often credited as one of the scientists responsible for the emergence of paleoecology as a distinct field of paleontological research. Bryan's research in the 1920's and 1930's focused on how to understand weather patterns in ancient environments by examining aspects of soil chemistry and the occurrence of fossilized pollen grains. Bryan went on to train a number of prominent students who later made major contributions to paleoecology and other branches of geological research.

Paleoecology utilizes data derived from the study of both paleozoology (the study of fossilized animal species) and paleobotany (the study of fossilized plant species) in an attempt to reconstruct assemblages of organisms that occupied ancient environments. In addition, paleoecologists utilize chemical analyses of soil and geological data regarding the movement of continents. Taken as a whole, these data allow paleoecologists to create models of extinct ecosystems that help them to understand how evolutionary drives and pressures create communities within specific environments. The study of these ancient ecosystems can provide insight for ecologists studying modern ecosystems and can help to create an understanding of how changes in environmental conditions alter the evolutionary path of an ecosystem.

Theoretical Concepts of Paleoecology

One of the most important theoretical underpinnings of paleoecology is the concept of substantive uniformitarianism, which is the theory that the physical and chemical properties of the earth have remained unchanged throughout the history of the planet. The concept of uniformitarianism was developed in the nineteenth century and popularized by scientists such as prominent Scottish geologist Charles Lyell and English scientist William Whewell, who believed that the properties and processes of the earth were universal constants that could be applied equally to both ancient and modern environments.

Since the nineteenth century, scientists have come to understand that the physical and chemical properties of the earth are not constant, and that they have changed considerably during the 3.4 billion years of life on the planet. Modern ecologists apply an altered version of uniformitarianism, which assumes that the properties of the earth have varied, but that they have remained relatively similar during most of life's history on the planet.

Applied to paleoecology, uniformitarianism holds that the materials that make up the earth and the forces that drive chemical reactions and geological processes are similar today to those active in ancient ecosystems. This assumption allows paleoecologists to tentatively apply modern theories about ecological relationships to ancient ecosystems. For example, paleontologists might assume that ancient microorganisms were able to gain energy by utilizing sulfur in the same way as modern sulfur-metabolizing bacteria, because the chemical reactions that drive the metabolism of sulfur functioned the same in the past as they do in the modern environment.

Similarly, the theory of actuopaleontology (also called analogy) holds that the features and behaviors of modern organisms are similar to the features and behaviors of ancient organisms. In general, actuopaleontology allows paleontologists to assume that extinct organisms behaved and functioned in similar ways to modern organisms. This applies not only to the structure and biological function of individual organisms but also to the formation of multispecies communities.

Therefore, when paleontologists observe a beak-like structure on a fossilized organism, they will often assume that the beak was used like the beaks of modern animals living in similar environments. Similarly, paleontologists often assume that organisms in ancient ecosystems filled the same roles as similar organisms in modern ecosystems of the same type. Paleontologists might therefore expect that organisms in an ancient marsh ecosystem would have displayed behaviors and environmental roles that are similar to organisms in modern marshes.

Research occasionally invalidates hypotheses developed using the principle of actuopaleontology. An extinct organism might, for instance, evolve a certain feature that is similar to features on modern organisms, but it might have used this feature for a different purpose. Similarly, ancient ecosystems that are superficially similar to modern ecosystems might be found to contain vastly different types of organisms with different sets of relationships not seen in modern ecosystems.

Whether or not to assume hypotheses based on the theories of actuopaleontology or uniformitarianism can be determined with the help of parsimony, defined as the idea that, given an unexplained phenomenon, it is best to assume the simplest possible explanation of the available data. Parsimony is a core concept that can be applied to all scientific thought because it discourages scientists from positing unlikely and overly complicated explanations.

Most scientists, for example, believe that life first originated on Earth about 3.4 billion years ago; however, some scientists have postulated that life came to Earth from elsewhere in the galaxy, carried on meteors from a distant planet. Although either

Brachiopod fossils in Devonian Period shale, from central New York. (Stephen P. Parker/Photo Researchers, Inc.)

explanation could be correct, the former is the simplest explanation in that it does not propose the extra assumption that life evolved elsewhere in the universe.

Applied to paleoecology, parsimony dictates that paleontologists utilize theories that encourage the simplest and most direct explanation. The principles of actuopaleontology and uniformitarianism are parsimonious, in that they assume that conditions in the past are similar to conditions in the present. While this may not always hold true, it is the simplest method for explaining similarities between extinct organisms and ecosystems and their modern counterparts.

Paleoclimatology

One of the most important steps to understanding extinct ecosystems is to determine the climate of the ecosystem in question. Geological data provide a variety of evidence regarding the climate of paleoenvironments. For instance, large deposits of coal are the result of decomposing woody plants, whose existence indicates the presence of a lush, humid forest system. By contrast, some ancient soils display characteristic signs of extended ice or frost coverage; these ancient soils include glacial tills, which are pulverized or striated rock segments that bear the scars of contact with moving glaciers. Rocks that show striations left by standing water as well as cracks resulting from heating and drying were exposed to an ancient monsoon system, in which hot, dry seasons alternated with periods of intense rainfall and flooding.

Fossils also can be used as indicators of climate conditions. Ecologists have noted that diversity of species is greatest near the equator and declines at higher latitudes; paleontologists have discovered the same trend in the diversity of fossil assemblages from different areas. This phenomenon can be partially explained by the warm and relatively stable climate near the equator, leading to the proliferation of plants and microorganisms that form the base of the food chain. Farther from the equator, the environment becomes more variable.

In the broad temperate zone, organisms are still plentiful, but diversity is lower overall than in equatorial zones. Here organisms are more generalized, filling multiple niches because food is less plentiful than in zones closer to the equator. The number of species continues to drop in the higher latitudes, where arctic weather conditions dominate and overall diversity is low.

By measuring the diversity of fossil species in a certain area, paleoecologists can begin to understand the relative climate of the region during the time that the fossils were deposited. By examining a diversity of fossils over time, paleoecologists also can begin to observe climate change in ancient environments.

An example of this approach is the research of paleontologist Warren O. Addicott, whose 1969 study of shallow-water mollusk fossils from the west coast of Mexico and the United States indicates that certain species of mollusk were gradually moving southward during the early portion of the Tertiary period (65 to 2.5 million years ago). Addicott believed that this movement indicates a pattern of global cooling that led to changing distributions of species requiring warm, tropical waters for feeding and breeding. Furthermore, Addicott's research also indicated that a pronounced period of global warming existed during the later part of the Tertiary, again demonstrated by the movement of mollusk species farther northward toward the upper edges of the United States.

Marine fossils are the most useful fossils for determining the specifics of ancient climate conditions. Brachiopods, which are members of a phylum of shelled, marine animals, first appear in the fossil record in the Cambrian period (542 to 488 million years ago) but are still represented by more than two hundred living species in modern marine environments. Brachiopods appear to be highly climate specific. Paleontologists have found that characteristic assemblages of brachiopod species appear at each climate zone. Because extinct brachiopod species appear to have changed little in millions of years of evolution, paleoecologists can examine the distribution of modern brachiopods and, by applying the theories of analogy, parsimony, and uniformitarianism, can assume that ancient brachiopods tended to exhibit the same climatic preferences as their modern counterparts. Paleontologists can therefore use the distribution of these characteristic brachiopod faunas to track changes in climate distribution in periods as distant as the late Cambrian.

Organisms that form shells are especially useful for studying marine climates because the process of shell formation is closely tied to the chemical composition of the marine environment. Oxygen isotopes occur naturally in marine environments, and they occur in multiple varieties. In warm environments, certain types of oxygen isotopes tend to disappear from the environment as they engage in chemical reactions and dissolve in the surrounding water. Therefore, in warmer waters, only the more stable types of oxygen isotopes remain. By examining the fossilized shells of ancient organisms, paleoecologists can measure the proportion of stable to unstable oxygen isotopes and can attempt to determine whether the organism lived in a warmer or cooler environment.

To determine climate in ancient terrestrial environments, paleoecologists often attempt to determine the closest living relatives of species represented in the fossil record of the paleoenvironment. For instance, paleontologists have used the distribution of fossil rodent species as an indicator of alternating periods of warm, productive climate and periods of cooler climate with less abundant food sources during the Pleistocene period (1.8 million to 10,000 years ago). In other cases, paleontologists have theorized that the presence of fossilized reptiles indicates a mean temperature in keeping with the temperature favored by similar-sized reptiles in the modern environment.

Paleontologists also have used the distribution of fossilized pollen grains to investigate climate change. Because plants produce pollen on an annual cycle, pollen grains may appear in layers in the sediment, with each layer representing a subsequent year in the history of the environment. Paleontologists can therefore retrieve a core sample of sediment from the environment of interest and then examine the distribution of pollen at each layer. In times of warmer weather and more rainfall, more pollen will appear in the sample, while cooler or drier temperatures will result in fewer pollen grains. By examining the distribution of pollen grains through longer periods, paleoecologists can begin to get an image of long-term trends in climate change.

Paleoecology of the Pleistocene and Holocene Epochs

Utilizing data from both geological sources and the distribution of fossil species, paleontologists have created a detailed model of the paleoecology of the Pleistocene period, when the earth experienced its most recent period of global cooling. The Pleistocene is characterized by extended glacial periods, when thick sheets of ice covered large portions of the temperate zone. These periods of intense cooling alternated with interglacial periods, when the climate grew warmer and the glaciers melted across much of the landscape.

Evidence for climate change in the Pleistocene has been derived partially from the distribution of fossil species that appear to be characteristic of certain climate zones. Single-celled organisms called foraminifera are some of the most important fossil organisms of the period. These tiny shelled microbes are associated with the growth of algae, and their distribution is tied closely to temperature and other climatic variables. The tiny shells of foraminiferans are abundant in many types of ocean sediment, and paleoecologists have used the proportion of oxygen isotopes, remaining from the pulverized shells of foraminiferans, to track shifting climatic patterns during the Pleistocene.

Paleontological knowledge of the Pleistocene ice age also has benefited from detailed investigations of fossilized pollen. The distribution of pollen grains shows distinct periods of increased productivity followed by periods of greatly reduced pollen production. Paleontologists have taken this as additional evidence that the Pleistocene experienced waves of glacial and interglacial conditions.

The Holocene epoch began approximately ten thousand years ago and continues into the twenty-first century. Paleoecologists believe that the Holocene represents an intermediate period between the last ice age of the Pleistocene and the next ice age. The climate has continued to fluctuate, and the earth experienced a period of cooling, beginning in the seventeenth century and ending at the beginning of the twentieth century. Sometimes called the little ice age, this period had lowered temperatures across most of the temperate zone, but it did not affect the globe as a whole.

Much evidence supports the assertion that human activity is accelerating the process of global warming across the planet, though scientists are not universally in agreement about the causes or scope of emerging climatic trends. Paleoecological research has helped scientists to understand how the environment will change if current trends in global warming continue. The study of ancient ecosystems indicates that further global warming will force many species to migrate to stay within comfortable climatic zones. Past periods of both global warming and cooling also have resulted in waves of extinction, during which thousands of species disappear from the earth as they are unable to adapt to changing environmental conditions.

Micah L. Issitt

Further Reading

Allmon, Warren, and David J. Bottjer. *Evolutionary Paleoecology.* New York: Columbia University Press, 2000. Thorough investigation of modern paleoecological principles with a focus on evolutionary theory and the evolutionary relationships existing in paleoenvironments. Focuses on speciation and other large-scale evolutionary changes that can be observed from aspects of the fossil record.

Benton, Michael J., and D. A. T. Harper. *Introduction to Paleobiology and the Fossil Record.* New York: Wiley-Blackwell, 2009. Introductory text covering a variety of topics in paleontological research. Chapter 4 covers theories in paleoecology and the reconstruction of ecological relationships from fossilized evidence.

Briggs, Derek E. G., and Peter R. Crowther, eds. *Paleobiology II.* New York: Wiley-Blackwell, 2001. Introduction to the study of paleobiology aimed at students of related disciplines. Contains a discussion of trace fossils, body fossils, and the fossilization process, and of methods used to evaluate biological activity from fossil remains.

Foote, Michael Joseph, et al. *Principles of Paleontology.* New York: Macmillan, 2007. Revised edition of a classic text on paleontology. Provides a basic introduction to field and laboratory methods used in the study of fossils and prehistoric life. Chapter 9 discusses the basic theories of paleoecology research.

Martin, Anthony J. *Introduction to the Study of Dinosaurs.* New York: Wiley-Blackwell, 2006. Introductory text covering a variety of concepts in paleontological research and in dinosaur behavior and classification. Includes a discussion of paleoecology in relation to dinosaur behavior.

Miller, William Charles. *Trace Fossils: Concepts, Problems, Prospects.* St. Louis, Mo.: Elsevier, 2007. Updated introduction to the history and modern methods used in trace-fossil research. Chapter 23 contains a discussion of using trace fossils to investigate paleoecology, including environmental conditions and animal behavior.

Shipman, Pat. *Life History of a Fossil: An Introduction to Taphonomy and Paleoecology.* Cambridge, Mass.: Harvard University Press, 1993. Thorough introduction to taphonomy and the use of fossils in reconstructing the paleoecology of extinct environments. Contains information on fossilized plants and animals and discussions of fossil assemblages and their use in reconstructing environmental variables.

Weishampel, David B., and Halszka Osmolska. *The Dinosauria.* Berkeley: University of California Press, 2004. Introduction to dinosaur species and the study of prehistoric environments. Chapter 26 contains a discussion of paleoecology in relation to dinosaur behavior, including detailed examinations of the ecology of herbivory and carnivory in dinosaurs.

See also: Biogeography; Biostratigraphy; Dinosaurs; Fish-Amphibian Transitions; Fossil Fish; Fossil Record; Fossil Reptiles and Amphibians; Geomicrobiology; Life's Origins; Mass Extinctions; Molecular Paleontology; Paleoclimatology; Paleogeography and Paleocurrents; Pangaea; Permo-Triassic Boundary; Snowball Earth; Trace Fossils; Uniformitarianism.

PALEOGEOGRAPHY AND PALEOCURRENTS

Paleogeography is both the geography of a past geologic time and the science of its determination. Paleocurrents are currents existing within a paleogeographic environment or region.

PRINCIPAL TERMS

- **azimuth:** degrees of arc measured clockwise from the north
- **correlation:** the determination of identity in age, fossil content, or physical continuity of rocks observed in different areas
- **endemic species:** species confined to a restricted area in a restricted environment
- **epicontinental sea:** a sea overlapping continental crust, as opposed to a sea underlain by oceanic crust
- **isotopes:** forms of an element having the same atomic number but differing atomic weights
- **planetary wind system:** a global atmospheric circulation pattern, as in the belt of prevailing westerlies
- **remanent magnetism:** the magnetic field imposed upon a rock at the time of its formation in accord with the global magnetic field

ANCIENT GEOGRAPHY

Paleogeography is the geography or the study of the geography of the past. Paleogeographic interpretations or maps may refer to any geographic phenomenon or group of phenomena, so there are many special types of paleogeographic maps. Paleobiogeographic maps show distribution of fossil species, ecological assemblages, floras, or faunas. Paleoclimatological maps plot ancient climatic realms, and paleophysiographic maps depict ancient landforms. Paleotectonic maps outline regions of deformation, including crustal compression, crustal extension, persistent uplift or depression, and stable areas. Paleobathymetric maps describe the depths of ancient bodies of water. Maps defining past patterns of circulation in the sea, a lake, or stream are called paleocurrent maps. Any geographic property, in fact, may be the subject of a specialized paleogeographic map. Paleogeographic maps are, however, interpretive maps; that is, they are not maps of existing conditions or objects but describe conditions inferred from the rock record. They cannot be verified by direct observation in nature. Paleogeographic maps and interpretations may illustrate the geography of the entire globe, of whole continents, or merely of specific regions or features.

Global maps generally show continents and ocean basins as they are inferred to have existed during a major interval of time, such as a geologic period. The continents shown on these maps may be parts of present-day continents or may be composed of the merged parts of several modern continents. Furthermore, the ancient continents are not likely to be located in their present position in reference to latitude and longitude. On many of these maps, ancient continental landmasses and archipelagos are plotted over the whole earth, with the outline of present continents and parts of continents superimposed. A series of such maps thus records the manner in which segments of continental crust have continually divided, split, merged, and moved about throughout geologic time. The maps for the Precambrian, which are least reliable, record the appearance of many "protocontinents," coalescing blocks of continental crust, lighter rocks of the composition of granite that form an incomplete outermost layer of the rocky earth.

These protocontinents eventually gathered to form perhaps five ancestral continents. During the Early Paleozoic, these gathered in two large continents: Laurasia in the Northern Hemisphere and Gondwanaland in the Southern Hemisphere. During the Late Paleozoic, Laurasia and Gondwanaland coalesced in a single continental mass, Pangaea. A large embayment, the Tethys Sea, indented the eastern side of Gondwanaland and is the remote ancestor of the Mediterranean Sea. During the Mesozoic and Cenozoic, this single large continent progressively separated into the modern suite of continents: North America, South America, Eurasia, Africa, Australia, and Antarctica. These continents gradually moved to their present positions relative to the poles and the equator. Global paleogeographic maps of this sort first appeared in the middle twentieth century, as the theory of plate tectonics and paleomagnetic concepts were discovered and accepted.

Past Positions of Continents and Ocean Basins

Knowledge of the global position, size, and interconnections of the continents and ocean basins is required to place past geologic events in a geographic context. It also is needed for full understanding of present global geologic relationships. The past position of crustal blocks, continents or parts of continents, and ocean basins may be inferred in part from the remanent magnetism of rocks of known age. Remanent magnetism is a magnetic field imprinted in a rock and conforming to the world's magnetic field existing at the time the rock was formed. This remanent magnetic field remains fixed, even if the rock is subsequently rotated or moved. It is, therefore, a permanent record of the original latitude and orientation of the rock. Thus, the azimuth direction of the remanent magnetism in a rock indicates the position of the pole with respect to the locality of a rock when it was deposited or intruded. The polarity of the remanent field reveals the polarity, normal or reversed, at the time the rock formed. The inclination of the vertical component of the remanent field gives the latitude at which the rock formed. Also, regions in which the remanent magnetic fields in rocks of common age are oriented uniformly may be considered parts of the same continental blocks or fragments. The remanent field is susceptible to destruction only by heating the rock to near-melting temperatures. The longitudinal position of a rock cannot be determined by paleomagnetic analysis, so other phenomena must be called upon to place ancient continental crustal blocks in their proper positions on the globe.

Former continuity between separated continental blocks may be inferred from their shape, as in the apparent match between the eastern coastline of the Americas and the western coasts of Europe and Africa. This pattern was noted and a relationship inferred as early as the first half of the nineteenth century. Such "jigsaw puzzle" techniques are a significant aid in continental placement. It should be noted, however, that the land area of a continent is not what is being matched; much of the continent is submerged, so the location of the continental margin and slope is essential.

Apparent continuity of ancient geologic features on the margin of separated continental masses suggested former connections before the advent of plate tectonics, however, and still reinforces paleogeographic reconstructions. For example, the apparent continuation of the Appalachian orogenic belt and the band of carboniferous coal deposits from Texas to Newfoundland, the British Isles, and Western Europe supports the idea of former continuity between the landmasses involved. Common occurrence of glacial deposits, coal beds, similar volcanic extrusives, and disjunct occurrences of fossil organisms incapable of crossing open oceans in Australia, India, Antarctica, South America, and Africa all support reconstruction of the Gondwanaland continent. Before the advent of plate tectonics, now-submerged continent-sized landmasses in the South Atlantic and Indian Oceans or long isthmian land bridges were postulated as crossing those oceans to explain the present distribution of the Gondwanaland fauna and flora.

Paleogeographic Maps

Paleogeographic maps of continental masses may show the distribution of ancient geographic features on modern-day continents, or they may refer to continents as they existed at some former time. Maps of the distribution of the land, sea, and ancient mountain ranges on present-day continents have been produced since the mid-nineteenth century. These maps typically show the epicontinental seas, seas consisting of marine waters overflowing continental crust, as opposed to seas in deep ocean basins. Such maps also show the continental shelf or inundated continental margin, the relatively steep continental slopes at the very edge of the continental crust, and dry-land areas. In addition, major belts of deformation, broad areas of uplift, and broad areas of subsidence, or basins, are frequently shown on these maps. More detailed, and generally more specific, maps will show such things as deltas, ancient volcanic tracts, coastal lagoons, estuaries, reefs, dune fields, mountains, and glaciers.

Regional maps may show the paleogeography of areas defined by geologic interest, or they may be bounded arbitrarily. Examples of the first would include a map of the Late Paleozoic Ancestral Rocky Mountains. A set of paleogeographic maps for a state would exemplify the second. Regional maps of smaller areas generally are more detailed than continental or global maps and are more directly derived from local geology and paleontology. Latitude and pole position may, however, be shown and are determined by the same techniques employed in producing global or continental scaled maps. At the regional level, paleogeography may be concerned with the distribution of land and sea, mountain ranges, volcanic regions, lakes, flora and fauna, and the climate.

Paleogeographic maps are constructed by correlating and establishing the extent of rocks of the age in question. The conditions under which contemporaneous rocks were deposited are then determined. In this way, areas of uniform or nearly uniform environment may be outlined and mapped. Areas lacking rocks of the age under consideration were regions of non-deposition or intrusion or were areas subsequently eroded to remove rocks of that age. Areas undergoing deformation and areas suffering igneous intrusion or extrusive volcanism also are defined and mapped.

Ralph L. Langenheim, Jr.

FURTHER READING

Berner, Robert A. *The Phanerozoic Carbon Cycle: CO_2 and O_2*. New York: Oxford University Press, 2004. Discusses climate and atmosphere of the Paleozoic, Mesozoid and Cenozoic Eras. Also covers aspects of weathering and erosion on the carbon cycle. This text is suited to undergraduates. Contains references for each chapter and indexing.

Dott, Robert H., Jr., and Donald R. Prothero. *Evolution of the Earth.* 8th ed. New York: McGraw-Hill, 2009. A geologic history, centering on North America. Includes a set of global paleogeographic maps, paleogeographic maps of North America for Precambrian through Cenozoic times, and regional paleogeographic maps of areas of particular interest. Maps are supported by summaries of the data upon which they are based. Paleocurrents are shown on many of these maps. Includes a comprehensive index and a useful glossary of technical terms. Designed for use as a beginning college geology text.

Grotzinger, John, and Tom Jordan. *Understanding Earth.* 6th ed. New York: W. H. Freeman, 2009. This comprehensive physical geology text covers the formation and development of the earth. Contains extensive sections on plate tectonics and its effect on paleogeography. Readable by high school students, as well as by general readers. Includes an index and a glossary of terms.

Koenigshof, P. *Devonian Change: Case Studies in Palaegeogrpahy and Palaeoecology.* London: Geological Society of London, 2009. Presents thirteen articles of studies conducted on the Devonian stratum. Case studies come from North and South America, Australia, China, and Europe. Covers historical geography, ecology, and climate.

McKinney, Frank. *The Northern Adriatic Ecosystem: Deep Time in a Shallow Sea.* New York: Columbia University Press, 2007. This text covers the paleogeography of the Adriatic Sea. Discusses the succession of the ecosystem as the sea's geography changed. Oceanography topics such as circulation and sedimentation are discussed. Topics are well described and logically ordered, to make this book accessible to undergraduates.

Shell Oil Company, Exploration Department. *Stratigraphic Atlas, North and Central America.* Houston: Author, 1975. Maps of the distribution and lithologic character of rocks belonging to the major subdivisions of each geologic system. Also paleogeologic maps, isopachous (thickness) maps, and lithofacies (regional changes in rock character) maps. Directly interpretable as paleogeographic maps.

Smith, A. G., D. G. Smith, and B. M. Funnell. *Atlas of Mesozoic and Cenozoic Coastlines.* New York: Cambridge University Press, 2004. Global paleogeographic maps showing past positions of Mesozoic and Cenozoic epicontinental seas.

Stanley, Steven M. *Earth System History.* 3rd ed. New York: Freeman, 2008. An introductory undergraduate level text integrating coverage of life and the environment.

Wanless, H. R., and Cynthia R. Wright. *Paleoenvironmental Maps of Pennsylvanian Rocks, Illinois Basin and Northern Midcontinent Region.* Boulder, Colo.: Geological Society of America, 1978. Detailed regional paleogeographic maps for individual subunits of the coal-bearing rocks of middle America.

See also: Biostratigraphy; Cenozoic Era; Cretaceous-Paleogene Boundary; Fossil Plants; Fossil Record; Geologic Time Scale; Glacial Deposits; Ice Ages; Land Bridges; Mesozoic Era; Microfossils; Paleobiogeography; Paleoclimatology; Paleoecology; Paleoseismology; Paleosols; Paleozoic Era; Proterozoic Eon; Stratigraphic Correlation; Transgression and Regression; Unconformities; Uniformitarianism.

PALEOSEISMOLOGY

Paleoseismology is the study of the evidence of past earthquakes. By studying the physical features of previous earthquakes, scientists gain knowledge of the forces operating within the earth, knowledge that may be used to develop a means of predicting future large earthquakes.

PRINCIPAL TERMS

- **earthquake:** a shaking or trembling of the earth caused by a break or rupture in the rocks of the earth's outer crust
- **elastic rebound:** the process of the buildup and release of geologic stress; the release of the strain in the rocks results in earthquakes
- **fault:** a fracture of the earth's crust along which rocks move
- **fault offset:** a feature such as a road, creek bed, or tree line across a fault that separates and becomes misaligned by the fault movement
- **fault scarps:** a steep cliff or slope created by movement along a fault
- **plate tectonics:** the theory that the earth's hard outer crust (lithosphere) is composed of sections, or plates, that are in constant motion
- **recurrence period:** the range of time between successive earthquakes
- **seismic gap:** a fault region known to have had previous earthquakes but not within the area's most recent recurrence period
- **tsunami:** a large ocean wave caused by an earthquake in the ocean floor

Ancient Earthquakes

Seismology is the science that deals with earthquakes and movement within the earth. Seismologists and geologists use historic and prehistoric information about earthquakes in order to make better calculations of the probability of earthquakes occurring in a particular area. However, historical evidence from earthquakes—that is, information acquired from observation and instrumental measurements—is short and incomplete. Paleoseismology is a relatively new branch of earthquake science that uses archaeological techniques to find physical evidence of large ancient earthquakes and thus to expand the seismic history of the earth.

In the 1960's, the theory of plate tectonics was introduced. According to this theory, the earth's outer crust is composed of sections, or plates, that are in constant motion. When these plates collide or spread apart, the resulting forces produce deformations, or tectonics, of the earth's surface that are identified with earthquakes. When plates collide and one plate moves beneath the other, the resulting fault is called a "subduction zone." Subduction zones are located under water and cause changes in land at the shoreline. Earthquakes at the subduction zones often cause tsunamis that are devastating to coastal areas. When the plates rub against one another, the resulting fault is known as a "plate interaction." The San Andreas fault in California is a well-known example of a plate-interaction fault.

The idea of elastic rebound is used to explain how earthquakes originate. Under normal circumstances, the rocks at a fault boundary move a few centimeters per year. If the rock cannot move, the force of the normal movement becomes stored as "strain energy." When the force becomes too great, the fault slips to release energy. As the plate boundaries shift or rupture, earthquakes occur. By studying the historic record, scientists have found that the plate boundaries rupture in segments until the entire section is broken, and the cycle then begins again. Scientists believe that future earthquakes are more likely to occur in places where the segments have not recently ruptured. These places are called "seismic gaps."

With the discovery of plate tectonics, scientists believed that long-term and short-term prediction of earthquakes would be possible. Long-term prediction is based on the probability that an earthquake will occur. Paleoseismic evidence of past earthquake recurrence periods allows scientists to determine when a large earthquake is "overdue" for a particular area.

Short-term prediction involves specifying the future place, magnitude, and time of the earthquake. Short-term prediction relies on data collected from monitoring equipment on known faults as well as precursory signals such as an increase in fault movement, change in deep-well water levels, changes in surface elevations, an increase in minor earthquake activity, the occurrence of a moderate foreshock, or an increase in the release of radioactive radon gas. Since other environmental factors can cause some of these signals, there are no definitive predictors of an imminent earthquake.

The advantages of knowing when, where, and how strong an earthquake will be are numerous. The cost of an earthquake in terms of loss of human life and property can be tremendous. In addition to the damage caused by the actual ground shaking from the earthquake, there may be loss of life and property caused by structural collapse and secondary effects such as landslides, fires, and tsunamis. The September 19, 1985, earthquake that struck Mexico City killed more than eight thousand people and injured thirty thousand people, destroyed or severely damaged more than one thousand multistoried buildings, and caused an estimated $5 billion in damage.

Study of Earthquakes

Scientists use several methods to study earthquakes. One of the methods is based on recorded observations. The earliest known scientifically collected observational data about earthquakes came from a survey taken after a strong earthquake, known as the All Saints Day earthquake, struck Lisbon, Portugal, on November 1, 1755. Local officials sent a list of questions to the affected areas asking about the duration of the shock, the times and intensities of aftershocks, the number of people killed, the structures destroyed, and the extent of damage from related fires. Firsthand accounts of death and structural damage, as well as obvious changes in the physical features of the land, allow scientists to judge the severity of an earthquake. This information is useful in identifying seismic gaps. Probability statistics indicate that an area is more likely to have a large earthquake if a large earthquake has occurred there before.

Another method for studying earthquakes is through instrumentation. Tiltmeters and strainmeters are used to measure changes in the crustal movement in earthquake areas. Hidden faults are located using a device that relies on how sound waves are reflected. Satellites, aerial photographs, and surveying equipment also provide clues to the location of underground faults. Seismic activity is monitored worldwide with seismographs, which record the occurrence, direction, severity, and duration of earthquakes. Modern seismographs are extremely sensitive and can detect ground motions of very small earthquakes. Scientists estimate that approximately eighty thousand small tremors and fifteen to twenty severe earthquakes occur around the world annually.

In the late 1970's, Kerry Sieh, a geologist at the California Institute of Technology, conducted the pioneering work in paleoseismology at Pallet Creek on the San Andreas fault. Scientists could now gather additional information about earthquakes to supplement and expand the existing observational and instrumentation knowledge. With the prehistoric evidence of earthquakes, scientists can better calculate the probability of another large earthquake.

Paleoseismic techniques may involve digging a trench across a fault and examining the rock strata, or layers, for evidence of previous ruptures. Scientists also look for evidence of flooded forests or sheets of sand along coastlines, which would indicate tsunami activity. The evidence can be direct, such as fault offsets, or indirect, such as landslides or sand deposits, which occur with strong ground shaking. By measuring the distance separating the two rock segments in a fault offset, scientists can determine how much the fault has slipped. The samples can be radiocarbon dated in order to determine when the earthquakes occurred. When the time gap between the earthquakes is determined, the recurrence period and average interval can be established. The recurrence period is a range of time between earthquakes; the interval is the average time between earthquakes. For example, if earthquakes happened in a particular region in 1700, 1775, 1825, and 1925, the recurrence range is fifty to one hundred years, with an average interval of seventy-five years.

Paleoseismic Data

Seismologists need information about past earthquakes in order to study and prepare for future earthquakes. Paleoseismology has been very useful in North America because of the short historical record. Additionally, the techniques have helped establish the accumulation of stress in faults all over the world.

Paleoseismic data enable scientists to determine higher levels of earthquake hazard in a geographic area. Engineers and public officials use the information from scientists to revise building codes and redraw maps of earthquake-shaking hazard zones. The U.S. Uniform Building Code, which includes nationwide standards for designing structures, defines six levels of earthquake-shaking hazard. By convincing people to reinforce existing structures and adopt earthquake-resistant construction standards, loss of life and property can be greatly reduced.

In California, a 1971 earthquake in the San Fernando Valley killed sixty-five people and caused an estimated $500 million in damages. California has had earthquake-resistant building criteria since 1933, which have been updated several times. In contrast, a 1972 earthquake of similar magnitude in Managua, Nicaragua, killed more than five thousand people and caused economic losses equal to the country's gross national product for that year. Managua had few buildings designed under earthquake-resistant standards.

Paleoseismology helps answer questions about faults, folds, secondary effects of prehistorical earthquakes, length of an earthquake cycle, the cycle's regularity or irregularity, and similarity of characteristics of an area's earthquakes. Scientists believe that earthquake processes of the past will probably be the same as those in the near future. The longer the record of earthquake occurrences in a particular area, the greater the predictive value of the knowledge. The ability to predict the likelihood and possible effects of future earthquakes enables people to take precautions to protect themselves against injury, death, and economic losses.

Study of Shorelines

Paleoseismology initially studied active faults to help establish the rates of movement. The technique was then applied to interpreting earthquake crustal deformations or shaking effects. Although paleoseismology is typically used to examine surface faults, the technique is also used to examine subduction-zone earthquakes indirectly. Since subduction zones are underwater faults, scientists examine changes in the land at the shoreline for evidence of past activity. The sudden lowering of a shoreline or a series of terraced beaches may be associated with major earthquakes.

In Alaska, scientists examined terraced beach lines near an island in Prince William Sound. This area was the site of the March 27, 1964, "Good Friday" earthquake, the second-strongest earthquake recorded. The earthquake killed 114 people and caused $350 million in property damage. The coastlines of nearby landmasses bowed by as much as 2.5 meters, flooding them with seawater and killing many square kilometers of forest. Scientists discovered that large earthquakes have occurred in the area an average of once every 850 years during the past 5,000 years. Because of this investigation, scientists believe that it is unlikely that a major earthquake will occur in this area in the near future.

Until the mid-1980's, scientists believed that the area off the coast of Washington, Oregon, and Northern California had a low probability of a massive earthquake, since there was no historical record of such an earthquake. Paleoseismic studies, however, show evidence of a sudden lowering of the shoreline associated with major earthquakes. In addition, scientists found sheets of sand deposited by tsunamis and sand-filled cracks caused by strong earthquake shaking. Researchers have thus concluded that three massive earthquakes have struck this area in the past three thousand years; the most recent great earthquake struck about three hundred years ago. Based on this knowledge, measures have been taken to reinforce existing structures and toughen earthquake design standards. Without the paleoseismic data, a powerful earthquake would have taken the area by complete surprise.

Scientists have been gathering much information about faults at plate boundaries, since an estimated 95 percent of all earthquakes occur at these boundaries. The probability of occurrence for some earthquakes is more difficult to predict, because they do not occur near plate boundaries. Often, these inner-plate earthquakes are devastating, because they occur so infrequently that there may be no historical record of an earthquake in that area.

On September 30, 1993, ten thousand people in Killari, India, died when an earthquake devastated the area. Prior to the earthquake, the area was included in the lowest level of the earthquake-hazard map of India. The seismologists were wrong at Killari because the hazard map relied mostly on the historical record of earthquakes. In India, this record is only about 150 years old. Investigation following the earthquake suggested that the area had been seismically quiet for 65 million years or more. Scientists believe that the earthquake may have been triggered by an artificial, or human-constructed, reservoir that changed the stress on a fault. Understanding how such construction affects the stress on the earth's crust that triggers earthquakes allows more considered and intelligent decisions about building nuclear power plants, dams, reservoirs, and cities.

In the United States, the area east of the Rocky Mountains has faults that are buried deep and accumulate stress slowly. Earthquake risk estimates for the eastern United States are not as advanced as those for the West, where the faults break the surface and earthquakes occur more often. Eastern faults may be quiet for hundreds, thousands, or millions of years.

One of the most destructive series of earthquakes in American history, however, occurred in 1811 and 1812 in New Madrid, Missouri. At least one earthquake per day took place in the area for more than a year, with three extremely powerful ones; these quakes were felt over approximately 2.6 million square kilometers and caused tremendous physical changes to the land. Destruction was severe, but casualties and economic losses were low because there were few settlements in the affected area. The earthquakes triggered landslides, submerged islands, opened fissures in the ground, flooded lowlands, and caused areas of land to rise or sink.

Large earthquakes have also occurred in Charleston, South Carolina, in 1886; Newfoundland, Canada, in 1929; and New York State in 1988. Numerous small earthquakes have occurred over the past two hundred years, with at least twenty damaging earthquakes in the central Mississippi Valley alone. Eastern earthquakes are puzzling because they do not generate visible surface faults. Because of this lack of physical evidence and the infrequent occurrences of eastern earthquakes, determining the probability and location of another large earthquake in the region is very difficult.

Virginia L. Salmon

Further Reading

Adams, John. "Paleoseismology: A Search for Ancient Earthquakes in Puget Sound." *Science* 258 (December 4, 1992): 1592-1593. Introduces a five-article series on paleoseismology fieldwork in the Pacific Northwest. Explains what scientists look for and how they interpret physical evidence of prehistoric earthquakes at coastlines. Briefly discusses the history of paleoseismology. Easy to follow.

Best, David M., and David B. Hacker. *Earth's Natural Hazards.* New York: Kendall/Hunt, 2010. Covers the geology necessary to understand natural hazards and provides extensive coverage of earthquakes, mass movement, and tsunamis. Intended for undergraduates with no previous geology background.

Bolt, Bruce A. *Earthquakes and Geological Discovery.* New York: Scientific American Library, 1993. Explains why earthquakes occur and how they are measured. Also discusses some of the most famous earthquakes and what can be done to minimize their destructive effects. Suitable for anyone interested in earthquakes.

Bull, William B. *Tectonic Geomorphology of Mountains: A New Approach to Paleoseismology.* Malden, Mass.: Wiley-Blackwell, 2007. This text provides technical coverage of topics in paleoseismology. Chapters contain information on uplifting, mountain formation, tectonic activity, and fault scarps. Recommended for graduate students and practicing paleoseismologists.

Chester, Roy. *Furnace of Creation, Cradle of Destruction.* New York: AMACOM Books, 2008. This text discusses the turbulent processes of the planet. It covers earthquakes, volcanoes, and tsunamis in reference to plate tectonics, natural disasters, and prediction and mitigation of these events. Multiple chapters explore seafloor spreading. The author also discusses hydrothermal activity. This text takes on an immense range of content, but still explains concepts clearly and with detail.

Guidoboni, Emanuela, and John E. Ebel. *Earthquakes and Tsunamis in the Past: A Guide to Techniques in Historical Seismology.* New York: Cambridge University Press, 2009. Discusses paleoseismology as a field of study, providing background information for topics to follow. The bulk of this text focuses on interpretation and analysis of historical data. Final chapters provide conclusions pulled from actual data. This text also contains a glossary, a bibliography and references, and an index.

Lay, Thorne, and Terry C. Wallace. *Modern Global Seismology.* New York: Academic Press, 1995. A complete text discussing seismology and its various applications. Includes a chapter dedicated to the earthquake cycle. Contains only brief information specifically on paleoseismology, but the overview of seismology is excellent.

Main, Ian. "Long Odds on Prediction." *Nature* 385 (January 2, 1997): 19-20. Brief article discussing the possibility of reliable earthquake prediction. Discusses why earthquakes are difficult to predict and suggests that efforts should concentrate on hazard mitigation.

McCalpin, James P., ed. *Paleoseismology*. 2d ed. New York: Academic Press, 2009. Includes an introductory chapter on paleoseismology and discusses field techniques and interpretation of data. Includes a bibliography and index. Suitable for more advanced readers.

Moseley, Charles. "Slow Progress in Earthquake Prediction." *Editorial Research Reports* 2 (July 15, 1988): 354-363. Well-written and easily understood overview of earthquake prediction and the role of paleoseismology. Includes examples of major earthquakes and a clear explanation of the need for and problems associated with prediction. Includes a bibliography.

Sintubin, Manuel, et al., eds. *Ancient Earthquakes*. Boulder, Colo.: Geological Society of America, 2010. This special paper discusses earthquakes from a pre-instrument time and their study using techniques in archaeology and paleontology. The text was written for use by seismologists, archaeologists, earth scientists, and historians.

Worsnop, Richard L. "Earthquake Research." *CQ Researcher* 4 (December 16, 1994): 1105-1128. A balanced discussion of the state of earthquake research. Discusses the issues of earthquake prediction and hazard-reduction preparedness; also gives background on faults and current risk areas and includes an explanation of how earthquake strength is measured. Includes a bibliography and suggestions for further reading. Suitable for high school readers.

Yeats, Robert S., Kerry Sieh, and Clarence R. Allen. *The Geology of Earthquakes*. New York: Oxford University Press, 1997. Detailed information about the geological explanations for and effects of earthquakes. Includes a discussion of tectonics and faults as well as an assessment of seismic hazards. Suitable for college-level readers. Includes a bibliography and informative illustrations.

See also: Biostratigraphy; Cenozoic Era; Continental Growth; Continents and Subcontinents; Cretaceous-Paleogene Boundary; Displaced Terranes; Faults; Geoclines; Geologic Time Scale; Gondwanaland and Laurasia; Ice Ages; Land Bridges; Mesozoic Era; Paleobiogeography; Paleoclimatology; Paleogeography and Paleocurrents; Paleosols; Paleozoic Era; Proterozoic Eon; Stratigraphic Correlation; Supercontinent Cycles; Transgression and Regression; Unconformities; Uniformitarianism.

PALEOSOLS

Paleosols are ancient soils that have been buried. Although natural acids (largely dissolved carbon dioxide) are almost entirely supplied by the atmosphere during modern soil formation, that does not seem to have been the case with all paleosols, particularly the most ancient paleosols and those that are associated with ore deposits. Paleosols that were produced by ancient atmospheric gases may record the environmental conditions of the ancient earth.

PRINCIPAL TERMS

- **caliche:** a type of soil or paleosol that contains a high proportion of calcium carbonate, calcium sulfate, or both
- **clay mineral:** a type of mineral that is the most common product of soil formation; it is composed of silicon, oxygen, hydrogen, usually aluminum, and possibly other elements
- **partial pressure:** the proportion of a gas mixture (for example, the atmosphere) that a particular type of molecule (for example, carbon dioxide) represents
- **soil:** all material that has been substantially altered at the earth's surface by interaction with the atmosphere, living things, or both, and that has not been laterally displaced subsequent to that alteration
- **soil horizon:** a distinct layer in a soil

Ancient Soils

Soil is one of the best-known, yet least precisely defined, geologic entities. Many definitions would not exclude beds of graded sediment; other definitions would not exclude sediment that has experienced only mild physical and chemical alteration. Additionally, many soil science textbooks restrict "soil" to that lying within the depth of plant roots, but such a definition is inappropriate for Precambrian and Early Paleozoic soils, which formed before the evolution of rooted plants.

The rock record of the earth covers the past 4.28 billion years (out of a total earth history of 4.6 billion years). Throughout the known rock record, soil formation (weathering) has involved more energy than other geologic processes, such as mountain building. Mountain building is driven by the internal energy of the earth (for example, by natural radioactivity), whereas soil formation is driven by solar energy and by chemical reactions between acidic atmospheric gases and exposed rock. Solar energy fuels photosynthetic plants, which concentrate the most abundant atmospheric acid (carbon dioxide); subsequent decay of these plants releases concentrated carbon dioxide into soil waters, which thereby become acidic and dissolve minerals. The downward solar flux of energy, which enhances weathering, currently is about 7,500 times greater than the upward flux of internal energy. The internal energy of the earth has progressively decreased with time as radioactive elements decay; solar radiation has progressively increased. In addition, the input of solar energy has exceeded the output of internal energy by at least a factor of 1,000 since the beginning of the rock record.

The collective volume of soils produced throughout earth history may have been comparable to the present volume of the continental crust, but only an insignificant volume of these soils has become preserved by burial within the crust. Paleosols now constitute a smaller proportion of continental crust than most other well-known rock types. The proportion of paleosols appears to be particularly small in the oldest rocks. In these rocks, the preserved portion is so small and the conditions for preservation of paleosols appear to have been so peculiar that it is dangerous to make sweeping interpretations of ancient earth environments based on paleosols.

Paleosols and Sedimentary Rock

Virtually all old soils have been eroded to become sediment rather than buried to become paleosols. Intense weathering causes a high proportion of rock to dissolve, and the remnant soil becomes rich in insoluble elements such as aluminum. Erosion of this soil produces aluminum-rich sediment, which usually accumulates in the shallow ocean as a clay-rich rock (mudrock), such as shale. Sedimentary rocks therefore may be environmental indicators, and the vastly greater volume of sedimentary rock has resulted in most interpretations of ancient environments coming from study of sedimentary rock rather than from study of paleosols. Environmental study of paleosols is less constrained than that of sedimentary rock, however, because the unweathered parent of

Caliche or nitrate salt is a hardened deposit of calcium carbonate. This calcium carbonate cements together other materials, including gravel, sand, clay, and silt. It is found in aridisol and mollisol soil orders and generally occurs worldwide in arid or semi-arid regions such as the Atacama Desert. (Stuart Fox)

the weathered material (soil) is observable beneath a paleosol, whereas the parent for a weathered sediment generally cannot be deduced with certainty; the parent rock for any given sediment may lie thousands of miles from where the sediment accumulates—for example, the Andean Mountain parent for much of the Amazon deltaic sediment.

Paleosols are known from all major divisions of earth history, from the Archean eon to the present epoch. The proportion of paleosols to other rock types roughly increases with time, consistent with the theory of continental growth through earth history; an increase in the area of exposed continents would lead to an increase in the volume of soil. Although the area of exposed continents probably was smallest in the Archean eon, well-preserved Archean paleosols are known from Canada and from South Africa, where Archean continental crust never has been so deeply buried that regional metamorphism could destroy evidence of the weathering processes that produced the paleosols. The area of exposed continents apparently increased dramatically from the Archean to the Proterozoic, so Proterozoic paleosols are correspondingly more abundant.

The role that plant life may have played in the weathering of these Proterozoic soils is unclear. Plants currently concentrate so much carbon in the upper portion of a typical soil that oxidation of this carbon provides more carbon dioxide to the underlying soil than does diffusion of carbon dioxide directly from the atmosphere. Much of this carbon dioxide comes from roots, but roots did not evolve until after the Proterozoic. Plant life in Precambrian (Archean plus Proterozoic) soils was limited to bacteria. Mats of photosynthesizing cyanobacteria (also called blue-green algae) lay at the surface, and other bacteria occurred deeper in the soil. Modern soil contains abundant bacteria below the surface, but the soil surface generally is occupied by complex photosynthesizing plants—for example, trees—instead of by cyanobacteria. A partial image of Precambrian soils may be obtained by studying cyanobacterial mats that grow in environments too harsh for more complex plants—for example, on salt flats and around geysers. These environments are not only harsh but also highly variable. The chemical composition of water on a modern salt flat may vary from hypersaline to nonsaline following a thunderstorm. The water temperature around a geyser may vary by several tens of degrees within a few minutes. The ability of cyanobacteria to withstand such variations may indicate that the environmental conditions of Precambrian soils either were consistently harsh or were more variable than those of more recent soils.

Geographic Variation

Ancient paleosols generally are too scarce for study of the geographic variation of soil types. Geographic variations in rainfall and temperature are the two prime variables that control the distribution of modern soils because the atmosphere mixes so rapidly that any local variation in the partial pressure of oxygen or carbon dioxide rapidly becomes globally homogenized. Molecular oxygen and carbon dioxide also would have been evenly distributed in ancient atmospheres, but their proportions of the atmosphere may have varied throughout earth history. The oldest paleosols therefore are studied to examine variation in the earth's atmosphere through earth history, whereas the youngest paleosols are studied to learn about geographic variation in climate. Paleosols and other indicators record dramatic variation in climates on the earth during the past 2 million years, related to the growth and melting of enormous continental glaciers.

Tropical climates produce distinct soils and thus distinct paleosols. Both wet and dry tropical climates characteristically produce yellow-to-red soils because of oxidation and retention of iron in the soil. Wet tropical conditions can produce soils that are so rich in aluminum that these soils may be mined profitably. In such soils, the more soluble elements have been leached away by groundwater draining lush vegetation. Decay of the vegetation produces acids that attack even quartz, leaving only aluminum-rich minerals. Aluminum-rich paleosols are among the oldest (Archean) paleosols on earth, so wet tropical conditions appear to have existed long ago, despite the fact that solar radiation should have been much smaller during the Archean eon.

Warm, dry climates may produce little soil of any kind. For example, more characteristic of the Sahara than the "sand seas" that are commonly illustrated in documentaries are vast stretches of bare rock. Caliche is a characteristic soil and paleosol produced in such a climate. Caliche generally forms by precipitation of calcium carbonate from upwardly moving groundwater as the groundwater approaches the earth's surface. The upward decrease in pressure may allow carbon dioxide to be released, just as carbon dioxide is released upon opening a bottle of soda pop. Release of carbon dioxide favors precipitation of calcium carbonate; this precipitation may be aided by evaporation of the groundwater—which points to a dry climate for the formation of an ancient caliche. A caliche paleosol generally may be interpreted to record an ancient dry climate, even if the precipitation of calcium carbonate was not related to evaporation within soil but was simply the result of release of carbon dioxide. The original excess of carbon dioxide could have been provided by the escape of gases to the earth's surface during metamorphism of carbon-bearing sedimentary rocks at great depth. The abundant infiltration of rainwater in a wet climate would dissolve calcium carbonate from soil, whatever its origin, so the preservation of calcium carbonate in a caliche paleosol generally records a dry climate, even if the precipitation of calcium carbonate were induced by deep crustal processes independent of climate.

Elemental Composition and Mineralogy

Interpretation of the elemental composition and mineralogy of a paleosol generally is controversial, especially for paleosols older than 544 million years (Precambrian paleosols), because paleosols potentially have experienced substantial modification (diagenesis) after burial. One of the most consistent chemical peculiarities of Precambrian paleosols is that they contain extreme ratios of potassium to sodium, unlike modern clayey soils. No known weathering process could fractionate sodium from potassium so severely. Precambrian paleosols commonly contain more than ten times as much potassium as sodium, whereas these two elements generally behave similarly under modern weathering conditions. This potassium in Precambrian paleosols mostly occurs in fine-grained, aluminum-rich mica. The potassium either is a record of pervasive diagenetic alteration of Precambrian paleosols or indicates that, unlike modern soils, they did not form as a result of atmospheric acid-forming gases. The majority of Precambrian paleosols could represent alterations on the ancient land resulting from exhalation of acid-containing mud from deep in the earth. Although paleosols that are older than 544 million years generally have the greatest ratios of potassium to sodium, paleosols that are 245 to 544 million years old (Paleozoic paleosols) also are more potassium-rich than are modern soils. In these Paleozoic paleosols, the potassium-bearing mineral typically is a clay mineral called illite. In illite-bearing paleosols, some investigators attribute the high potassium content to peculiar weathering conditions, whereas others attribute it to precipitation of potassium from through-flowing groundwater long after burial of the soil.

Michael M. Kimberley

Further Reading

Bloom, Arthur. *Geomorphology: A Systematic Analysis of Late Cenozoic Landforms*. 3rd ed. Long Grove, Ill.: Waveland Press, 2004. This college-level text covers the basics of geomorphology. Includes a section on the development of soils. Index and bibliography.

Brady, N. C., and R. R. Weil. *Nature and Properties of Soils*. 11th ed. Englewood Cliffs, N.J.: Prentice Hall, 1996. This advanced text requires some background in geology and geomorphology. However, it contains a wealth of information on the development of soils.

Chernicoff, Stanley, and Donna Whitney. *Geology: An Introduction to Physical Geology.* Upper Saddle River, N.J.: Prentice Hall, 2006. This is a good overview of the scientific understanding of the geology of the earth and surface processes. Includes a section on the development of paleosols. Includes a link to a Web site that provides regular updates on geologic events around the globe.

Kapur, Selim, and George Stoops, eds. *New Trends in Soil Micromorphology.* New York: Springer, 2010. This text is a compilation of articles suited for graduate students and researchers. Discusses the identification and analysis of paleosols, soil micromorphology of the Quaternary period. Articles address vertisols, andosols, and other paleosols.

Kimberley, M. M., and D. E. Grandstaff. "Profiles of Elemental Concentrations in Precambrian Paleosols on Basaltic and Granitic Parent Materials." *Precambrian Research* 32 (1986): 133-154. All papers in this issue are devoted to Precambrian paleosols. This particular paper compares Precambrian paleosols that developed on basalt with those that developed on granitic rocks.

Reinhardt, J., and W. R. Sigleo. *Paleosols and Weathering Through Geologic Time: Principals and Applications.* Special Paper 216. Boulder, Colo.: Geological Society of America, 1988. Several paleosols in the United States (from Pennsylvania, Kentucky, Georgia, Alabama, Colorado, and California) are described in this volume, along with field methods and theoretical models for the development of ancient paleosols.

Retallack, G. *Laboratory Exercises in Paleopedology.* Eugene: University of Oregon Press, 1985. Practical exercises are outlined for the study of paleosols. Examples of paleosols are described from several localities around the world.

__________. *Soils of the Past: An Introduction to Paleopedology.* 2d ed. New York: Blackwell Science, 2001. The best and most comprehensive coverage of this subject. This text is an excellent source of information on paleosols, discussing soil classification and formation, identifying paleosols, and their characteristics, and pointing out the factors involved in paleosol formation. The author covers the soil fossil records with discussion of ancient landscapes and ecology. Written for undergraduate students.

Rutter, Nat. "International Year of Planet Earth 4: Utilizing Paleosols in Quaternary Climate Change Studies." *Geoscience Canada* 36 (2009): 65-77. This article discusses paleosols as valuable tools for determining ancient environments. The author describes the process of identifying paleosols to determine past vegetation and climate. The article also provides an explanation of paleosol dating techniques. Best suited for researchers and graduate students.

Samama, J. C. *Ore Fields and Continental Weathering.* New York: Van Nostrand Reinhold, 1986. Some paleosols are ore deposits for aluminum or nickel; part of this book reviews the characteristics and metal-concentrating processes that produce such paleosols. Another part reviews a different class of paleosol-related ore in which subsurface holes in limestone coincidentally or subsequently become filled with metal-precipitating solutions. Research subsequent to this book has shown that the dissolution of ore-bearing limestone probably shortly precedes precipitation of the lead-zinc minerals and that both the limestone-dissolving and metal-precipitating fluids arise from deep in the earth.

Schaetzl, Randall J., and Sharon Anderson. *Soils: Genesis and Geomorphology.* New York: Cambridge University Press, 2005. This text provides a large amount of fundamental content for studying soils. Part I discusses soil morphology and classification. Part II discusses the formation of soil and other soil dynamics. In the final part the authors discuss paleosol characteristics and reconstruction of environments through the study of paleosols.

Wright, V. P., and Alfred Fischer, eds. *Paleosols: Their Recognition and Interpretation.* Princeton, N.J.: Princeton University Press, 1986. This volume emphasizes paleosols that occur outside the United States, particularly those in the British Isles, Spain, and New Zealand. Chapter 6 reviews selected paleosols in the western United States.

See also: Biostratigraphy; Cenozoic Era; Cretaceous-Paleogene Boundary; Geoarchaeology; Geologic Time Scale; Ice Ages; Land Bridges; Mesozoic Era; Paleoclimatology; Paleogeography and Paleocurrents; Paleoseismology; Paleozoic Era; Proterozoic Eon; Stratigraphic Correlation; Transgression and Regression; Unconformities; Uniformitarianism.

PALEOZOIC ERA

Paleozoic stratigraphy concerns the study of rock sequences that date from 544 to 245 million years before the present. The primary method by which the stratigraphic relationships of the Paleozoic are established is stratigraphic analysis of distinctive types of fossils.

PRINCIPAL TERMS

- **biostratigraphy:** defining rock layers on the basis of their fossil content
- **correlation:** matching rock units of equivalent age
- **index (guide) fossil:** the remains of an ancient organism that are useful in establishing the age of rocks; index fossils are abundant and have a wide geographic distribution, a narrow stratigraphic range, and a distinctive form
- **series:** a time-rock unit representing rock deposition during a geologic epoch
- **stage:** a time-rock unit representing rock deposition during a geologic age
- **stratigraphy:** the study of rock layers (strata)
- **system:** a time-rock unit representing rock deposition within a geologic period

CAMBRIAN SYSTEM

The Paleozoic is the oldest era of the Phanerozoic eon, ranging from approximately 544 to 251 million years before the present. In North America, it is divided into seven periods. Beginning with the oldest, these are the Cambrian, Ordovician, Silurian, Devonian, Mississippian, Pennsylvanian, and Permian. Outside North America, the Mississippian and Pennsylvanian are combined to form the Carboniferous period. In stratigraphy, strata laid down during a particular period are referred to as a system. Systems are subdivided into series (the time-rock unit equivalents of epochs), which may be further subdivided into stages (the time-rock unit equivalents of ages). These subdivisions are very important in subdividing time-rock units of the Paleozoic and other eras of earth history.

The Cambrian is the oldest system in the Paleozoic, ranging from approximately 542 to 488 million years before the present. It was named by Adam Sedgwick in 1835 on the basis of rock exposures in northern Wales. The Cambrian system is subdivided globally into a Lower (Early), Middle, and Upper (Late) series. The earliest rocks of the Phanerozoic eon are stratigraphically difficult to interpret. One reason is that although most rocks of the Paleozoic are subdivided primarily on the basis of their included fossils, the record of Precambrian and Early Cambrian life is poorly known. By Middle Cambrian times, the composition of the fossil communities had changed greatly; this proliferation of forms preserved in the rock record has enabled paleontologists to establish fairly well the stratigraphy of the Middle and Upper Cambrian series. The primary tool used in the establishment of Middle and Upper Cambrian stages is the stratigraphic distribution of trilobites, marine arthropods that somewhat resembled modern pill bugs. Acritarchs—tiny planktonic algae that form a resistant, fossilizable covering—were another group that have become useful index fossils for the Proterozoic (Upper Precambrian) through Devonian. Another common Cambrian fossil group is the brachiopods; these two-valved suspension feeders were of relatively simple form during the Cambrian, a fact that lessens their utility as stratigraphic indicators. During the Late Cambrian, a series of events caused many types of trilobites to become extinct. These events of extinction and evolutionary radiation are especially useful in precise definition of Cambrian stratigraphy. The last of these extinction episodes, which occurred at the very end of the Cambrian, eliminated numerous trilobite species. After this event, the trilobites never rebounded; in post-Cambrian strata they cannot be widely used as index fossils.

ORDOVICIAN, SILURIAN, AND DEVONIAN SYSTEMS

The second Paleozoic system is the Ordovician. It ranges from approximately 488 to 444 million years before the present. The Ordovician was named in 1879 by Charles Lapworth, who combined portions of the Cambrian and Silurian systems as first defined in Wales, thus ending a debate that had developed concerning those sequences of strata. The Ordovician may also be subdivided into Early, Middle, and Late series, and for these the stages are much better established than in the Cambrian.

Strata of Ordovician age are often characterized by an abundance of carbonate (limestone and dolomite) rocks, which were deposited upon shallow seaways extending over many of the continents. Although trilobites declined in number throughout the Ordovician, many other animal groups became abundant or appeared for the first time. Especially important for biostratigraphy are graptolites, primitive hemichordate animals that lived in colonies; often they are preserved as black marks that resemble tiny hacksaw blades on rocks. Also of biostratigraphic importance are the acritarchs and the nautiloids. The latter group were predatory mollusks that at this time typically had straight shells, with tentacles emanating from one end. The chitinozoans are another group used in biostratigraphy for rocks of Ordovician through Devonian age. Although they are often classified as algae, the precise relationships of these microscopic, typically vase-shaped organisms are unknown. Finally, the evolution of more complex, hinged-shell groups of brachiopods with distinctive form enables stratigraphers to utilize this group in biostratigraphy from Late Ordovician through Early Pennsylvanian series rocks. The end of the Ordovician is marked by another mass extinction event, in which approximately one hundred families of marine animals were wiped out.

The Silurian system was named by Roderick Murchison in 1835 on the basis of rock exposures in southern Wales. The age of the Silurian ranges from 444 to 416 million years before the present. Worldwide, an Early and a Late series are recognized. In the best-known exposures of Silurian rocks, however, various series designations are used. Silurian rocks are common on all continents and were deposited in continental or marine environments. The most common rocks of the Silurian are limestones and dolomites, although many dark, clay-rich marine rocks containing graptolites are present. As in the Ordovician, this group is a useful index fossil for the Silurian. Brachiopods, acritarchs, and chitinozoans are other important fossils used in determining the age of Silurian strata.

The fourth system of the Paleozoic is the Devonian, ranging from 416 to 359 million years before the present. The Devonian system was named by Murchison and Sedgwick in 1839 for rock exposures in Devon and Cornwall, in southwest England. Globally, the Devonian is subdivided into Early, Middle, and Late series. Devonian rocks are very common; they are present on all continents, where they formed in a great variety of environments. Thick sequences of Devonian-age sediments accumulated along the edges of ancient continents and record a time of changing climate and environments. On land, red bed sequences, colored by oxidation of iron-bearing minerals, are widely distributed for the first time. Of these units, the Old Red Sandstone of the British Isles is best known. As in the Ordovician and Silurian, graptolites, brachiopods, and chitinozoans are important marine index fossils for the Devonian. The proliferation of different groups of organisms also adds to the list of fossils used in stratigraphic correlation. Here, for the first time, pollen and spores can be used to define strata; they are especially important in correlating Devonian continental rocks. Also of biostratigraphic utility are the strange, tiny, toothlike conodont fossils (from animals of unknown affinity) and ostracods (minuscule, two-valved crustaceans). In addition, for the first time, ammonoids (cephalopods with typically coiled shells) are of widespread use in biostratigraphy. Goniatite ammonoids are of major importance in defining rocks of the Devonian through Permian age.

Mississippian, Pennsylvanian, and Permian Systems

The term "Carboniferous" was first applied to rocks within the area of north-central England by William Conybeare and William Phillips in 1822. Yet no type section (the area in which rock units are first described) was ever designated. The Carboniferous lasted from approximately 359 to 299 million years before the present. Although the system is named for coal-bearing strata, only its upper portion contains large amounts of coal; the lower portion typically consists of carbonates. In North America, a different classification is employed, based largely on these rock differences. In 1870, American geologist Alexander Winchell proposed the name "Mississippian" for Lower Carboniferous exposures in the Mississippi River valley between southeastern Iowa and southern Illinois. Rocks of the Mississippian range from 359 to 299 million years before the present. In 1891, Henry Williams named Upper Carboniferous exposures the Pennsylvanian, after the widespread coal-bearing strata in that state. The Pennsylvanian ranges from 318 to 299 million years before the present. The widespread application of these names to Carboniferous-age rocks in North America led to the formal recognition of the Mississippian and Pennsylvanian as systems by the U.S. Geological Survey in 1953.

The Mississippian system (Lower Carboniferous series) is characterized by the widespread distribution of limestone and dolomite formed by marine incursions over many continents. Cyclic successions are developed in many places within Mississippian-age strata, especially within the upper portion of the system, in which sequences of limestones, sandstones, and shale were repetitively deposited one upon the other. Goniatites, brachiopods, spores, and pollen are all utilized in the Mississippian-age strata for correlation purposes.

Within North America, there is a widespread unconformity (a gap in the rock record) between the Mississippian and Pennsylvanian (Upper Carboniferous). The Pennsylvanian was a time of extensive coal deposition within low-lying swamps within North America, Europe, and many portions of Asia. Coal deposits are often found capping cyclical sequences consisting of limestone, shale, sandstone, and more shale (with the associated coal). Within the marine sequences, goniatite ammonoids are utilized for determining age and, at the base of the system, brachiopods are used for correlation. In addition, fusulinid foraminifers (small fossils resembling grains of rice but giants of this one-celled group) become important for Upper Pennsylvanian correlation. With the abundance of coal-producing plants, fossil spores and pollen serve as valuable index fossils for continental sediments of the Upper Carboniferous.

The Permian system was named for the region of Perm in the western Ural Mountains of Russia. The system, first proposed by Murchison in 1841, roughly corresponds to a period of earth history between 299 and 251 million years before the present. The Permian may be subdivided into a Lower and Upper series. Permian deposits are thick and widespread in many parts of the world and indicate a time of climatic complexity. Permian stratigraphic sequences are quite varied. Coal was still present but was no longer being extensively formed in North America or central Europe. Continental deposits of the Permian are primarily known for extensive red bed sequences created through the oxidation of iron-rich sediments. Extensive evaporites are also present, especially in Upper Permian-age strata, with thick layers of gypsum and salt. Large reefs are found, and, on the Gondwana continents, tillite deposits give evidence for extensive Permian glaciation. Major index fossils for the Permian include goniatite ammonoids and fusulinids for the marine sequences and spores and pollen for continental deposits.

The end of the Permian is marked by an extinction event of major significance. Fusulinids, tabulate and rugose corals, and trilobites became extinct; brachiopods, bryozoans, and echinoderms sustained huge losses, and substantial numbers of bivalve and gastropod mollusks also became extinct. On land, many groups of the larger coal-swamp trees became extinct, and several major groups of vertebrates also died out. These huge losses in plant and animal types mark the stratigraphic boundary between the Paleozoic era ("the era of ancient life") and the subsequent Mesozoic era.

Study of Paleozoic Stratigraphy

Studies of Paleozoic stratigraphy primarily utilize analyses of rock units, paleontological studies, and radiometric dating techniques. Lithostratigraphy defines strata (rock layers) on the basis of rock, or lithologic, characteristics. Distinctive rock layers are very useful in determining the position and relative age of local outcroppings of Paleozoic strata. Most Paleozoic

Trilobite fossils are used to identify Middle and Upper Cambrian stages in rock formatons. (National Geographic/Getty Images)

systems tend to be dominated by specific rock types. For example, carbonate rocks (limestones and dolomites) are characteristic of many Ordovician, Silurian, and Mississippian sequences. Clastic rocks, including sandstones and mudrocks with their included coals, are diagnostic of Pennsylvanian-age deposits. Red bed deposits, consisting of oxidized iron-rich sediments, are often associated with clastic sequences in Devonian- and Permian-age strata. Unconformities—gaps in the rock record representing periods of erosion or nondeposition—often separate the Paleozoic systems and form natural boundaries for their separation. A major unconformity also typically separates Precambrian rocks from those of the earliest Paleozoic.

Biostratigraphy defines rocks on the basis of their fossil content. The basic unit of biostratigraphic classification is the biozone, in which strata are characterized by the occurrence of a certain fossil or fossils. These index or guide fossils are abundant, distinctive forms with a wide geographic distribution but narrow stratigraphic range and are therefore useful in correlation and in determining the age of strata. The fossils most often used are planktonic (floating) or nektonic (swimming) forms. Of special importance for the study of Paleozoic stratigraphy are trilobites, brachiopods, and graptolites. For certain portions of the Paleozoic, acritarchs, conodonts, chitinozoans, fusulinids, pollen, and spores, as well as nautiloid and ammonoid cephalopods, are widely utilized in biostratigraphic studies.

Lithostratigraphic and biostratigraphic studies can establish only the relative ages of strata. Determination of the absolute age of Paleozoic rocks is done primarily through radiometric dating techniques. These studies utilize the known decay rates of radioactive isotopes to establish the age of rocks in terms of years. Isotopes used for studies of Paleozoic rocks are uranium-235, uranium-238, and thorium-232, all of which decay to lead. Also used are potassium-40, which decays to argon-40, and rubidium-87, whose daughter isotope (product of decay) is strontium-87. Igneous and metamorphic rocks are generally utilized in radiometric dating techniques.

Significance

Probably the most important application of Paleozoic stratigraphy is in the search for natural resources. Without knowing the age of the rocks being evaluated or the economic products potentially included within them, the exploration geologist's job would be impossible. Of particular significance is the widespread presence of coal in rocks of Pennsylvanian and Permian age. Most of the coal produced in North America and central Europe has been mined from Pennsylvanian-age rocks. As these coal beds often occur within the same portion of a cyclical sequence of sediments, the search for them is made easier through knowledge of Paleozoic stratigraphy. Permian-age coal sequences are found in Australia, China, the former Soviet Union, India, and South Africa. There are also great quantities of oil and natural gas within rocks of Paleozoic age. Significant finds of hydrocarbons have been made in Devonian- and Carboniferous-age rocks, and discoveries of huge oil and gas fields in the Permian Basin of west Texas were a result of exploration of Upper Paleozoic marine strata. Arid conditions, especially common during the Permian, resulted in huge deposits of sodium and potassium salts as well as gypsum and anhydrite. Mountain-building events during the Paleozoic also created tremendous deposits of both precious metals and metals that are important for industry. Without an understanding of the stratigraphic occurrence of these valuable resources, the recovery of these materials would be impossible.

Phillip A. Murry

Further Reading

Briggs, D. E., and R. A. Fortey. "Wonderful Strife: Systematics, Stem Groups, and the Phylogenetic Signal of the Cambrian Radiation." *Paleobiology* 31 (2005): 94-112. The authors discuss the unresolved issues of systematics in reference to the morphological disparity and divergence times of the Cambrian period. The article provides specific examples to support the broad concepts. Best suited for graduate students and professionals interested in theories of evolutionary biology.

Brookfield, Michael E. *Principles of Stratigraphy*. Hoboken, N.J.: Wiley-Blackwell, 2004. Written for undergraduate students, this text provides an overview of the principles and applications of stratigraphy. Includes stratigraphic techniques and case studies. It is organized into three sections, beginning with foundational material, followed by data collection and research topics, and completed with interpretations and analysis.

Bruton, David L., ed. *Aspects of the Ordovician System.* New York: Oxford University Press, 1984. A symposium volume from the Fourth International Symposium on the Ordovician system, held in Norway in 1982. The introductory paper and a few of the papers on the stratigraphic framework and sea-level changes of the Ordovician would be understandable to a college-level student with some training in geology. The sections on Ordovician environments and their included faunas are written by and for specialists in those areas.

Dineley, D. L. *Aspects of a Stratigraphic System: The Devonian.* New York: John Wiley & Sons, 1984. A well-written and well-illustrated book covering many aspects of Devonian stratigraphy, sedimentology, and fossils. Because it is a single-authored text, it possesses much more continuity than do most symposium volumes. The text would be understandable and very readable for a college student with some training in geology.

Grotzinger, John, and Tom Jordan. *Understanding Earth.* 6th ed. New York: W. H. Freeman, 2009. This comprehensive physical geology text covers the formation and development of the earth. Readable by high school students, as well as by general readers. Includes an index and a glossary of terms.

Levin, Harold L. *The Earth Through Time.* 9th ed. Hoboken, NJ: Wiley, 2009. An introductory college-level text that covers aspects of historical geology. A large section is devoted to Paleozoic-age rocks. Well written and profusely illustrated.

Ogg, James G., Gabi Ogg, and Felix M. Gradstein. *The Concise Geologic Time Scale.* New York: Cambridge University Press, 2008. This book is a complete overview of the geological time scale, including stratigraphy topics such as chronostratigraphy and magnetic stratigraphy. The book is organized by geological time periods. Also contains a reference appendix, the geological time scale table, and indexing.

Prothero, Donald R. *Bringing Fossils to Life.* 2d ed. Boston: McGraw-Hill, 2004. This well-illustrated and entertaining text covers a broad range of paleontological topics, including biostratigraphy and the use of fossils in the study of Paleozoic rocks. Glossary, bibliography, and index.

Prothero, Donald R., and Robert H. Dott, Jr. *Evolution of the Earth.* 8th ed. New York: McGraw-Hill, 2009. This basic textbook on historical geology is aimed at students of geology. However, it is very readable by anyone with a background in science. Presents an up-to-date account of the earth's history from the viewpoint of plate tectonics. Includes a glossary.

Stanley, Steven M. *Earth System History.* 3rd ed. New York: Freeman, 2008. An introductory undergraduate-level text integrating coverage of life and the environment. Good coverage of the Mesozoic.

Whittington, H. B. *The Burgess Shale.* New Haven, Conn.: Yale University Press, 1985. Whittington reviews the beautifully preserved animals from the Middle Cambrian of the Rocky Mountains in Canada and gives a detailed look at the often strange animals from that famous locale. A review of the stratigraphy and sedimentology of the locality is given, along with comments on the evolutionary significance of the Early Paleozoic animals. The work is written for general audiences.

Wicander, Reed, and James S. Monroe. *Historical Geology.* Belmont Calif.: Brooks/Cole, Cengage Learning, 2010. This textbook is accessible to university students with or without a background in geology. It covers earth's geological systems and plate tectonics, physical and biological history, geological time, and evolution. There are many diagrams, images, and charts along with a CD-ROM to supplement the text.

See also: Biostratigraphy; Cenozoic Era; Continental Growth; Cretaceous-Paleogene Boundary; Dinosaurs; Evolution of Life; Geologic Time Scale; Gondwanaland and Laurasia; Ice Ages; Island Arcs; Land Bridges; Mass Extinctions; Mesozoic Era; Microfossils; Ocean Basins; Oceanic Crust; Paleobiogeography; Paleoclimatology; Paleogeography and Paleocurrents; Paleoseismology; Paleosols; Proterozoic Eon; Stratigraphic Correlation; Transgression and Regression; Unconformities; Uniformitarianism.

PANGAEA

Pangaea was a supercontinent that existed between 270 and 200 million years ago. It contained most of the world's current continents in a single landmass. Supercontinents form because of tectonic forces within Earth that lead to the movement of continents and the expansion of oceans. Pangaea was the birthplace of the dinosaurs and other species that later colonized the entire planet.

PRINCIPAL TERMS

- **continental drift:** the hypothesis that continents shift position relative to one another rather than remaining fixed in a single location; this concept was first developed in the early nineteenth century
- **craton:** a stable portion of the continental crust, as differentiated from the more malleable section vulnerable to transformation during continental movement
- **Gondwana:** a supercontinent that emerged from the breakup of Pangaea in the Jurassic period and contained the continents that now constitute Africa, South America, Australia, and India
- **Laurasia:** a northern supercontinent that broke away from Pangaea in the Jurassic period and contained the landmass that formed into most of Asia and North America
- **lithosphere:** the outer shell or surface of Earth consisting of rocks, soil, and other sedimentary material and connected physically to Earth's molten inner core
- **Panthalassa:** a single ocean that existed during the Permian and Triassic periods and that was surrounded by Pangaea
- **plate tectonics:** a scientific theory used to explain shifting patterns in the distribution of land across Earth
- **subduction:** the process by which portions of Earth's crust are sucked into the mantle, creating a vacuum that is largely responsible for the movement of the continents
- **supercontinent:** a continental structure containing two or more cratons
- **Tethys Ocean:** the ocean that formed between Gondwana and Laurasia during the initial breakup of Pangaea

THE PANGAEA THEORY

In the early twentieth century, scientists determined that the continents of the world were not always in their present state, but were once united in a series of much larger landmasses known as supercontinents. Paleontologists and geologists have given the name Pangaea, derived from the Greek words meaning "all land," to the supercontinent that existed during the Paleozoic and Mesozoic periods, between 270 and 200 million years ago, and that contained most of the world's current continents in a single landmass. German meteorologist Alfred Wegener was the first to suggest the existence of Pangaea, in his 1915 book *Die Entstehung der Kontinente und Ozeane* (*The Origin of the Continents and Oceans*, 1922).

The Pangaea theory was controversial in its time, but it is now accepted by the scientific community and has been supported by evidence from many different branches of research. The Pangaea theory helps scientists to understand the distribution of both fossils and living species. When the continents were joined to form Pangaea, species could migrate throughout most of the globe. This ancient connection between continents is made clear by the global distribution of species found today.

PLATE TECTONICS AND THE FORMATION OF SUPERCONTINENTS

The Pangaea hypothesis was part of Wegener's theory of continental drift, which held that the continents were in a state of constant motion and drifted across the surface of the ocean in response to forces within Earth's core. The theory was based on evidence from early-twentieth-century paleontological and geological research.

One of the most obvious and compelling pieces of evidence for Pangaea is that the shapes of some of the continents—notably Africa, South America, and Australia—appear to be complementary, like the pieces of a puzzle. Additionally, fossils of the same species and rocks of the same age and type are often found on continents separated by ocean basins. Before the arrival of the idea of continental drift, geologists and paleontologists could not explain the distribution of these fossil remains. Additional evidence came from close inspection of the structure of intercontinental mountain passages and from irregularities in Earth's magnetic polarity.

Wegener's continental drift hypothesis was the subject of intense debate until the 1950's, when new evidence derived from studies of the ocean floor bolstered the idea that the continents could move. The theory of seafloor spreading, proposed by meteorologist Harry Hess in 1962, holds that the ocean surface is in a constant cycle of recycling, as sediment is pulled into ocean trenches while new ocean sediment is released from trenches in other locations. Portions of Earth's floor are sucked into deep ocean trenches, creating a force called subduction, in which Earth's continents are pulled toward active trenches. The discovery of seafloor spreading, subduction, and related processes led to the science of plate tectonics, which studies the movement and formation of the continents.

According to plate tectonics, each of the continents is associated with a stable portion of Earth's lithosphere called a craton. As the sea floor moves in response to forces from Earth's core, the cratons are pushed or pulled on currents of moving earth, thereby shifting the position of the attached continents on Earth's surface. A supercontinent is a landmass that contains two or more of Earth's cratons in a single structure.

Pangaea was composed of all Earth's cratons, with the exception of those containing portions of what is now China, which existed during that time as a separate island continent. Geologists now believe that Earth's cratons also formed into a single supercontinent on other occasions. Some geologists have suggested that these periods of unification may be part of a supercontinent cycle, in which the continents form into a single unit and then split apart every 400 to 500 million years. While scientists have a great deal of evidence to support the Pangaea theory, evidence for earlier supercontinents remains a subject of serious debate among geologists.

The World Before Pangaea

Geologists believe that about 1.2 billion years ago, before the time of Pangaea, the Earth's continents were united in a single supercontinent known as Rodinia. Geological traces of this ancient union are still present in formations around the world. Evidence suggests that life on Rodinia was confined to marine environments, while the terrestrial environment was barren. The dissolution of Rodinia, between 100 and 300 million years later, was a major milestone in the evolution of life. As the environment changed in response to the geological forces that caused the breakup of the supercontinent, the environment was then able to support terrestrial life; it was at this time that cells colonized the land of each continent. Geologists are uncertain of what happened next, but eventually the continents of Earth began to draw together again.

The initial formation of Pangaea began more than 500 million years ago, during the Ordovician period, when two supercontinents in the Northern Hemisphere, known as Baltica and Laurentia, collided to form a massive supercontinent known as Euramerica. By this time, Australia, Africa, India, Antarctica, and South America were already unified into a southern supercontinent known as Gondwanaland. Geologists believe that the collisions of cratons to form Euramerica and Gondwanaland caused the formation of some of today's mountain ranges in North America, Asia, and Africa.

Formation and Development

Near the end of the Permian period (290 to 248 million years ago), Gondwanaland and Euramerica collided to form Pangaea. This occurred through millions of years and involved prolonged periods of intense volcanic and seismic disturbances. Mountains and other geological features that formed in this collision attest to the force generated as the continents united. Pangaea was fully formed between 250 and 225 million years ago, though continued movement caused the development of seismic and volcanic zones within the supercontinent. Paleontologists have given the name Panthalassa, from the Greek terms meaning "all ocean," to the single massive body of water that surrounded most of Pangaea during the Permian and Triassic. To the east of Pangaea was a second, smaller ocean known as the Tethys.

The formation of Pangaea had a far-reaching effect on global conditions and on organisms living both in the oceans and on Earth's surface. During this time in the late Permian, Earth experienced the largest mass extinction in its history, resulting in the loss of some 90 percent of species living on the planet. A variety of causes may have played a part in the extinction event, but paleontologists believe that the formation of Pangaea was an important contributing factor.

Because Earth's cratons were united in a single mass, there was less available land to support coastal shelf habitats, which support the majority of ocean life. This loss of habitat may have been one of the factors that led to high extinction rates for marine organisms. Part of the supercontinent covered one of Earth's poles, leading to glaciation and a reduction in available habitat for some species. Land near the coasts was far lower than it is today because of rising ocean levels, which also may have contributed to a loss of habitat and habitat diversity. Additionally, because Pangaea was an exceptionally large continent, much of the inland area was a great distance from the ocean. This inland zone developed into an arid region, a massive central desert that wreaked havoc on weather patterns.

The supercontinent existed for less than 100 million years, during which time life on Earth changed dramatically. The Permian-Triassic extinction left the planet with a variety of unoccupied habitats, and new species soon evolved to fill these empty niches. Among the most famous groups of animals to evolve on Pangaea were the dinosaurs, which first appear in the fossil record in sediment from the late Triassic period. The dinosaurs evolved from small reptilian ancestors that survived the worst of the Permian extinction. The dinosaurs possessed key evolutionary adaptations that allowed them to thrive in the late Triassic environment.

Because the continents were unified, early dinosaur species could spread across the globe. By the time Pangaea began to separate, in the Jurassic period, dinosaurs had spread into many habitats around the globe. Dinosaurs were not the only new group of reptile descendants to colonize Pangaea; they were joined by the prehistoric marine and flying reptiles, filling the niches that are today occupied by marine mammals and birds. Like the dinosaurs, these groups also spread across the globe, leaving descendants in every continent. While the dinosaurs and other reptiles dominated the terrestrial environment of Pangaea, the ancestors of modern mammals also appeared at this time and spread across the supercontinent. Unlike the reptiles, however, mammals remained small throughout the Mesozoic period.

The Breakup of Pangaea

Pangaea began to break apart during the late Triassic period, possibly as much as 235 million years ago. Geologists have identified several distinct stages that occurred as the supercontinent separated. The separation began with the formation of rift zones, as tectonic pressures from Earth's interior caused the development of unstable fissures in its surface. Magma and superheated minerals were forced to the surface along these areas, which were frequently inundated by volcanic eruptions and earthquakes.

The southern and northern sections of Pangaea began to separate first, leading to the formation of a central ocean, called the Tethys, between the two areas. This occurred as the eastern part of what is now North America separated from the western part of Africa. As the Tethys Ocean expanded, so did the diversity of life on Earth, especially in terms of sea life. The Tethys Ocean provided a variety of shallow oceanic habitats that supported large numbers of fish and prehistoric marine reptiles. The terrestrial portions of Earth also began to flourish, as the ocean stabilized temperatures and created a vast forested zone inland from the coast.

In the middle of the Jurassic, the rift between Africa and North America continued to spread, detaching the southern portion of North America from the northern part of South America. This resulted in the formation of two supercontinents: Gondwana in the south and Laurasia in the north. For the dinosaurs and other animals living on Earth's surface, the split between the supercontinents led to divergent evolution, as unique species formed on each of the continents in response to different environmental conditions. This was the height of the age of dinosaurs, as southern and northern dinosaur faunas evolved into spectacular arrays of species.

The continents continued to pull farther apart, forming a new ocean, the Atlantic. Tectonic activity along the Atlantic Ocean trench widened the ocean and continued to push the continents along their trajectories. By the middle of the Cretaceous, about 150 to 130 million years ago, the supercontinent of Gondwana had broken farther apart, leading to the formation of Australia, India, South America, and Africa. These fissures led to the formation of the Indian Ocean and eventually closed most of what had originally been the Tethys Ocean. It was not until the Cenozoic period, about 65 million years ago, that Laurasia split into North America and Eurasia, giving rise to the Norwegian Sea and eventually resulting in the formation of the current continental structure.

Micah L. Issitt

Further Reading

Evans, David A. D. "The Palaeomagnetically Viable, Long-Lived and All-Inclusive Rodinia Supercontinent Reconstruction." *Geological Society* 327 (January 2009): 371-404. Detailed discussion of the lesser-known supercontinent of Rodinia and its relation to the well-known supercontinents of the Mesozoic period.

Foote, Michael Joseph, et al. *Principles of Paleontology.* 3rd ed. New York: Macmillan, 2007. Excellent introduction to the science of paleontology that includes information on continent formation, drift, plate tectonics, and the distribution of life across the supercontinents.

Nield, Ted. *Supercontinent: Ten Billion Years in the Life of Our Planet.* Cambridge, Mass.: Harvard University Press, 2007. A comprehensive look at the changing environments of Earth, with discussions of the processes and stages between continental stages. Presents information about Pangaea, including wildlife patterns on the planet during the Triassic and Jurassic periods.

Rogers, John J. M., and M. Santosh. *Continents and Supercontinents.* New York: Oxford University Press, 2004. Provides a detailed look at continental development and provides comprehensive information on the geologic processes that led to the formation of the modern continents and the supercontinents.

Weishampel, David B., Peter Dodson, and Halszka Osmolska, eds. *The Dinosauria.* 2d ed. Berkeley: University of California Press, 2004. A basic reference book containing information about many subjects relevant to modern paleontology, including the formation and breakup of the supercontinents. The authors also describe the movement of dinosaur groups and species among the continents from the Triassic to the end of the Cretaceous.

See also: Continental Crust; Continental Drift; Continental Growth; Continental Rift Zones; Continental Shelf and Slope; Continental Structures; Continents and Subcontinents; Dinosaurs; Faults; Gondwanaland and Laurasia; Hydrothermal Vent Faunas; Ocean Ridge System; Permo-Triassic Boundary; Seafloor Spreading; Supercontinent Cycles.

PERMAFROST

Permafrost is a thermal condition existing in any type of earth material that is perennially frozen. It occurs in ground in which the temperature remains below 0 degrees Celsius for at least two consecutive years. Affecting about 25 percent of the earth's surface, the condition hampers construction in the far north in regions such as Siberia and Arctic Canada.

PRINCIPAL TERMS

- **periglacial:** originally restricted to regions modified by frost weathering such as tundra
- **Pleistocene:** the geologic era between 2.58 million and 11,000 years ago, characterized by extensive continental glaciation and a colder climate
- **subsidence:** a downward vertical movement of the ground, commonly caused by land-cover changes of the surface
- **talik:** a layer of unfrozen ground in permafrost; water under pressure frequently occurs in talik layers
- **thermokarst:** a group of landforms in flat areas of degrading permafrost and melting ice

Formation and Distribution

Permafrost, or permanently frozen ground, is a thickness of earth material such as soil, peat, or even bedrock, at a variable depth beneath the ground surface, in which a temperature below 0 degrees Celsius exists continuously for a number of years (from at least two years to several thousands of years). Permafrost is thus not a process or an effect, but rather a temperature condition of earth material on land or beneath continental shelves of the Arctic Ocean. Its occurrence is conditioned exclusively by temperature, irrespective of the presence of water or of the composition, lithology, and texture of the ground. However, unfrozen water horizons may exist within a body of permafrost as the result of various conditions such as pressure or impurities in the water.

With regard to the climatic history of the earth, cold conditions and glaciers are the exception, not the rule. Because they are so rare, they are referred to as "climatic accidents." During the Pleistocene, advances of continental glaciers covered vast areas of Europe and North America. Glaciers continue to be significant landscape features, since they lock up as ice some 2.15 percent of the earth's waters. Permafrost is a relatively new condition introduced with the onset of the Pleistocene glacial events, and it will be maintained as long as the climate remains cold.

As the earth's climate has warmed during the past 20,000 years and the continental ice has receded, the location, distribution, and thickness of permafrost have changed. Today, the area underlain by permafrost is vast. It has been estimated that about 25 percent of the earth's surface is underlain by permafrost. One reason that vast areas of the earth possess permafrost is that the continents widen toward higher latitudes in North America and Eurasia. In the Northern Hemisphere, Mongolia, Russia, and China contain about 12.2 million square kilometers of land underlain with permafrost. North America and Greenland have 8.8 million square kilometers of permafrost. In the Southern Hemisphere, Antarctica accounts for 13.5 million square kilometers of permafrost. During the Pleistocene, perennially frozen ground had a different distribution. Permafrost produces soil and sediment structures that remain after the climate warms. Relict features such as ancient ice wedges can be identified, and former permafrost areas can be mapped. Evidence of preexisting permafrost conditions, and thus a record of climate change, has been extensively documented in Great Britain, southern Canada, and the central United States.

In terms of aerial distribution and continuity, permafrost may be classified as either continuous or discontinuous. At higher-latitude land circling the Arctic Ocean, the lateral extent of permafrost is uninterrupted nearly everywhere, except in areas of recent deposition and under large lakes that do not freeze to the bottom. Discontinuous permafrost includes areas of frozen ground separated by nonpermafrost areas. A transect from the Arctic Ocean southward, into Canada or Russia, reveals more and more permafrost-free areas. Farther south, in central Canada and central Russia, only patchy or sporadic zones of perennially frozen ground are to be found.

Permafrost is a unique condition because it is maintained very close to its melting point. Several thousand square kilometers are warmer than −3 degrees Celsius. If global warming continues, as many scientists predict, most discontinuous permafrost will eventually degrade. Russian scientists, in fact, have documented the northward retreat of permafrost near Archangel at a rate of 400 meters per year since 1837. In Canada, permafrost has retreated northward more than 100 kilometers since 1945, as ground temperatures have increased by 2 degrees Celsius in the northern prairie provinces. Changes in temperature may result from a local microclimate change, alteration of land cover, or changes in the atmosphere. Conversely, permafrost not only is degraded by global warming, but may also contribute to atmospheric change. Scientists suggest that thawing of Arctic terrain, which is rich in peat accumulation, may affect the chemistry of the atmosphere. It has been determined that one-fourth of the earth's terrestrial carbon is stored in organic matter in the permafrost and active layer. Long-term warming would release enormous quantities of greenhouse gases—methane and carbon dioxide in particular—accelerating global warming.

Vertical Characteristics

A cross section of excavated earth in the high Arctic reveals the vertical characteristics of permafrost. The thickness of permafrost is variable because of differences in air temperatures, occurrence of water, composition and texture of earth materials, and many other factors. The permafrost in Siberia may be at least 5,000 meters thick, whereas in Alaska thicknesses of 740 meters have been recorded. Farther south, thicknesses decrease as discontinuous permafrost becomes more common. The thickness of permafrost is in part determined by a balance between the increase of internal heat with depth and the heat loss from the earth's surface. If the thickness and depth of permafrost remain unchanged for many years, the heat loss at the surface and the heat at the base of the frozen ground are in equilibrium, and a steady state is maintained. At the surface, the ground seasonally thaws and freezes. This section of the permafrost is known as the "active layer." Its thickness is controlled by numerous variables in addition to seasonal temperature variations, and it exhibits great variability in thickness. Vegetation, snow cover, albedo, water distribution, and human alteration of the surface can all act as insulators and affect the thickness of the active layer. However, in areas not disturbed by people, the ground will typically thaw to a depth of from 1 to 12 meters during the summer.

The deeper boundary separating the active layer from the permafrost is known as the "permafrost table" or "zero curtain." Within the permafrost, unfrozen lenses known by the Russian term *talik* occur. These unfrozen bodies frequently contain water and function as protected water sources or aquifers in the Arctic environment. Talik lenses may be isolated bodies completely enclosed in impermeable permafrost and may result from a change in the thermal regime in permafrost. Conversely, talik may be laterally extensive even in continuous permafrost and is frequently a valuable source of groundwater. Talik lenses are larger in discontinuous permafrost, and the active layer thickens at the expense of the permafrost.

Below the permafrost table, the perennially frozen ground has low permeability. Maps and photographs of the Arctic reveal many lakes, swamps, and marshes, suggesting high precipitation. However, evaporation rates are low because of the cold air temperatures, and standing water occurs because the ground is frozen at depth. Throughout the region, precipitation is probably less than 35 centimeters annually, equivalent to that of marginal deserts. The abundant water and poor drainage

Exposed permafrost in the Katherine River Valley, Torngat Mountains, Labrador. (Geological Survey of Canada)

result from the occurrence of permafrost at depth, not to high atmospheric precipitation. As seasonal warming and cooling occur, the land surface expands and contracts, particularly in continuous permafrost areas. In essence, the lowering of the ground temperature causes thermal contraction of the ground, and vertical fissures or frost cracks occur. Water filling and freezing in these cracks creates ice wedges in the permafrost that widen with seasonal expansion and contraction. Ice wedges are often 3 to 4 meters wide and extend 5 to 10 meters into the perennially frozen ground below the active layer. In areas of degradation, ice wedges in the thawing permafrost begin to melt, creating thermokarst features characterized by numerous depressions similar to karst plains. As temperatures rise above freezing, the active layer extends deeper and deeper and eventually may intercept an ice wedge. If large ice wedges begin to thaw, the ground subsides, creating steep-sided conical depressions. The perimeter of a depression, often composed of very moist sediment, will slump into the depression, thus making it wider. Water then fills the depression, creating a thermokarst, or thaw lake. Averaging about 3 meters across, the lakes occur in clusters, are elliptical in shape, and are all oriented at right angles to the prevailing wind direction.

C. Nicholas Raphael

Further Reading

Anderson, Bjørn G., and Harold W. Borns, Jr. *The Ice Age World.* Oslo: Scandinavian University Press, 1994. The well-illustrated text details the Quaternary history of North America and northern Europe over the last 2.5 million years. Contains an extended glossary, references list, and index. For the general reader as well as the serious student.

Ballantyne, C. F., and C. Harris. *The Periglaciation of Great Britain.* New York: Cambridge University Press, 1994. A good upper-level description of permafrost in Britain. The discussion ranges from basic freeze-and-thaw processes to permafrost conditions and distribution in the geologic past.

Bloom, Arthur. *Geomorphology: A Systematic Analysis of Late Cenozoic Landforms.* 3rd ed. Long Grove, Ill.: Waveland Press, 2004. This college-level text covers the basics of geomorphology. Includes three chapters on glaciation, cold climates, and permafrost. Index and bibliography.

Chernicoff, Stanley, and Dona Whitney. *Geology: An Introduction to Physical Geology.* Upper Saddle River, N.J.: Prentice Hall, 2006. This is a good overview of the scientific understanding of the geology of the earth and surface processes. Includes a section on the development and effects of permafrost. Includes a link to a Web site that provides regular updates on geologic events around the globe.

Clark, M. J., ed. *Advances in Periglacial Geomorphology.* New York: Wiley InterScience, 1988. A series of state-of-the-art reviews of numerous processes and landforms in cold climates. Emphasizes processes of formation; very technical. Suitable for advanced earth science students as a reference and literature source.

Grosse, G., et al. "Why Permafrost Is Thawing, Not Melting." *EOS* 91 (2010): 87. Discusses terminology used in past and recent scientific communications and suggestions for future communications. The authors' recommendations provide the basis for an interesting debate over proper word usage to convey information beyond the scientific community. The article also provides a good description of the process of thawing, contrasting it to the physical process of melting.

Harris, S. A. *The Permafrost Environment.* Totowa, N.J.: Barnes & Noble, 1986. A primer on the consequences of construction in permafrost regions. Topics include the soil properties and methods of permafrost analysis. Written for contractors and environmental specialists.

Romanovsky, V., et al. "Permafrost Temperature Records: Indicators of Climate Change." *EOS, AGU Transactions* 83 (2002): 589-594. Discusses the changes to the ecosystem brought on by permafrost thawing. The article provides information on permafrost zones and the methods used to study their degradation.

Schuur, Edward A. G., et al. "The Effect of Permafrost Thaw on Old Carbon Release and Net Carbon Exchange from Tundra." *Nature* 459 (2009): 556-559. This article discusses the permafrost soils of the Arctic ecosystem. Provides a good understanding of how permafrost thaw alters the carbon cycle. Somewhat technical, so having a background in science is helpful for readers.

Williams, P. J., and M. W. Smith. *The Frozen Earth: Fundamentals of Geology*. New York: Cambridge University Press, 1991. A college-level textbook for science majors, emphasizing the physics of cold regions. Equations are used to illustrate significant physical ideas.

See also: Continental Glaciers; Geomorphology and Human Activity; Geomorphology of Dry Climate Areas; Geomorphology of Wet Climate Areas; Glacial Landforms; Grand Canyon; Karst Topography; Marine Terraces; River Valleys.

PERMO-TRIASSIC BOUNDARY

The Permo-Triassic boundary is that time in Earth's history marked by the transition between the Permian extinction period and the beginning of the Triassic period. The Permian extinction was the largest extinction in the history of Earth, resulting in the loss of more than 90 percent of species on the planet. Evidence for the Permian extinction includes geological sediment indicating a lack of oxygen and much paleontological data indicating the disappearance of large numbers of species from the fossil record. Reduced species concentrations and diversity in the wake of the Permian extinction allowed new species to flourish in both the terrestrial and marine environments, contributing to an overall increase in species diversity throughout the Mesozoic period.

PRINCIPAL TERMS

- **anoxia:** the conditions caused by an absence of oxygen
- **archosaur:** a reptilian branch that colonized much of the terrestrial environment during the Permian and early Triassic periods and gave rise to the earliest mammalian ancestors
- **chert:** a mineral sediment that often contains the pulverized remains of diatoms, shelled microbes that inhabit marine and aquatic ecosystems
- **flood basalt:** the sediment of liquid rock; associated with ancient patterns of volcanic eruption
- **mass extinction:** a period in Earth's history in which the species extinction rate rises sufficiently to cause large numbers of extinctions on a global scale and at a rate faster than the normal level of extinction for Earth as a whole
- **plate tectonics:** a branch of geology concerned with studying the movement and formation of continents
- **radiometric dating:** a scientific dating method that uses the decay of naturally occurring radioactive elements in the strata to provide an approximate date for the deposition of a mineral sample
- **stratigraphy:** a branch of geology concerned with studying rock layers and the process involved in sedimentary layering
- **supercontinent:** a combination of two or more of Earth's continents into a larger landmass

Geologic Time and Mass Extinction Theories

The geological discipline of stratigraphy involves studying the layers of Earth to discern patterns of deposition for clues to ancient geological events. Paleontologists have discovered patterns to the appearance and disappearance of certain types of fossils. These patterns can be used to mark the division between geologic time periods. The disappearance of a certain type of fossil may indicate the extinction of that species. The end of one geologic stage is usually marked by a significant change in the stratigraphy and associated fossil assemblages.

Paleontologists have noted portions of the geological record wherein many types of fossils seem to vanish from the sediment within a relatively short span of geologic time. Paleontologists believe that these major changes in the fossil record represent mass extinctions, when environmental conditions changed with sufficient rapidity to drive extinction rates far higher than their normal level. Paleontologists have discovered five major extinction events in Earth's history: the Triassic-Jurassic, Permian-Triassic, Devonian-Carboniferous, Ordovician-Silurian, and Cretaceous-Tertiary. Each of these mass extinctions represents changes in the biota significant enough to warrant a division between geologic periods.

Evidence for the Late Permian Extinction

Paleontologists have found significant evidence of a mass extinction at the end of the Permian period (290 to 248 million years ago), marking the division between the Permian and the Triassic periods. The extinction occurred about 251 million years ago and led to the loss of more than 90 percent of species on Earth, including nearly 95 percent of marine species and more than 70 percent of terrestrial plants and animals. In addition, paleontologists have estimated that more than 50 percent of species families disappeared during the Permian extinction, representing a major reduction in biodiversity on a global scale.

Evidence of the Permian extinction has been growing since English paleontologist John Phillips published a series of papers in the 1840's describing evidence that the end of the Permian coincided with a major turnover in marine species. Since that time, paleontologists have found evidence of similar changes in species composition from strata uncovered around the globe. With the advent of radiometric dating in the 1950's and 1960's, paleontologists were able to estimate the absolute age of geological samples and, in turn, to refine the geologic time scale. Using radiometric dating, paleontologists created a more accurate time line for the Permian extinction.

Fossil evidence of the Permian extinction is widespread and includes the disappearance of thousands of species worldwide. One recognizable fossil example is the chert gap, observed in global deposits from former marine habitats. Chert, a mineral formed from the bodies of millions of microscopic shelled organisms that accumulate on the ocean floor, is absent in the strata representing the shift between the Permian and the Triassic. Paleontologists believe that this gap was caused by a mass extinction that affected the microscopic shelled organisms whose bodies contribute to the buildup of chert on the sea floor. Similarly, paleontologists studying fossilized pollen have noted that many species common in the Permian period seem to have disappeared at the Permo-Triassic boundary. Fossilized pollen from early Triassic sediment is largely representative of smaller, weed-like species that were presumably better able to colonize nutrient-poor environments.

In addition to global shifts in fossil assemblages, geologists noted the presence of unusual mineralogical samples at the Permo-Triassic boundary. These geological indicators include a layer of dark, iron-rich sediment at the boundary layer, which indicates the presence of an anoxic environment, wherein oxygen is absent or present in very low levels. Geologists have found similar evidence of shrinking oxygen levels from the ratios of oxygen and carbon isotopes found at the Permian boundary. Sediment from the early Triassic also contains evidence indicating that there was a lack of foliage in many terrestrial environments immediately following the Permian extinction. This lack of plant life may have been responsible for the observed anoxia in the local environment.

Geological evidence therefore indicates that, at the end of the Permian, most of the plant life on Earth disappeared and oxygen levels plummeted worldwide within a relatively short time. Paleontologists now believe there was no single cause for the Permian mass extinction; instead, they believe a combination of factors, occurring through a few million years, changed the environment in relatively sudden and drastic ways, leading to a global loss of species.

Plate Tectonics and the Permian Extinction

During the Permian and the early Triassic, the continents of the world were united in a single supercontinent known as Pangaea. Earth's continents change position gradually in response to forces generated within its core. Geologists are now evaluating a new theory that the continents unite into a single, unified landmass on a repeating cycle of about 500 million years. During the formation of Pangaea, both terrestrial and marine environments experienced major environmental changes that may have had significant effects on species in both environments.

The vast majority of marine organisms live in the shallow areas off the coast of continental landmasses, where nutrients are in much greater supply than in deeper waters. When the continents joined to form Pangaea, less of this continental shelf would have been available, possibly leading to an overall loss of species. As diversity fell in the shallow coastal areas, local food chains may have collapsed, eventually affecting species farther from the continental shelf and precipitating the collapse of the entire marine ecosystem.

Paleontologists also have noted evidence of seafloor regression occurring during the Permian extinction. Seafloor regression occurs when portions of the sea floor become exposed at sea level, leading to a reduction in habitat available to marine organisms. The causes of seafloor regression are poorly understood, but some geologists theorize that the phenomenon may be related to changes in the rates of seafloor spreading. Seafloor regression may have been another factor contributing to the loss of shallow marine environments, thus placing further pressure on the biota in these normally species-rich environments.

In terrestrial environments, the formation of Pangaea had major effects on habitat diversity and weather patterns. The central regions of Pangaea were isolated from the tempering effects of the ocean, and would therefore have developed into vast deserts with little vegetation and with violent shifts in temperature. Coastal areas were subject to flooding, and many formerly tropical and temperate zones would have developed into floodplains, contributing to lower levels of diversity.

In addition, the formation of Pangaea may have reduced species diversity because a single species, or group of closely related species, might have been able to colonize vast areas; when the continents were separated, separate species would evolve to fill similar niches. Paleontologists studying Permian sediment have found that certain species were exceedingly widespread, appearing in vast areas across the globe. This lack of species diversity, coupled with a drop in habitat diversity, could have helped to make the biosphere vulnerable to environmental changes.

EXTRATERRESTRIAL CAUSES

Some paleontologists have suggested that Earth may have suffered a meteor impact around the time of the Permo-Triassic extinction. Evidence suggests that a meteor impact was one of the contributing causes to the extinction at the end of the Cretaceous period that resulted in the loss of all dinosaur species, with the exception of the birds. This evidence suggests that extraterrestrial causes occasionally play an important role in mass extinction events.

In 2001, geophysicist Luanne Becker and colleagues published an article in the journal *Science*, reporting on the presence of unusual carbon molecules, called fullerenes, in sediment from the Permo-Triassic boundary. These Permo-Triassic fullerene molecules contained trapped molecules of helium and argon with isotopic compositions similar to helium and argon isotopes uncovered from extraterrestrial sources, leading Becker and colleagues to theorize that an extraterrestrial impact occurred near the boundary layer. In 2004, Becker and colleagues published the results of a study indicating that the Bedout crater, off the northwest coast of Australia, appears to have been caused by a meteor impact that occurred at the end of the Permian with sufficient size and apparent composition to be the source of the unusual sediments found at the boundary layer.

The Bedout crater is roughly 120 miles across, indicating that it may have been formed by a meteor similar in size to that which impacted Earth at the end of the Cretaceous period. Paleontologists believe that an impact of this magnitude would have vaporized most living organisms within a wide area surrounding the impact site and would have further ejected particulate matter into the atmosphere, significantly altering temperature and weather patterns on a global scale. For hundreds of years following the impact, patterns of vegetative growth would have been significantly reduced, leading to a global collapse in food chains and to the extinction of thousands of species. Changes in weather patterns brought about by the Bedout impact would have led to the loss of sufficient vegetation to cause the global drop in oxygen levels observed in sediment from the Permo-Triassic boundary.

VOLCANIC CAUSES OF EXTINCTION

Another potential factor contributing to the Permo-Triassic extinction may have been a global increase in volcanic activity. Geologists have discovered a set of large flood basalts, known as the Siberian traps, in sediment from the Permo-Triassic boundary. Flood basalts are sediments created by liquid lava floes, and the Siberian traps indicate a continuous period of eruptions that lasted for a minimum of four million years and covered with lava an area larger than Europe.

The eruptions at the Siberian traps would have devastated the environment for hundreds of miles surrounding the eruption site, while gases and particulate matter ejected into the atmosphere from the plume of the active volcanoes might have led to worldwide devastation in the biosphere. Sulfur dioxide released from a volcanic eruption of this type would react with water to form sulfuric acid, which may have rained on the earth thousands of miles from the eruption site for millions of years. Sulfuric acid would devastate plant life across the planet, and it also would poison oceanic environments, accounting for a precipitous drop in global oxygen levels.

In addition, carbon dioxide released in volcanic plumes, especially through such a long period, could have had a major effect on global warming. Isotope levels measured from sediment at the Permo-Triassic boundary indicate a minimum of a 9 degree Celsius increase in temperature during the period. Geologists theorize that the volcanic eruptions at the Siberian traps may have been sufficient to account for the observed increase in temperature.

METHANE GAS INCREASES

Paleontologist James Kennett has suggested that the Siberian traps may have destabilized global deposits of methane hydrate, or clathrate, which is a lattice of ice-like molecules that forms when water combines with gaseous methane. Methane hydrate exists in large quantities in the modern environment within the polar oceans and under layers of permafrost in terrestrial environments. Though the substance resembles sheets of frozen water, it is highly reactive to heat.

Benton has theorized that global warming may have raised the temperature of the oceans during the Permian extinction. As the temperature of the oceans increased, large deposits of methane hydrate may have destabilized and returned to gaseous form. As this occurred, the methane gas would have bubbled to the surface of the oceans in large quantities, contributing to the buildup of greenhouse gases. When methane released from the ocean reacted with oxygen in the atmosphere, it would form carbon dioxide, thereby further contributing to the warming trend.

Benton and others have suggested that the carbon dioxide effect could have led to the establishment of a feedback cycle, wherein global warming caused the release of large deposits of methane, which then further increased warming and led to the release of additional greenhouse gases. Taken as a whole, this warming cycle could have caused the extinction of most plant life in terrestrial environments, and it could therefore account for the anoxic conditions observed in Permo-Triassic sediment.

The Triassic Recovery

In the wake of the Permian extinction, both the terrestrial and marine environments were sparsely populated. Research indicates, however, that life was quick to respond to the empty environment. A 2011 study from the University of Rhode Island indicates that, in some environments, genetic diversity among the terrestrial biota returned to levels similar to those before the Permian mass extinction within only five million years. Paleontologists believe that Earth recovered quickly because those species that remained were forced to explore new habitats and resources, producing evolutionary pressures that led to a rapid explosion of species in a relatively short period of evolutionary time.

During the early Triassic, the few marine species that survived the extinction colonized the continental shelf around Pangaea. As the supercontinent began to break apart in the Jurassic, marine diversity increased exponentially. Whereas in the Permian many oceanic species were bottom-dwelling organisms, few of these species survived; the Triassic seas became home to a larger proportion of organisms that were able to swim and travel to obtain necessary resources. Overall, marine diversity increased globally because of the Permian extinction.

Some paleontologists have theorized that the radiation of species in the newly empty environment created a situation in which a large number of organisms were competing for a limited supply of materials. This competitive regime then fueled the evolution of new species that were able to capitalize on alternative food sources and types of habitats.

In terrestrial environments, the Permian extinction allowed for an explosive radiation of conifers as one of the dominant forms of terrestrial vegetation. The archosaurs, a group of reptiles that would eventually give rise to the first mammals, underwent an evolutionary explosion in the wake of the Permian extinction, spreading across the world and evolving into a variety of species, ranging from small herbivores to large predators. The reptile lineage in general experienced a significant advantage in the Triassic, including the evolution of the first dinosaur species.

On the whole, the Permian mass extinction led to higher levels of diversity in all environments on Earth. Paleontologists have found similar changes in the biota accompanying other mass extinction events. As such, mass extinction has been a driving force in the evolution of life and biological diversity on the planet.

Micah L. Issitt

Further Reading

Elewa, Ashraf M. T. *Mass Extinction.* New York: Springer, 2008. Comprehensive discussion of mass extinctions and their role in shaping the history of life on Earth. Chapter 6 provides a detailed discussion of the Permian mass extinction with a focus on how it stimulated an increase in the biodiversity of life on Earth.

Erwin, Douglas H. *Extinction: How Life on Earth Nearly Ended 250 Million Years Ago.* Princeton, N.J.: Princeton University Press, 2006. Detailed discussion of the Permian extinction and associated issues in biology, paleontology, and geology. Contains detailed discussions of the leading theories regarding the cause of the Permian mass extinction, including volcanism, sea-level changes, and extraterrestrial impact theories.

Foote, Michael Joseph, et al. *Principles of Paleontology.* New York: Macmillan, 2007. Revised edition of a classic text on paleontology giving a basic introduction to field and laboratory methods used in the study of fossils and of prehistoric life. Chapter 10 provides a discussion of the Permian mass extinction and its potential effects on the evolution of life. Also contains information about the Mesozoic radiation of life.

Lieberman, Bruce M., and Roger Kaesler. *Prehistoric Life: Evolution and the Fossil Record.* Hoboken, N.J.: Wiley-Blackwell, 2010. Overview of research into paleontology and the study of macroevolution. Contains a section detailing current extinction theories and the effect of extinctions on the history of life.

Raup, David. *Extinction: Bad Genes or Bad Luck?* New York: W. W. Norton, 1992. Overview of extinction theories and the debate surrounding mass extinctions presented by a leading scientist in the field. Contains a discussion of the Permian extinction and its significance to the evolution of life in the Mesozoic.

Taylor, Paul D. *Extinctions in the History of Life.* New York: Cambridge University Press, 2004. Academic discussion of extinction and extinction theories from the perspective of how extinction fuels the evolution of life on Earth. Contains a discussion of the Permian extinction that includes an overview of leading theories regarding the potential causes leading to the extinction event.

Ward, Peter Douglas. *Out of Thin Air: Dinosaurs, Birds, and the Earth's Ancient Atmosphere.* Washington D.C.: National Academies Press, 2006. Detailed discussion about the evolution of Earth's atmosphere and how it facilitated and guided the evolution of life on Earth. Chapter 7 discusses the evolution of endothermy in the wake of the Permian extinction and its effects on the biosphere.

See also: Biogeography; Biostratigraphy; Cambrian Diversification of Life; Dinosaurs; Evolution of Birds; Evolution of Flight; Evolution of Life; Fish-Amphibian Transition; Fossil Fish; Fossil Record; Fossil Reptiles and Amphibians; Geomicrobiology; Life's Origins; Mass Extinctions; Mesozoic Era; Molecular Paleontology; Paleoecology; Pangaea; Radiometric Dating; Seafloor Spreading; Trace Fossils.

PETRIFIED WOOD

Petrified wood is the fossilized remains of ancient forests that were saturated with mineral-filled water, which converted the woody tissues into stone. Such preserved trees are studied by scientists interested in learning about prehistoric plants and their environments.

PRINCIPAL TERMS

- **Miocene:** geological epoch of the Tertiary period in the Cenozoic era, beginning about 26 million years ago
- **Oligocene:** geological epoch about 38 million years ago in the Tertiary period of the Cenozoic era
- **Pliocene:** geological epoch in the Tertiary period of the Cenozoic era that began about 12 million years ago
- **Pteridophyte:** plants, such as ferns, that have leaves, stems, and roots but no seeds or flowers
- **silicification:** a process in which molecules of silica that are dissolved in groundwater replace the organic material in fallen tree trunks and branches
- **strata:** layer of earth where petrified fossils are located according to the time period in which the organism lived
- **Triassic:** period of time about 225 to 195 million years ago at the beginning of the Mesozoic era, when dinosaurs lived
- **xylem:** plant cells through which water and nutrients move

Petrifying Ancient Forests

The Latin word *petros,* meaning rock, is the source for the scientific term "petrification." Petrified wood is actually stone remnants from prehistoric trees. During the Triassic era, gymnosperms—seed-producing trees without flowers, such as gingkos and conifers—grew over much of the earth's landmass. Volcanic eruptions triggered tremors, lightning, and heavy rains, which washed trees from higher elevations down to swampy valleys. As they were pushed downhill, the trees were stripped of their bark, branches, and roots from the force of the water's impact, and they also broke into pieces. Usually, trees soaking in deep, muddy water would decay, but silt rapidly and completely covered these trees, preventing exposure to oxygen and inhibiting aerobic decomposition. Volcanic ash in the floodwater consisted of inorganic compounds such as magnesium carbonate and iron sulfide, and the trees also absorbed silicon dioxide (called silica) that had dissolved in groundwater. The minerals filled the spaces between cells in the tree trunks and branches. Molecules of these inorganic materials replaced molecules of organic tissues. During the next millions of years, wood gradually became stone in the process of silicification. Assisted by extreme pressure and temperatures, the silica that was lodged in the wood was transformed into quartz. Plants that have undergone petrification are also referred to as being permineralized.

The trees remained preserved under the soil for millions of years until soil erosion and shifting plates exposed them. Manganese, lithium, copper, and iron created patterns of bright colors as wood fossilized. Some petrified wood displays varying rings of vivid colors, resembling agates. Other pieces are brown and look like driftwood. More significant than its beauty, petrified wood reveals information about the history of plants on the earth. Unlike other fossils that are seen as an impression or compression, petrified wood is a three-dimensional representation of its organic material that preserves its external shape and internal structure. The preserved tree trunk sections also indicate the size of the Triassic forest. Scientists have even seen chromosomes and stages of nuclear division in petrified cells. A termite nest was discovered in one petrified log, offering clues about that insect's communal evolution.

Identifying Petrified Wood

Geologists and paleobotanists analyze petrified wood samples to specify the type of ancient tree that became fossilized. A piece of petrified wood from a hardwood tree is carved into a small cube approximately 2.5 centimeters square. The cell structure of hardwood fossils is more diverse than that of softwood trees, causing its source to be more easily identified. Scientists choose pieces that exhibit an intact cell structure. Using a saw and grinding wheel as needed, they flatten the cube's surfaces and sand them to preserve the natural rings of the tree and to remove any marks made while shaping the cube.

Researchers then examine the specimen with a microscope at ten times, forty times, and one hundred times magnification to scrutinize how the wood's bands and pores are arranged, both of which are crucial identifying characteristics. Softwood samples require greater magnification (four hundred times) of thin slivers only one or two cells thick in order to permit enough light to shine through when mounted on a slide.

Wood anatomists then describe the sample's cell structure, which they compare to records of previously identified petrified wood and existing trees. North Carolina State University and the International Association of Wood Anatomists created a computer program known as the General Unknown Entry and Search System (GUESS) , which efficiently checks its databases for matching cell patterns. In addition to databases that contain information about existing hardwood and softwood trees, another database for fossil hardwoods has information on at least 1,356 types of trees of more than 1,200 species that have been identified. Codes direct researchers to scientific journal citations for detailed illustrations.

A petrified tree trunk, c. 1890, on Specimen Ridge, Yellowstone National Park. (U.S. Geological Survey)

Early identification of petrified wood relied on seeking similarities with existing trees even though the petrified sample may have represented an extinct tree type. Because most petrified wood lacks fruit or leaves to aid categorization, paleobotanists had to focus on unique aspects of cell structure to distinguish fossil trees. As petrified wood samples were described in journals, scientists developed models for comparison to determine the geological age and geographical source of fossil wood. Identification remained difficult, however, because the cell structures of some trees resembled those of other species more than trees in their own families.

Three types of petrified wood are found in the Tertiary strata: nondescript silicified wood, petrified palm wood, and massive silicified wood. Nondescript silicified wood has undergone silicification but still appears to be woody structurally. Difficult to identify because of its generic structure, nondescript silicified wood requires expert authentication for accurate labeling. Petrified palm wood has rod structures that reinforced the tissue strength before the wood grain became silicified and that look like spots or lines when the wood is cut. Popular among rock collectors, this type of petrified wood is also Arizona's state fossil. Massive silicified wood is difficult to recognize because the tree's grain was destroyed during silicification, thus making identification reliant on awareness of the area in which the tree was located and comparison to other petrified wood in adjacent territory.

Petrified Wood in the United States

Arizona's Petrified Forest National Park is one of the best-known areas with petrified wood. This wood originated from a forest of giant conifer-like trees that grew from Texas to Utah. Located in northeastern Arizona, the park's 93,533 acres is home to one of the largest and most varied assortment of petrified wood in the world. The desert is dotted with stone log fragments. Although some visitors expect to see rock trees standing in clumps similar to a natural forest, these petrified trees rest where they fell individually or in groups.

The park features four primary petrification zones. At Blue Mesa, petrified logs have emerged from an area of blue- and gray-colored clay. The

Jasper Forest consists of a valley scattered with petrified wood. At one time it was covered with logs, but nineteenth-century prospectors collected the stones to sell. Nearby, a petrified tree trunk called Agate Bridge crosses a stream. The Rainbow Forest contains the largest amount of petrified wood in the national park, as well as some of the largest logs, some of which are several meters long. The Agate House is a hut that sixteenth-century Native Americans constructed completely from petrified wood. The Crystal Forest's petrified wood represents what was once a vast landscape of logs sparkling with quartz and purple amethyst crystals. By the twentieth century, prospectors and tourists had taken the most beautiful pieces; as a result, local residents sought government protection against further theft. In 1906, the petrified wood site was declared a national monument, finally achieving national park status in 1962.

The National Park Service tries to protect petrified wood by preventing the nearly 1 million annual visitors from seizing samples. Rangers patrol sites and ask tourists to report any thefts they witness. Despite these precautions, several tons of petrified wood disappear annually from the Arizona park. As a deterrent, local superstitions have been encouraged to warn people that illegally acquired fossils will bring them bad luck. Privately owned sites adjacent to the park offer collectors opportunities to search for petrified wood without any restrictions. Stores sell small pieces that have been artistically cut and polished at affordable prices, which people use to decorate their homes, gardens, and even aquariums. Petrified wood is both Arizona's state fossil and its state rock.

Because the ancient forest lived simultaneously with the dinosaurs, archaeologists look for dinosaur fossils near petrified wood in the Arizona park. Although collection and thefts have greatly reduced the number of petrified logs, authorities believe that some areas of the park may shelter petrified wood buried as much as 100 meters beneath the surface.

Petrified wood has been discovered in many other regions of the United States, especially in areas in the West where volcanic activity occurred, such as Yellowstone National Park, and in areas where rivers and streams deposited large amounts of sand, such as Louisiana and Texas. Washington State is home to the Ginkgo Petrified Forest State Park, which contains petrified logs that began fossilizing during the Miocene epoch. This petrified wood is unique because it includes petrified ginkgo, an indigenous tree that no longer grows naturally there. An unusual type of petrified wood that resembles pebbles and that originated in the Chehalis Valley is sometimes seen on the state's beaches.

The Calistoga, California, petrified forest is considered one of the best sources of Pliocene fossils similar to existing redwoods. Measuring more than 2 meters in diameter, the fossil logs reveal gray stone veins of quartz. Petrified wood in New Mexico's Bisti Badlands is not as colorful as neighboring Arizona's fossil wood. In Utah, petrified wood has been found near the Escalante River and the Coyote Buttes region near the Paria River. Petrified wood has also been located in the southern states of Mississippi and Alabama.

Petrified Wood Worldwide

Scientists have also found petrified forests in other parts of the world. A petrified forest on the Greek island of Lesvos was named a protected site by presidential decree. A museum was established to encourage scientific research about this Greek forest's origins. Scientists have determined that the petrified wood began fossilizing during the Late Oligocene to Lower-Middle Miocene epochs. Unlike other sites of petrified forests, the Lesvos fossil trunks are erect and still have intact roots penetrating into fossilized soil, as well as fossilized branches, leaves, cones, and seeds.

Scientists realize that these trees were fossilized where they grew, offering insight about the environment and climate of ancient Lesvos. The ancient trees of Lesvos were well preserved during the petrification process, and such details as rings indicating age and growth patterns are visible. Paleobotanists have determined that both gymnosperms and angiosperms (flowering trees) grew in the ancient forests, along with pteridophytes such as ferns. Many of these ancient species no longer grow in the Mediterranean. Instead, they are found in Asian and American tropical and subtropical regions, indicating that Lesvos's petrified forest presents information about how the earth evolved as the continents moved apart.

Other sites of fossilized wood also reveal details about the planet's development. Argentina's Petrified Forest on the Central Steppes was created after the formation of the Andes Mountains. Larger than American wood fossils, Argentinean petrified wood includes pieces 27 meters long and 3 meters in diameter. The petrified forest in Namibia on the African continent contains giant tree trunks as long as 30 meters. Petrified wood has also been discovered in Australia, India, England, Turkey, and Switzerland.

Petrified Wood's Legacy

Paleobotanists research petrified wood to determine how the earth and the plants that have grown on its surface have changed since ancient geological periods. They study how plants are related and descended from similar ancestors. In addition to compiling lists and descriptions of thousands of modern tree species, scientists have noted existing trees that are represented by petrified wood and have cataloged types of ancient trees identified solely by fossil samples. Paleobotanists realize that during the earth's existence many more tree species have existed than are documented and that many species, existing and extinct, await discovery. They note that pieces of fossil wood are often the oldest known specimens of a tree species and might be the predecessors of living trees.

Petrified wood helps researchers comprehend how trees have evolved to adapt to environmental and climatic conditions. For example, researchers have hypothesized that the cell structure of tree xylem has not changed as much as fruit and leaves. Fossil wood also sometimes reveals the reason for the demise of extinct trees. The study of petrified wood has altered how scientists perceive both ancient geological ecosystems and modern environments by offering perspective on which plants lived on the earth and survived, adapted, or died according to changes in the atmosphere and crust.

Elizabeth D. Schafer

Further Reading

Agashe, Shripad N. *Paleobotany: Plants of the Past, Their Evolution, Paleoenvironment, and Application in Exploration of Fossil Fuels.* Enfield, N.J.: Science, 1997. A discussion of how ancient trees became petrified and how they are beneficial to environmental research. Illustrated with photographs of fossil chromosomes and diagrams of cell structure. Also includes an index and bibliographical notes.

Andrews, Henry Nathaniel. *The Fossil Hunters: In Search of Ancient Plants.* Ithaca, N.Y.: Cornell University Press, 1980. Contains biographical information about paleobotanists who have studied petrified wood. Illustrated and indexed, with bibliographical references. Useful reference for high school students wishing to supplement more technical discussions.

Baldridge, W. Scott. *Geology of the American Southwest: A Journey Through Two Billion Years of Plate Tectonic History.* New York: Cambridge University Press, 2004. Provides a sweeping overview of the geology of the southwestern United States, beginning with Precambrian events and continuing through modern time. Covers plate tectonics and orogeny. Accessible to undergraduate and graduate students.

Daniels, Frank J., Richard D. Dayvault, and Brooks B. Britt, eds. *Petrified Wood: The World of Fossilized Wood, Cones, Ferns, and Cycads.* Grand Junction, Colo.: Western Colorado, 1998. This elaborately illustrated book covers petrified wood found around the world. The text explains how wood petrifies, what its scientific applications to paleobotany and geology are, and how to collect petrified wood. Contains an index and bibliographical sources. Suitable for readers of all ages.

Hickey, Leo J. *The Forest Primeval: The Geologic History of Wood and Petrified Forests.* New Haven: Yale Peabody Museum, 2010. This text discusses the properties of wood and explains how the properties of wood have evolved. The author highlights the history of humans using wood. This text also discusses how petrified forests developed and where petrified wood has occurred in the fossil record.

Lubick, George M. *Petrified Forest National Park: A Wilderness Bound in Time.* Tucson: University of Arizona Press, 1996. Lubick discusses the scientists and custodians who have studied and protected fossil wood in Arizona's Petrified Forest National Park. Illustrations, maps, index, and bibliography. Suitable for high school readers.

Prothero, Donald R. *Bringing Fossils to Life: An Introduction to Paleobiology.* 2d ed. Boston: McGraw-Hill, 2004. Prothero explains how fossil wood offers clues to the evolution of existing trees from ancient species. Includes photographs and drawings, bibliographical sources, and index. An excellent reference for students.

Stewart, Wilson N., and Gar W. Rothwell. *Paleobotany and the Evolution of Plants.* 2d ed. New York: Cambridge University Press, 2010. This text provides fundamental information about the science of studying plant fossils to identify species past and present and to determine how they are related. Bibliographical references and index.

Wilson, Stewart, and Gar Rothwell. *Paleobotany and the Evolution of Plants.* 2d ed. New York: Cambridge University Press, 2009. Explains in a clear way the origin and evolution of plants. Well illustrated and written at the level of undergraduate students.

See also: Archaebacteria; Biostratigraphy; Cambrian Diversification of Life; Catastrophism; Colonization of the Land; Dinosaurs; Ediacaran Fossils; Eukaryotes; Evolution of Birds; Evolution of Flight; Evolution of Humans; Evolution of Life; Fossilization and Taphonomy; Fossil Plants; Fossil Record; Fossils of the Earliest Life-Forms; Fossils of the Earliest Vertebrates; Geoarchaeology; Gondwanaland and Laurasia; Ice Ages; Life's Origins; Mammals; Mass Extinctions; Microfossils; Neanderthals; Paleobiogeography; Paleobotany; Paleoclimatology; Prokaryotes; Stromatolites; Systematics and Classification.

POPOCATÉPETL

The Popocatépetl volcano is located in central Mexico, 40 kilometers east of the outskirts of Mexico City. It is in sight of 30 million people, most of whom are living on the debris from previous eruptions. The volcano has erupted explosively in the past, and it became active again during the 1990's, raising major concerns about the safety of people in this heavily populated area.

PRINCIPAL TERMS

- **andesite:** a volcanic rock with intermediate silica content that contains the minerals plagioclase, biotite, and hornblende; because the molten equivalent of andesite is often viscous, gases can build up in it and produce explosive eruptions
- **ashfall:** fine-grained volcanic material that settles out of the atmosphere
- **ash flow:** a moving mass of hot ash and gas that flows from a volcano; the hot material is deposited as it spreads out on flatter areas
- **lahar:** a volcanic mudflow formed when volcanic ash and minor coarse volcanic particles mix with the abundant rainfall that often accompanies eruptions
- **Plinian eruption:** a rapid ejection of large volumes of volcanic ash that is often accompanied by the collapse of the upper part of the volcano
- **pumice:** volcanic glass, usually light in color, that contains so many bubbles that it may float in water
- **subduction:** a tectonic process in which one lithospheric plate is underthrust below another plate into the mantle

VOLCANISM IN MEXICO

The Popocatépetl volcano, located in central Mexico, rises to a height of 5,465 kilometers above sea level. Although there is usually snow on its summit, oranges, mangoes, bananas, and palms grow at its base.

The volcano is the result of tectonic activity in the region, where the Cocos oceanic plate is underthrust or subducted below the Caribbean or Americas continental plates. The subduction process produces melting along or above the subducted plate, resulting in the creation of fairly viscous melted rocks such as andesites. These stiff or viscous melts do not readily flow up through the volcano, often resulting in a plug of partially solidified lava that causes a buildup of gases. The gas pressure may eventually become so great that it explodes out of the volcano at its weakest point.

Such an explosion may take the form of a Plinian eruption, in which copious amounts of volcanic ash, gas, and larger particles are blown upward out of the central cone. This type of eruption has occurred at Popocatépetl several times throughout history. As the particles settle out of the atmosphere, they produce ashfalls. Sometimes the ash and gas may be blown out laterally and flow down the sides of the volcano as ash flows. Heavy rains produced by hot updrafts over the hot, central eruptions may mix with the abundant volcanic ash, producing lahars that move rapidly and for many kilometers downstream and river channels.

ERUPTION HISTORY

Scientists believe that Popocatépetl has experienced major Plinian eruptions every one thousand to three thousand years since 4000 B.C.E. Some of the most recent of these events have destroyed human settlements. Large eruptions occurred sometime between 3195 and 2830 B.C.E., between 800 and 215 B.C.E., and between 675 and 1095 C.E. Each of these eruptive cycles began with ashfalls and ash flows in small volumes, producing a number of 2- to 10-centimeter-thick volcanic layers. The subsequent phase of Plinian eruptions produced a thick sequence of coarser-grained pumice fragments. It is estimated that the eruption column of fragments and gas would have reached heights of more than 25 kilometers. This would have had a major impact on world climate as the particles spread high into the atmosphere. Some of these major eruptions also produced lahars and hot ash flows that moved rapidly down valleys and were deposited on the plains.

The area around Popocatépetl has long supported fairly large populations because of the fertile soils, the presence of water, and the climate. For example, agricultural and ceremonial towns were established in the area between 1000 and 100 B.C.E. Pottery shards, building foundations, and agricultural furrows dating to this era have been found in pumice deposits, suggesting that the settlements suffered widespread destruction as the result of volcanic activity. Subsequently, areas near the eruptive center declined in population between 100 B.C.E. to

100 c.e., while areas to the north became more populated.

By 750 to 800 c.e., several large cities containing as many as 150,000 people each had been established in the region surrounding Popocatépetl. Two of these, Cholula and Teotihuacán, were large religious centers. Archaeologists have determined that Cholula was abandoned around 800 c.e., a phenomenon that was accompanied by a significant decrease in population in the general area. A variety of reasons were initially proposed for this decline, including drought, invasion, food shortages, and the exhaustion of soil nutrients. However, this time period also correlates with a major Plinian eruption in which more than 3,000 square kilometers of the area was covered by deposits from the eruption. The remains of lahars at the base of the pyramid in Cholula were found to contain pieces of pottery and volcanic glass artifacts. Several researchers who have studied these eruptions have suggested a scenario in which the area was covered by a muddy plain of lahars, with only the great pyramids projecting above the mud.

Popocatepetl volcano erupting. This volcano (5,452 meters elevation) is the second-highest peak in Mexico. It had been dormant for around sixty years when it became active again in 1994. Moderate eruptions and earthquakes have occurred every few months, as of early 2001. (Wesley Bocxe/Photo Researchers, Inc.)

Modern Activity

After a long period of dormancy, Popocatépetl became active again during the early 1920's. The first stages of activity began with steam and ash emissions during the summer of 1920. A lava dome formed on the floor of the crater in the latter part of the same year, then stopped growing. The dome eventually exploded during a minor eruption in January of 1922.

Relatively little activity occurred at Popocatépetl between 1922 and December of 1994. During that time, researchers noted a gradual increase in earthquake activity and gas emissions enriched in sulfur dioxide. On December 21, 1994, ash emissions also began, which continued into the middle of 1995. The activity then quieted until emissions resumed on March 5, 1996. Ash was observed covering snow and glaciers on the summit, and ash and gas were thrown as high as 800 meters above the crater. A viscous dome of lava about 400 square meters in area was observed on the central crater floor on March 29, 1996, indicating that the lava phase of the eruption had begun. Lava continued to flow until most of the inner crater was filled.

On April 30, a large explosion blew a hole in the surface of the lava dome. This event resulted in the deaths of five climbers who had defied warnings to stay away from Popocatépetl. A rain of warm volcanic debris was deposited in nearby towns. Particles 0.5 centimeter in diameter were observed 12 kilometers away from the volcano, while sand-sized material was observed at a distance of 60 kilometers.

Volcanic activity continued sporadically through late 1997 as lava domes periodically built up on the crater floor, only to be destroyed by often spectacular explosions. An explosion on October 28, 1996, at 9:05 A.M., emitted large amounts of ash and gas. This was followed by a lesser explosion the following day at 10:11 P.M. Minor explosions also occurred on December 21, 1996, and between February 17 and June of 1997. An eruption on April 24, 1997, produced an ash and gas plume 13 kilometers high and 8 kilometers wide. It eventually spread out to a plume that was 65 kilometers long. Meanwhile, the lava body first observed within the crater in March 1996 remained unchanged.

The largest ash emission since the beginning of the eruption cycle in 1994 occurred on June 30, 1997. It began with several earthquakes related to the volcanic activity at 4:56 P.M. The first of the ash eruptions began at 5:11 P.M. and lasted 135 minutes. The second ash eruption began at 7:26 P.M. and lasted for 90 minutes. The ash plume rose to more than 13 kilometers within a few minutes, and ash began falling in towns around the volcano two to three hours later. The ashfall shut down the airport in Mexico City for twelve hours, leading to numerous press reports. A red alert was posted, but no evacuations were ordered. No casualties were reported, but minor lahars were reported in an area with heavy rain, causing the flooding of one house. Several lava tongues up to 2 kilometers long flowed south to southeast down the volcano on July 3 and 4.

More small, sporadic eruptions were reported through the end of 1997, with the most significant eruptions occurring on August 12, in late September, in early October, and in early to late December. The biggest eruption during this period occurred on December 24, during which incandescent material was thrown from the volcano and grass fires were started to the east. Intermittent activity continued through 1998 and 1999, with several significant eruptions. A thirty-second explosion sent ash 5 kilometers into the atmosphere on January 1, 1998. Larger explosions began in late November of 1998, producing ash plumes 3 kilometers high. Ashfalls were reported on the southwest side of Mexico City. Explosions on November 25 broke windows and walls in several villages 60 kilometers away, and large volcanic blocks landed 5 kilometers away. People were warned to stay at least 7 kilometers away from the volcano. Large explosions in December 1998 shattered windows, fractured walls, opened doors, and knocked tiles off roofs. Popocatépetl continued to be active through 1999. Significant eruptions occurred on March 17, March 23, June 16, and June 22. The eruption on March 17 was large enough to result in evacuation centers being put on alert.

Significance

Much has been learned about the long-term eruptive cycles of volcanoes through the study of ancient volcanic deposits. If authorities can understand the potential danger of Popocatépetl in terms of its historic eruption cycle, they can set up better evacuation procedures in case of a catastrophic eruption. The authorities have such evacuation and emergency procedures in place should a large eruption cycle begin.

Scientists are also interested in the effects of volcanic eruptions on the global climate. Several of Popocatépetl's major eruptions appear to have been powerful enough to send volcanic ash into the stratosphere, allowing it to spread worldwide. For example, analysis of ice cores from the Greenland ice sheet has suggested that a major eruption occurred somewhere in the world around 822 or 823 C.E., which corresponds to the approximate date of a major eruption of Popocatépetl. Such large eruptions appear to affect the climate of the world by lowering the penetration of sunlight, thus causing a cooling of the surface of the earth.

Robert L. Cullers

Further Reading

Bullard, F. M. *Volcanoes of the Earth.* 2d ed. Austin: University of Texas Press, 1984. A nontechnical and readable discussion of the location and origin of many volcanoes. It contains sections on Plinian eruptions, ash flows, and lahars. A glossary is included.

Lockwood, J. P., and Richard W. Hazlett. *Volcanoes: Global Perspectives.* Chichester: Wiley-Blackwell, 2010. An up-to-date and well-illustrated undergraduate level text that is also accessible to the interested reader.

Macdonald, G. A. *Volcanoes.* Englewood Cliffs, N.J.: Plinian Press, 1972. Macdonald gives detailed accounts of volcanoes and volcanic eruptions. There is a lot of terminology, but it is generally defined. The book contains sections on Plinian eruptions, ash flows, and lahars, and includes an appendix listing active volcanoes and their locations.

Oppenheimer, Clive. *Eruptions That Shook the World.* New York: Cambridge University Press, 2011. Describes the mechanics of volcanoes and types of eruptions. The author also outlines the effects of eruptions on the atmosphere, humans, and other organisms. Specific volcanoes are examined and an appendix of major eruptions is provided.

Siebe, C., M. Abrams, J. L. Macias, and J. Obenholzer. "Repeated Volcanic Disasters in Prehispanic Time at Popocatépetl, Central Mexico: Past Key to the Future?" *Geology* 24 (1996): 399-422. This article is in a technical geology journal, but it has a minimum of jargon. It summarizes the main eruptive cycles of Popocatépetl in terms of age and archeological studies.

Stuart, G. E. "Teotihuacan." *National Geographic* 188 (1995): 2-35. This article, which contains numerous photographs, discusses the excavation of Teotihuacan, as well as evidence for its decline.

White, S. E. "Popocatépetl: The Ever-Burning Torch." *Volcano News* 17 (1984): 1-3. An easily read overview of Popocatépetl.

See also: Eruptions; Forecasting Eruptions; Krakatau; Mount Pelée; Mount Pinatubo; Mount St. Helens; Mount Vesuvius; Recent Eruptions; Shield Volcanoes; Stratovolcanoes; Volcanic Hazards.

PROKARYOTES

Prokaryotes are primitive, one-celled organisms that have left an extensive fossil record in the form of sedimentary structures produced by physiological activity of cell communities. For 80 percent of the earth's history, communities of prokaryotes made up the biosphere of the earth. They are a well-defined group of organisms and occupy a highly diverse variety of habitats.

PRINCIPAL TERMS

- **aerobes:** prokaryotes (usually bacteria) that live in the presence of elemental oxygen
- **anaerobes:** prokaryotes that can live only in an atmosphere that is free of elemental oxygen
- **deoxyribonucleic acid (DNA):** a molecule made up of two strands of nucleotides arranged in a double helix; the molecular basis of heredity
- **grazing and cropping eukaryotes:** single-celled (protists) or multicelled (metazoans) eukaryotes (cells with a definite nucleus), which appear in the fossil record during the close of the Precambrian eon, about 1 billion years ago
- **nucleotide:** a molecule made up of a series of amino acids that, when linked together, are capable of carrying genetic information
- **Precambrian eon:** the first 3.5 billion years of the geologic record; it is followed by the Paleozoic era of the Phanerozoic eon
- **ribonucleic acid (RNA):** a complex compound made up of nucleotide bases that acts as a template, or "messenger," in the replication of DNA
- **ribosome:** a large multienzyme found associated with cell nuclear materials, composed of RNA and protein molecules

Characteristics of Prokaryotes

From about 3.5 to 1 billion years ago, life on Earth, as determined from the fossil record, consisted entirely of one-celled organisms that have a cell morphology and a metabolism different from those of all other life-forms. These organisms, the prokaryotes, are characterized by their lack of a cell nucleus, their lack of sexual reproduction (meiosis), the small size of the prokaryotic cell, and their distinctive biochemistry. Prokaryotes are neither plants nor animals, although the aerobic photosynthetic forms, often called blue-green algae, have in the past been placed with the plants. Eukaryotes, organisms with a cell nucleus and a larger, more complex cell, make up the animals, plants, fungi, and protists. Prokaryotes are thus quite separate from all other life-forms in terms of their cell biology. Prokaryotes and eukaryotes are the two most basic categories of living things, exhibiting basic differences in their biologic processes that are greater than those that exist between animals or plants or between any of the other kingdoms.

Prokaryotes are mainly single-celled organisms, usually found living together in "colonies" consisting of immense numbers of cells. Their deoxyribonucleic acid (DNA) is distributed throughout the cell; it is not, as in the case of the eukaryotes, localized in a cell nucleus surrounded by a nuclear membrane. The prokaryotic cell is smaller by a factor of ten than the average eukaryotic cell. It lacks chloroplasts and mitochondria and consequently is considered primitive when compared with the eukaryotic cell.

Prokaryotes constitute the kingdom Monera, one of five kingdoms in modern taxonomy. (The other kingdoms are the protists, fungi, animals, and plants, all of which have more complex eukaryotic-type cells.) Phyla, or categories, within the kingdom Monera include the bacteria, the cyanobacteria (or blue-green algae), the archaebacteria, and the prochlorophytes. The bacteria, as well as the other moneran phyla, are further subdivided into a number of classes. Bacterial classes of the Monera include the eubacteria, photosynthetic bacteria, myxobacteria (slime bacteria), actinomycetes (moldlike bacteria), and other groups, each characterized by its own distinctive metabolism and biochemistry. The bacteria consist of obligate or strict anaerobes and facultative anaerobes; the former include the photosynthetic bacteria, which differ from the cyanobacteria not only in their ability to function, if required, under anaerobic conditions and low light levels, but also in their different photosynthetic pigment.

The archaebacteria are considered by some to be the most primitive and ancient of the monerans. Archaebacteria have a number of biochemical and metabolic characteristics that allow them to live under very adverse conditions—conditions such as those that appear to have existed during the early history of the earth. The archaebacteria are defined from their ribosomal ribonucleic acid (RNA), which in sequencing is quite different from that of all other monerans. The archaebacteria differ fundamentally from the other bacteria classes in structural and biochemical aspects as well.

Geological Significance

Fossil prokaryote cells of great antiquity have become widely known from the fossil record. They were first reported in the 1910's by C. D. Walcott from 1.5-billion-year-old strata of western Montana (Belt series). However, the authenticity of these fossils was doubted until the discovery, in 1954, of one of the oldest known paleontological "windows" on life of the past, the 2-billion-year-old Gunflint biota. Since then, many occurrences of prokaryote cell fossils have been reported, most from very fine-textured flinty cherts associated with stromatolites of the Proterozoic (latter part of the Precambrian) eon and dating from as far back as 3.5 billion years.

The geologic significance of the prokaryotes is great: Not only do they (at present as well as in the geologic past) play an important part in the recycling of many chemical elements, but they also have a role in basic geologic processes such as weathering and other alteration of rocks. For example, prokaryotes are involved in the formation of stromatolites. Stromatolites are layered organosedimentary structures, frequently found fossilized in rock strata of many different geologic ages. Stromatolites come in a considerable variety of shapes and sizes; the different types have often been given Linnaean biological names because, when originally discovered, they were thought to be fossil organisms like corals or sponges. Most stromatolites are dome-shaped, finger-shaped, or laminar structures that have a characteristic "signature." They can form significant parts of rock strata, particularly in limestone and dolomites. Stromatolites are found in rock strata as ancient as the Archean (former part of the Precambrian) eon and are particularly diverse and abundant in strata of the Proterozoic. Locally, they can be quite common in early Paleozoic marine strata as well.

The origin of stromatolites was debated for many years; as late as the 1950's, many paleontologists seriously doubted their biogenic origin. This doubt stemmed, at least in part, from the fact that stromatolites occur so much further in the geologic past than do any other fossils. Through thousands and thousands of meters of Precambrian strata, they are the only fossils that can be found. Early workers on stromatolites, such as Walcott, suggested a cyanobacterial origin for them. The discovery of the well-preserved cells of prokaryotic type in digitate (fingerlike) stromatolites of the Gunflint Chert of Ontario in the 1950's led to a gradual acceptance by most geologists and paleontologists of the organic origin of the majority of stromatolites. It became clear that, under the right conditions, small, fragile cells could be preserved in very ancient strata.

During the 1970's and 1980's, studies on Precambrian stromatolites and the prokaryotic organisms responsible for them became widespread. Stromatolite occurrences going as far back as 3.5 billion years have been documented. These ancient stromatolites yield not only morphological information but also carbon isotope ratios indicative of a biogenic origin. They sometimes supply biochemical information in the form of hydrocarbons, amino acids, and porphyrins (the latter is apparently a degradation product of original photosynthetic pigment). In 1999, studies of Proterozoic rocks from western Australia confirmed the presence of chemicals that could only have been synthesized by cyanobacteria.

The morphology of a prokaryotic organism is simple. Unlike fossils of eukaryotic organisms, fossils of most prokaryotes provide little specific information about the actual living organism. Prokaryotic cells can be single coccoid (spherical) forms, or they can be elongate chains of cells, as with the filaments, or trichomes, of the cyanobacteria.

Study of Prokaryotes

A standard petrographic thin section mounted on a glass slide is the common mount for observing cells preserved in a stromatolite. Oil immersion is usually required if fossil prokaryote cells are to be observed. Thin slivers of a stromatolite can also be examined under oil immersion; however, the best results are generally with well-made thin sections. Often considerable trial and error is involved in finding stromatolites that preserve cells and then in actually locating those cells; different parts of a particular stromatolite specimen usually have varying degrees of cell preservation. Very fine-grained sediments, such as those that occur with stromatolites preserved by black cherts or finely crystalline limestones, generally give the best results.

In this section under high optical magnification, a stromatolite may exhibit fossil cyanobacterial cells as either filaments or rod-shaped forms. If preservation of these small prokaryote cells is excellent, as in the stromatolites of the Gunflint formation, the biogenic origin of the cells will be clear, and distinct cell types can be observed. When most stromatolites are examined in thin section under high magnification, however, the biogenicity of the small objects seen is usually not so certain. Often, small black globules of carbon, suggestive of macerated cells, are evident, but their origin usually cannot be proved. Contaminants such as spores, pollen grains, bacteria cells, and fungi fragments can be a problem, particularly in examination of suspected fossiliferous rocks when thin sections are not used. Even with most thin sections, the unequivocal verification of a biogenic origin for fossil cells is rare. In the case of the Gunflint prokaryotes, the detail preserved in these fossil cells is highly remarkable; some of them show internal cell structure and cells in the process of division.

The earliest stromatolites that yield these fossil cells are generally either broad domes or laminar forms. Associated cells either are single-cell coccoid forms or consist of probable chains of photosynthetic bacteria. Chains of cells of filamentous cyanobacteria generally first appeared about 2.3 billion years ago, and this appearance of filaments agrees fairly well with the first appearance of branched or digitate stromatolites, for which filamentous cyanobacteria seem to be responsible.

Chemical Signatures

Often more significant than single-cell morphology or the megascopic morphology of a stromatolite is the chemical signature left by a group of prokaryotes as a consequence of their metabolic activity. Prokaryotes are classified according to their type of metabolism; some prokaryotes have a metabolism that enables them to occupy a wider variety of ecological niches than do eukaryotic organisms. Anaerobic and aerobic forms are the two fundamental forms of prokaryotic metabolism. In these two categories are the autotrophs and the heterotrophs; heterotrophic prokaryotes require previously formed organic material on which to live, while autotrophs do not. The autotrophs obtain their energy from their environment either in the form of sunlight (photoautotrophs) or through chemical reactions such as oxidation, as in the sulfur-oxidizing bacteria; such bacteria are called chemoautotrophs. This type of metabolism is unique in the organic world, for all other life-forms obtain their energy from photosynthesis or through utilization of the chemical energy contained in previously formed organic compounds. The cyanobacteria are photoautotrophs and are responsible for the formation of the various types of stromatolites. The process of photosynthesis changes the microenvironment around the photosynthesizing prokaryote; the mineral precipitation that results is responsible for the formation of stromatolite layers.

Some stromatolites contain oxidized manganese, cobalt, or other "transitional" elements, possibly incorporated into these fossil communities by oxidative metabolism of bacteria. Chemoautotrophic prokaryotes, which are various types of bacteria, may leave a chemical signature in the form of these oxides and precipitate their production of a layered stromatolite-like structure containing these oxidized metals. A number of bacteria oxidize manganese to higher oxidation states so that it is precipitated; deep-sea manganese nodules presently being formed are believed to have such an origin. Sectioning of these nodules shows a finely layered, stromatolite-like structure. Some of the heavy-metal-bearing stromatolites of the early Precambrian may reflect a similar chemoautotrophic metabolism. Analysis of organic residues present in many stromatolites in small quantities can sometimes shed light upon the specific organisms responsible for forming them. This technique, however, has met with only limited success, although degradation products of the photosynthetic pigment present in cyanobacteria have been identified, supporting the cyanobacterial origin of many ancient stromatolites.

The earliest stromatolites, those of Archean age (about 3.5 billion years old), exhibit certain distinctive morphological and chemical aspects. Some of these early stromatolites may be products of anaerobic assemblages of photosynthetic bacteria rather than of cyanobacteria communities. Geochemical evidence suggests that the atmosphere in the Archean may have been anoxygenic (oxygen-free) and that the photosynthetic bacteria, not being obligate aerobes, would have been favored by such an environment.

Bruce L. Stinchcomb

Further Reading

Broadhead, T. W., ed. *Fossil Prokaryotes and Protists: Notes for a Short Course.* Knoxville: University of Tennessee, 1987. Information on a broad range of fossil prokaryotes and protist microfossils is presented. This work is the text of one of a series of "short courses" sponsored by the Paleontological Society, but it can be useful to anyone interested in the various fossil groups covered.

Fedonkin, Mikhail A., James G. Gehling, Kathleen Grey, Guy M. Narbonne, and Patricia Vickers-Rich. *The Rise of Animals: Evolution and Diversification of the Kingdom Animalia.* Baltimore: Johns Hopkins University Press, 2007. An up-to-date account of all the known Ediacaran sites that is extensively and beautifully illustrated. Accessible to all interested readers.

Gunde-Cimerman, Nina, Aharon Oren, and Ana Plemenita. *Adaptation to Life at High Salt Concentrations in Archaea, Bacteria, and Eukarya.* New York: Springer, 2011. This text provides information on high-salinity habitats and the halophilic organisms found there. The environment overview is followed by sections devoted specifically to Archaea, bacteria, fungi, protozoa, and viruses.

Margulis, Lynn, and Michael J. Chapman. *Kingdoms and Domains: An Illustrated Guide to the Phyla of Life on Earth.* 4th ed. Boston: Elsevier, 2009. Provides a comprehensive overview (current at the time of publication) of life's classification and phylogeny. An improvement on Margulis's previous work, *Five Kingdoms,* this text includes more recent discoveries and systematics. The classification is a modified version of Linnaean taxonomy.

McMenamin, Mark. *Discovering the First Complex Life: The Garden of the Ediacara.* New York: Columbia University Press, 1998. This entertaining study of the earliest complex life-forms on the planet details the author's work on these organisms. Written for the interested student but understandable by the general reader.

Nisbet, Evan G. *The Young Earth: An Introduction to Archean Geology.* Boston: Unwin Hyman, 1987. A highly comprehensive coverage of geologic phenomena of the earth's earliest geologic time span, the Archean eon. Included in this work is information on both crustal evolution and biosphere. The book's sections vary considerably in technical coverage and terminology, some parts being readily comprehensible to the lay reader while others are quite technical and require considerable background in trace element geochemistry, isotope geochemistry, and petrology.

Schopf, J. William, ed. *Major Events in the History of Life.* Boston: Jones and Bartlett, 1992. An excellent overview of the origin of life, the oldest fossils, and the early development of plants and animals. Written by specialists in each field but at a level that is suitable for high school students and undergraduates. Although technical language is used, most terms are defined in the glossary.

Yaacov, Davidov, and Eduard Jurkevitch. "Predation Between Prokaryotes and the Origin of Eukaryotes." *BioEssays* 31 (2009): 738-757. Discusses how eukaryotes evolved as a result of parasitism or predation occurring between prokaryotes. Covers the topic of mitochondrial symbiosis.

See also: Archaebacteria; Cambrian Diversification of Life; Catastrophism; Cenozoic Era; Colonization of the Land; Dinosaurs; Ediacaran Fossils; Eukaryotes; Evolution of Birds; Evolution of Flight; Evolution of Humans; Evolution of Life; Fossilization and Taphonomy; Fossil Plants; Fossil Record; Fossils of the Earliest Life-Forms; Fossils of the Earliest Vertebrates; Geoarchaeology; Life's Origins; Mammals; Mass Extinctions; Mesozoic Era; Microfossils; Neanderthals; Paleobiogeography; Paleobotany; Petrified Wood; Stromatolites; Systematics and Classification.

PROTEROZOIC EON

The Proterozoic eon is the interval between 2.5 billion and 542 million years ago. During this period in the geologic record, many of the geological processes presently active on earth first appeared. This includes the first clear evidence for plate tectonics. Rocks of the Proterozoic eon also document changes in conditions on Earth, particularly an apparent increase in atmospheric oxygen.

PRINCIPAL TERMS

- **Archean eon:** the period of geologic time from about 3.6 billion to 2.5 billion years ago
- **Hadean eon:** the period of time before the Archean and lasting from 4.5 to 3.6 billion years ago.
- **orogenic belt:** a belt of crust that has been severely compressed, deformed, and heated, probably by convergence of crustal plates
- **paleomagnetism:** the study of magnetism preserved in rocks, which provides evidence of the history of Earth's magnetic field and the movements of continents
- **Phanerozoic eon:** the period of geologic time with an abundant fossil record, extending from about 544 million years ago to the present
- **plate tectonics:** the theory that the outer surface of the earth consists of large moving plates that interact to produce seismic, volcanic, and orogenic activity
- **Precambrian:** the collective term for all geologic time before the Phanerozoic—that is, before about 544 million years ago
- **radiometric dating:** the use of radioactive elements that decay at a known rate to determine the ages of the rocks in which they occur
- **terrane:** a structurally distinct block of crust added to a continent by plate tectonic processes

Earth's Four Eons

The largest subdivision of geologic time is the eon. Four eons constitute the known history of earth: Phanerozoic, Proterozoic, Archean, and Hadean. The Phanerozoic, whose abundant fossil record permits fine subdivisions of geologic time and intercontinental correlation of strata, is the interval about which the most is known. The Proterozoic, by contrast, is generally characterized by a sparse fossil record of simple life-forms that do not permit the sort of stratigraphic subdivision possible in the Phanerozoic. Ages of rocks in the Proterozoic can be established only by radiometric dating. For rocks as old as the Proterozoic, the inherent uncertainty in even the best dating is on the order of 10 million to 20 million years, comparable in length to some Phanerozoic periods.

The Proterozoic has now been subdivided into eras and periods, which are comparable to those in the Phanerozoic. The Hadean is not divided but the Archean now comprises the Eoarchean (3.6 to 3.2 billion years ago [bya]), the Mesoarchean (3.2 to 2.6 bya), and the Neoarchean (2.8 to 2.5 bya). This is followed by the Proterozoic, which is divided into three eras. The Paleoproterozoic (2.5 to 1.6 bya) consists of the Siderian, Rhyacian, Orosirian, and Statherian periods. It is followed by the Mesoproterozoic (1.6 to 1.0 bya), comprising the Calymmian, Ectasian, and Stenian periods, and the Neoproterozoic (1.0 to 0.542 bya), comprising the Tonian, Cryogenian, and Ediacaran periods.

Much of the significance of the Proterozoic derives from the first appearance of processes that are still operating on earth. The Archean, in contrast, may have experienced a quite different set of processes. It is therefore impossible to discuss the Proterozoic without also discussing the Archean to some extent. There are many unanswered questions about major earth processes in the Archean. In particular, the Archean lacks structures that closely resemble Phanerozoic orogenic belts. Whether plate tectonics operated in its present form during the Archean is still unresolved. Dynamic processes within the earth may have undergone a significant change about 2.5 billion years ago.

The contrast between Archean and Proterozoic geology is striking, if still imperfectly understood. The Archean is dominated by two principal types of regional structure: greenstone belts and gneiss-migmatite terrains. Greenstone belts are troughs of volcanic rocks and deep-water sedimentary rocks intruded by elliptical granite bodies. Gneiss-migmatite terrains are bands of very highly metamorphosed and deformed rocks. Structures similar to both the greenstone belts and the gneiss-migmatite terrains have formed throughout the history of the earth and can form through conventional plate tectonic processes, but the almost complete lack of other types of structures in Archean time is perplexing. It is widely (though not universally) believed that the Archean crust was more mobile than at later times and that large masses of continental crust did not exist then. The first extensive shallow-water rocks deposited in a continental-shelf or stable continent setting occur in the latest Archean rocks of South Africa and are abundant throughout the Proterozoic and Phanerozoic. The appearance of such rocks may mark the first appearance of stable continental crust.

Crustal and Atmospheric Events

During the Proterozoic, there is the first widespread evidence for plate tectonics and orogeny (mountain building) comparable with that of the Phanerozoic. There seem to have been two major periods of plate collision during which large continents were assembled out of small continental plates. North America was assembled between about 1.9 and 1.6 billion years ago by the collision of several smaller Archean blocks plus the accretion of many smaller terranes. This episode is known as the Trans-Hudson Event. The overall process was very similar to the accretion that is known to have added large areas to western North America during the Phanerozoic. Between about 900 and 600 million years ago, South America and Africa also appear to have been assembled in much the same way to form part of the early Phanerozoic supercontinent of Gondwanaland. The sequence of events that assembled South America and Africa is called the Pan-African Event.

Some Proterozoic crustal events are still poorly understood. In particular, there was a widespread heating event that resulted in the intrusion and eruption of silica-rich igneous rocks (granite and rhyolite) between about 1.5 and 1.3 billion years ago across much of North America. This interval also resulted in the intrusion of large bodies of anorthosite, an igneous rock made mostly of feldspar minerals, on many continents. Anorthosite is otherwise uncommon, and the reason it was formed so extensively in the Proterozoic is not known.

A number of significant developments took place on the surface of the Proterozoic Earth. There is evidence for increasing oxygen content of the atmosphere, as indicated by the appearance of the first "red beds"—red sandstone and conglomerate colored by iron oxide. Also, there were a number of major ice ages during the Proterozoic. One occurred about 2.3 billion years ago, and there is widespread evidence for a series of ice ages between about 900 and 600 million years ago.

The early Earth probably lacked free oxygen, except in minor amounts. Free oxygen is not a component of the raw materials that formed the planets, and oxygen is a highly reactive gas that would rapidly have combined with other substances. Oxygen originally was given off as a waste product by early organisms and is actually toxic to some organisms even today. How and when the present oxygen level of the atmosphere was attained is controversial. Many scientists consider that increasing oxygen was related to the abrupt expansion of life at the start of the Phanerozoic. There is also evidence that a significant threshold in oxygen level was crossed about 2.0 to 1.8 billion years ago.

Sedimentary Deposits

Several types of sedimentary deposit, all closely related to the availability of free oxygen, either ceased or began to be deposited about 1.8 billion years ago. Detrital uranium deposits and banded iron formations, both of which are believed to indicate low oxygen levels, become very uncommon in the geologic record after that time, while red beds appear at roughly the same time. Detrital uranium deposits are sandstone deposits in which dense minerals were concentrated by current action. This process is common today, but the Proterozoic deposits include uranium minerals, which are highly susceptible to weathering, and even pyrite (iron sulfide), which oxidizes extremely quickly and which is almost never found in recent sedimentary deposits.

A detrital uranium deposit at Okolo, in the African nation of Gabon, is remarkable for another reason. The uranium ore from the deposit was found to be greatly depleted in uranium-235, the isotope used in nuclear reactors. Further investigation showed that the ore deposit had been a natural nuclear reactor. Present-day uranium is too poor in uranium-235 to sustain a nuclear chain reaction except under very controlled conditions. Yet 2.2 billion years ago, when the Okolo deposit formed, uranium-235 was far more abundant—abundant enough for a sustained chain reaction to begin when sedimentary processes collected enough uranium minerals in one place. So far the remarkable deposit at Okolo is unique; other detrital uranium deposits were studied for effects of natural chain reactions, and no other examples were found.

Another type of ore deposit common in early Proterozoic rocks is banded iron formations. Banded iron formations consist of layers of iron oxides and silicates interbedded with fine-grained silica, or chert. The significance of these ore deposits, apart from their great importance as sources of iron, lies in the degree of oxidation of the iron. Iron in nature has two oxidation states. Ferrous iron, the less oxidized state, consists of iron atoms that have lost two electrons and thus have a positive electric charge of +2. Ferric iron atoms have lost three electrons and have a +3 charge. Common rust is mostly ferric iron oxide, as would be expected in the earth's presently oxygen-rich atmosphere. The iron in banded iron formations, however, is mostly ferrous, even though it was deposited on the surface of the earth. The ferrous iron in banded iron formations presents one of the strongest arguments for an oxygen-poor atmosphere on the early Earth. The iron was originally dissolved in seawater and was probably precipitated by microorganisms. Indeed, some of the best-preserved Proterozoic fossils are those of microorganisms preserved in the chert of banded iron formations. Ferrous iron is far more soluble in water than is ferric iron, so the Proterozoic seas may have been richer in dissolved iron than are the present oceans. Iron deposits precipitated by microorganisms have also formed in the Phanerozoic, but the iron is mostly ferric.

Both detrital uranium deposits and banded iron formations become rare in the geologic record after 1.8 billion years ago. At about the same time, red sandstones appear in abundance. Red sandstones owe their color to ferric (highly oxidized) iron and are most commonly deposited on land or in shallow water. One of the earliest extensive red bed deposits is in northwestern Canada. These rocks were deposited shortly after the crust of the region had experienced a major orogeny about 1.8 billion years ago and represent debris eroded from the newly formed mountain range. A deposit of this sort is termed molasse. These red beds furnish not only evidence for the oxygen level of the atmosphere but also some of the first evidence for topographically high mountains.

GLACIATION

The Proterozoic experienced a number of major glacial events about 2.3 billion years ago and again from 900 to 600 million years ago. The intervening period appears to have been largely ice-free. The glaciation 2.3 billion years ago is best documented in the Gowganda formation of Ontario, where virtually every type of glacial deposit occurs. Other deposits in Wyoming and Quebec have been interpreted as glacial deposits of the same age, but the Gowganda formation remains the clearest evidence for a 2.3-billion-year-old glaciation.

The glacial deposits 900 to 600 million years old are far more problematical. Many of these rocks are diamictites, fine-grained sedimentary rocks with scattered large pebbles. Diamictites cannot have been simply deposited by running water. Slow-moving water could not have transported the pebbles, while sediment deposited by fast-moving water would have a much higher proportion of coarse material than do diamictites. Only a few plausible ways of forming diamictites exist. One way is for floating glacial icebergs to melt and drop trapped rocks into otherwise fine-grained sediment. When the sediments are finely layered, the occurrence of so-called dropstones is generally taken as clear evidence of glaciation. Diamictites can also form from nonglacial processes. For example, submarine landslides can mix a small amount of coarse debris into a large amount of fine sand and silt. The glacial origin of many diamictites has been controversial, and some of the late Proterozoic diamictites may not be glacial.

An additional problem with the late Proterozoic glaciation is the geographic distribution of the deposits. Glacial deposits by themselves indicate cold climates, but in many cases there are nearby carbonate rocks (dolomite and limestone) of similar age that typically form in warm climates. Also, paleomagnetic studies indicate that some areas with glacial deposits, notably Australia, were at low latitudes during the late Proterozoic. A number of possible explanations, none entirely satisfactory, have been proposed. A few geologists find a flaw in the evidence, either in the climatic indicators or the paleomagnetic evidence. Others propose that the late Proterozoic Earth was abnormally cold, still others that the temperature decrease with altitude was much greater than at present, allowing warm climates at sea level but glaciers at even moderate elevations.

Throughout most of the Proterozoic, simple life-forms dominated on Earth. The first life-forms, already established in the Archean, were prokaryotes, organisms without a cell nucleus, such as bacteria and blue-green algae. About 2 billion years ago, eukaryotes, or organisms with a cell nucleus, appeared. The only common Proterozoic fossils of large size are stromatolites, domelike masses of calcium carbonate or silica deposited by colonies of algae in shallow water. The first widespread evidence for large multicelled organisms appears in rocks about 800 million years old. These organisms are rare as fossils because they lacked hard body parts but seem to have resembled jellyfish and marine worms. Many are unrelated to present-day organisms and seem to represent extinct evolutionary lines.

Steven I. Dutch

FURTHER READING

Dott, Robert H., Jr., and Donald R. Prothero. *Evolution of the Earth.* 8th ed. New York: McGraw-Hill, 2009. This basic textbook on historical geology is aimed at students of geology. However, it is very readable by anyone with a background in science. Presents an up-to-date account of the earth's history from the viewpoint of plate tectonics. Includes a glossary.

Fedonkin, Mikhail A., et al. *The Rise of Animals: Evolution and Diversification of the Kingdom Animalia.* Baltimore: Johns Hopkins University Press, 2007. Discusses the study of Ediacaran biota from initial identification through recent studies of the ecology and evolution of these species. Although the title suggests otherwise, this text focuses on Ediacarans, their diversity, and distribution.

Grotzinger, John, and Tom Jordan. *Understanding Earth.* 6th ed. New York: W. H. Freeman, 2009. This comprehensive physical geology text covers the formation and development of the earth. Readable by high school students, as well as by general readers. Includes an index and a glossary of terms.

Kusky, T. M., ed. *Precambrian Ophiolies and Related Rocks.* San Diego: Elsevier, 2004. The book is divided into four sections covering Proterozoic ophiolites, Archean ophiolites, oceanic crust models, and Precambrian analogs. The introduction alone provides an excellent summary of the collective knowledge of ophiolites. Background in geology is necessary when reading this text, as this is a compilation of scientific studies rather than a general textbook.

Hoffman, Paul F. "United Plates of America, the Birth of a Craton: Early Proterozoic Assembly and Growth of Laurentia." *Annual Review of Earth and Planetary Sciences* 16 (1988): 543-603. Probably the most comprehensive general account of the Trans-Hudson Event that assembled North America between 1.9 and 1.6 billion years ago. Uses fairly technical language. Suitable for advanced college students.

Kasting, James B., O. B. Toon, and J. B. Pollack. "How Climate Evolved on the Terrestrial Planets." *Scientific American* 258 (February 1988): 90-97. This article compares the evolution of Earth, Venus, and Mars. Its relevance to the Proterozoic is that the early sun was fainter than is the present sun, and the earth's atmosphere must have been different to have trapped enough heat to allow liquid water to exist on the early earth.

McMenamin, Mark. *Discovering the First Complex Life: The Garden of the Ediacara.* New York: Columbia University Press, 1998. This entertaining study of the earliest complex life-forms on the planet details the author's work on these organisms. Written for the interested student but understandable by the general reader.

Meyer, Charles. "Ore Deposits as Guides to Geologic History of the Earth." *Annual Review of Earth and Planetary Sciences* 16 (1988): 147. A synthesis of a lifetime of experience as a mining geologist. Meyer documents the changing patterns of ore deposits through geologic time, with strong emphasis on the Precambrian. Uses fairly technical language. Suitable for college students.

Ogg, James G., Gabi Ogg, and Felix M. Gradstein. *The Concise Geologic Time Scale.* New York: Cambridge University Press, 2008. This book is a complete overview of the geological time scale, including stratigraphy topics such as chronostratigraphy and magnetic stratigraphy. The book is organized by geological time periods. Also contains a reference appendix, the geological time scale table, and indexing.

Schopf, J. William, ed. *Major Events in the History of Life*. Boston: Jones and Bartlett, 1992. An excellent overview of the origin of life, the oldest fossils, and the early development of plants and animals. Written by specialists in each field but at a level that is suitable for high school students and undergraduates. Although technical language is used, most terms are defined in the glossary.

Stanley, Steven M. *Earth System History*. 3rd ed. New York: Freeman, 2008. An introductory undergraduate-level text integrating coverage of life and the environment through time.

Van Schmus, W. R., and W. J. Hinze. "The Midcontinent Rift System." *Annual Review of Earth and Planetary Sciences* 13 (1985): 345-383. A comprehensive description of a major rifting event in central North America about 1.1 billion years ago. Uses fairly technical language. Suitable for advanced college students.

Wicander, Reed, and Monroe, James S. *Historical Geology*. Belmont Calif.: Brooks/Cole, Cengage Learning, 2010. This textbook is accessible to university students with or without a background in geology. It covers Earth's geological systems and plate tectonics, physical and biological history, geological time, and evolution. There are many diagrams, images, and charts along with a CD-ROM to supplement the text.

See also: Biostratigraphy; Cenozoic Era; Cretaceous-Paleogene Boundary; Fossil Plants; Geoarchaeology; Geologic Time Scale; Geomorphology of Dry Climate Areas; Geomorphology of Wet Climate Areas; Ice Ages; Land Bridges; Mesozoic Era; Paleoclimatology; Paleogeography and Paleocurrents; Paleoseismology; Paleosols; Paleozoic Era; Stratigraphic Correlation; Transgression and Regression; Unconformities; Uniformitarianism.

R

RADIOMETRIC DATING

Radiometric dating is based on the principle of exponential decay, by which certain atoms called radionuclides undergo nuclear fission, changing their elemental identities in specific ways. Identification of the relative amounts of specific radionuclides permits determination of the amount of time that has passed since the formation of the particular sample.

PRINCIPAL TERMS

- **beta (β) particle:** a subnuclear particle emitted during nuclear fission as an electron (β–) from the decomposition of a neutron
- **exponential decay:** a process of decomposition, particularly nuclear fission, that obeys the time-related function $A = A_o e^{kt}$
- **fissible:** capable of undergoing nuclear fission
- **fission:** the splitting of a nucleus through the emission of nuclear particles to form atoms of a different element
- **fusion:** the combination of two atomic nuclei to produce a single nucleus of a higher element; the opposite of fission
- **half-life:** the length of time required for one-half of an amount of material to decompose or be consumed through an exponential decay process
- **isotope:** an atom of a specific element that contains different numbers of neutrons in its respective nucleus
- **labeling:** the incorporation of atoms of a detectable radionuclide into the normal molecular structure of a material
- **radionuclide:** a specific radioactive isotope of an element
- **transmutation:** the conversion of one element into another by natural or artificial fission or fusion processes

Atoms and Isotopes

According to modern atomic theory, the structure of atoms consists of three basic particles in a specific structural arrangement. These particles are the positively charged proton, the negatively charged electron, and the electrically neutral neutron. Protons and neutrons are of essentially equal mass, being 1,834 and 1,835 times as massive as an electron, respectively.

Atoms are described according to quantum mechanics as having a small dense nucleus containing almost all of the mass of the atom, as protons and neutrons, surrounded by a diffuse cloud of electrons. The identity of the atom is determined solely by the number of protons contained in its nucleus. Each neutral atom has an equal number of electrons and protons. The identity of the atom also determines the identity and chemical properties of the chemical element.

The nucleus of any atom may contain a different number of neutrons from other atoms of the same element. Such atoms are called isotopes, meaning atoms that have the same elemental identity according to the number of protons but a different number of neutrons. Some isotopes are entirely stable, but others are not stable and spontaneously break down by the process of nuclear fission to produce either the stable isotopic form of the same atom or atoms with different chemical identities.

Exponential Decay

The rate at which a radioisotope breaks down follows a specific exponential decay pattern that is described mathematically by the equation $A = A_o e^{kt}$. The symbol A represents the amount of a specific isotope that is present at the time t, whereas the symbol A_o is the amount of that material that was present originally and k is the rate constant for the particular process. A special relationship exists for this formula when A is exactly one-half of A_o, which occurs after a specific time called the half-life has elapsed.

The relationship is constant for exponential decay in that it requires exactly the same amount of time for one-half of an amount of material to decompose no matter how much of the material is present. That is, the same amount of time is required for one kilogram of a material to decompose to one-half kilogram as it does for one gram of the material to decompose to one-half gram of the material. Because the breakdown begins and ends with specific materials, the exact amounts of each are also related to the elapsed time for the process. Thus the ratio of the materials that are present is an important factor in determining age by radiometric methods.

Time Scales, Identifiability, and Mass Spectrometry

Isotopic fission takes place at very different rates according to the isotopes involved. Some isotopes undergo fission quickly, while others require billions of years. Tables of isotopes and their half-lives, or lifetimes (as are commonly listed in reference books such as the *CRC Handbook of Chemistry and Physics*), show isotopic half-lives ranging from 10^{-16} seconds to more than 10^9 years. This range provides the opportunity to employ an appropriate radiometric dating method for different ages.

Each method requires that the fission process ends with the formation of stable elements so that the proportion of starting radionuclide to final element can be determined. The method also will provide the best results when the final element is uniquely identifiable within the matrix in which it is contained. These conditions of time range and identifiability have greatly limited the number of useful isotopic changes.

Mass spectrometry is the most widely used and exact method of determining the ratios of radionuclides. The technique of mass spectrometry relies on the ratio of the mass of an ion to the electrical charge on that ion as it determines the trajectory of the ion traveling through a magnetic field. According to precise mathematical relations, the ion moves along a circular path having a specific radius determined by the strength of the applied magnetic field and the charge/mass ratio of the ion. In typical techniques, a sample containing the nuclides to be measured is injected into the high-vacuum inlet of the mass spectrometer, where it is given an electrical charge and directed into the magnetic field. The detector section determines the relative numbers of ions of each mass that are present in the sample, which provides the nuclide ratio.

Nuclide Pairs

Fission of a radionuclide occurs by distinct steps governed by the law of conservation of mass. The emission of a single electron or positron, termed a "β-particle" (beta particle) from a nucleus does not reduce the isotopic mass of those atoms, but may alter their elemental identity. However, the emission of an α-particle (alpha particle), a helium nucleus containing two protons and two neutrons, changes both the mass of the atom and its elemental identity. The change is specific, and the identification of corresponding nuclide pairs as starting material and product of a fission process is fundamental to the process of radiometric dating.

Three of the most commonly used nuclide pairs for radiometric dating are carbon-14 and nitrogen-14 ($C^{14} - N^{14}$), potassium-40 and argon-40 ($K^{40} - Ar^{40}$), and uranium-238 and lead-206 ($U^{238} - Pb^{206}$). In the transition from C^{14} to N^{14}, it is believed that one of the extra neutrons in C^{14} decomposes into a proton by emitting a β- particle (an electron) to produce the more stable N^{14} atom. The half-life of C^{14} is 5,730 ± 30 years. The $C^{14} - N^{14}$ transition is thus most useful for the radiocarbon dating of materials from organic sources and carbonaceous minerals.

The $K^{40} - Ar^{40}$ transition has a half-life of 1.28×10^9 years and presumably occurs by a similar but opposite process in which a proton and an electron combine to produce the extra neutron of the resulting highly stable argon atom. The $U^{238} - Pb^{206}$ transition is a more complicated process, occurring by α-particle decay through rapid intermediate stages resulting in the formation of lead-206. The common key feature of these transitions is that the atoms of the elements that end the corresponding chain are entirely stable and do not decompose further.

Radiocarbon Dating

The factor that makes radiocarbon dating by the $C^{14} - N^{14}$ transition feasible originates in the constant bombardment of cosmic ray neutrons that impinge upon Earth's atmosphere. These neutrons convert some of the normal N^{14} of atmospheric nitrogen (N_2) into the C^{14} radionuclide. This labeled carbon is then taken up along with normal CO_2 in various biological and mineralization processes. Photosynthesis is the most common process by which C^{14} is incorporated into organic materials, while the formation of carbonate minerals is the predominant means of incorporation of C^{14} into inorganic materials. The rate of incorporation does fluctuate, but is essentially constant in the time in which plants and animals are functionally alive and ceases immediately upon the death of the organism. The amount of C^{14} remaining in the material, compared with the amount in living organisms, provides the number of half-lives that have occurred since the replenishment process ceased.

The radiocarbon dating of inorganic materials is somewhat more demanding in principle, because no living system is involved in the process and because the rates of incorporation are more speculative. Stalactites taken from caves containing human and other remnants may be dated absolutely by radiocarbon methods by measurement of the ratio of C^{14} in specific samples of the material and can be dated relatively by comparison with the dating of any accompanying organic materials.

Radiocarbon dating is most appropriate for materials between one hundred years and fifty thousand years old. Before one hundred years, insufficient conversion of C^{14} has taken place to allow an accurate determination of age, while after fifty thousand years, insufficient C^{14} remains to permit comparison. In both cases, the error limits of the determination exceed the precision of the measurement.

Mineralogical Dating

The K^{40} – Ar^{40} transition and the U^{238} – Pb^{206} transition overlap extensively, and they are most useful for mineralogical dating, including determination of Earth's age. Potassium is incorporated into biological systems, particularly animal systems, according to its natural abundance for the duration of the living process; as in C^{14} dating the ratio of Ar^{40} to K^{40} in the remnant is used to calculate the age of a material such as a fossil. The half-life of K^{40} and the ready identifiability of Ar because of its chemical nature as an inert gas allow the K^{40} – Ar^{40} dating of inorganic materials and minerals from 100,000 years of age to the formation of the planet some 4.6 billion years ago. This makes it useful in geological studies involving muscovite, biotite, and hornblende minerals, and intact volcanic rock.

The U^{238} – Pb^{206} transition has a half-life of some 4.51 billion years, with the end product atoms of lead being both common, stable, and easily recovered. The precision of analyses involving the U^{238} – Pb^{206} transition is adversely affected more by the scarcity of U^{238} and the presence of several other uranium nuclides than by any difficulties with the end product, the lead atoms. The extreme longevity of the half-life, however, makes the transition suitable for radiometric dating of materials from ten million years of age to the formation of the planet. Accordingly, it is useful in geological studies involving zircon, uraninite, and pitchblende, or uranium ore.

Relative Age Versus Absolute Age

Radiometric dating provides valuable chronological data regarding the age of materials, which allows other observed phenomena and properties to be placed into a chronological context. One example of the utility of this method is in determining the ages of fossil materials and the ages of much more recent archaeological findings. An important distinction is made between the relative age and the absolute age of a material.

Relative age, as the name implies, is determined in relation to the occurrence of other materials. The applicable assumption in geology is that underlying strata are older than the strata above them. Accordingly, strata having the same or similar characteristics and material content, but at a different location, can be assigned, at least tentatively, approximately the same age. Relative dating thus does not relate the age of an object or a material to the present time.

The validity of radiometric age determination is based on assumptions about the original state of the material. The half-life of a material, for example, assumes that the material was in its pure form or was otherwise not contaminated by the presence of its breakdown nuclides at the outset. Such an assumption can be validated for many materials. For example, if an ore contains a measurable amount of U^{238} and an appropriate amount of Pb^{206}, one can conclude that the age of that ore is at least 4.51 billion years, the half-life of the original amount of U^{238} that was present in the ore. This value cannot be used to determine whether or not the ore is older than 4.51 billion years, however, because there is no comparative standard for the older age.

Absolute age requires a phenomenon or characteristic that links the age of a material directly to a known point in time and is most useful when that link is made to the present day. The best example of a mechanism for absolute age is found in annual growth rings. These provide an unequivocal determination of the age of the tree, fish scales, or other structure in which they are found. In the same way, the relative proportions of nuclides in a material, according to a specific nuclide pair transmutation, provide a direct link from the present time to the date of origin of the material within the practical error limits to which such a determination can be made.

Richard M. Renneboog

Further Reading

Dalrymple, G. Brent. *Ancient Earth, Ancient Skies: The Age of Earth and Its Cosmic Surroundings.* Stanford, Calif.: Stanford University Press, 2004. Chapter 4, "How Radiometric Dating Works," is devoted to the principles of radiometric dating and also provides a list of isotopic materials used for dating rocks and minerals.

De Pater, Imke, and Jack J. Lissauer. *Planetary Sciences.* 2d ed. New York: Cambridge University Press, 2010. This book discusses radiometric dating in the context of Earth and planetary sciences. It also provides detailed breakdown sequences for some important nuclide pairs.

Dossetto, Anthony, Simon P. Turner, and James A. van Orman, eds. *Timescales of Magmatic Processes from Core to Atmosphere.* Hoboken, N.J.: Wiley-Blackwell, 2011. A series of specialist papers includes chapters discussing radiometric dating in the context of the geological time scale of magmatic processes.

Hunt, Roy E. *Characteristics of Geologic Material: A Field Guide for Geotechnical Engineers.* Boca Raton, Fla.: CRC Press, 2007. A minor section describes basic principles of radiometric dating, but the book itself provides useful data and information on geological formations and their age.

Leddra, Michael. *Time Matters: Geology's Legacy to Scientific Thought.* Hoboken, N.J.: Wiley-Blackwell, 2010. This book contains a thorough examination of the methods and meaning of geological time measurement.

Walker, Mike. *Quaternary Dating Methods.* Hoboken, N.J.: John Wiley & Sons, 2005. Describes various dating techniques used for materials from the Quaternary period, explains how each method works, and outlines where and how errors are likely to occur.

See also: Carbonate Rocks; Fossil Record; Geomicrobiology; Life's Origins; Minerals; Molecular Paleontology; Paleoecology.

RECENT ERUPTIONS

Volcanoes and earthquakes are dramatic expressions of the fact that the earth is a living, dynamic planet. Unfortunately, because of previously erroneous concepts regarding them as well as inadequate descriptive techniques, most written records of specific eruptions undertaken before 1800 are now of limited value. Since then, however, a number of particularly instructive examples have been witnessed.

PRINCIPAL TERMS

- **ash:** rocky, unconsolidated ejecta of sand-grain size
- **eruption:** volcanic activity of such force as to propel significant amounts of magmatic products over the rim of the crater
- **lava:** molten rock (as opposed to ash) erupted by a volcano
- **nuée ardente:** a sudden basal surge of incandescent, heavier-than-air gas

Although volcanic eruptions are a normal part of the earth's workings, they have always been regarded with awe because of their unpredictability and power. Only within the last two hundred years has scientific understanding of them been achieved. During that time, certain specific eruptions (and recurring activity of certain volcanoes) have been especially influential.

TAMBORA

Located on the island of Sumbawa, about 160 kilometers east of Java, Tambora volcano erupted in April 1815—the exact date and preliminary activity are uncertain—creating the greatest explosion and the most powerful volcanic eruption of modern times. There were no surviving eyewitnesses, as virtually the entire population of Sumbawa died. The noise of the explosion was heard 1,500 kilometers away, and its atmospheric effects were felt worldwide. As clouds of volcanic ash circulated throughout the stratosphere, 1816 became in America and Europe "the year without a summer," during which normal harvests were greatly curtailed.

GRAHAM ISLAND

Another otherwise routine Vesuvian eruption of October 22-26, 1822 (together with its predecessor of 1818), prompted George Julius Poulett Scrope to study the volcanoes of Europe. His *Considerations on Volcanoes* (1825) inaugurated the modern science of volcanology. Graham Island was a submarine volcano that appeared off the southern coast of Sicily for a few months in 1831, only to disappear by year's end. The first volcano of any kind to be observed almost daily throughout its existence, it was considered by such European geological theorists as Scrope and Sir Charles Lyell to be an example of fundamental importance in their debates regarding the formation of volcanoes. The forceful but otherwise routine eruption of Vesuvius on April 24-30, 1872, stimulated Luigi Palmieri, director of the Volcano Observatory on the slopes of Mount Vesuvius, to formulate his influential theory that volcanic eruptions are often characterized by predictable phases. This eruption was also the first major eruption to be photographed.

KRAKATAU

Krakatau was a three-part volcanic island located between Java and Sumatra. After preliminary activity beginning on May 20, 1883, it erupted explosively on August 26 and 27, the reports being heard nearly 5,000 kilometers away. Tsunamis generated by the eruption killed more than 36,000 people. As with Tambora, subsequent atmospheric effects were worldwide, generating much scientific and popular interest. Dutch and British investigators published important studies.

TARAWERA

Following some preliminary earthquakes, the top of domelike Mount Tarawera, on the North Island of New Zealand, split apart on June 10, 1886, exposing the internal structure of the dome and creating a series of steam-blast craters. When fissuring extended beneath Lake Rotomahana, further steam explosions followed, destroying two world-famous hot-spring formations, the White and Pink terraces. Ash and mud from the eruption buried several Maori villages and killed more than 140 people.

Bandai-san

After more than one thousand years of relative inactivity, Bandai-san volcano (on the island of Honshu) erupted suddenly on July 15, 1888, for less than two minutes initially. The north slope of the mountain was demolished, burying the villages and people of an entire valley with avalanches and mudflows. Although only a steam explosion, it was the worst volcanic disaster in the history of Japan.

La Soufrière

La Soufrière at the northern end of the island of St. Vincent in the West Indies became active on May 6, 1902, expelling its crater lake in a series of mudslides. Its major eruption, on May 7, was the first modern one in which glowing avalanches (nuées ardentes) were recognized. They sped down all sides of the mountain, causing little damage. The thrust of the eruption had been vertical, with the glowing avalanches on its slopes a secondary phenomenon caused only by fallout and gravity. Activity continued into March 1903.

Mount Pelée

Nearby Mount Pelée, on the island of Martinique, was active from April 24, 1902, to October 5, 1905 (and again from September 16, 1929, to December 1, 1932). After giving ample warning, it erupted violently on May 8, 1902—one day after Soufrière on St. Vincent—but the nueés ardentes this time were ejected straight down the western slope of the volcano. Thus, searing, suffocating clouds of gas and ash rushed like an avalanche to engulf the city of St. Pierre, destroying it and all but two (some sources say four) of its citizens—about thirty thousand people in all. Second only to the destruction of Pompeii by Vesuvius in 79 C.E., this volcanic disaster was the most famous of modern times. It greatly stimulated interest in volcanoes, attracted two major figures (Thomas A. Jagger and Frank A. Perret) to their study, established a new type of eruption (the Peléan), and initiated a dispute regarding nuées ardentes that evidence from its own later eruptions resolved.

Alaskan Eruptions

Accounts of a major Alaskan Peninsula eruption from June 6 to June 9, 1912—the most powerful modern one on the North American continent—differ with age. Current understanding is that the primary effusion radiated from a newly formed volcano named Novarupta on the flank of Mount Katmai rather than from Katmai itself. (There were no nearby eyewitnesses, and no scientific investigations before 1916.) In any case, explosions were heard 1,000 kilometers away, ash fell thickly at a distance of 160 kilometers, and an adjacent valley floor was transformed into a steaming fumarolic basin, the so-called Valley of Ten Thousand Smokes (with rapidly decreasing activity). The old crater of Mount Katmai later collapsed, perhaps because the liquid rock (magma) underlying it had been withdrawn. Aside from its forcefulness, Katmai is significant for being the only example of an ignimbrite (fused rhyolite tuff) eruption in modern times.

Lassen

Lassen Peak in Northern California is, like Tarawera, a volcanic dome, and one of the largest known. Long dormant, it resumed activity on May 30, 1914, becoming particularly violent one year later, from May 19 to May 22, 1915, when directional nuées ardentes coursed down its shattered slopes to devastate whole forests and create powerful mudslides. Some less important lava flows then followed, with activity continuing into 1917. Prior to the eruption of Mount St. Helens in 1980, the 1915 eruptions of Mount Lassen were the most famous recent ones to have taken place within the United States.

Parícutin

Parícutin emerged overnight from a cornfield in central Mexico on February 20, 1943. This fragile but rapidly growing cinder cone reached a height of 50 meters the next day; lava began flowing from its base rather than its crater, destroying a nearby farm. In June 1944, it buried an entire village. After years of spectacular but relatively harmless activity, Parícutin ended as abruptly as it had begun, on February 25, 1952, its height being 410 meters. Parícutin was the first wholly new volcano to arise in North America since nearby Jorullo (1759; poorly observed) and attracted unprecedented scientific attention.

Hekla

Hekla, the best-known Icelandic volcano, erupted memorably on March 29, 1947, after more than one century of dormancy. Large amounts of ash and other debris were ejected from a fissure, and lava flows poured down the volcano's slopes from a series of fissure-related vents. After nearly thirteen months of continuous activity (ending April 21, 1948), approximately 1 billion cubic meters of lava had been extruded, in addition to immense volumes of ash. Eruptions of this type do not usually endanger human life, but Iceland's livestock and farming industries were severely affected, both from the destruction of arable land and from the widespread emissions of toxic gases.

Lamington

Prior to 1951, Mount Lamington, on the northern side of Papua New Guinea, was regarded as extinct. It consisted of a deeply eroded complex of volcanic domes, with heavy forestation and no record of previous activity. On January 15, however, earthquakes and emissions began, and then gradually intensified. The major eruption took place on May 21, featuring a huge eruption cloud and glowing avalanches of the St. Vincent type. These avalanches devastated an area of more than 230 square kilometers and killed about six thousand people. Later eruptions were of the Vulcanian type and did not feature glowing avalanches but generated mudflows. The volcano's activity, which continued into 1956, then culminated in dome-building.

Bezymianny

Bezymianny ("no name"), in Kamchatka, was also thought to be extinct. After one month of seismic activity, eruptions began on October 22, 1955. The most violent blast, on March 30, 1956, destroyed the top 200 meters of the volcano and most of one side. Nuées ardentes, followed by huge ash and pumice flows, rushed down adjacent valleys. Dome formation then concluded the volcano's activity, all of which was meticulously recorded by Soviet volcanologists at a nearby observatory.

Ilha Nova, I and II

The Azores are a group of volcanic islands in the north Atlantic Ocean. A new vent adjacent to the island of Fayal opened on September 27, 1957, creating the island volcano of Capelinhos. Called Ilha Nova I ("new island") in this manifestation, it lasted only one-half month before disappearing (as Graham Island did in 1831). On November 7, Ilha Nova II appeared, soon joining the island of Fayal as a new peninsula. It then fell into dormancy after little more than one year of activity. The resulting cone resembles others, such as Diamond Head in Hawaii, formed by underwater explosive eruptions.

Surtsey

The best-studied submarine eruption of all is that of Surtsey, a newly created island in the Westman Islands off the southern coast of Iceland. Named for a fire giant in Norse mythology, Surtsey announced its presence with undersea turmoil on November 14, 1963. An island appeared the next day and grew rapidly, with every step in its progress being closely observed. The main vent became inactive for a time (beginning May 7, 1965), but two associated ones then spewed up smaller independent islands (Syrtlingur and Jolnir) of their own; both subsequently disappeared. All activity ended on June 5, 1967, yet Surtsey continues to be of scientific interest as ecologists note its increasingly diverse complement of plant and animal life.

Heimay

Heimay ("home island") is the largest and most permanent of the Westman Islands, the same group to which Surtsey belongs. It is approximately five thousand years old but had never been known to erupt. On January 23, 1977, however, a 2-kilometer fissure opened at the edge of the major town, whose five thousand inhabitants were then evacuated. The new vent, Eldfell, quickly developed a cone and poured lavas down into the deserted village, eventually destroying more than three hundred homes. Accumulations of heavier-than-air noxious gases made salvage operations hazardous and endangered farm animals, many of which had to be destroyed. While various attempts to impede the encroaching lava achieved only limited success, the volcano stopped of its own accord on June 28.

Mount St. Helens

After erupting nine times between 1831 and 1857, Mount St. Helens, in the state of Washington, was dormant until March 27, 1980, when significant activity began. The major eruption of May 18 blew off the top of the mountain and was said to be the most forceful in the contiguous forty-eight states during historic times. Mudslides triggered by the eruption then destroyed bridges, logging camps, homes, and thousands of trees. About sixty people perished in one way or another. Further eruptions followed on May 25, June 12, July 22, and August 7, with occasional activity thereafter. These volcanic eruptions were the first in the contiguous United States since those of Mount Lassen. They attracted immense popular interest and were closely monitored by scientists.

El Chichón

In a now-familiar pattern, El Chichón volcano, in northern Chiapas, Mexico, awoke from a prolonged dormancy to erupt without warning on March 28, 1982. At least ten people who fled into a local church died when earthquakes demolished their haven; others were killed by falling debris and molten rock. A long ash cloud rained cinders across southern Mexico. Later eruptions in April left more people dead and homeless; survivors were then evacuated. Little had previously been known about this hitherto inactive killer.

Mount Pinatubo

Until 1991, Mount Pinatubo in the northern Philippines had no recorded eruptions. This was suddenly changed by the second-largest eruption of the twentieth century. More than three hundred people were killed by the eruption, but loss of life would have been much greater if the region had not been evacuated.

Soufrière Hills

The Soufrière Hills of the island of Montserrat in the Caribbean Sea consist of a stratovolcano, the first eruption of which began in 1995. Pyroclastic flows were produced, and more than six thousand people were evacuated to the north of the island.

A Small Sampling

Although these eruption episodes are among those most likely to appear in discussions of modern volcanism—either because of their historical significance or because of the thoroughness with which they were observed—it would be extremely misleading to suggest that they include most, or even the most typical, examples. Thousands of eruptions take place every year, most of which scientists know nothing about because the volcanoes remain hidden within the depths of the oceans. There are still remote areas on the earth, moreover, in which even a moderate land-based eruption might go unnoticed and therefore unrecorded.

Among the volcano records that do exist, all of the following have erupted twenty times or more since 1800: the three classic Italian volcanoes, Vesuvius, Stromboli, and Etna; Nyamuragia in Africa; Karthala (Grand Comoro) and Piton de la Fournaise (Reunion) in the Indian Ocean; White Island, Tarawera, Tongariro, Ngauruhoe, and Ruapehu in New Zealand; Bam and Manam, New Guinea; Ambrym and Lopevi, New Hebrides; Marapi and Dembo, Sumatra; Anak ("son of") Krakatau, between Sumatra and Java; Gedeh, Gunter, Slamet, Merapi, Semeru, Tengger, Lamogan, and Ruang, Java; Batur, Bali; Soputan and Lokon-Empung, Sulawesi; Apu Siau, Sangihe Islands; Gamalama, Halmahera; Mayon and Taal, the Philippines; Sakurajima, Kirishima, and especially Aso, Kyushu; Yake-Dake, Asama, Zao, and Iwake, Honshu; Oshima, Izu Islands; Komaga-Take and Taramai, Hokkaido; Karymsky and Kliuchevskoi, Kamchatka; Akatan and Shishadlin, Aleutian Islands; Pavlof, Alaskan peninsula; Kilauea and Mauna Loa, Hawaii; Colima, Mexico; Fuego, Guatemala; Izalco and San Miguel, El Salvador; Cerro Negro and Masaya, Nicaragua; Poas and Irazu, Costa Rica; Purace, Colombia; Reventador, Cotopaxi, and Sangay, Ecuador; Fernandina, Galápagos; and Rupungatito, Llaima, and Villarrica, Chile. It will be noted that a number of the most famous volcanoes do not appear; they have erupted either less frequently or not at all since 1800.

Study of Volcanic Eruptions

Broadly speaking, data relevant to any given volcanic eruption fall into three separate but related categories, depending upon whether they were collected prior to, during, or after the eruption. Data collected prior to an eruption include the previous eruptive history of the volcano: the types, the frequency, and the magnitude of its known eruptions. Illuminating as this information can be, however, it has not been shown to have reliable predictive value. Yet a volcano on the verge of eruption will often signal its upcoming behavior in a variety of ways. These include increased fumarolic activity, changes in the chemistry of associated waters, seismic disturbances, and tumescence (swelling). Certain changes in electrical and magnetic phenomena have been noted as well. Volcanologists, therefore, monitor all of these indicators carefully.

To ensure objectivity, precision, and thoroughness, much of this monitoring is accomplished through instruments. Pyrometers and thermometers, for example, measure the heat of volcanic rocks (the temperature of the cone rises before an eruption) or of associated waters, as in a crater lake. As rocks adjacent to the vent heat up, they become less powerfully magnetic, as surface or aerial surveillance with a magnetometer will establish. In some cases, rising temperature can also be detected through a series of aerial infrared photographs. Unfortunately, none of these methods predicts eruptions definitively.

The two most commonly used instruments (and, on the whole, the most reliable) are the seismometer and the tiltmeter. The seismometer measures and records vibrations within the earth. When strong enough to be detected by humans, these vibrations constitute earthquakes. It has been known for centuries that seismic activity often precedes volcanic eruptions. Since the invention of the seismometer in the latter nineteenth century, however, it has also been possible to detect fainter, more deep-seated vibrations, signaling the subterranean rise of molten lava within the volcanic vent. Typically, such vibrations become stronger and more frequent as the time of eruption nears. Seismometry, therefore, is one of the most useful observational techniques available to the volcanologist.

As molten lava rises within the vent of an existing volcano, the cone of that volcano subtly changes shape. Although the changes are generally too small to be obvious, they can be ascertained and verified through the use of a tiltmeter. This instrument can be in any of several forms (a pendulum, for example) but most commonly depends upon the displacement of a liquid, as in a carpenter's level. Laser beams have been used to measure tilt with great precision. In addition, some tumescence takes place so rapidly and conspicuously (generally along a seacoast) that detection by the naked eye is possible. Subsidence usually follows in any case, regardless of whether there has actually been an eruption.

Actual volcanic eruptions can last either seconds or days; a period of activity, characterized by frequent eruptions, can continue for years. During all or part of the times involved, it may well not be possible to inspect the volcano closely. First at Vesuvius, then at Hawaii and elsewhere, a number of the most active (or the most dangerous) volcanoes have been monitored constantly, by instruments and by humans, at volcano observatories. During a very violent eruption, instruments, observers, and even the observatory itself may be lost. In any case, volcanologists have always relied upon whatever eyewitness data they could gather to supplement their understanding of a volcanic event. Any eruption is a contribution to history. In this regard, the most significant new technique has been the use of photography.

Although photography was invented in 1839, early techniques did not permit the recording of subjects in motion. Thus, no photographs of any eruption exist before 1872, when on April 26, memorable views of Vesuvius were obtained. The first eruption to be photographed in the Western Hemisphere was that of Izalco, El Salvador, in 1894. There were then striking images of nuées ardentes at Mount Pelée by Tempest Anderson and Alfred Lacroix in 1902. Anderson's *Volcanic Studies in Many Lands* (1903) became the first picture book of volcanic phenomena. Eruptions of Mount Lassen in 1914-1915 were the first within the continental United States to be photographed. B. F. Loomis's *Pictorial History of the Lassen Volcano* (1926) is perhaps the earliest example of its kind.

Even in more modern times, it is possible for a major eruption to take place without being observed by humans; Katmai, 1912, is a well-known example. The only recourse then available to volcanologists is that of hypothetical reconstruction based upon the physical evidence. This evidence

Houses crushed by an ash-slide following the 1912 eruption of Katmai in Alaska. (U.S. Geological Survey)

generally consists of changes in the volcano's cone, adjacent geological effects, and the type and pattern of its discharges, insofar as they can be ascertained.

Significance

In the early days of geology, it was believed by some that volcanoes were only local phenomena, deriving their power from the ignition of coal beds. If so, then they were of little consequence in the history of the earth and hardly more than a curiosity within it. Since the latter eighteenth century, when this view was promulgated by Abraham G. Werner, scientists have come to realize how mistaken it is.

In his *Theory of the Earth* (1788), James Hutton first emphasized that the earth's internal heat was the fundamental driving force underlying its surface permutations. Volcanoes for him were safety valves intended to prevent his heat-engine earth from blowing up. Unlike certain predecessors, however, he did not emphasize their major constructive role in creating the surface of the earth. The first theorist to hold essentially modern views regarding volcanoes was George Julius Poulett Scrope in *Considerations on Volcanoes* (1825). As more volcanoes were studied, three facts strikingly emerged. First, volcanoes are not randomly placed on the earth but are instead grouped into narrow belts—one surrounding the Pacific, in particular. Second, volcanoes are of different types, differing in origin, appearance, and behavior. Third, eruptions are also of different types, though individual volcanoes often produce more than one type, even during the same eruptive episode. By the beginning of the twentieth century, therefore, the understanding of volcanism and its role in the history of the earth had become much more sophisticated.

Two of the most important twentieth-century discoveries about volcanoes were that they are even more numerous in ocean basins than they are on land and that those on land are almost always associated with established or incipient plate boundaries. With very few exceptions, continental volcanoes occur as a byproduct of subduction, the process through which one tectonic plate slides underneath another and is melted. Many theorists believe that the earth is less thermal now than it was in the past—that its geological activity in this respect is slowing down. Others insist that the most powerful episodes of volcanicity with which humans are familiar have taken place in relatively recent geological times. However this debate is eventually to be resolved, scientists regard volcanoes as the most visible indicators of the earth's still-active interior processes.

Dennis R. Dean

Further Reading

Decker, Robert, and Barbara Decker. *Volcanoes.* 4th ed. New York: W. H. Freeman, 2005. This readable and well-illustrated text is geared toward the general public. The authors explain geologic phenomena in a clear and simple manner. The appendices list the world's 101 "most notorious" volcanoes, as well as Internet sources for volcano information.

Francis, Peter, and Clive Oppenheimer. *Volcanoes.* 2d ed. New York: Oxford University Press, 2004. This text covers volcanic activity ranging from lava flows to pyroclastic currents. Provides many examples from Hawaii, Italy, Ethiopia, Japan, Mount St. Helens, and Krakatau. Written for the layperson, but some knowledge of geology is helpful.

Lockwood, J. P., and Richard W. Hazlett. *Volcanoes: Global Perspectives.* Chichester: Wiley-Blackwell, 2010. An up-to-date and well-illustrated undergraduate-level text that is also accessible to the interested reader.

Siebert, Lee, Tom Simkin, and Paul Kimberly. *Volcanoes of the World.* 3rd ed. Berkeley: University of California Press, 2011. A book to be consulted rather than read, this one lists all known eruptions for all known volcanoes.

Sigursson, Haraldur, ed. *Encyclopedia of Volcanoes.* San Diego, Calif.: Academic Press, 2000. This book contains a complete summary of the scientific knowledge of volcanoes. It contains eighty-two well-illustrated overview articles, each of which is accompanied by a glossary of key terms. Although this is a college-level text, it is written in a clear and comprehensive style that makes it generally accessible. Cross-references and index.

Sigursson, Haraldur, and Steven Carey. "The Far Reach of Tambora." *Natural History* 97 (June 1988): 67-73. A nontechnical, well-illustrated account of the 1815 eruption and its worldwide effects.

Trombley, R. B. *The Forecasting of Volcanic Eruptions.* Lincoln, Neb.: iUniverse, 2006. The author has a PhD in volcanism, providing a sense of authority to his writing. The book examines probability and distribution models used to predict volcanic activity. Data analyses using current software are discussed. The final chapter addresses the concept of supervolcanoes. Each chapter contains algorithms and references.

See also: Eruptions; Forecasting Eruptions; Hawaiian Islands; Iceland; Krakatau; Lava Flow; Mass Wasting; Mount Pelée; Mount Pinatubo; Mount St. Helens; Mount Vesuvius; Popocatépetl; Shield Volcanoes; Stratovolcanoes; Volcanic Hazards.

RING OF FIRE

The Ring of Fire is a relatively narrow band of land composed of parts of the continents and island chains that surround the Pacific Ocean. The region is characterized by extensive volcanic and earthquake activity. The ring represents the surface expression of a tectonic boundary in which oceanic crust is forced into the upper mantle, producing molten rock that rises to the surface and forms volcanoes. The stresses imparted to the crustal rocks as they are forced into the mantle provide the energy for numerous earthquakes. Geologic investigations of the Ring of Fire have helped scientists refine the theory of plate tectonics.

PRINCIPAL TERMS

- **continental crust:** near-surface rocks primarily composed of low-density, light-colored minerals and constituting the bulk of continental land masses
- **mantle:** the zone of the earth immediately underlying the crust; the uppermost region of the mantle is composed of partially melted and melting rocks
- **oceanic crust:** near-surface rocks primarily composed of dense, dark-colored minerals that underlie the ocean floors
- **plate:** a relatively thin slab of crustal rock, either continental or oceanic, that moves over the face of the globe, driven by currents of circulating molten rock in the underlying mantle
- **plate tectonics:** a theory that explains the distribution of geologic features and phenomena across the face of the earth based on the interactions of crustal plates
- **subduction:** the process occurring at colliding plate boundaries when denser oceanic crust is forced down into the upper mantle by overriding continental crust
- **trench:** a very deep portion of the ocean where oceanic crust is being forced beneath continental crust

PLATE TECTONICS

The "Ring of Fire" is the dramatic name of a narrow band of the earth's surface surrounding the Pacific Ocean. This band extends from the islands of the South Pacific, through Japan, to the Aleutian Islands between Asia and North America, down the West Coast of North America, and through Central America and the western portions of South America. The name was applied to the roughly circular region because of its many active volcanoes. In addition to volcanic activity, this region of the earth's surface exhibits a combination of geologic features that have intrigued earth scientists for years The region is characterized by mountain chains with many active and dormant volcanoes that parallel coastlines. Another characteristic of the area is its numerous strong earthquakes. In addition, some of the deepest portions of the world's oceans lie just offshore the volcanically active mountain chains of the region's land masses. Geologists had noticed the correlation between the land masses surrounding the Pacific, the volcanic and earthquake activity, and the oceanic depths long before they had an explanation relating these phenomena. The advent of the theory of plate tectonics provided the mechanism to explain the relationships among these geologic features.

The theory of plate tectonics, developed during the 1960's, views the earth's surface as divided into a number of crustal plates. Like the pieces of a jigsaw puzzle, the plates fit together at their edges and cover the surface of the earth. However, these pieces are in motion. Each plate moves slowly over the surface of the earth, riding on currents of molten rock in the upper mantle, the layer within the earth immediately underlying the crustal plates.

Geologists have long known that rock temperatures increase with depth in the earth. For example, temperatures at the bottom of the deepest mines are sometimes uncomfortably hot. Deeper into the earth—in the upper mantle and beyond—temperatures are so high that most rocks begin to melt. However, sources of heat within the mantle are not evenly distributed, and circulation systems exist. In areas of high heat, the molten rocks are less dense because they are hotter, and they rise. In cooler areas of the mantle, the molten rocks become denser and sink. This circulation system acts as a conveyor belt, propelling the crustal plates across the surface.

The two major types of crustal plates are composed of either continental or oceanic crust. Because the types of rocks making up these crustal types are chemically different, they also differ in density, or the mass per unit volume of the rock. Continental crust is less dense than oceanic crust. Tectonic plates may be composed of oceanic crust (for example, the Pacific plate), continental crust (for example, the Arabian plate), or some combination of the two types.

Plate Boundaries

There are three basic types of plate boundaries. Boundaries where two plates are moving away from each other are called divergent boundaries and are probably associated with the areas of rising molten rock within the mantle. A good example of a divergent plate boundary is found in the middle of the Atlantic Ocean, where an underwater mountain chain, the Mid-Atlantic Ridge, runs the length of the sea floor. Plate boundaries where two plates are moving past each other are called transverse boundaries and are characteristically marked by a major fault. An example of a transverse boundary is found in the San Andreas fault zone of Southern California.

The third type of plate boundary occurs where two plates are colliding. These are known as "convergent plate boundaries." This type of plate boundary is characteristic of the Ring of Fire. At a convergent plate boundary, two plates being moved along on mantle currents collide. At the point of collision, something must give. In many cases where the plates are composed of different crustal types (such as continental and oceanic), the denser crustal type (oceanic) will tend to be pushed beneath the lighter, less-dense continental crust and into the upper mantle. An example of this type of convergent boundary exists along the West Coast of South America. There, the denser oceanic plate is subducted beneath the lighter continental plate. The subduction zone, where the oceanic plate is pushed beneath the continental plate, runs parallel to the coastline. Immediately offshore, the subduction zone is characterized by extremely deep ocean water. As the oceanic plate is subducted, it is bent down into the mantle, and the bottom of the ocean in the subduction zone becomes very deep. These deep linear features in the ocean, called "submarine trenches," are some of the deepest parts of the ocean.

As the oceanic crustal plate sinks deeper into the upper mantle, temperatures rise, and the rocks begin to melt. The molten rock, which is much hotter than the solid oceanic crust it came from, is less dense than the surrounding mantle material, and it tends to rise, like a beach ball held under water. As the molten rock from the oceanic plate rises, it hits the bottom of the overlying continental crust. Working its way through cracks and other openings in the continental crust, the molten rock often forces its way to the surface, where it pours out from a volcano. The volcanic activity tends to occur in a narrow band within the continental plate some distance inland from where the oceanic plate is being subducted. Mountain chains on the continent, composed of many volcanic peaks, are characteristic of this zone.

Another type of convergent plate boundary occurs where two similar crustal types collide. In the case of the Ring of Fire, this type of convergence is found in the South Pacific, where plates of oceanic crust are colliding. The result is that one of the plates again sinks beneath the other. The subducted plate is heated and melts within the upper mantle, and rising melt finds its way through the overlying oceanic crust to the surface of the ocean floor. The molten rock that builds these volcanoes continues to spill onto the ocean floor until the volcano rises above sea level to form a volcanic island. Volcanic island chains, called "island arcs," form with a deep submarine trench offshore.

Hot Springs and Geysers

Another feature of volcanic areas in the Ring of Fire is the presence of hot springs and geysers at the earth's surface. The abundance of hot and molten rock at relatively shallow depths above a subduction zone provides a perfect setting for hot-spring formation. Water from surface precipitation enters the earth's surface to become groundwater. Some of this groundwater will penetrate deeply into the earth's crust and be heated by the rocks and melt associated with the Ring of Fire. When the water is heated, like the heated, molten rock rising from the subducted plate, it becomes less dense and rises toward the surface through cracks and fractures in the earth's crust to emerge as a hot spring or geyser.

In addition to absorbing heat from the surrounding rocks and melt, the circulating water may dissolve many chemical elements from the rocks. These dissolved metals are carried along with the water through the cracks in the earth's crust toward the surface. However, as the water rises, it cools, and its ability to keep the elements dissolved decreases. As a result, minerals containing specific elements, especially metals such as lead, zinc, and copper, are deposited along the path of the water. Mineral deposits being mined today are thought to have been deposited from just such hot (hydrothermal) waters.

Earthquakes

Earthquakes are another common feature of the Ring of Fire. Rocks at plate boundaries are subjected to the stresses at that boundary. Rocks are being forced apart, are sliding past each other, or are colliding. Under these pressures, all rocks tend to break along planes called faults. Sometimes the movement along a fault is not smooth. The rocks along the fault get stuck, and the pressures driving the movements build up. Eventually, the built-up pressures become so great that the sticking point along the fault breaks free, and the stored energy is released. This energy travels through the surrounding rocks as vibrations—similar to the vibrations from a bell after it is struck—producing an earthquake. The intensity of the vibrations is related to the amount of energy released along the fault. The vibrations from a strong earthquake can do severe damage to buildings and roads.

Earthquakes are common in the Ring of Fire because areas of converging plates are ideal locations for rocks to break and form faults. In addition, the pressures of collision are constant and can build up if movement along the faults ceases. Deep earthquakes are characteristic of the Ring of Fire. Faults only occur in solid rocks. In a subduction zone, solid rocks are forced deep into the upper mantle, an area usually composed of molten or partially melted rock. The faults in the relatively cold subducted plate are the only places in the upper mantle where pressures can build up and be released as an earthquake.

Ocean Surveys

The explanation for the Ring of Fire has depended, in large part, on investigations into the processes of plate tectonics. Geologists have used a broad range of technologies to understand how plate tectonics operates, and that understanding has helped explain the association of geologic phenomena composing the Ring of Fire.

Geologists have mapped the ocean floor with tools as crude as a weighted rope and with more sophisticated tools such as sonar. Sonar uses sound waves generated within the water to measure the distance to any object that can reflect some of the sound waves back to a receiver. The time that elapses between when the sound is made and when the reflected sound is received can be used to calculate the distance to the reflecting object. A map of the ocean floor can thus be created by towing a sound source and sound receiver behind an oceanic research vessel as it crisscrosses the ocean. Data from many such surveys have been compiled into maps of the ocean floor, which clearly show the mid-ocean ridges and the submarine trenches characteristic of divergent and convergent plate boundaries.

Oceanic surveys also often involve the taking of heat measurements. Using a sensitive heat sensor towed near the ocean floor behind a research vessel, scientists are able to map the amount of heat entering the ocean through the oceanic crust. The results of these surveys show that the warmest oceanic crust and the areas where the most heat enters the oceans are associated with the mid-ocean ridges. Conversely, the coolest portions of the ocean floor are those associated with submarine trenches. However, some of the highest heat-flow measurements taken on land are associated with the areas of volcanic activity in island arcs and continental mountain chains within the Ring of Fire.

Seismology

Seismologists are geologists who specialize in the study of earthquakes. Seismologists use seismographs, sensitive instruments that measure vibrations in the earth's crust, to identify where an earthquake has occurred. Seismographs can be used to identify the time earthquake vibrations arrive at a seismic measurement station. Knowing the velocity of earthquake vibrations through rocks, seismologists can calculate the distance from their measuring point to the earthquake source. With only a single measurement point, a seismologist would know only that an earthquake could have been centered anywhere on a circle of that radius. However, with many seismographs around the world, seismologists at each station combine measurements from other stations to locate the point on the earth's surface directly over the source of the earthquake. This point is called the epicenter of the earthquake. Measurements showing the locations of many epicenters over the years were used to identify the Ring of Fire as an area of high earthquake activity.

Another characteristic of the Ring of Fire is the depth of the earthquake source. Earthquakes occur where two bodies of solid rock are moving past each other, separated by a fault plane, when that movement is temporarily halted. The rocks must be solid; within the earth's depths, however, temperatures increase rapidly, and solid rock is not thought to exist below a certain depth. The implication is that earthquakes cannot occur below the depths predicted for melting of continental or oceanic crustal rocks. However, seismologists can use seismograph data to determine the depth of an earthquake source. Examination of seismograph data from many earthquakes in the Ring of Fire clearly indicates that, in addition to relatively shallow earthquakes, there are many very deep earthquakes. The epicenters of the very deep earthquakes are centered inland away from coastlines and submarine trenches. In the context of plate tectonics theory, the data suggest that deep earthquakes are centered within the relatively cold, and solid, rocks of the subducting plate; researchers have theorized that the plate remains solid and able to generate earthquakes because it is being driven into the earth's depths rapidly enough that the rock does not heat up and melt, as geologists might otherwise predict. Additional analysis of the seismic data has allowed seismologists to differentiate between subduction zones where the subducting slab is entering the upper mantle at a sharp angle, which produces a relatively narrow zone of earthquake activity, with deep earthquakes closer to a submarine trench, and those entering the upper mantle at a shallow angle, which produces a broader earthquake zone and deep earthquakes farther inland.

Richard W. Arnseth

Further Reading

Chester, Roy. *Furnace of Creation, Cradle of Destruction.* New York: AMACOM Books, 2008. This text discusses the turbulent processes of the planet. It covers earthquakes, volcanoes, and tsunamis in reference to plate tectonics, natural disasters, and prediction and mitigation of these events. Multiple chapters explore seafloor spreading. The author also discusses hydrothermal activity. This text takes on an immense range of content, but still explains concepts clearly and with detail.

Decker, Robert, and Barbara Decker. *Volcanoes.* 4th ed. New York: W. H. Freeman, 2005. This readable and well-illustrated text is geared toward the general public. The authors explain geologic phenomena in a clear and simple manner. Includes a chapter on the Ring of Fire. The appendices list the world's 101 "most notorious" volcanoes, as well as Internet sources for volcano information.

Erickson, Jon. *Plate Tectonics: Unraveling the Mysteries of the Earth.* Rev. ed. New York: Facts on File, 2001. A comprehensive, technical review of plate tectonics theory, from the initial development of the theory to more recent areas of investigation. The material is presented in the manner of a classroom textbook. Illustrated; includes an extensive bibliography for each chapter subject and a comprehensive index.

Francis, Peter, and Clive Oppenheimer. *Volcanoes.* 2d ed. New York: Oxford University Press, 2004. This text covers volcanic activity ranging from lava flows to pyroclastic currents. Provides many examples from Hawaii, Italy, Ethiopia, Japan, Mount St. Helens, and Krakatau. Written for the layperson, but some knowledge of geology is helpful.

Harris, Stephen L. *Fire Mountains of the West: The Cascade and Mono Lake Volcanoes.* 3rd ed. Missoula, Mont.: Mountain Press Publishing Company, 2005. Although this book focuses mostly on the Cascades, it includes information on Mount St. Helens, Mammoth Mountain, and Crater Lake. Chapter 2 covers the Ring of Fire and hot spots. Written in a manner accessible to anyone with some interest in volcanoes.

Levy, Matthys, and Mario Salvadori. *Why the Earth Quakes.* New York: W. W. Norton, 1995. A general, readable discussion of the forces creating earthquakes and volcanoes. The text is supplemented throughout with helpful illustrations. The last few chapters discuss case histories of earthquakes, some of the engineering aspects of building in an earthquake-prone area, and the social consequences of earthquake predictions. Includes a glossary of terms and an index.

Lockwood, J. P., and Richard W. Hazlett. *Volcanoes: Global Perspectives.* Chichester: Wiley-Blackwell, 2010. An up-to-date and well-illustrated undergraduate-level text that is also accessible to the interested reader.

Ritchie, David. *The Ring of Fire.* New York: Atheneum, 1981. A mixture of science and history devoted to the Ring of Fire. The author combines the history of the development of explanations for the Ring of Fire with descriptions of some of the dramatic, and catastrophic, events associated with the area. Contains a glossary of terms, a bibliography, and an index.

Vogel, Shawna. *Naked Earth: The New Geophysics.* New York: Dutton, 1995. A readable discussion of the role of modern geophysics in exploring the mysteries of the earth. The author focuses on the roles of individual scientists and the array of technologies used in solving geologic mysteries. The text reads like a mystery, with just enough details to satisfy the more knowledgeable reader. Includes an index.

See also: Calderas; Faults; Flood Basalts; Geysers and Hot Springs; Hawaiian Islands; Hot Spots, Island Chains, and Intraplate Volcanism; Island Arcs; Lava Flow; Spreading Centers; Volcanic Hazards; Yellowstone National Park.

RIVER VALLEYS

The valleys in which streams flow are produced by those streams through long-term erosion and deposition. The landforms produced by fluvial action are quite diverse, ranging from spectacular canyons to wide, gently sloping valleys. The patterns formed by stream networks are complex and generally reflect the bedrock geology and terrain characteristics.

PRINCIPAL TERMS

- **aggradation:** the process by which a stream elevates its bed through deposition of sediment
- **base level:** the theoretical vertical limit below which streams cannot cut their beds
- **fluvial:** of or related to streams and their actions
- **prokaryotic:** an upland area between valleys
- **stream equilibrium:** a state in which a stream's erosive energy is balanced by its sediment load such that it is neither eroding nor building up its channel
- **underfit stream:** a stream that is significantly smaller in proportion than the valley through which it flows
- **valley:** part of the earth's surface where stream systems are established; it includes streams and adjacent slopes

Formation of River Valleys

River valleys consist of valley bottoms and the adjacent valley sides. Between valleys are undissected uplands known as interfluves. Valley floors may be quite narrow, as in the case of the Black Canyon of the Gunnison River, or they may be quite wide, as in the case of the Hwang Ho or the Brahmaputra. Similarly, valley sides may have very gentle to rolling slopes, or they may be nearly sheer, as in the case of the Arkansas River's Royal Gorge. In many areas, the interfluves are simply divides between adjacent valleys, but on tablelands such as the Colorado Plateau, they may be tens of kilometers wide in places.

River channels and river valleys are products of the streams that flow through them. As a stream erodes a channel for itself in newly uplifted terrain, it eventually carves a valley whose form is determined by the erosive power of the stream, by the structural integrity of the rock and debris of the valley walls, by the length of time that the stream has been operating on its surroundings, and by past environmental conditions. These past environmental conditions are attested by stream channel and valley profiles that have not entirely erased landforms produced during the most recent episodes of glaciation and climate change. The valley of the Mississippi River, for example, is formed in the complex deposits of what was a much larger, more heavily laden glacial meltwater stream that existed only 15,000 years ago. River channels and valleys may be referred to as palimpsests, a term originally used to describe parchment manuscripts that had been partly scraped clean, then reused. On the fluvial (river-carved or river-deposited) landscape, previous landscape elements are seen, just as old words show through on a recycled piece of parchment.

In many parts of the world, streams are found flowing through valleys that appear to have been formed by a far larger stream. The valley width, amplitude of meanders, and caliber of coarse sediment are proportional to far larger stream courses. Such streams are said to be underfit, and the valleys are largely remnant features from times of wetter and

The Columbia River Gorge. (PhotoDisc)

cooler climates that accompanied glaciation, or the streams were glacial meltwater channels during deglaciation.

The fact that stream channels and valleys are not chance features on the landscape was first noted by British geologist John Playfair in 1802. Playfair suggested that instead of being isolated features on the landscape, streams are part of well-integrated networks. More important, these networks are finely adjusted to the landscape and to one another in such a way that tributaries almost always join the main trunk stream at the same level as that stream. Streams that must plunge over waterfalls to join a larger stream are quite rare. Such discordant streams are found primarily in recently glaciated terrain, where the stream-valley system has not yet become fully adjusted, such as in Yosemite Valley. This remarkable consistency of stream accordance is the strongest evidence indicating that streams carve their own valleys.

Streams develop into network patterns that are strongly influenced by bedrock structure. Where the bedrock is relatively uniform and without strong joints and faults, a dendritic pattern of drainage develops. In this pattern, a stream system is branched like a tree. On inclined plains such as the Atlantic coastal plain, the stream pattern is often parallel, with major streams flowing directly down the topographic slope to the sea. Inclined mountain systems such as the Sierra Nevada also produce parallel drainage. Where structural folding of the terrain has produced linear ridge and valley topography, main trunk streams occupy the linear valleys and are quite long, with short, steep tributaries feeding them off the flanks of the hills or mountains. In areas such as the Great Valley of Virginia, this type of pattern has developed and is known as trellis drainage. Volcanoes often produce a radial pattern of drainage. Mount Egmont in New Zealand is often cited as an example.

Base Level

The depth of a river valley is a function of the height of the land above base level, the length of time that has passed since the stream began to erode, the resistance of the bedrock to erosion, the load of

A river gorge at Fall Creek, Ithaca, Thompkins County, New York. The gorge wall shows erosion partially controlled by joints. (U.S. Geological Survey)

sediment that the stream is carrying, and the spacing of adjacent streams. "Base level" refers to the theoretical lower limit to which a stream can cut. Ultimate base level for all streams is sea level, but local base level is significantly higher for many streams. Streams require a minimum slope to transport their sediment to the sea; this limits the depth to which a stream may cut. The upper section of the Hwang Ho near Tibet flows at elevations of more than 3,000 meters, but the river must still flow more than 4,000 kilometers to the sea. In order to carry its heavy load of sediment over that distance, the river must maintain its channel at a rather high elevation.

Valleys that are deep and relatively narrow are called canyons. Valleys that are especially narrow are called gorges. Streams require time to erode great canyons, although the rate of cutting can be quite rapid compared with most geologic processes. The process of downwearing of interfluves is far slower, so young (recently uplifted) landscapes produce the deepest canyons. As downcutting by the main stream ceases, upland weathering and erosion lower the local relief. The deepest, narrowest canyons are eroded in strong, homogeneous rock such as granite and quartzite. The Royal Gorge of the Arkansas River and the Colca Gorge of Peru are excellent examples; both are cut in recently uplifted masses of resistant rock formations. The Colca Gorge is far narrower than is the Grand Canyon and, at more than 3,000 meters, it is nearly twice as deep.

As a mountain mass or tableland is uplifted, streams often develop that flow directly down the initial slope of the land. Such streams are called consequent streams because their course is a direct consequence of the terrain slope. Because streams seek the path of least resistance, their courses often follow the outcrop pattern of weak rocks such as shale, producing what is known as a subsequent stream pattern. In many cases, however, streams seem to ignore the terrain and structural slope of the land entirely, flowing through mountains of quite resistant rock. An excellent example is the Black Canyon of the Gunnison River in Colorado, where the river carves a deep canyon through a high plateau, with its channel cut in resistant gneisses and igneous intrusives. What makes this so surprising is that much lower terrain, underlain by thick sequences of weak shale, lies only 3 kilometers to the west of the head of the gorge, which would seem to provide a much easier path to the sea. The reason for this course—and many other anomalous stream courses—is that the course of the Gunnison was established prior to the uplift of the plateau. As the land rose beneath the river, it maintained its position, carving an ever-deepening gorge in the rocks as they rose. Far older and more extensive is the anomalous course of the New River, which cuts directly across the structure of the Appalachians, flowing a great distance to the Ohio River rather than the shorter, more direct route to the Atlantic. Once again, this river is one that was established prior to the uplift of the land in which it is now entrenched—in this case, the ancient Appalachians. Thus, the river is rather ironically named, considering that its course is older than the Appalachians.

At any given time, a stream may either erode or aggrade (build up) its bed. This vertical change in stream profile is determined by many factors internal to the stream, as well as outside environmental factors. Over geological time, the tendency of streams is to erode their beds deeper and deeper. Over shorter time intervals, however, a stream may reach a state of equilibrium between its erosive energy and the load of sediment that the stream is carrying. A delicate balance is maintained within the stream channel. If stream energy is increased by an increase in flood volume or frequency, for example, the stream is likely to respond by eroding its bed. Conversely, if stream power remains constant and the sediment load is significantly increased, the stream will respond by aggrading. This channel aggradation increases the channel slope, giving the stream more energy to transport the sediment load, and a new equilibrium is reached.

Michael W. Mayfield

Further Reading

Bloom, Arthur. *Geomorphology: A Systematic Analysis of Late Cenozoic Landforms.* 3rd ed. Long Grove Ill.: Waveland Press, 2004. This college-level text covers the basics of geomorphology. Includes three chapters on fluvial systems and the development of river valleys. Index and bibliography.

Chorley, R. J., S. A. Schumm, and D. E. Sugden. *Geomorphology.* New York: Methuen, 1985. An advanced text in geomorphology that is best used as a reference. Provides detailed citations of the important scholarly works on each subject addressed and is quite thorough. Many diagrams, graphs, and line drawings supplement the material, but there are few photographs. Suitable for college students who have already gained some background in geomorphology.

Dobrazanska, Halina, Erzsebet Jerem, and Tomasz Kalicki, eds. *The Geoarchaeology of River Valleys*. Budapest: Archaeolingua, 2004. A collection of papers that discuss the development of societies along rivers and the ways in which river valleys influence culture. Also covers anthropogenic effects on river valley evolution.

Fletcher, Charles. *Physical Geology: The Science of Earth*. Hoboken, N.J.: John Wiley & Sons, 2011. This text deals with all aspects of physical geology. It is intended for undergraduate students but accessible to the interested reader.

Hunt, C. B. *Natural Regions of the United States and Canada*. San Francisco: W. H. Freeman, 1974. This book is primarily a physiography text, aimed at describing and explaining the pattern of landforms across North America. The author produced it in the light of "the general public's interest in the natural environment and the need for an authoritative account of it in language as nontechnical as possible." In both regards, the book is quite successful. Well illustrated with diagrams, line drawings, and photographs. Introductory chapters provide an explanation of the general phenomena, and the remaining chapters provide a very thorough, understandable coverage of the regional expression of landforms. Suitable for high school students and above.

McKnight, T. L., and Darrel Hess. *Physical Geography: A Landscape Appreciation*. 9th ed. Englewood Cliffs, N.J.: Prentice-Hall, 2007. This physical geography text provides very thorough but understandable coverage of the concepts of stream networks, stream channel history, and fluvial dynamics. Richly illustrated with color diagrams and photographs. Suitable for high school students and above.

Ollier, Cliff, and Colin Pain. *The Origin of Mountains*. London: Cambridge University Press, 2000. This text encompasses the geographic, geological, and geophysical concepts of mountains. Topics include fault-block mountains, fold belts and folding, gravity structures, rift valleys and river valleys, plateaus and mountain erosion. Begins with a general overview of subduction and plate tectonics. Mountain ranges discussed in detail include the Appalachians, Andes, Cascades, Himalaya, and the Rockies.

Restrepo, Juan D., Sergio A. Lopez, and Juan C. Restrepo. "The Effects of Geomorphic Controls on Sediment Yield in the Andean Rivers of Colombia." *Latin American Journal of Sedimentology and Basin Analysis* 16 (2009): 79-92. A technical and detailed study on the sedimentation within the Andean basin. Compares Colombian runoff of the Pacific system and the Caribbean system. Hydroligic, climatic, and morphometric data were collected to build a model for predicting sediment yield. A background in geology or hydrology is recommended.

Skinner, Brian J., and Stephen C. Porter. *Physical Geology*. New York: John Wiley & Sons, 1987. A consistently excellent text in introductory physical geology, richly and generously illustrated in full color. Very thorough and quite extensive (more than seven hundred pages). Chapter 11, "Streams and Drainage Systems," is one of the best chapters. Suitable for advanced high school students and above.

See also: Alpine Glaciers; Fluvial Processes; Geomorphology and Human Activity; Geomorphology of Dry Climate Areas; Geomorphology of Wet Climate Areas; Glacial Landforms; Grand Canyon; Karst Topography; Marine Terraces; Mass Wasting; Permafrost.

ROCKY MOUNTAINS

The various mountain ranges and basins that make up the Rocky Mountains represent a region of tremendous mineral wealth that has played an essential part in the economic development and prosperity of the United States and Canada. In addition, the ranges constitute a recreational resource of inestimable value. Finally, the Rockies have proved to be an invaluable key to understanding the earth's past as well as the processes that continue to transform it.

PRINCIPAL TERMS

- **Ancestral Rockies:** the mountain ranges that preceded the Rockies by many millions of years and no longer have surface topographic expressions of their former existence
- **cordillera:** a major mountain chain or system of such chains, especially one that is a dominant feature of a continent-sized landmass
- **craton:** a large, tectonically stable core area of a continental-type tectonic plate, usually of great age; also termed a continental shield
- **epeiric sea:** a shallow sea that temporarily (in geologic terms) covers a portion of a craton; also termed an epicontinental sea
- **Laramide revolution:** an orogeny that occurred between the Late Cretaceous and the Early Tertiary periods of geologic time, instrumental in producing the ranges termed the Rocky Mountains in the United States
- **orogeny:** a mountain-building episode or event that extends over a period usually measured in tens of millions of years; also termed a revolution
- **plate tectonics:** a widely accepted theory in geology that contends that the earth's crust is composed of about twelve moving, rigid plates
- **subduction:** a process, according to the theory of plate tectonics, whereby one crustal plate is thrust beneath another at a convergent boundary
- **transgression:** the flooding of a large land area by the sea either by a regional downwarping of continental surface or by a general global rise in sea level

MOUNTAIN FORMATION

Numerous mountain ranges are sometimes grouped under the general definition of the Rocky Mountains by various geologists and geographers. Canadian scientists often define the Rockies differently from the way their counterparts in the United States define the Rockies. During the past few centuries, the defined extent of the Rockies has expanded or contracted to include first one range or feature, then yet another. To define the Rockies in a geologically meaningful way, it is necessary to include those features that are surface expressions of particular systems of formative processes. The Rockies and their geologic history are products of long-term and very large-scale processes that shaped not only the North American continent but also the configuration of the earth's continents and oceans as a whole. Thus, the term "Rockies" will mean the more easterly of those mountain ranges and closely associated features that constitute the North American cordilleran region.

The cordillera is a long, basically continuous highland structure extending from eastern Alaska southward to northern New Mexico. For the majority of their length, the Rockies front the Great Plains to the east and, in the United States, border a geophysical feature called the Basin and Range Province to the west. The Rockies owe much of their present conformation to the Cordilleran orogeny, of which the Laramide orogeny was one of the last great pulses. Thus, a geologic history of the Rocky Mountains is, for the most part, a few localized acts in the greater drama of the Cordilleran orogeny and the formation of North America.

Orogenies are mountain-building episodes or events, occurring over periods of time measured typically in tens of millions of years. In modern geologic theory, they are considered as consequences of large-scale forces operating in the realm of plate tectonics. Plate tectonics theory, aspects of which were once called continental drift, holds that the crust of the earth is composed of about twelve major rigid units called lithospheric plates. Some of these plates are oceanic while others are continental in nature. All are involved in some type of motion relative to one another.

Plate motion can take various forms: Plates can move away from a common center, slide alongside each other, or collide with each other with various effects. One collision type involves a process called subduction, in which a heavier plate is forced to dive beneath the edge of a lighter plate. This effect is believed to be taking place along the western coast of South America as the edge of a heavier oceanic plate is subducted beneath the lighter continental plate, on which South America rides. The Andes mountain chain is considered a direct result of this subduction. Geologists point to the Andes and the associated subduction as a model for what probably occurred in relation to the formation of the Rockies. In both cases, a long, high belt of mountains possessing similar rocks, minerals, and geologic structures was formed as a product of plate collision.

Much of the evidence of the way the Rockies formed exists in the findings of two subbranches of geology: stratigraphy and sedimentology. Both sciences deal with the layers of rock termed strata and the story they tell about past environments in which deposition and erosion took place. Findings indicate that the ranges known as the Rockies did not exist in their present incarnation before the close of the Mesozoic era, 65 to 70 million years ago. Across great durations of time previous to that period, however, evidence exists for other incarnations of the Rockies occupying somewhat similar geographic orientations. The incarnation just prior to the present range has been termed the Ancestral Rockies. Before the appearance and elimination through erosion of this range, it appears that other Rocky Mountain-like ranges existed at least two other times in earth history.

These very ancient pre-Ancestral Rockies are evidenced by recrystallized, tightly folded rocks, which represent orogenies estimated to have occurred during the Late Precambrian eon, also termed the Proterozoic eon, which ended about 600 million years ago. Very general ages for two orogenies of these Precambrian Rockies have been estimated at about 1.5 billion and 2.5 billion years before the present. After about 1 billion years before the present, orogenic activity seems to have terminated, and a very lengthy period of crustal stability and erosion appears to have dominated. During this period, the last of the Precambrian Rockies seems to have been totally beveled down to the roots by erosion during a time referred to as the Lipalian interval, which lasted minimally for hundreds of millions of years.

Paleozoic and Mesozoic Eras

With the advent of the Paleozoic era about 600 million years before the present, stratigraphic evidence indicates marine conditions dominated in the region formerly occupied by mountains. Sandstones, limestones, and shales indicative of quiet, marine depositional environments prevailed. This situation continued during much of the Early Paleozoic over much of what later would become the North American cordilleran region. Beginning with the Late Devonian period in the Late Paleozoic, about 380 million years before the present, and extending into the following Mississippian period, tectonic disturbances began to have effect again. An orogeny called the Antler took place that deposited at least 1,000 meters of sandstone and conglomerate sediments found in such places as present Nevada. These sediments accumulated as the mountains began to erode. Other evidence of this orogeny exists in the form of the clastic (broken rock) layers lying unconformably above older marine strata.

Unconformable stratigraphic rock sections indicate that a break in sedimentary deposition occurred, often as a result of erosional processes outpacing depositional ones, which is exactly what one would expect if a mountain range were rising in an area. Although the Antler orogeny is not directly related to the formation of the present Rockies, it is an event representative of the collage of many smaller orogenic events that contributed, over time, to the Cordilleran orogeny and, thus, to the formation of the Rockies. Like most, if not all, orogenesis, plate tectonic activity is believed to be the driving force behind the Antler event. It is theorized that either a relatively small lithospheric plate called a microcontinent or another tectonic feature called a volcanic arc was being subducted somewhere to the west.

Following the Antler orogeny by tens of millions of years, another orogenic event occurred in the area now part of the western United States termed the Colorado orogeny. This mountain-building event was responsible for uplifting the Ancestral Rockies. During the Pennsylvanian period, about 300 million years ago, two large, mountainous islands or island chains arched up in the Colorado area. They rose above a shallow sea that covered, with the exception of the Appalachia cordilleran area to the east, most of the North American continental landmass. A sea of this epeiric, or epicontinental, type represents a

Looking south to the Sidenius thrust fault in the Northern Rocky Mountains, Halfway River area. A broad hanging-wall syncline is visible to the left. (Geological Survey of Canada)

transgression, or advance, of marine waters over a core area of a continental tectonic plate, known as a craton.

The Colorado orogeny produced what are believed to have been two distinct subranges: Front Range and Uncompahgre. A narrow, linear, shallow seaway is thought to have separated them. As the two island ranges eroded, they deposited extensive aprons of sediment into the surrounding seas. These sediments accumulated particularly in the narrow intervening region and are represented by the Fountain and Maroon geologic formations of Colorado. Evaporite basins also formed in the intermontane region as some arms of the surrounding seas became trapped or isolated. Sedimentary environments typical of such basins are represented by gypsum (calcium sulfate) and salt deposits in the west-central Colorado area. Erosion of the Ancestral Rockies continued for tens of millions of years, resulting eventually in their complete leveling. Remnants of these ranges are thought to have persisted in the form of minor hilly topographic relief until the middle of the following geologic era—the Mesozoic, famous for its dominant reptilian fauna such as the dinosaurs.

Tectonic forces had configured the arrangement of the continents into one great landmass by the Late Paleozoic era. Epeiric seas virtually disappeared for a time from North America by the close of the Paleozoic—the Permian period, 280 to 240 million years ago. Arid conditions prevailed for millions of years in western North America and in other areas that made up the supercontinent, which geologists call Pangaea. By the Triassic period, at the onset of the mid-Mesozoic, Pangaea slowly began to break up into smaller units. The following Jurassic period witnessed not only the formation of a large seaway between North America and Eurasia and the southern continents, termed Gondwanaland, but also an intensification of tectonic activity along the western margin of North America. This intensification culminated in the long process of mountain building in the west termed the Cordilleran orogeny.

As the North American continental plate progressively overrode various microcontinents, also known as suspect terranes, as well as volcanic arcs and sea floor to the west, an Andean-type plate margin was initiated. This convergent-type tectonic margin generated the various cordilleran ranges. Widespread crustal deformation and orogenic activity began during the Jurassic near the Pacific Coast with such events as the Nevadan orogeny. Orogenic activity spread eastward, initiating more uplifting activity such as the Sevier event in the Late Cretaceous period, near the close of the Mesozoic era. Finally, the Laramide event was generated as one of the last, most eastward pulses of the Cordilleran orogeny, starting in the Late Cretaceous period. This orogeny lasted for about 20 million years, from roughly 65 million to 45 million years ago, up into the Eocene epoch of the early Cenozoic era. Its end product was the easternmost ranges of the North American Cordillera—the ranges known as the Rocky Mountains.

Cenozoic and Pliocene Eras

By the onset of the Laramide orogeny, an epicontinental sea had formed by the transgression of marine waters from the Arctic to the north and the Gulf of Mexico to the south. These seas joined in midcontinent to form a long, wide but shallow, epeiric sea, which left marine sediments behind known generally as the Sundance formation. To the west of these marine strata, a famous, nonmarine sequence of strata formed, termed the Morrison formation, famous for numerous dinosaur fossils. As the land still farther to the west began to uplift, erosion accelerated, depositing continental sediments to the east, prograding, or extending out, into the epeiric, marine sediments. These sediments are termed the Morrison clastic wedge and are the first evidence, in Late Jurassic times (about 180 million years ago), of the eastward spreading effects of the Cordilleran orogeny. Lower Cretaceous period deposits in the central Rocky Mountain area exhibit a much thicker clastic wedge of up to 10,000 meters. This wedge's accumulation coincides with the draining of the epeiric sea from the continental interior and immediately presages the Laramide event.

By the Paleocene and Eocene epochs of the Early Cenozoic, the Laramide orogeny had created the long, high ranges of mountains of the Rockies. Between the ranges, intermontane basins had formed as lowlands between the north-south trending, linear uplifts. These basins began to fill with sediments eroded off the adjacent mountains as well as with river and lake sediments. Numerous basins became large lakes, which, over millions of years, gradually filled completely. During their existence as active lake environments, the larger lakes were hosts to thriving ecosystems of plants and animals. The economically important deposits of oil shale found widely in Colorado, Wyoming, and Utah formed from the petrochemical mineral kerogen (which yields oil when heated) trapped within the shale. The kerogen is believed to have formed primarily from the decay of lake-dwelling planktonic organisms during the lifetime of the larger lake bodies. Indications suggest the lakes were in existence for at least 5 to 8 million years.

The filling of the intermontane basins and lakes is believed to have been completed by the end of the Eocene or perhaps the Early Oligocene epoch, about 35 million years ago. At roughly the same time, regional uplift had come to a halt and the effects of the Laramide orogeny had definitely dwindled to a standstill. Erosion by Oligocene times had made great inroads toward leveling the landscape of a generally stable tectonic region. For something like 5 to 10 million years, the eastern Rocky Mountain region saw no new uplift, and erosion proceeded to wear the mountains away. A huge, sloping apron of eroded sediments began to develop out onto the Great Plains to the east, forming a deep clastic wedge during the Oligocene and most of the following Miocene epochs of the Cenozoic. The remnants of the once-lofty Laramide summits stood only as minor prominences, sticking out of huge, surrounding aprons composed of their own eroded debris. As erosion proceeded still further, these aprons began to coalesce into a solid sheet of sediment termed the Tertiary pediment, which was once erroneously called the Rocky Mountain peneplain.

This trend was thwarted, however, during the Late Miocene by a great regional upwarping of the crust over most of the Rocky Mountain region as well as adjacent areas. To the west, the huge area of the present Colorado Plateau began to lurch upward toward its present altitude during this period. As the regional uplift proceeded into the Pliocene, rivers and streams were rejuvenated by the increased gradients they now flowed down, and stream-cut erosion accelerated. A lengthy period of canyon cutting ensued, which has continued. This phenomenon of stream rejuvenation by regional uplift explains much of the unusual entrenched meander and superimposed drainage found in the Rocky Mountain region and in the western states at large. The Miocene-Pliocene regional uplift added 1,000 to 2,000 meters of new height to the peaks of the Rockies and also lifted surrounding terrain such as the Tertiary pediment so that these former lowlands have been considerably elevated and seem like anomalous terrain.

Finally, as the cooling of the Pliocene gave way to the intermittently frigid period of continental glaciation dominating the Pleistocene epoch of the past 2 million years, large alpine glaciers formed at higher altitudes in the ranges of the Rockies. At least three episodes of alpine glaciation are believed represented in places such as the Colorado Rockies of the United States. Glaciation helped to scour sedimentary layers from mountaintops, exposing the harder and older crystalline rocks lying beneath. Glaciation also widened and deepened downslope valleys, giving some a U-shaped cross section, characteristic of glacially influenced topography. Glaciers also left their mark on the Rocky Mountain landscape in the form of myriad examples of glacially sculpted terrain: landforms such as cirques (steep-walled basins), moraines (accumulations of rock debris), and razor-sharp ridges called arêtes. The end of the Pleistocene epoch's cycle of glacial advance and retreat is not certain, however, and the few thousand years since the waning of widespread glacial effects has not been enough time for significant geologic processes to leave a new page in the history of the Rocky Mountains.

Frederick M. Surowiec

Further Reading

Baldridge, W. Scott. *Geology of the American Southwest: A Journey Through Two Billion Years of Plate Tectonic History.* New York: Cambridge University Press, 2004. Provides a sweeping overview of the geology of the southwestern United States, beginning with Precambrian events and continuing through modern time. Covers plate tectonics and orogeny. Accessible to undergraduate and graduate students.

Caputo, Mario V., James A. Peterson, and Karen J. Franczyk, eds. *Mesozoic Systems of the Rocky Mountain Region, USA.* Denver: SEPM, 1994. This multiauthored volume deals with a variety of aspects of the Rocky Mountains, including tectonics, sedimentation, and paleontology. Written for the specialist.

Chronic, Halka. *Roadside Geology of Colorado.* 2d ed. Missoula, Mont.: Mountain Press, 2002. One of the more comprehensive treatments of the landforms found in Colorado that is designed for the general reading public. Like other titles in this series, it is written for the geologically aware or curious traveler or reader. The author assumes no prior geology education or training. Profusely illustrated throughout with numerous maps, diagrams, and photographs. These visual aids in combination with the text effectively explain the geologic history of the middle Rockies of the United States.

__________. *Time, Rocks, and the Rockies: A Geologic Guide to Rocks and Trails of Rocky Mountain National Park.* Missoula, Mont.: Mountain Press, 1984. This source provides numerous illustrations and descriptions pertaining to Rocky Mountain National Park, but with the general intent geared toward explaining the general story of the Rockies through time to both park visitor and nonvisitor alike. Suitable for readers at all levels of geologic knowledge, high school and above.

Crowley, John. "Types of Mountain Ranges in The Rocky Mountain Region." *American Geographical Society's Focus on Geography* 50 (2007): 24-29. Describes the five basic types of mountain ranges within the Rocky Mountain region. The author discusses geographical features and environmental differences among the range types. This article does not address the geology of the region and is thus most useful to the layperson.

Ehlers, J., and P. L. Gibbard. *Quaternary Glaciations—Extent and Chronology: Part II.* New York: Elsevier, 2004. One of a three-part volume discussing the Quaternary glaciations of the world. Discusses the glaciation of Sierra Nevada, Rocky Mountains, the Hawaiian Islands, and specific states associated with the Appalachians.

Ehlers, J., P. L. Gibbard, and P. D. Hughes, eds. *Quaternary Glaciations—Extent and Chronology: A Closer Look.* New York: Elsevier, 2011. One volume in a series of texts discussing the Quaternary period. This book covers Pleistocene and Quaternary glaciation events around the world, including in the Alps, Andes, Appalachians, Sierra Nevada, and Rocky Mountains among others. Contains an extensive index and a CD containing instruction and maps for use with ArcView, ArcGIS, and GoogleEarth.

Kearey, Philip, Keith A. Klepeis, and Frederick J. Vine. *Global Tectonics.* 3rd ed. Hoboken, N.J.: Wiley-Blackwell, 2009. This well-illustrated and clearly organized undergraduate text discusses the principles of geology in relation to plate tectonics. Index and bibliography.

Lageson, David R., and D. R. Spearing. *Roadside Geology of Wyoming.* Missoula, Mont.: Mountain Press, 1988. A fascinating and informative introduction to the geologic history and processes still at work in this state, which possesses a number of outstanding examples of Rocky Mountain ranges and related geomorphologist features. Well illustrated throughout, like other titles in this series, with highly explanatory geologic and geographic maps, diagrams, and photographs. A useful sourcebook for the general reader on the Rockies and their geologic history.

McPhee, John. *Rising from the Plains.* New York: Farrar, Straus & Giroux, 1986. One of about three books on the subject of American geology by the well-known, "new" or "literary" journalist, McPhee. The author uses a novel approach to discussing geologic concepts by involving the reader in the lives of working field geologists. Excellent introduction to the geologic history and processes of the entire western United States, focusing on the middle and northern Rockies. Suitable for all readers high school and above.

Ollier, Cliff, and Colin Pain. *The Origin of Mountains.* London: Cambridge University Press, 2000. This text encompasses the geographic, geological, and geophysical concepts of mountains. Topics include fault-block mountains, fold belts and folding, gravity structures, rift valleys and river valleys, plateaus and mountain erosion. Begins with a general overview of subduction and plate tectonics. Mountain ranges discussed in detail include the Appalachians, Andes, Cascades, Himalaya, and the Rockies.

Orr, Brendon B., et al., eds. *Rocky Mountain Geology.* Laramie, Wyo.: *GeoScience World,* 1962-current. An academic journal presenting articles quarterly on geology and paleontology topics related to the Rocky Mountains. An excellent resource for earth scientists specializing in the Rocky Mountain region.

Plummer, Charles C., Diane H. Carlson, and Lisa Hammersley. *Physical Geology.* 13th ed. Boston: McGraw-Hill, 2009. This is a straightforward, easy-to-read introduction to geology intended for those with little or no science background. Chapter 20 discusses the development of mountain belts such as the Rockies. Includes many excellent illustrations.

Prothero, Donald, and R. H. Dott, Jr. *Evolution of the Earth.* 8th ed. New York: McGraw-Hill, 2009. An in-depth survey of North America from the viewpoint of historical geology. Each geological era and period is treated chronologically with respect to the processes and effects that influenced various regions of the continent. Significant formations and deposits are explained in terms of geologic structure and stratigraphy. Suitable for college readers or others with a good, basic knowledge of geologic principles.

See also: Alps; Andes; Appalachians; Basin and Range Province; Cascades; Himalaya; Minerals; Mountain Belts; Sierra Nevada; Transgression and Regression; Transverse Ranges.

S

SEAFLOOR SPREADING

Seafloor spreading is the mechanism of continental drift. Convection currents above magmatic plumes in Earth's mantle pull the solid crust apart, allowing new material to well up and solidify. Driven by magmatic convection, the sea floor spreads until it encounters a continental plate and is forced into the mantle. The motion of seafloor material pushing sections of continental plates together leads to the formation of exotic terranes along the western coast of North America and elsewhere, characterized by seismic faults and volcanism.

PRINCIPAL TERMS

- **asthenosphere:** a thin boundary layer between the earth's solid crust material and the molten magma of Earth's mantle
- **craton:** a region of unaltered original magmatic crustal material, typically as continental shield structures
- **hot spot:** the crustal region above a large rising plume of magmatic material in the mantle, presumed to be caused by an intense prominence of the outer core
- **magmatic:** having its origin in the melted rock, or magma, of the planet's mantle layer
- **paleomagnetism:** the planet's magnetic polarity signature embedded in molten rock as it cools and solidifies
- **plutonic:** having its origin and focus in the planet's mantle in the asthenosphere below the level of the crust
- **seamount:** a mountain formed under the sea through volcanic processes
- **subduction:** the movement of dense crustal material back into the mantle as it is directed downward against the less dense material of the continental plates
- **terrane:** a coastal subsection of a continent, formed by the accretion of a small continental mass or plate to a larger one under the influence of seafloor subduction movement; generally, a fault-ridden rock mass relocated from its point of origin and unrelated to adjacent rock structures
- **volcanic:** having its origin within the crust, above the level of the asthenosphere

ORIGINS OF THE THEORY OF SEAFLOOR SPREADING

The idea of seafloor spreading was preceded by German geophysicist Alfred Wegener's 1910 theory of continental displacement, or continental drift, in which Wegener posited that Earth's continents had once been conjoined into a single supercontinent, which he called Pangaea. In time, the segments of Pangaea drifted apart, to eventually take up their current positions. Wegener's theory was met with disdain and ridicule until well after his death. The primary failing of the theory was that it contained neither a notion of a mechanism whereby such a process could take place nor a notion of a driving force for such displacement.

Not until the 1950's was the first evidence to support Wegener's theory observed. U.S. Navy and civilian field studies observed the paleomagnetic signature embedded in seafloor bedrock formations. The Navy's mission, which crisscrossed sections of the Pacific Ocean to map the topography of the seabed, was accompanied by a civilian crew who mapped the Pacific seabed's paleomagnetic signature. This mapping revealed numerous geomagnetic bands with opposite magnetic polarity. While this was not a surprise in itself, because magnetic pole reversal in rock formations had been observed before, what was surprising was that the bands could be made to match precisely across seismic fault lines, indicating that the bands had shifted in time from the demarcation point of that fault line.

Oceanic topography mapping in the Atlantic Ocean revealed the presence of the Mid-Atlantic Ridge, a line of volcanically active sea floor that essentially bisected the Atlantic Ocean. Magnetic anomaly mapping across this feature revealed a symmetrical pattern of geomagnetic bands cataloging some 171 magnetic pole reversals in 76 million years. In 1960, American geologists Harry Hammond Hess and Robert S. Dietz, working independently, concluded from these geomagnetic observations that new seafloor material was being generated by the continuous upwelling of magma from Earth's mantle at mid-oceanic ridges. Coupled with a theory put forward by British geologist Arthur Holmes in 1945, which suggested magmatic convection currents as the means by which the new crustal material was carried away from ridges and subducted back into the mantle at deep ocean trenches, the overall mechanism was termed "seafloor spreading." This discovery provided the final vindication of Wegener's theory of continental displacement by revealing the mechanism that had been missing in Wegener's hypothesis.

The Structure of the Earth and the Mechanism of Action

The theories of continental drift and seafloor spreading are intimately bound into plate tectonics and concepts of the interior structure of the earth. The inner core of the earth is believed to be a sphere of solid nickel-iron at a temperature well above its melting point but kept in the solid state by gravity and the pressure of the surrounding material. The inner core is surrounded by an outer core of liquid nickel-iron. Seismic investigations show that the outer core may not be spherical but rather covered with radial prominences. The remaining material, extending practically to the planet's surface, is the mantle, composed of molten rock in a plastic state. The final, outer layer of the planet's structure is the crust, a mere 70 to 80 kilometers (40-50 miles) thick across the sea floor and nearly twice that size across the continental plates. The thickness of the crust is only about 1 percent of the radius of the planet, roughly the same proportion as the shell of an egg.

Hot spots are formed in the mantle above the various prominences that dot the outer core, increasing the plasticity of the magma in those locations and giving rise to volcanic activity. The Hawaiian Islands, the South Pacific island chains, and the infamous volcano Krakatoa in Indonesia are formed atop such hot spots. Elsewhere, as at the Mid-Atlantic Ridge and the mid-Pacific rise, the convective movement of magma within the mantle drives the material upward with sufficient force to break the crustal surface. At those locations, magma continually and slowly oozes against the pressure of thousands of meters of seawater, forming new crustal material as it freezes. The convective motion of the underlying magma drives the new seafloor material from the ridge structure where it was formed and eventually into the continental plates on either side.

The mechanism is slightly different at the mid-Pacific rise because of the more intense hot spot beneath the Hawaiian Islands and the much more extensive, relatively thin crust that makes up the Pacific Ocean floor. The rate of spreading may be limited to some extent in the Atlantic Ocean by the relative nearness of the American and African continents to either side. This effect is not present in the Pacific Ocean, and the rate of seafloor spreading is as much as ten times greater there. This rate indicates that considerably more magmatic material is being extruded from the mid-Pacific rise; also, the lack of restraining pressure allows the magma over the Hawaiian hot spot to roll out almost continuously, building up volcanic shield seamounts in the process. The effect of seafloor spreading in this area is evidenced by the chains of volcanic islands in different regions of the Pacific Ocean.

The Hawaiian Islands comprise a series of seamount islands that have formed over the stationary hot spot that currently provides the energy for the activity of Kilauea, Maunaloa, and other active volcanic sites in Hawaii. The island chain began to form from the movement of the active vents away from the hot spot as the sea floor moved westward from the mid-Pacific rise; the curve of the chain was caused by the slow turning movement of the Pacific plate. Other Pacific island chains resulting from the combination of these movements include the Javanese island system, home to Krakatoa, and the east Pacific chain, which includes the volcano Pinatubo in the Philippines.

When the seafloor material approaches the continental plates, it meets the resistance of the more massive, though less dense, material of the continental mass. The natural tendency is for the lighter material to float atop the denser material in isostatic equilibrium. This is augmented by the now-downward convective movement of the mantle magma. As a result, the spreading sea floor is subducted beneath the continental plate to become part of the magmatic mantle once again. Subduction zones are characterized by deep oceanic trenches where ocean-side sea floor trends downward, into the mantle, and continent-side sea floor is pushed upward by the material sliding beneath it.

Terranes and Fault Zones

The movement of seafloor material from mid-oceanic source ridges and toward subduction zones has had some significant effects on the structures of the continents. It can be assumed that the breakup of Pangaea and its daughter segments Laurasia and Gondwana was accompanied by several small subcontinental fragments that separated from the main mass. Inevitably, these relatively small fragments were carried toward the major continental plates by the seafloor spreading mechanism, to locations far removed from their points of origin.

At the subduction zone, the fragments' greater size brought them up against the continental plate instead of into subduction. The force of the moving sea floor inexorably drove the smaller fragments against the continental mass, as it still does, essentially welding them and driving up high mountain ranges. These regions, called terranes, are notable for their essential geological difference from the main continental mass, sometimes being composed of entirely incongruous structural features. The junctures of these smaller fragments and continental plates are characterized by active seismic faults and volcanic activity fueled by the displacement of material in the subduction zone. The fault lines that permeate terranes tend to run parallel to the motion of seafloor spreading, with the San Andreas fault in California being a notable example.

Plate tectonics, however, does not only involve seafloor spreading. As it relates to the entire crust of the earth, a number of tectonic plates other than the continental plates exist and are in motion, driven by the same magmatic convection forces that drive seafloor spreading. The largest of these plates, the Pacific, is also affected by the rotation of the planet and, correspondingly, has a rotational motion of its own. This additional motion of tectonic plates creates transform faults because of the sheering force between moving plates. At these locations, new seafloor material is neither produced nor removed.

Fault boundaries are classified as either convergent or divergent. The Mid-Atlantic Ridge and mid-Pacific rise are divergent boundaries, where solid plate material, such as the sea floor, is driven to separate, allowing new material to form from upwelling magma. The deep trenches of subduction zones are convergent boundaries, where solid plate material is forced down into the mantle; here it is melted again and then circulated into the overall motion of the magma. Convergent boundaries therefore destroy crustal material and divergent boundaries create crustal material or sea floor.

Subduction Zone Earthquakes

The difference in the direction of movement of tectonic plates and of the sea floor as it spreads gives rise to a different form of seismic activity: the earthquake. Plate edges and shapes are irregular and do not slip smoothly past each other. The lateral relative movement of plates is interrupted as plate edges bind, generating lateral torsional stress. Earthquakes and related seismic activity in fault zones is generated when the torsional stress exceeds the resistance of the binding sites and the plates surge forward in their normal movement. This type of plate movement generates continuous minor earthquakes, with few being sufficiently powerful to cause damage.

Seafloor spreading generates seismic activity in an entirely different manner because of the movement of the sea floor toward the mantle in a subduction zone. The normal motion of the sea floor is orthogonal to the edge of a plate, rather than with the parallelism of lateral slippage. Accordingly, there is a buildup of pressure between the advancing sea floor and the leading edge of the continental plate when binding occurs. This type of binding is not relieved by lateral slippage, but by the sudden release of pressure as the sea floor suddenly breaks free and plunges to resume its normal motion. At the same time, the edge of the continental plate lurches vertically to ride up on top of the seafloor plate, then downward and toward a new isostatic balance. The energy released by the movement is often orders of magnitude greater than that which is released by horizontal slippage earthquakes.

Complicating the effect is the mass of water lying above the active fault. The sudden movement of the subducting plate is accompanied by a corresponding displacement of the water mass, generating a series of waves of mass equal to the mass of water that was displaced. This wave, and its echoes, comes to shore as a tsunami that can result in a great deal of damage. To maintain the energy of the wave, its height increases as the depth of water below it decreases. Depending on the topography of the wave's path, a tsunami can attain a height of several meters (more than 6 feet), with the geological record of some locations indicating that tsunamis more than 300 meters (1,000 feet) in height have been produced.

The movement of seafloor spreading in the Atlantic Ocean threatens the collapse of one of the Canary Islands along an existing fault line. If that collapse occurs, it is expected to generate a tsunami of 200 meters (more than 650 feet) or more in height that will strike the eastern seaboard of North America, obliterating many major cities.

Richard M. Renneboog

Further Reading

Chester, Roy. *Furnace of Creation, Cradle of Destruction: A Journey to the Birthplace of Earthquakes, Volcanoes, and Tsunamis.* New York: Amacom Books, 2008. Discusses the mechanics of seafloor spreading at some length, clearly describing its dynamic role with regards to earthquakes, volcanoes, and tsunamis.

Karner, G. D., G. Manatschal, and L. M. Pinheiro. *Imaging, Mapping, and Modelling Continental Lithosphere Extension and Breakup.* London: Geological Society of London, 2007. A series of learned papers relating current methods and theory with regards to seafloor spreading and tectonics.

Kearay, Philip, Keith A. Klepeis, and Frederick J. Vine. *Global Tectonics.* 3rd ed. Hoboken, N.J.: John Wiley & Sons, 2009. An intermediate geology textbook that provides a good technical description of seafloor spreading in the context of tectonic plate theory.

Merrill, Ronald T. *Our Magnetic Earth: The Science of Geomagnetism.* Chicago: University of Chicago Press, 2010. This volume provides a detailed description of the phenomenon of geomagnetism and its role in the discovery and proof of seafloor spreading.

Monroe, James S., Reed Wicander, and Richard Hazlett. *Physical Geology: Exploring the Earth.* 6th ed. Belmont, Calif.: Thomson Higher Education, 2007. A textbook for college-level programs that provides a readable, detailed discussion of plate tectonics, continental drift, and seafloor spreading set in a question-and-answer format.

Pinet, Paul R. *Invitation to Oceanography.* 5th ed. Sudbury, Mass.: Jones and Bartlett, 2009. This introductory college-level textbook provides a clear discussion of continental drift, seafloor spreading, and plate tectonics in the context of being the source of ocean basins.

See also: Coastal Processes; Continents and Subcontinents; Continental Drift; Continental Rift Zones; Faults; Geoclines; Hydrothermal Vent Faunas; Iceland; Mass Wasting; Oceanic Basins; Ocean Ridge System; Oceanic Crust; Pangaea.

SHIELD VOLCANOES

Shield volcanoes are the products of eruptions of low-viscosity materials. They have been identified on five solar system bodies. The most active volcanoes on the earth are shields.

PRINCIPAL TERMS

- **coronae:** ring structures on Venus consisting of alternating concentric ridges and valleys, higher than the external terrain; possibly volcanic in origin
- **hot spots:** areas of anomalously hot mantle
- **ignimbrite:** rock formed by widespread deposition and consolidation of block, pumice, and ash flows
- **monogenetic:** pertaining to an eruption in which a single vent is used only once
- **paterae:** inverted, saucer-shaped features considered to be of volcanic origin
- **polygenetic:** pertaining to volcanism from several physically distinct vents or repeated eruptions from a single vent punctuated by long periods of quiescence
- **pyroclastic materials:** broken rock formed by volcanic explosion or aerial expulsion from a volcanic vent
- **tholus:** an inverted, bowl-shaped feature considered to be of volcanic origin

Volcano Classification

The shield volcano is so called because of its similarity to the ornate shields of the ancient Nordic warriors. More specifically, the name is derived from the Icelandic word *dyngja* (shield). The typical shield volcano is circular to ellipsoidal when viewed from above. Its "ornamentation" can take the form of superposed cinder cones, radiating lava flows, a summit or flanking crater or craters, and rift valleys. Viewed from the side, the shield can take forms ranging from inverted saucers to inverted mixing bowls. In this respect, shields differ from the familiar volcano shape represented by the symmetrical, conical Mount Fuji in Japan.

Most shields are basaltic in composition and formed primarily of lavas (more than 98 percent). Exceptions are found in shields that have lavas that are richer in silica and of andesitic composition, such as Hayli Gub in Ethiopia, and those that are poorer in silica, such as the carbonatitic volcanoes of the East African Rift zone. There are also shields composed primarily of pyroclastic materials, although frequently they are separately classified.

One of the earlier attempts to produce a systematic classification of volcanoes was made by Swiss volcanologist Alfred Rittmann in 1936. This scheme did not achieve wide recognition until the second German edition of *Vulkane und ihre Tätigkeit* (1960; *Volcanoes and Their Activity*, 1962) was translated into English. In this text, Rittmann recognized a Hawaiian type and an Icelandic type of shield volcano. A similar subdivision was recognized by Sir Charles Cottonin his text *Volcanoes as Landscape Forms* (1944). In the most influential of the English geomorphology textbooks of the twentieth century, *Principles of Physical Geology* (1944), Arthur Holmes classified the shield volcanoes as "domes of external growth" to distinguish them from the domes of viscous lava built by the addition of material to the interior.

Aerial view of snow-covered Mauna Loa, a shield volcano on the big island of Hawaii, with several craters visible. (U.S. Geological Survey)

With the acquisition of images of volcanoes on other bodies in the solar system, there began a new phase in volcano classification. These new images provided views of volcanoes produced on bodies with different gravities, atmospheres, compositions, and other important parameters. The volcanologist now needed a database with which to make objective comparisons of the morphologies of the volcanoes on the different bodies. One such attempt was made by British-born American geologist James Whitford-Stark in the book *Volcanoes of the Earth, Moon, and Mars* (1975). This classification was made on the basis of the basal diameter versus the height of the shield volcanoes and resulted in the identification of four classes of terrestrial shield volcanoes: Hawaiian, Icelandic, Galápagos, and scutulum. (The last is from the Latin *scutum*, or "shield.") Lunar domes were shown to be similar to Icelandic shields, whereas Martian volcanoes had dimensions exceeding those of most terrestrial volcanoes.

A more detailed analysis was undertaken by British chemist Richard Pike and American geophysicist Gary Clow, who employed a statistical technique called principal component analysis. They first distinguished between polygenetic and monogenetic shields and then identified subgroups on the basis of size, shape, composition, and location. The monogenetic shields, shields produced by only one eruption phase, included the large Icelandic shield, a smaller steep shield, and a low shield. The polygenetic shields included a Hawaiian type, a Galápagos type, and a locally varied type, all of which are characterized by tholeiitic basalts; an oceanic and continental type dominated by alkali-rich basalts; and a group represented by seamounts, or submarine volcanoes. The more silicic of the shields were grouped into alkalic and calcalkalic ashflow plains categories. Pike and Clow also recognized three distinct groups of Martian volcanoes and a group of lunar domes. Volcanic landforms have since been recognized on the Jovian satellite Io and on Venus. Unfortunately, there is a paucity of topographic data for these features, so objective comparisons cannot yet be made.

Types of Shield Volcanoes

Small low and small intermediate-slope shields are monogenetic and variously described as scutulum-type shields, lava cones, or low shields. These shields are abundant in the Snake River plain volcanic field of the western United States and are represented by the volcano Mauna Iki, which erupted between 1919 and 1920 on the flanks of Kilauea. Typically these shields have lower slopes of about 0.5 degree and steepen to about 5 degrees near the summit. This slope change has been attributed to a change from extensive, thin, fluid, pahoehoe flows early in the eruption to more viscous, aa flows during the waning eruption. Many have summit craters formed by collapse, and a few have spatter ramparts around the vent. In the Snake River plain, the shields are aligned along fractures and probably represent initial fissure-type eruptions that subsequently contracted to point-source eruptions.

Small steep shields are typified by Mauna Ulu, which was produced in an eruption lasting from 1969 to 1974 and punctuated by a three-and-one-half-month period of quiescence in late 1971 and early 1972. The early phase was marked by lava fountains reaching heights of up to 540 meters, whereas in the second phase, the fountains rarely exceeded 80 meters. Nearly 350×10^6 cubic meters of lava contributed to the final structure. The eruption was characterized by a summit lava lake. Sustained overflows of this lake led to the production of tube-fed lava flows that traveled up to 12 kilometers from the vent. Short-duration overflows resulted in growth of the shield structure. The final structure rose 121 meters above the pre-1969 base and had a basal diameter of slightly more than 1 kilometer. The smaller Icelandic and seamount shields also fall within this category.

Medium low shields are polygenetic constructs represented by calcalkalic ash-flow plains and the smaller of the alkalic ash-flow plains. The calcalkalic ash-flow plains are composed primarily of rhyolitic or dacitic ignimbrites, rather than lavas, surrounding a caldera ranging from 7 to about 60 kilometers in diameter. Toward the caldera, lavas and pyroclastic fall deposits tend to dominate, and the flank slopes increase from about 1 degree to about 8 degrees. These structures are the products of extremely powerful eruptions that, fortunately, have not occurred within historic times. The Venusian coronae, if of volcanic origin, would probably fall within this group. Medium intermediate-slope shields consist of a wide variety of the groupings established by Pike and Clow. Included are the steeper of the ash-flow sheet shields, the majority of the continental alkalic basaltic shields, and the smaller of the oceanic and Martian shields. Medium steep shields are represented primarily by the Icelandic shields, the larger of the seamounts, the steeper continental shields, and the small steep oceanic shields. The small-volume members can be monogenetic, whereas the large-volume members are invariably polygenetic.

Large low shields are represented by the terrestrial Toba volcano, the Martian paterae, and the larger of the Venusian volcanoes, such as Colette and Sacajewea. Large intermediate-slope shields include the majority of the Hawaiian volcanoes, other terrestrial oceanic shields, and the Martian tholii. The mature Hawaiian shields typically have radiating rifts, a summit collapse caldera, and superposed small shields and cinder cones of a more evolved composition than the mass of the volcano. The large steep shield group includes the Galápagos volcanoes and other oceanic shields. The Galápagos shields differ from the Hawaiian shields in being much steeper in the summit region, having concentric rifts, and having a greater portion of more alkaline basalts.

The very large low shield category currently includes only the Martian volcano, Alba Patera. It is probable that a number of the Io volcanoes will be found to fall into this category. Very large intermediate shields are represented by the giant Mons volcanoes on Mars (Olympus Mons is the largest volcano in the solar system). Data suggest that these very large shields are dominated by lava flows.

Study of Shield Volcanoes

The morphologies of terrestrial shield volcanoes are readily obtainable from topographic maps, their compositions are obtained via standard geochemical techniques, their ages can be ascertained by radiometric dating, and their interior structures can be determined by observation of eroded structures. On other bodies in the solar system, these quantities are less easily obtained. Topography can be determined by measuring shadow lengths or determining the time delay between the sending and receiving of electromagnetic radiation by such instruments as laser altimeters. Since there are rock samples only from the moon (excluding an exotic group of igneous meteorites found on the earth), the compositions of other extraterrestrial materials have to be determined by analysis of their spectral characteristics. Relative ages of the outermost layers of extraterrestrial volcanoes can sometimes be determined from the density of superposed impact craters—the more craters, the older the surface. The internal structure of these volcanoes cannot be determined, since none of their parent bodies has as powerful agents of erosion as those found on the earth.

A lava fountain, or fire fountain, on Hawaii, 1983. (U.S. Geological Survey)

One of the reasons for assembling morphometric data for terrestrial volcanoes is to ascertain any relationships between the size and shape of the volcano and the forces that led to its construction. Then, by making corrections for the different gravities, atmospheres, and other parameters on the nonterrestrial bodies, one can infer the conditions that led to the construction of their shield volcanoes.

In spite of the wide range in the volumes of shield volcanoes, all appear to be constructed of low-viscosity material, whether it be basaltic lava or silicic ignimbrite. A major difference exists in the volume eruption rates between the lava and ignimbrite shields. The lava shields are formed primarily by an eruption style called Hawaiian, which is characterized by volumetric eruption rates of about 100 cubic meters per second for short durations but less than 1 cubic meter over time spans of a few years for the smaller structures, to about 1 million years for the larger Hawaiian shields. The ignimbrite shields are products of a plinian or ultraplinian eruption style and are thought to have had volumetric eruption rates of the order of 1 million cubic meters per second for eruptions lasting less than one day.

A feature that serves to distinguish the majority of terrestrial shield volcanoes from cone volcanoes is that the shields are commonly not found at plate margins; that is, they are predominantly intraplate volcanoes. It has been suggested that many shields are associated with hot spots, or areas of anomalously hot mantle. The paucity of terrestrial continental shields with heights in excess of 2 kilometers implies that the earth's oceans are acting as buttresses to prevent the collapse of the oceanic shields. The great heights of the Martian Mons volcanoes are probably, in part, a function of the lower gravity on that planet.

Significance

Shield volcanoes include the most active volcano on Earth (Kilauea), the tallest volcano in the solar system (Olympus Mons on Mars), and perhaps some of the highest eruption columns (more than 200 kilometers high, on Io). They have undergone some of the most violent eruptions ever to take place on Earth, such as the eruption at Toba Caldera, Sumatra. On Earth, shields are present on the continents and on the ocean floor, and they occur as oceanic islands. From the sea floor to its summit, Mauna Loa is the largest mountain on Earth, rising some 10 kilometers above its base. Small shields resembling those of Iceland are found on the moon. Gigantic shields have been identified on Mars. Active shield volcanoes are present on Io, and numerous shieldlike structures occur on Venus.

Chains of terrestrial shield volcanoes extending across a plate and formed by a single, fixed hot spot have been employed to determine the direction of motion of the plate, and the ages of the various volcanoes along the chain have been used to infer the plate's rate of motion. At the large and very large end of the shield volcano scale, inferences can be drawn as to the nature of the planet's interior, such as the internal heat transfer as a function of time and the strength of the outer layers of the planet required to support such huge masses. The heights of shield volcanoes have been employed to determine lithosphere thicknesses, and the horizontal separations between volcanoes have been used to infer the depths in the interior at which magma originated.

The high eruption frequency of terrestrial shield volcanoes, such as those of Hawaii, is of importance to those living in the paths of lava flows, which may destroy their homes. The enormous scale of eruptions associated with the formation of the ash-flow shields has fortunately not been witnessed within historic times, but such eruptions have been relatively frequent throughout geologic history and may recur in the future. On a more positive note, shield volcanoes can provide a source of tourist revenue; furnish new, rich arable land; and endow geothermal energy.

James L. Whitford-Stark

Further Reading

Decker, Robert, and Barbara Decker. *Volcanoes.* 4th ed. New York: W. H. Freeman, 2005. This readable and well-illustrated text is geared toward the general public. The authors explain geologic phenomena in a clear and simple manner. Includes a good section on Hawaiian shield volcanoes. The appendices list the world's 101 "most notorious" volcanoes, as well as Internet sources for volcano information.

Francis, Peter, and Clive Oppenheimer. *Volcanoes.* 2d ed. New York: Oxford University Press, 2004. This text covers volcanic activity ranging from lava flows to pyroclastic currents. Provides many examples from Hawaii, Italy, Ethiopia, Japan, Mount St. Helens, and Krakatau. Written for the layperson, but some knowledge of geology is helpful.

Lockwood, J. P., and Richard W. Hazlett. *Volcanoes: Global Perspectives.* Chichester: Wiley-Blackwell, 2010. An up-to-date and well-illustrated undergraduate-level text that is also accessible to the interested reader.

Lunar and Planetary Institute, Houston, Texas. *Basaltic Volcanism on Terrestrial Planets.* Elmsford, N.Y.: Pergamon Press, 1982. A compendium, by several dozen expert authors, of the majority of information published in English relating to basaltic volcanism. Chapter 5 is particularly pertinent to shield volcanoes. Aimed at a college-level audience.

Siebert, Lee, Tom Simkin, and Paul Kimberly. *Volcanoes of the World.* 3rd ed. Berkeley: University of California Press, 2011. A summary of locations, dimensions, and eruption histories of all terrestrial historic eruptions. For the reader desiring a detailed source.

Sigurdsson, Haraldur, ed. *Encyclopedia of Volcanoes.* San Diego, Calif.: Academic Press, 2000. This book contains a complete summary of the scientific knowledge of volcanoes. It contains eighty-two well-illustrated overview articles, each of which is accompanied by a glossary of key terms. Although this is a college-level text, it is written in a clear and comprehensive style that makes it generally accessible. Cross-references and index.

See also: Calderas; Eruptions; Forecasting Eruptions; Hawaiian Islands; Hot Spots, Island Chains, and Intraplate Volcanism; Krakatau; Lava Flow; Mount Pelée; Mount Pinatubo; Mount St. Helens; Mount Vesuvius; Popocatépetl; Recent Eruptions; Stratovolcanoes; Volcanic Hazards.

SIERRA NEVADA

The Sierra Nevada preserve rocks that formed several kilometers below the earth's surface during subduction of the Pacific Ocean below North America more than 100 million years ago. Erosion and deposition by streams 50 million years ago concentrated gold in the western foothills discovered in the 1849 California gold rush.

PRINCIPAL TERMS

- **batholith:** body of rock crystallized from magma, or molten rock; compositions are typically granitic
- **crust:** the outermost layer of the earth; continental crust is 30 to 35 kilometers thick; oceanic crust is 5 to 10 kilometers thick; the greater density of oceanic crust relative to continental crust forces it to subduct
- **faulting:** the process of fracturing the earth such that rocks on opposite sides of the fracture move relative to each other; faults are the structures produced during the process
- **granite:** an igneous rock that is composed of mostly pink and white (quartz) minerals, which frequently has a speckled appearance
- **metamorphic rocks:** rocks deposited initially as sediments and subsequently exposed to extensive heat, resulting in their recrystallization
- **normal fault:** a steep fault that forms during extension, or stretching, of the earth's surface; the rock above the fault moves down relative to the rock below
- **quartz:** one of the most common minerals on the earth's surface; it occurs in many different forms, including agate, jasper, and chert
- **vein:** a mineral deposit that fills a crack; veins form by precipitation of minerals from fluids

Overview

Much of what is admired in the Sierra Nevada—the face of the eastern slope, the deep canyons of Yosemite National Park, the peaks where the interaction of light and shadow inspired the name the "range of light," and the undulating topography of the western foothills—is what makes the range geologically significant. The uplift and erosion that created the Sierran landscape over the last 20 million years exposed the roots of a giant volcanic chain active between 225 million and 80 million years ago. During that interval, the Sierra Nevada resembled the modern volcanic belts of the Andes in South America and the Cascade Range of the Pacific Northwest; however, only the bowels of the Sierran volcanoes remain. The origins of the ancient and modern volcanic chains are similar; all three formed during the important process of subduction, or descent, of an oceanic plate below a continent. Subduction is fundamental to plate tectonics, the basic theory of the earth sciences. Exposure in the Sierra Nevada of rocks originally formed several kilometers below the surface of the earth allowed scientists to investigate processes at depth in continental crust above subduction zones and to assess the validity of plate tectonics theory in explaining events in the geologic past. Furthermore, the processes inferred for the Sierra Nevada probably occur beneath the Andes and the Cascade Range.

The description of the Sierra Nevada includes the morphology, or physical characteristics that define it, and the geology, or the rocks that compose it. The Sierra Nevada represent the longest continuous mountain range in the conterminous United States, extending 640 kilometers north-northwesterly along the eastern border of California. The mountains rise gently in the west from the agricultural belt of central California's Great Valley to the glaciated crest. From the crest, the range drops precipitously to the deserts of the Basin and Range Province in Nevada to the east. To the northwest, the Sierra merge with the Klamath Mountains of Northern California and Oregon. The rocks and faults of the Sierra Nevada can be identified in the Klamath Mountains, suggesting that the two ranges were once one. To the north, the Sierra Nevada terminate against the Cascade Range of northern California, Oregon, Washington, and British Columbia. The southern Sierra Nevada bend westward into the Transverse Ranges. The Sierra Nevada form a barrier to storms that enter California from the Pacific Ocean and create a rain shadow east of the range crest. Thus, the shallow western slope has pine forests and alpine lakes, whereas the steep eastern slope is barren and dry.

Different processes created the landscapes of the northern, central, and southern Sierra Nevada. In the north, much of the morphology was formed by volcanoes that spewed lava and ash. Glaciers dominated in the central portion, carving great U-shaped valleys. The highest peaks, including the tallest mountain in the lower forty-eight states, Mount Whitney, are in the southern Sierra, where elevations approach 4,250 meters. Here, glaciers and streams were not very active. What is remarkable about the entire Sierra Nevada, however, is their flat top. Looking from a distance toward the crest, the Sierra Nevada appear to be a tableland cut by numerous deep canyons. This flatness led scientists to conclude that the mountains represent uplift of an area that was previously eroded to a uniform level. The Sierra Nevada consist of a series of long belts of metamorphic rocks that were intruded by a large batholith. The belts parallel the axis of the mountain range and grow younger from east to west. Their boundaries are faults across which rocks change age and composition. The metamorphic belts of the eastern Sierra Nevada contain sediments that accumulated on the continental shelf of North America between 500 million and 225 million years ago. In contrast, the sediments in the metamorphic belts of the western Sierra (which are the same age as those of the eastern Sierra Nevada) collected on the floor of the old ocean basin that bordered the ancient continental shelf. Slivers of the old sea floor, called ophiolites, occur with the sediments in the western metamorphic belts.

The batholith that intruded the mosaic of fault-bound belts appears to be a monolith of pink and gray speckled rock. Closer inspection, however, reveals that it consists of numerous small bodies of rock, each of which is a few kilometers across. The bodies are between 135 and 85 million years old and were derived from tens of kilometers depth. The pink and gray rock is granite, a chief constituent of continental crust, and is readily distinguished from the dark-colored metamorphic rocks. The large size of the mineral crystals in the granite suggests that the magma cooled slowly below the surface of the earth. Granite is estimated to make up more than 60 percent of the Sierra Nevada. Other rocks in the Sierra Nevada include glacial deposits, volcanic rocks, and stream gravels that are younger than the metamorphic belts and the batholith.

Half Dome, in the most famous section of the Sierra Nevada, Yosemite National Park. A glacier that moved over the Sierra Nevada batholith formed the Yosemite Valley and other landmarks during the last ice age. (© Mick Roessler/Corbis)

Formation

To understand the evolution of the Sierra Nevada, one must have a grasp of the theory of plate tectonics, which states that the surface of the earth is composed of rigid plates between 5 and 100 kilometers thick. The plates, either oceanic or continental, float on a partially molten layer that allows them to move relative to one another at velocities of a few centimeters per year. Oceanic plates are much younger than continental plates and, in fact, are created continuously along the ridges that line the world's ocean basins. Magma (molten rock) from deep in the earth erupts at the mid-oceanic ridge and cools, forming new sea floor. To make room for the new material, old sea floor is removed from the ocean basins at trenches. Trenches are on the opposite edges of oceanic plates from ridges and are the sites where the sea floor subducts, or descends, below another plate, either continental or oceanic. Ridges and subduction zones, therefore, define two types of plate boundaries: divergent (where plates move apart) and convergent (where plates come together), respectively.

The geology and morphology of the Sierra Nevada tell a story that spans 0.5 billion years—of which subduction dominated the last 200 million years. For the first few hundred million years, however, the area of the Sierra Nevada was quiet. Sediments were deposited in shallow water on the continental shelf of Nevada to the east and in deep water on the Pacific Ocean floor to the west. (The edge of the North American continent during this time was several hundred kilometers farther east.) Approximately 225 million years ago, the scene changed. The Pacific Ocean began to subduct below the North American continent. Sediments on the sea floor scraped against and accreted to the overriding continent in long, linear belts. Pieces of sea floor occasionally accreted with the sediments. Addition of material to its edge forced the continent to grow progressively westward. As subduction continued, the oceanic plate reached depths in the earth below the continent where the temperature was high enough to cause melting of the interface between the overriding and descending plates. The melting produced small bodies of magma that ascended to the surface. Significant quantities of subducted sediments and continental crust also melted and mixed with the magmas to produce liquids of granitic composition. Most of the magmas solidified at shallow depths between 135 million and 85 million years ago, leading to the large batholith now exposed in the Sierra Nevada. Some of the magmas, however, erupted on the surface as volcanoes. Emplacement of the magmas near the surface raised the temperature sufficiently to recrystallize the sediments into metamorphic rocks.

About 80 million years ago, the magmas cooled, the volcanoes stopped erupting, and the sea floor ceased accreting to the North American continent. For the next 50 million years, erosion dominated. Streams stripped the volcanoes from the surface, exposing the batholith and metamorphic rocks that lay below. The removed material was deposited as soil and gravel in the rolling hills between the Sierra Nevada and the Pacific Ocean. Eventually, the landscape was eroded to a uniform level. Subduction below California stopped completely approximately 25 million years ago. The rise of the modern Sierra Nevada began 20 million years ago. The continental crust between the Sierra Nevada and Utah stretched and broke apart, producing normal faults oriented in an almost northerly direction. This stretched province of narrow mountains and wide valleys is the Basin and Range Province. The westernmost of the normal faults in the Basin and Range Province defines the eastern edge of the Sierra Nevada. Rocks west of the Sierra fault moved up to create the modern mountains, while those to the east moved down to form a valley. Eruption of lava and ash accompanied the stretching of the continental crust as spaces formed to allow the passage of molten rock from deep in the earth. Lavas frequently flowed down the old stream channels. New streams circumvented the lavas, leaving them as high grounds in the landscape. Finally, glaciers of the Wisconsin glacial epoch (60,000 to 90,000 years ago) carved the spectacular valleys of the crest, including Yosemite Valley.

The combination of subduction followed by stretching built and is building the modern Sierra Nevada. The Sierra Nevada are still active; large earthquakes on the eastern edge of the range reveal that the mountains are rising and the fault is moving. Magma lies within a few kilometers of the surface below Mammoth Lakes, ready to feed the volcanoes at Mono Craters, which last erupted 6,500 years ago. Streams continuously erode the high peaks trying to bring the land to sea level, thereby restoring the landscape of 0.5 billion years ago.

Study of the Sierra Nevada

A variety of techniques from all disciplines in the earth sciences are used to study the Sierra Nevada. Two, however, are particularly important: fieldwork and isotope geochronology. By doing fieldwork, investigators map the distribution of rock types, document structures such as faults, and analyze the relative ages of the rocks, primarily by painstakingly describing each rock type and walking throughout the range. In contrast, isotope geochronology allows scientists to ascertain the absolute ages of the different rock types from field samples taken back to the laboratory.

To determine the evolution of ancient rocks, geologic fieldwork is necessary. In the field, the geologist systematically notes the different types of rocks that exist and the contacts among them. Features in the rock that may be chosen for evaluation include the kinds of minerals, the sizes of the different types of particles, and the nature of the boundaries between grains. Contacts are evaluated to determine positions of rock types to establish their relative ages. Generally, rocks at the top of a sequence are assumed to be the youngest; those at the bottom, the oldest. The geologist also watches carefully for faults that may indicate that two adjacent sequences of rocks have had dissimilar histories. Such faults define the boundaries of the western metamorphic belts in the Sierra Nevada. Although the rocks on either side of these faults were deposited in deep water, they have different compositions and ages, leading scientists to conclude that the rocks now juxtaposed in the field were not deposited next to one another but were brought together during faulting. They originally may have been separated from one another and North America by great distances. The term "exotic terrane" frequently is applied to such rock sequences. Accretion during subduction explained this phenomenon. Sediments could arrive at North America from far-off sites, carried by the moving oceanic plate.

Isotope geochronology, or radiometric dating, allows scientists to determine the "absolute" age of a rock by taking advantage of the property of radioactivity. The dates derived from this method differ from those calculated by fossils and the superposition of strata. The latter are "relative" ages based on the geologic time scale. Prior to radiometric dating the number of years represented by divisions on the time scale were unknown. Radioactivity is the process by which the nuclei of certain chemical elements such as uranium, radium, thorium, rubidium, and potassium break apart and emit radiation. The process yields an atom of a different element. Uranium, for example, changes to lead. Isotopes are species of the same chemical element that differ only in the number of neutrons in their nucleus. Some isotopes are stable; some are unstable and break apart by radioactive decay. Each isotope has a characteristic rate of decay measured as a half-life (the time in years for half the nucleus to decay). Examples of half-lives include the 47 million years for rubidium to change to strontium and the 1.35 million years for potassium to become argon. Because half-lives can vary by several orders of magnitude, isotopes used for young rocks are different from those used for old rocks. By accurately determining in a mineral the amounts of the radioactive isotope and the atom into which it decays, geologists can calculate the age at which the mineral crystallized. The technique, therefore, is very useful for dating igneous and metamorphic rocks. Minerals crystallize in these rocks when the rocks form. In contrast, sedimentary rocks contain fragments of minerals that were eroded from older areas. Thus, radiometric dating of minerals in sediments yields only the ages of the region from which the minerals were derived.

The ages of and the relationships between the batholith and the metamorphic rocks in the Sierra Nevada were greatly elucidated by isotope geochronology. Intrusion of the batholith was discovered to have spanned tens of millions of years as individual bodies rose one by one. Despite the similar appearance of two granites, their age can differ by as much as 50 million years. Field observations suggested that heat from the batholith metamorphosed the accreted sediments; dating of the metamorphic belts corroborated the field evidence. In addition, isotopic studies established the age of the lavas that postdated magmatism.

Pamela Jansma

Further Reading

Adams, Ansel. *Yosemite and the Range of Light.* Reprint. New York: Little Brown & Company, 1992. This collection of photographs was taken by the noted photographer Ansel Adams, after whom a wilderness has been named in the high Sierra Nevada near Mammoth Lakes. Adams spent many years hiking the Sierra Nevada early in the twentieth century prior to the advent of cars in the alpine country. His images are beautiful, and this book is suitable for everyone.

Alt, David D., and Donald W. Hyndman. *Roadside Geology of Northern and Central California.* Missoula, Mont.: Mountain Press, 2000. This book is an excellent guide to the geology of Northern California. Discussion of the Sierra Nevada and the Klamath Mountains occupies more than one-third of the book. Other sections discuss the Coast Ranges, the Cascade Range, and the Great Valley. The section on the Sierra Nevada includes an introduction followed by summaries of the geology along the major federal and state highways that pass through the region. The illustrations are very good, and the text is easy to read. The book is recommended for anyone interested in the geology of Northern California and in the Sierra Nevada, specifically. If the reader enjoys this book, several more are available for the western United States.

Ehlers, J., P. L. Gibbard, and P. D. Hughes, eds. *Quaternary Glaciations—Extent and Chronology: A Closer Look.* New York: Elsevier, 2011. One volume in a series of texts discussing the Quaternary period. This book covers Pleistocene and Quaternary glaciation events around the world, including in the Alps, Andes, Appalachians, Sierra Nevada, and Rocky Mountains among others. Contains an extensive index and a CD containing instruction and maps for use with ArcView, ArcGIS, and GoogleEarth.

Glazner, Allen F., and Greg Stock. *Geology Underfoot in Yosemite National Park.* Missoula, Mont.: Mountain Press Publishing Company, 2010. Written for the layperson, this book describes the geology and history within the park. Discusses structures and processes occurring in and around Yosemite.

Grotzinger, John, and Tom Jordan. *Understanding Earth.* 6th ed. New York: W. H. Freeman, 2009. This comprehensive physical geology text covers the formation and development of the earth. Contains extensive sections on the deformation of the continental crust and the development of the Sierra Nevada. Readable by high school students, as well as by general readers. Includes an index and a glossary of terms.

Harden, Deborah R. *California Geology.* 2d ed. Upper Saddle River, N.J.: Prentice Hall, 2003. Using plate tectonics as its main theme, this book focuses on the geological highlights of California. A good source for anyone interested in the geology of California.

Hatcher, Robert D., Jr. *Structural Geology: Principles, Concepts, and Problems.* 2d ed. Englewood Cliffs, N.J.: Prentice-Hall, 1995. This undergraduate textbook provides an overview of the structural development of the Sierra Nevada. Intended for the more advanced reader.

Hill, Mary. *Geology of the Sierra Nevada.* Rev. ed. Berkeley: University of California Press, 2006. Covers fundamental topics in geology with a focus on the Sierra Nevada region. This book contains a rock identification key, a table of geological features, and maps. Very clear writing accessible to anyone with some science background and an interest in geology.

Oakeshott, G. B. *California's Changing Landscapes: A Guide to the Geology of the State.* New York: McGraw-Hill, 1978. This book has excellent illustrations and a fine glossary. Geological concepts are introduced prior to the discussion of the geology of the state. Short sections on geochronology and the geologic time scale are particularly good. Recommended for the college-level student.

Seyfert, Carl K., ed. *The Encyclopedia of Structural Geology and Plate Tectonics.* New York: Van Nostrand Reinhold, 1987. As its title suggests, this reference is a good source for learning more about plate tectonics. Faulting and folding also are discussed. The text is not technical and is recommended for anyone interested in the subject.

Stanley, Stephen M. *Earth System History.* 2d ed. New York: W. H. Freeman, 2004. A comprehensive overview of the development of the earth. Chapter 19 deals with the development of the Sierra Nevada. Illustrations, glossary, bibliography, and CD-ROM. Readable by high school students.

Webster, Paul. *The Mighty Sierra: Portrait of a Mountain World.* Palo Alto, Calif.: American West, 1972. This book is a pictorial voyage through the entire Sierra Nevada. The morphology is discussed extensively; the geology is mentioned briefly. Quite a large amount of space is devoted to plant and animal life. Recommended for anyone who wants to get a feel for the high alpine country but does not want to be limited to geological information.

Wise, James M. *Mount Whitney to Yosemite: the Geology of the John Muir Trail.* Scotts Valley, Calif.: CreateSpace, 2008. Provides an account of the rock formations within the Sierra Nevada following the John Muir Trail. Although written as a guide to mountaineers and arm-chair explorers, the detailed geology referenced makes this an interesting starting point for students of geology as well. The extensive first chapter covers a variety of geology topics. Following chapters address specific sections of the trail.

See also: Alps; Andes; Appalachians; Basin and Range Province; Cascades; Continental Growth; Himalaya; Mountain Belts; Rocky Mountains; Transverse Ranges.

SNOWBALL EARTH

Snowball Earth is the theory that, at certain times in the earth's history, the planet was covered by a global glacier that reached from the poles to the equator. Evidence supporting the snowball Earth theory includes the discovery of widespread glacial sediments in equatorial areas. Global glaciation might have played a role in facilitating the complex evolution of life by refining the gene pool, as only adaptive organisms were able to survive periods of glacial coverage.

PRINCIPAL TERMS

- **ablation:** the removal of snow through either melting or sublimation
- **agglutination:** the clumping of particles to form a shell around a particular organism or object
- **albedo:** a portion of light from the sun reflected by a certain surface
- **evaporation:** the transformation of a liquid to a gas that occurs at the surface of a liquid
- **glacial erratic:** a rock sample that differs from an area's typical rock size and type and is thought to have been carried with moving glacial ice
- **glacial till:** the ground and deeply scarred rocks believed to originate from the movement of ancient glacial ice across underlying layers of sediment
- **glacier:** an area of solid ice that forms from snow in areas where accumulation of snow exceeds the speed of melting and sublimation
- **paleomagnetism:** the study of the magnetic field in Earth's sediment and of the record of Earth as recorded in the magnetic field of rocks
- **sublimation:** a chemical process involving transformation from a solid to a gaseous state

Development of the Snowball Earth Hypothesis

The snowball Earth hypothesis holds that at several times in Earth's history, the planet has been covered by glaciers that stretched from the poles to the equator. Evidence for global glacial periods has been building since the 1940's and became a major subject of debate among geologists following the discovery of new evidence in the 1990's. This evidence provided strong support for the theory, as well as additional evidence to suggest that the earth may have experienced more than one global glacial period.

Geologist and explorer Douglas Mawson discovered glacial sediments in southern Australia in 1949 that were later linked to similar glacial deposits discovered in Europe and China. In 1964, geologist W. Brian Harland discovered evidence for the existence of glaciers near the equator. Harland's findings were initially dismissed, as most geologists believed that glacial deposits near the equator came from rocks that were gradually moved to the equatorial zone by perhaps ocean or river currents.

In the late 1980's, paleomagnetist Joseph Kirschvink discovered sediment samples in Adelaide, Australia, that provide some of the best evidence for the snowball Earth theory. Kirschvink's rock samples clearly had been affected by glacial activity, yet bore distinct magnetic signatures that pinpointed their origin as being in the equatorial zone, convincing many geologists that glacial ice at one time covered part of the equator. Kirschvink coined the term "snowball Earth" in a 1992 paper to refer to these global glacial periods in Earth's history.

Geologist Paul F. Hoffman popularized the snowball Earth theory in a series of works published in the late 1990's. Hoffman discovered distinct carbon caps from geological samples in Namibia, Africa, which indicated a rapid decrease in photosynthetic activity during the period of global glaciation. Hoffman's data bolstered the case for the snowball Earth theory and sparked more research into glacial cycles.

Since the 1980's, scientists have discovered evidence for a minimum of three distinct periods of global glaciation. The oldest is the Makganyene glaciation, which appears to have occurred about 2.2 billion years ago. Evidence for this event, reported by Kirschvink, David Evans, and Nik Beukes in 1997, includes sedimentary samples bearing signs of glacial deposition and paleomagnetic traces.

Geologists sometimes refer to the period from 850 to 630 million years ago as the Cryogenian, during which the earth experienced two or more periods of global glaciation. The first of the Cryogenian ice ages, called the Sturtian glaciation, occurred sometime between 720 and 710 million years ago. Geologist Francis Macdonald and colleagues refined the timeline for the Sturtian glaciation with the discovery of sedimentary deposits indicating that the equator was covered with ice about 716.5 million years ago. Macdonald and colleagues believe that the glacial period lasted for 5 million years or more, during which time most of the earth's surface was covered with ice.

The latest glaciation event, called the Marinoan glaciation, occurred between 640 and 635 million years ago. Geologists discovered glacial sediment from the Marinoan on every continent, and evidence suggests that the glaciation may have lasted 6 to 12 million years before retreating to the poles.

Evidence for Snowball Earth Events

Glacial tills are samples of rock that have been ground or scarred by the force of glaciers moving across their surface. By dating the sediment within glacial tills, geologists can determine the approximate time of the glacier's formation. In addition, moving glaciers can carry rocks as they move across the surface of the earth. These rocks, called erratics, might differ from the surrounding sediment in which they are deposited. The appearance of erratics has provided further evidence for glacial activity in the past. Magnetic data from glacial deposits can indicate the origin of the sediment, as sediment carries distinct magnetic signatures based on a given locale. Glacial sediment with magnetic signatures indicating equatorial origin is considered strong evidence that glaciers reached into the temperate and equatorial zones.

Additional evidence comes from measurements of carbon-13, a by-product of photosynthesis, in samples of sediment above and below glacial deposits. Hoffman's research indicates that sediment below glacial deposits is rich in carbon-13. This evidence shows that photosynthesizing bacteria were abundant in the environment before the glacial deposits formed. The amount of carbon-13 drops sharply in the sediment approaching the glacial deposits, indicating that changing environmental conditions were hindering photosynthesis. Hoffman's carbon-13 data seem to show that photosynthesis slowed to low levels during the glacial period and then rebounded as the glaciers retreated.

Causes of Global Glaciation

Scientists do not know the exact chain of events that led to increasing glaciation during the three snowball Earth periods. The overall cause seems to have been a reduction in greenhouse gases, most notably methane and carbon dioxide, which resulted in a period of global cooling.

The portion of solar radiation that is reflected from an object's surface is called the albedo, and different surfaces have different albedo levels depending on their color, chemical makeup, and other factors. The albedo of ice and fresh snow is much higher than that of unfrozen water. The albedo effect is responsible for snow-temperature feedback, in which the presence of snow on the ground further reduces the temperature of the ground by reflecting more sunlight, which in turn leads to the accumulation of more snow.

In the 1960's, climatologists calculated the theoretical extent of the snow feedback phenomenon and found that if more than one-half of the earth was covered in ice, the albedo effect would create a runaway feedback loop, which would rapidly cover the rest of the planet with ice. Runaway feedback occurs when the accumulation of snow and ice occurs faster than ablation, which is the combined effect of sublimation of evaporation in removing snow from the surface. Geologists believe that each snowball Earth period occurred when global cooling initiated a runaway cooling cycle. Global cooling may have been related to a massive reduction in carbon dioxide and methane immediately preceding the glaciation.

In 2009, geochemist Kent Condie published the results of an extended study of ancient sediment samples suggesting that the Makganyene glaciation, which occurred more than 2 billion years ago, may have been precipitated by a global reduction in tectonic and volcanic activity. According to Condie, evidence indicates that volcanic activity came nearly to a halt and remained at low levels for more than 200 million years. Condie suggests that this would have led to a massive reduction in greenhouse gases and could have precipitated the formation of a global glaciation event.

Some geologists also note that glacial sediment from the Sturtian and Marinoan glacial periods appears close to a layer of igneous rock, which is the result of lava flows that hardened over millions of years. Some geoclimatologists suggest that a major period of widespread volcanic eruptions might have cooled the climate by ejecting smoke and particulate matter into the sky. Alternatively, some scientists suggest that these ancient volcanic deposits may have been caused by a period of intense volcanism that precipitated the end of a snowball Earth period. Intense volcanism would create additional greenhouse gases and lead to global warming.

The Possible End of Global Glaciation

During a snowball Earth period, tectonic activity continues under the earth's surface, producing volcanic material and large amounts of carbon dioxide. Under the ice-covered earth, chemical mechanisms that remove greenhouse gases from the atmosphere would be greatly reduced in scope. Taken together, this might lead to a buildup of carbon dioxide, methane, or other greenhouse gases, which would eventually be released into the atmosphere, countering the albedo effect.

Hoffman discovered evidence of thick limestone overlaying glacial sediments in many areas. Limestone is deposited only in areas where the temperature is warm and conditions are wet. This finding led Hoffman to assume that the average temperature of the earth grew rapidly warmer immediately after the global glaciers began melting. Furthermore, evidence suggests that the earth remained in a superheated state for some time, allowing the oceans to melt and return to liquid form. Hoffman believes that each glacial period may therefore have been linked to an ensuing period of global heating lasting millions of years.

Geologist Martin Kennedy proposed an alternative theory to explain the end of the Marinoan glaciation. His theory assumes the production of methane, a common greenhouse gas that is more effective in trapping solar radiation than carbon dioxide. Kennedy and colleagues believe that methane slowly accumulated under the ice layers that covered the earth. Tectonic movement eventually caused some of the fragile ice layers to split, releasing trapped methane into the atmosphere. The methane trapped a portion of the solar radiation that would otherwise escape into space, causing the temperature to rise and leading to additional splits in the ice.

Effects on Life

Most climatologists believe that even during the height of a global glaciation, the earth would have been able to support life. First, geologists believe that ice and snow covering the earth would not have been continuous, but likely would have been dotted with areas containing liquid water.

In the glaciers of modern Antarctica, for instance, biologists have found rich deposits of algae and microorganisms within gaps in the sea ice. Sedimentary geologist Daniel Le Heron reported in 2011 on the discovery of sedimentary evidence from the Cryogenian period, indicating that ocean currents continued to flow in some portions of the sea even at the heart of the glacial period. Some climatologists suggest that continued volcanic activity would have littered some areas with blankets of dark dust and debris. These areas would have heated faster because of their lower albedo levels, and thus may have become pockets of water allowing life to subsist during the height of the snowball effect.

Second, in 2011, researchers found fossil evidence from the end of the Sturtian glaciation, indicating the presence of the first "shelled" microbes known to science. Like some modern microbes, these ancient single-celled organisms used a process called agglutination to build shells around their body, insulating themselves from environmental hazards. Researchers conducting the study suggested that the harsh environment of the glaciated Earth led to a situation in which only organisms with advanced survival tools, like agglutinated shells, could survive. This paved the way for the evolution of stronger, hardier organisms in subsequent generations and may have been a major step in the evolution of life.

Third, the end of the Marinoan glaciation came just before another major turning point in the history of life: the Cambrian explosion, during which the animal fauna of the earth rapidly evolved into many forms that could colonize Earth's surface. Geologists Daniel Schrag and Hoffman suggested that life forms living during the last glacial period would have been separated into small, isolated populations divided by sheets of thick ice, in which few organisms could survive. These isolated pockets of microorganisms evolved in different ways, leading to an overall leap in diversity beyond what existed in the marine environment before the glacial period. As the glaciers melted and the earth rebounded into a period of high average temperatures, these microorganisms proliferated and spread across the globe, exploiting many environments left empty by species that became extinct during the snowball Earth period.

Micah L. Issitt

Further Reading

Barry, Roger, and Thian Yew Gan. *The Global Cryosphere: Past, Present, and Future.* New York: Cambridge University Press, 2011. Detailed academic text discussing all forms of ice and snow on Earth, past and present. Chapter 9 discusses the snowball Earth theory and includes an overview and comprehensive discussion of the ice systems in Earth's distant past. Also provides a detailed discussion of the future of glacial systems on Earth.

Benton, Michael J. *The History of Life: A Very Short Introduction.* New York: Oxford University Press, 2008. Basic history of life on Earth since the evolution of the first microorganisms. Provides a short discussion of the snowball Earth theory accompanying the evolution of advanced microbes in the Neoproterozoic.

Condie, Kent C. *Earth as an Evolving Planetary System.* San Diego, Calif.: Academic Press, 2011. Introduction to the geological history of life on Earth with a focus on the cycles and processes that change the planet's lithosphere and atmosphere in geological time. Discusses the evidence surrounding the snowball Earth theory and the relevance of global glaciation to the history of the atmosphere and the evolution of life.

Macdonald, Francis A., et al. "Calibrating the Cryogenian." *Science* 327 (2010): 1241-1243. Attempts to provide a more exact time frame for the Cryogenian period and provides an overview of evidence on the snowball Earth theory.

Macdougall, Douglas. *Frozen Earth.* Berkeley: University of California Press, 2006. Overview of research regarding glaciation in both the ancient and modern world. Provides information on ice age cycles, the evolution of life during glacial periods, and the potential future of glaciation on Earth.

Stanley, Steven M. *Earth System History.* New York: Macmillan, 2004. Introductory discussion of geological processes and Earth systems science. Chapter 12 examines the snowball Earth theory as part of an overall description of geological developments in the Precambrian period. Also discusses the movement of continents during the theorized periods of global glaciation.

Walker, Gabrielle. *Snowball Earth.* New York: Bloomsbury Press, 2004. Provides a detailed biographical account of Paul Hoffman's research on the snowball Earth theory. Also discusses expeditions to uncover evidence of glaciation.

See also: Biogeography; Coastal Processes; Continental Glaciers; Fluvial Processes; Geomicrobiology; Glacial Deposits; Ice Ages; Mass Wasting; Molecular Paleontology; Paleoecology; Permafrost; Radiometric Dating; Weathering.

SPREADING CENTERS

Volcanism occurs along spreading centers, the places where the earth's crust is rifting apart. This volcanism has formed the mid-ocean ridge system, the largest mountain chain on the earth. It also contributes to the dense oceanic crust, allowing ocean basins to form. Associated hydrothermal activity produces valuable deposits of copper, zinc, and other metals.

PRINCIPAL TERMS

- **basalt:** a fine-grained, dark, heavy rock of volcanic origin, primarily composed of calcic feldspars and pyroxene
- **convection currents:** the transfer of material caused by differences in density, usually brought about by heating
- **crust:** the thin, rigid outer layer of the earth, extending generally 35 kilometers under the continents and 10 kilometers under the oceans
- **fissure:** a fracture or crack in rock along which there is a distinct separation
- **hydrothermal:** related to hot water, particularly involving the production or dissolution of minerals
- **lava:** the fluid rock issued from a volcano or fissure and the solidified rock it forms when it cools
- **magma:** molten rock below the earth's surface that has not erupted and therefore retains its gaseous components
- **mantle:** the layer of the earth's interior between the crust and core, either semisolid (lower mantle) or plastic and nonbrittle (upper mantle)
- **seamount:** a submarine mountain 1,000 meters or higher, often a volcanic cone

Mid-ocean Ridges

Plate tectonics explains that the earth's crust is composed of relatively thin, rigid plates that float on top of the denser, less rigid mantle. Where the individual plates of crust meet, three types of motion are possible: Plates may collide with each other, forming a convergent boundary; they may slide past each other, forming a lateral, offset boundary; or they may move away from each other, forming a divergent boundary. Along divergent boundaries, cracks or rifts in the crust are formed between the plates. Lavas made of basalt seep up and fill these cracks from below, forming new crust. Where the new basalt erupts, great ridges form.

As spreading continues, the newly formed volcanic crust is slowly forced away from the ridge, as if carried on a great conveyor belt. This motion seems to be driven by large-scale convection currents within the mantle, fueled by the release of heat from the earth's interior. As basaltic crust moves away from the spreading center, it gradually subsides. The basaltic crust produced by volcanism at spreading centers along divergent boundaries is heavier than the granitic continental crust. Because it is denser, it sits lower in the earth's surface and forms the great topographic depressions of the ocean basins.

The ridges that form along the spreading centers are interconnected and make up the mid-ocean ridge system. This is a vast mountain chain that extends for more than 60,000 kilometers and spans the globe. The mid-ocean ridge system covers approximately 33 percent of the ocean floor, a surface equal to that of all the earth's continents. The segment of the mid-ocean ridge system that occurs in the Pacific basin is referred to as the East Pacific Rise. The Atlantic segment of the ridge is known as the Mid-Atlantic Ridge and bisects almost the entire Atlantic Ocean from the Arctic to the Antarctic Circle. Most of the ridges are totally submerged, though a few of the highest peaks (notably Iceland and the Azores) extend above sea level as islands.

The typical mid-ocean ridge stands about 2 to 3 kilometers above the surrounding sea floor. The ridges are broader than most mountain ranges. Individual ridges are typically 1,000 to 4,000 kilometers wide at their base. Unlike most continental mountains, the ridges typically have a central rift that is a few kilometers wide and about 1 kilometer deep. Volcanic activity associated with spreading occurs within this central rift.

Undersea Eruptions

The basaltic lava that erupts along the spreading centers is exceptionally hot (1,000 to 12,000 degrees Celsius) and, therefore, very fluid compared with other types of volcanic flows. As a consequence, these basaltic eruptions are less violent and less explosive than are other types of volcanic events, and the lava is spread over large areas, similar to eruptions recorded on land in Iceland and, to a lesser extent, those of Hawaii.

At depths of more than 2,000 meters below the sea, volcanic eruptions are considerably less explosive than they are on land. At these depths, the water pressure is so high that the gases released during volcanic eruptions do not explode but immediately go into solution with the seawater. Consequently, below 2,000 meters, there are no explosions, only effusive outpourings of lavas. This concept is supported by the low percentage of vesicles, which are the spherical voids formed around gas bubbles found in deep mid-ocean ridge basalts.

At depths of less than 2,000 meters, steam, ash, and sulfuric gases may be released. Unless the volcanism occurs at very shallow depths, however, the steam, ash, and gases will be absorbed by the seawater before reaching the surface. Many of these eruptions go completely unnoticed by humans, unless an ocean traveler happens to note a yellow tinge to the sea, the characteristic evidence of volcanism at depth. If the eruption occurs at shallow water depths (less than a few hundred meters), the eruption may be more explosive than if it had occurred on land, because of the secondary explosions caused by the violent vaporization of the water. The eruptions of steam propel the volcanic projectiles higher and faster through the air than they would normally travel. Spectacular ash plumes known as "cock's tail" explosions may arise.

Pillow Lava and Pahoehoe

The eruptions along the mid-ocean ridge system are episodic and frequently emanate from conical vents of basalt along the central rift. These vents have been measured up to 20 meters high. In some areas, there is evidence of multiple lava flows erupted in quick succession. Many of these flows form thick layers of basaltic "pillows," which are structures that form only under water. Pillow lavas are cemented masses of bulbous, slightly squashed ovoids of rock. They form as the molten lava contacts the cold water (which, at the depths of the mid-ocean ridges, is

Pillow lava rocks, such as these off Hawaii's shore, are cemented masses of bulbous, slightly squashed ovoids that form when molten lava contacts ocean water. As the magma oozes out of undersea vents, its surface is chilled by the cold water and quickly forms a bulge of dark, glassy rock. (National Oceanic and Atmospheric Administration)

barely above freezing, at 2 degrees Celsius). As the lava flows, its surface is chilled by the cold water and quickly cools into a bulge of dark, glassy rock. The still-molten interior of the mass may burst under pressure and stretch or may open the pillows, allowing small buds or toothpaste-like bulges to protrude.

Along the axis of the central rift, flows of pillow basalts have been recorded up to 200 meters high and 500 to 1,000 meters wide, particularly along the Mid-Atlantic Ridge. These flows form steep-sided, flat-topped ridges similar to the table mountains or möberg mountains formed by eruptions under glaciers.

Smooth or ropy flows of lava known as pahoehoe also occur within the central rift. These flows form when the slopes are too gentle for pillows to take shape. Under water, a flow's outer 10 to 30 centimeters quickly solidify, allowing the lava beneath to advance under this hardened shell. These flows form sheets or lava plains. Single flows emanating from spreading centers have been measured up to 20 kilometers from their source.

Sometimes, a single eruption may create such a rapid outpouring of basalt that a lava pond or lake may form. Some of these lakes have been measured up to hundreds of meters long and up to 5 meters deep. Lava lakes may display a number of interesting features. In places, the surface may have caved in, forming a collapse pit. Often, there are residual hollow basalt pillars, which upheld the roof of the lava lake as it drained and collapsed. Horizontal ribs, like the rings around a bathtub, occur repeatedly on the pillars and on the edges of the lake. These ribs apparently record changes in the lava level.

In addition to the flows, large, circular volcanic seamounts sometimes form near the ridge crests, particularly along the Mid-Atlantic Ridge. These may have been formed from lava with a slightly lower temperature (and therefore lower fluidity), which allowed the basalt to pile up. Some seamounts occur as pairs equally spaced from the central rift, indicating that they were probably once part of a single volcano that split and was separated.

Eruption Frequency and Volume

Along spreading centers, the amount of basalt generated and the time between eruptions may vary considerably. The East Pacific Rise is spreading at a relatively fast rate (averaging 6 centimeters per year). The frequency of eruptions at a given location is probably about once every few thousand years and the volume of material per eruption may reach up to 200 cubic kilometers. The spreading rate along the Mid-Atlantic Ridge is approximately three times slower. Along parts of this ridge, eruptions are expected to occur with an average frequency of once every 14,000 years. Estimates of the amount of material erupted vary considerably. At one submarine location, the rate of lava being generated has been speculated to be 8,700 cubic meters per kilometer per year. Another study near Iceland suggested that ten times that amount of fresh material was being released. To put this amount into perspective, that would be enough material to cover the entire landmass of Great Britain with a meter of new basalt every year.

Although volcanism has not been directly observed on the ocean floor, scientists have seen heated fluids (over 350 degrees Celsius) spewing from chimneylike vents along the rifts in the mid-ocean ridges. These chimneys rise from layers of basaltic lava and stand up to 30 meters tall. In some areas, they coincide with the source of the most recent lava flows. Sometimes the fluids emanating from the chimneys are black because of large quantities of sulfide precipitates. The chimneys are called "black smokers" or "white smokers" depending on the clarity of their fluids. They are apparently produced as seawater circulates down through cracks in the basalt to the molten magma beneath. As the water heats up, it dissolves elements such as zinc, iron, copper, and lead as well as silver and cadmium. The hot, mineral-rich water rises and is recycled back into the ocean. Upon contact with the cold water, sulfide minerals precipitate, forming chimneys around the rising jets.

This hydrothermal activity supports an incredible concentration of previously unknown animal life, including foot-long clams, fish, crabs, and giant tube worms up to a meter and a half long. These creatures thrive on bacteria that digest the sulfide provided by the vents. This is the only known ecosystem that is not directly or indirectly based on solar energy.

Mary D. Albanese

Further Reading

Ballard, Robert B. *Exploring Our Living Planet.* Rev. ed. Washington, D.C.: National Geographic Society, 1993. The chapter "Mountains of the Sea" gives an excellent firsthand account of some of the most illuminating journeys to the mid-ocean ridge system. The chapter "Discovery of Planet Earth" provides a highly readable, informative treatment of plate tectonics. The text is well illustrated with large color photographs and graphics. Contains a glossary and is completely indexed. Excellent reading for anyone interested in volcanism.

Chernicoff, Stanley, and Donna Whitney. *Geology: An Introduction to Physical Geology*. Upper Saddle River, N.J.: Prentice Hall, 2006. This is a good overview of the scientific understanding of the geology of the earth and surface processes. Includes sections on the ocean ridge system. Includes a link to a Web site that provides regular updates on geologic events around the globe.

Chester, Roy. *Furnace of Creation, Cradle of Destruction*. New York: AMACOM Books, 2008. This text discusses the turbulent processes of the planet. It covers earthquakes, volcanoes, and tsunamis in reference to plate tectonics, natural disasters, predicting and mitigating. Multiple chapters explore seafloor spreading. The author also discusses hydrothermal activity. This text takes on an immense range of content, but still explains concepts clearly and with detail.

Decker, Robert, and Barbara Decker. *Volcanoes*. 4th ed. New York: W. H. Freeman, 2005. This readable and well-illustrated text is geared toward the general public. The authors explain geologic phenomena in a clear and simple manner. The first chapter discusses the development of the ocean ridge system. The appendices list the world's 101 "most notorious" volcanoes, as well as Internet sources for volcano information.

Dott, Robert H., Jr., and Donald R. Prothero. *Evolution of the Earth*. 8th ed. New York: McGraw-Hill, 2009. This basic textbook on historical geology is aimed at students of geology. However, it is very readable by anyone with a background in science. Presents an up-to-date account of the earth's history from the viewpoint of plate tectonics. Includes a glossary.

Fletcher, Charles. *Physical Geology: The Science of Earth*. Hoboken, N.J.: John Wiley and Sons, 2011. This text deals with all aspects of physical geology and has good coverage of spreading centers. It is intended for undergraduate students but accessible to the interested reader.

Fowler, C. M. R. *The Solid Earth: An Introduction to Global Geophysics*. 2d ed. New York: Cambridge University Press, 2004. Includes bibliographical references and index. An introductory textbook to the topics of geophysics, focusing on tectonics and earth's forces and motions. Contains color plates.

Frisch, Wolfgang, Meschede, Martin, and Blakey, Ronald C. *Plate Tectonics: Continental Drift and Mountain Building*. New York: Springer, 2010. This textbook provides a basic overview of continental drift and plate tectonics focusing on the resulting changes in the earth's surface.

Garrison, Tom S. *Oceanography: An Invitation to Marine Science*. Belmont, Calif.: Brooks/Cole, Cengage Learning, 2010. Chapter 3 provides excellent coverage of plate tectonics, including ocean formation, seafloor spreading, and island arc formation. Chapter 4 covers topography and bathymetry of deep-ocean basins. Abundant diagrams aid readers from the layperson to the advanced undergraduate.

Haymon, Rachel M., et al. "Manifestations of Hydrothermal Discharge from Young Abyssal Hills on the Fast-Spreading East Pacific Rise Flank." *Geology* 33 (2005): 153-156. This article discusses oceanic hydrothermal heat released via black smokers. The authors comment on the research possibilities available due to fluid expulsion from the abyssal seafloor. This article is best suited for advanced undergraduates and graduate students. It provides a view of where research of the sea floor is heading.

Rona, Peter A., K. Boström, L. Laubier, and K. L. Smith, Jr., eds. *Hydrothermal Processes at Seafloor Spreading Centers*. New York: Plenum Press, 1983. For those readers interested in a technical treatment of the subject delineated in the book's title, this volume contains a compilation of thirty papers ranging from hydrothermal mineralization to the biology of hydrothermal vents. The book is well indexed, and each paper contains its own bibliography.

Rosbacher, L. A. *Recent Revolutions in Geology*. New York: Franklin Watts, 1986. Chapters 3 and 4, "Exploring the Ocean Floor" and "Riches from the Sea," respectively, provide a good treatment of the subject for the younger reader interested in the topic of undersea exploration.

Singh, Satish C., et al. "Discovery of a Magma Chamber and Faults Beneath a Mid-Atlantic Ridge Hydrothermal Field." *Nature* 442 (2006): 1029-1032. This article provides an example of current seafloor exploration and research. It discusses the use of seismic analysis for developing models of mid-ocean ridges. Also provides some information on hydrothermal activity and submarine volcano structure. Understood by the layperson with careful reading.

See also: Calderas; Eruptions; Flood Basalts; Geysers and Hot Springs; Hawaiian Islands; Hot Spots, Island Chains, and Intraplate Volcanism; Iceland; Island Arcs; Lava Flow; Ocean Basins; Ocean-Floor Exploration; Oceanic Crust; Ocean Ridge System; Ring of Fire; Yellowstone National Park.

STRATIGRAPHIC CORRELATION

Stratigraphic correlation is the process of determining the equivalence of age or stratigraphic position of layered rocks in different areas. It is critical to understanding earth history because stratigraphic correlation is one of the principal methods by which the succession and synchrony of geological events are established.

PRINCIPAL TERMS

- **biostratigraphy:** the identification and organization of strata based on their fossil content and the use of fossils in stratigraphic correlation
- **correlation:** the determination of the equivalence of age or stratigraphic position of two strata in separate areas; more broadly, the determination of the geological contemporaneity of events in the geological histories of two areas
- **index fossil:** a fossil that can be used to identify and determine the age of the stratum in which it is found
- **lithology:** loosely used by many geologists to refer to the composition and texture of a rock
- **sedimentary rocks:** rocks formed by the accumulation of particles of other rocks, organic skeletons, chemical precipitates, or some combination of these
- **stratigraphy:** the study of layered rocks, especially of their sequence and correlation
- **stratum (plural strata):** a single bed or layer of sedimentary rock

METHODS OF STRATIGRAPHIC CORRELATION

Stratigraphic correlation is the process of determining the equivalence of age or stratigraphic position (position in a vertical sequence of rock layers) of strata in different areas. Strata thus determined to be of the same age or in the same stratigraphic position are called "correlative" or "correlated."

The most obvious method of stratigraphic correlation is to demonstrate that strata are continuous over a given area. This is achieved by tracing the layers (often by walking along them) throughout the area to demonstrate their physical continuity. In areas where strata are continuously exposed (the Grand Canyon in Arizona is a good example), this direct method is simple and best. Over much of the earth's surface, however, the strata that scientists wish to correlate are covered by younger rocks or vegetation, or erosion has removed them over some portion of their geographic extent, making this method impractical.

A commonly used method of stratigraphic correlation that is not subject to this drawback is correlation by lithologic similarity, or similarity in rock composition and texture. Thus, strata of the same lithology (rock type) in different areas may be considered correlative. In effect, this method attempts to identify the same layer in different regions without demonstrating the continuity of (that is, without tracing) the layer. The major problem with this method, however, is that there are many layers in the earth's crust that are of similar lithology. In eastern Colorado, for example, it is not easy to tell a 90-million-year-old sandstone layer from a 70-million-year-old sandstone from lithology alone. Thus, mistakes in determining age equivalence are easy to make when correlating only by lithologic similarity.

A folded mountain (Grassy Mountain, Alberta), with Jurassic and Lower Cretaceous clastic rocks exposed in the pit below. Layers of coal, sandstone, schist, and carbon have been deformed from their original horizontal positions due to deformations in the earth's crust. (Geological Survey of Canada)

A third method of stratigraphic correlation relies on the idea that two layers in different areas that occupy the same stratigraphic position are correlative. This method can be referred to as correlating by position in the stratigraphic sequence. For example, if a limestone layer overlies sandstone A in one region and a shale layer overlies sandstone A elsewhere, scientists may conclude that the limestone and shale are correlative. This method, however, is not without its pitfalls. In the example, it might be that the limestone layer at the first location was deposited right after sandstone A was deposited, whereas at the second location, the shale layer was deposited many millions of years after sandstone A. To correlate the limestone and shale in this case is to recognize their equivalence of stratigraphic position but not to demonstrate their age equivalence.

Age equivalence in stratigraphic correlation is almost always based on fossils. Stratigraphic correlation by fossils depends on biostratigraphy, the recognition of strata by their fossil content. If a fossil represents an organism characteristic of a particular interval of geologic time, then it is referred to as an index fossil. Thus, if an index fossil is collected from a stratum, then that stratum is the same age as strata elsewhere that contain the index fossil.

Besides fossils, there are some other, less often used, methods of determining the age equivalence of strata. The most common of these methods is to obtain a numerical estimate (usually in millions of years) of the age of the strata. This can rarely be undertaken unless the strata contain volcanic material, which contains chemical elements needed to obtain a numerical age. These elements are not normally present in sedimentary rocks; volcanic ash beds are not nearly as common in sedimentary rocks as are fossils. That is why fossils have been and continue to be the primary means for determining the age equivalence of strata. To undertake stratigraphic correlation, it thus is necessary to define the lateral extent of strata in a given region, characterize their lithology, determine their stratigraphic sequence, and collect and establish the stratigraphic ranges of fossils. Once these data have been collected in one or more regions, a geologist can attempt to correlate the strata within an area or between disparate areas.

Development of Stratigraphic Correlation

The development of stratigraphic correlation began in 1669 when Nicolaus Steno, a Danish naturalist, recognized that the sequence of strata is directly related to their relative ages. Steno's principle of superposition thus identified the oldest layers as those at the bottom and the youngest strata as those at the top of a sequence of strata. Steno's principle and the recognition of fossils as the remains of past life allowed William Smith in England and Georges Cuvier and Alexandre Brongniart in France to undertake the first stratigraphic correlations based on fossils during the early nineteenth century. Smith's correlations resulted in the first geologic map of England (1815). Cuvier and Brongniart were able to correlate stratigraphically the rock layers in the vicinity of Paris and thereby reconstruct the geological and biological history of this area. Previously, during the late eighteenth century, the German mining engineer Abraham Werner and his students had laid the basis for stratigraphic correlation by lateral continuity and lithologic similarity by arguing for the continuity of strata of a particular lithology over a broad area. Thus, by the early nineteenth century, stratigraphic correlation by lateral continuity, lithologic similarity, position in the stratigraphic sequence, and fossils was already being practiced by European geologists.

Applying Stratigraphic Correlation

Stratigraphic correlation plays a central role in understanding geological history. By allowing geologists to determine the synchrony or diachrony of strata in different areas, the geological and biological events recorded in these strata can be ordered in time. This ordering is the basis of the chronology of geological and biological events during the last 3.9 billion years of earth history. Stratigraphic correlation is thus one of the methods by which the relative geological time scale of eons, eras, periods, epochs, and ages was constructed.

A good example of stratigraphic correlation helping to decipher geological history and leading to the discovery of a giant oil field comes from the Guadalupe Mountains of western Texas and eastern New Mexico. In the years before World War II, geologists and paleontologists there used stratigraphic correlation (especially biostratigraphy) to unravel the complex geological history of the strata that form these mountains. This history revealed that huge barrier reefs had developed as the sea encroached from the south about 260 million years ago. When it was later learned that hydrocarbons often accumulate around reefs, interest in the petroleum potential of these rocks was aroused. Because drilling the rocks in the rugged Guadalupe Mountains was impractical, an effort was made to trace the reef strata into the nearby lowlands. There, drill cores brought up rocks and fossils from deep beneath the surface. Stratigraphic correlation, by lithologic similarity, position in stratigraphic sequence, and fossils, was used to identify the strata adjacent to the reef in what is now called the Delaware basin, south and east of the Guadalupe Mountains. Successful drilling for petroleum soon turned the Delaware basin into one of the world's giant oil fields, thanks to the identification of petroleum source rocks through stratigraphic correlation.

By ordering events in the history of the earth, stratigraphic correlation has been critical to the development of a global geologic time scale. Such a time scale provides geologists with a shared temporal framework within which to view their observations. A good example of the use of stratigraphic correlation in the development of the global geological time scale comes from the concept of the Cambrian period. The British geologist Adam Sedgwick coined the term "Cambrian" (from "Cambria," the ancient Roman name for Wales) in 1835 to refer to rocks in northern England that he believed contained the oldest fossils. Geologists now recognize the Cambrian period (544 to 505 million years ago) worldwide because stratigraphic correlation proved that an interval of geologic time that corresponds to Sedgwick's original Cambrian rocks can be identified across the globe. One of the most distinctive aspects of Cambrian strata is the abundance and types of trilobites they contain. Correlation of the Cambrian strata of England with strata elsewhere has been mostly undertaken on the basis of these trilobite fossils. Indeed, Sedgwick himself first conducted such stratigraphic correlation in continental Western Europe.

By the 1880's, when American geologist Charles Doolittle Walcott identified Cambrian trilobites in North America, the Cambrian was well on its way to becoming a geologic time period recognized worldwide. Stratigraphic correlation, especially using fossils (biostratigraphy), thus played a significant role in the recognition of the Cambrian. All the geologic time periods now recognized worldwide are similarly rooted in stratigraphic correlation.

Spencer G. Lucas

Further Reading

Ager, Derek V. *The Nature of the Stratigraphical Record.* 3rd ed. Hoboken, N.J.: John Wiley & Sons, 1993. A witty and unabashed look at stratigraphy. Much of the discussion focuses on problems of stratigraphic correlation. The bibliography is extensive, there is an index, and a few, well-chosen illustrations illuminate the text.

Amos, Salvador, ed. *International Stratigraphic Guide.* 2d ed. Hoboken, N.J.: John Wiley & Sons, 2008. The international "rule book" for stratigraphy. It sets procedures and standards to be met when naming stratigraphic units. It also defines many terms used in stratigraphy and has an extensive bibliography. Chapter 6 is devoted to biostratigraphy.

Berry, William B. N. *Growth of a Prehistoric Time Scale.* Rev. ed. Palo Alto, Calif.: Blackwell Scientific Publications, 1986. Provides an excellent review of the principles of stratigraphic correlation but is mostly devoted to the history of how the global geological time scale was formulated. As such, it well relates many of the stratigraphic correlations upon which the time scale is based. Well illustrated, with a good bibliography and an index.

Brenner, Robert L., and Timothy R. McHargue. *Integrative Stratigraphy Concepts and Applications.* Englewood Cliffs, N.J.: Prentice-Hall, 1988. Chapters 11-13 provide a comprehensive look at all facets of stratigraphic correlation. Well illustrated, with extensive reference lists and an index.

Brookfield, Michael E. *Principles of Stratigraphy.* Hoboken, N.J.: Wiley-Blackwell, 2004. Written for undergraduate students, this text provides an overview of the principles and applications of stratigraphy. Includes stratigraphic techniques and case studies. It is organized into three sections, beginning with foundational material, followed by data collection and research topics, and completed with interpretations and analysis.

Dott, Robert H., Jr., and Donald R. Prothero. *Evolution of the Earth.* 8th ed. New York: McGraw-Hill, 2009. This basic textbook on historical geology is aimed at students of geology. However, it is very readable by anyone with a background in science. Presents an up-to-date account of the earth's history from the viewpoint of plate tectonics. Includes a glossary.

Grotzinger, John, and Tom Jordan. *Understanding Earth.* 6th ed. New York: W. H. Freeman, 2009. This comprehensive physical geology text covers the formation and development of the earth, as well as stratigraphic correlation. Readable by high school students, as well as by general readers. Includes an index and a glossary of terms.

McGowran, Brian. *Biostratigraphy: Microfossils and Geological Time.* New York: Cambridge University Press, 2008. This text expands the application of biostratigraphy from the original use in paleontology to the fields of petroleum exploration and deep-ocean drilling. Addresses the relatively new methodology of studying microfossils and absolute aging.

Miall, Andrew D. *The Geology of Stratigraphic Sequences.* 2d ed. New York: Springer, 2010. The text covers stratigraphic sequences and the geological processes that form them. The author refers to stratigraphic cycles of different lengths, discussing the characteristics and succession of the stratigraphy within these cycles. It provides a technical and detailed overview of current and future practices in the study of stratigraphy and solidification of the geological time scale and dating.

Ogg, James G., Gabi Ogg, and Felix M. Gradstein. *The Concise Geologic Time Scale.* New York: Cambridge University Press, 2008. This book is a complete overview of the geological time scale, including stratigraphy topics such as chronostratigraphy and magnetic stratigraphy. The book is organized by geological time periods. Also contains a reference appendix, the geological time scale table, and indexing.

Prothero, Donald R. *Bringing Fossils to Life.* 2d ed. Boston: McGraw-Hill, 2004. This well-illustrated and entertaining text covers a broad range of paleontological topics, including stratigraphy. Glossary, bibliography, and index.

See also: Archaebacteria; Biostratigraphy; Cenozoic Era; Clastic Rocks; Cretaceous-Paleogene Boundary; Ediacaran Fossils; Eukaryotes; Geologic Time Scale; Ice Ages; Land Bridges; Mesozoic Era; Paleoclimatology; Paleogeography and Paleocurrents; Paleoseismology; Paleosols; Paleozoic Era; Prokaryotes; Proterozoic Eon; Transgression and Regression; Unconformities; Uniformitarianism.

STRATOVOLCANOES

A volcano that erupts both cinder and lava is a composite volcano, or stratovolcano. These volcanoes are found at subducting tectonic plate margins and are the most abundant of the large volcanoes. The tallest and most famous stratovolcanoes of the world, such as Vesuvius, Fuji, and Mount St. Helens, have produced the most devastating and violent eruptions.

PRINCIPAL TERMS

- **andesite:** a gray volcanic rock with a silica content of about 60 percent
- **ash:** fine volcanic ejecta less than 2 millimeters in diameter
- **caldera:** a large circular basin with steep walls that resulted from the collapse of a summit or of an earlier volcanic cone
- **cinder cone:** a small volcano composed of cinder or lumps of lava containing many gas bubbles, or vesicles; often the early stage of a stratovolcano
- **pumice:** a solidified volcanic froth of a glassy texture light enough to float on water
- **pyroclastics:** fragmented ejecta released from a volcanic vent
- **rhyolite:** a light-colored volcanic rock composed of a viscous lava containing about 70 percent silica
- **tephra:** all pyroclastic materials blown out of a volcanic vent, from dust to large chunks
- **viscosity:** the resistance of lava to flow; it depends upon the chemical composition, temperature, and crystalline nature of the magma

Characteristics of Stratovolcanoes

The most abundant type of large volcano on the earth is the stratovolcano, also referred to as a composite volcano. These classic "poster volcanoes" constitute 80 percent of all active volcanoes, including Mount Fuji, Mount Vesuvius, Mount St. Helens, and Krakatau. Stratovolcanoes vary in elevation from a few hundred to 4,000 meters or more above base levels, with diameters approaching 40 kilometers. Their slopes vary from 10 to 35 degrees, with an increase toward the summit because of larger amounts of volcanic material deposition there. In humid regions, much of this volcanic material, in the form of tephra, is washed downslope in mudflows, but when transported aerially, very little tephra actually reaches the base.

The shapes of most of these volcanoes are symmetrical, with eruptions from a single pipe vent at the center, although some build up around long fissure vents. Mount Hekla in Iceland is an example of the latter, with a 5-by-10-kilometer oval shape. In the former case, a central crater, usually a funnel-shaped hole, is found on the summit. The central vent tends to remain in the same position despite the explosive events. Often, the crater may be enlarged by wall collapse because of magma withdrawal. A flat central floor in the crater may be more prevalent in humid areas, where heavy rains wash fine sediments downward.

Stratovolcanoes may reach great heights, but these large, symmetrical cones require a riblike structure of lava for support. Pure cinder cones, in contrast, rarely exceed 500 meters in elevation because their symmetry tends to be destroyed by slumping. One eruptive cycle may be an effusive flow of lava, followed in some later period by an explosive event that produces cinder and ash layers. The relative proportions of tephra and lava in stratovolcanoes' structures vary considerably, from pure ash or cinder to pure lava flows. Rarely do volcanoes exhibit classic "layer-cake" geology—that is, alternating layers of lava and ash. Some mountains (for example, Taal in the Philippines) consist of 70 to 80 percent pyroclastics, but others are dominated by lava eruptions.

Eruption Patterns

Eruption patterns vary from one stratovolcano to another; three basic types are described here. The Vulcanian type of eruption, characteristic of many active volcanoes, produces a solid crust that lasts until the next eruption. Gas pressure builds up in the magma column, eventually blowing out the solidified obstruction. A great explosion may follow, accompanied by a large, dark, cauliflower-shaped eruption cloud. With the reduction of pressure, gas-charged magma is replaced by pumice and ash. After the vent has cleared, lava flows may issue forth from the crater. Vesuvius, the stratovolcano east of Naples, falls within this group of eruptions.

The Strombolian eruption is distinguished by glowing fragments of lava accompanied by a white eruption cloud. In contrast to the Vulcanian eruption, which contains much ash, the Strombolian cloud contains very little ash. The crust that forms over the lava column is very thin, allowing for frequent, mild eruptions. Stromboli, a stratovolcano located on an island west of Italy, is the classic example, but some volcanoes may exhibit Strombolian activity during some portion of their history.

The Peléan eruption gives rise to the expulsion of nuées ardentes. Pumice and ash are characteristic; no liquid lava develops. The magma is under such intense pressure that it is shattered into a fine dust and mixed with superheated steam. This mixture rolls over the top of the crater as an avalanche of red-hot dust. In the final stages of eruption, the gas content of the magma is greatly reduced, no longer breaking the surface but instead pushing upward to form a dome with spinelike projections, as in the case of Mount Pelée on the island of Martinique.

Stratovolcanoes tend to be positioned at the margins of descending or subducting masses of ocean floor or tectonic plates beneath the continents. The western United States has thirty-five volcanoes that have erupted and may erupt again in the future. One region of volcanic activity is found in the Cascade Range, extending from Washington and Oregon to Northern California, with a continuation into Nevada and through Idaho to Yellowstone National Park. Another volcanic zone winds through southeastern Utah into Arizona and Mexico.

In terms of frequency of eruption, volcanoes are divided into two groups. The first comprises those that have experienced eruptions on average every two hundred years and have last erupted within that period of time. Examples in this group are Mount St. Helens, Lassen Peak, and Mounts Shasta, Rainier, Baker, and Hood, all in the Cascade Range. The other group includes those whose last eruption was more than one thousand years ago; their frequency of eruption is greater than one thousand years. Examples are Crater Lake Volcano and Mounts Adams and Jefferson, also located in the northern Cascade Range.

The stratovolcano gains its heat from subducting-plate friction. The magma tends to have a lower temperature and a higher viscosity than does the basaltic magma of shield volcanoes (found at rift zones and located over hot spots). The eruptions are explosive because of the high silica and gas content and high viscosity. Rhyolite and andesite are often present in the flows, along with volcanic products called pyroclastics. Not all stratovolcanoes, however, erupt lava containing such high amounts of silica; Mount Etna in Sicily and Mount Fuji in Japan tend to have a silica composition of 50 to 60 percent in a basaltic lava.

Stratovolcanic Eruptions

Several stratovolcanic eruptions have had significant consequences for people living nearby. Located in the southeastern state of Chiapas, Mexico, El Chichón experienced a series of explosive eruptions on March 28, 1982, killing 187 people and leaving another 60,000 homeless. A large dust cloud, which was monitored by satellite, rose to an altitude of 25 kilometers and lasted for one month after the eruption. Scientists recorded 50 centimeters of ashfall 16 kilometers away and 20 centimeters of ashfall 80 kilometers away. In Palenque, 120 kilometers east of the volcano, ashfall was measured at 40 centimeters.

On the eastern shore of Sicily, Mount Etna, the largest and highest of the European volcanoes, rises 3,320 meters above sea level. Small subsidiary cones around its flanks, up to 1,000 meters high,

Stromboli Volcano, Aeolian Islands, Italy. (Connie Bransilver / Photo Researchers, Inc.)

erupt rather frequently, but only rarely does sufficient energy accumulate to cause an eruption from the summit crater. Early eruptions, several centuries B.C.E., were recorded by the Greeks. Thousands of lives have been lost and several towns destroyed during Etna's history. In 1669, 20,000 died in an eruption. The town of Mascati was destroyed in 1853, and in 1928 the village of Nunziata was nearly leveled. Explosive eruptions near the summit in 1979 claimed nine lives. Mount Etna is classified as a transitional volcano, transforming from shield to stratovolcano, and has had more than 150 eruptions since the first recorded one in 1500 B.C.E.

On June 6, 1912, one of the greatest eruptions of the twentieth century occurred on the Alaska Peninsula in the Valley of Ten Thousand Smokes. Mount Katmai is a complex stratovolcano, with both a caldera and a lake on its summit. The explosion was heard in Juneau, 1,200 kilometers away. The valley was flooded with 25 cubic kilometers of pumice, ash, and gas and was filled to a depth of nearly 200 meters. In Kodiak, some 160 kilometers away, pumice and ash nearly blocked out the sunlight, reducing visibility to about 2 meters. The upper 1,900 meters of the mountain collapsed, forming a giant caldera more than 2 kilometers wide, which filled with water and became a crater lake. A new vent, Mount Trident, on the west flank of Katmai, became active in 1949 and has since experienced several thick lava flows, the last erupting in 1974.

Stromboli, one of the most active volcanoes, lies on a Liparian island 60 kilometers north of Sicily. It has been in a continuous state of eruption its entire history, although the intensity varies considerably. It was known for centuries to sailors as the Lighthouse of the Mediterranean because eruptive flashes from its summit were visible far out to sea. Even during World War II, it was used by Allied bombers as a navigational aid. The more explosive eruptions occur at about fifteen-minute intervals, with moderate activity in between. The principal crater, located 600 meters up the 900-meter mountain, is always full of semifluid lava and thus does not greatly resist magma pressure from beneath.

Mount Tambora, on the island of Sumbawa, Indonesia, began a series of eruptions on April 5, 1815, that did not end until July of that year. The early explosions sounded like cannon fire and could be heard 720 kilometers away. The greatest eruptive activity occurred on April 11 and 12, when explosions were heard in Sumatra, 1,600 kilometers to the west. There were only a few survivors of a population of 12,000 on the island, and 45,000 other lives were lost on surrounding islands. Volcanic ash was so heavy that there was darkness at noon in Java, 480 kilometers distant. The dust sent into the atmosphere obscured sunlight more than had any other volcanic dust within the previous four hundred years. The year 1816 was known as the "year without summer."

Vesuvius is the only active volcano on mainland Europe. The famous eruption of August 24, 79 C.E., was described by Pliny the Younger from a distance of 30 kilometers. His account describes a strange cloud that rose up and away from the mountain as well as roars and explosions coming from deep within the mountain. The residents of Pompeii were overcome with searing ash and toxic gases. Some 16,000 lives were lost, mainly by suffocation; the residents of nearby Herculaneum, for example, were overcome by boiling mudflows. Another violent eruption took place in 1631, causing the loss of 18,000 lives. During World War II, on March 18, 1944, an eruption forced the evacuation of the city of Naples, and lava flows caused extensive damage at an air base close to the mountain. Mount Vesuvius is a complex stratovolcano, with lava flows generally following explosive eruptions.

Volcanology

A major goal in volcanology is the prediction of the time, place, and nature of an eruption to minimize loss of lives and property. A variety of methods are employed, for no single factor is sufficient to warn of impending eruption.

Geophysicists monitor the earthquakes and tremors that accompany volcanic eruptions with seismographs that measure the intensity of earthquake activity. Almost every eruption or increase in volcanic activity is preceded by earthquakes—often very small earthquakes, or microearthquakes, that occur in swarms of hundreds or thousands. A volcanic tremor is almost always present during an eruption and often begins before the surface outbreak. This motion has a natural frequency from 0.5 to 1.0 hertz and produces a very low hum. The source of this noise is not clearly understood, but it may be related to the formation of gas bubbles in the lava.

Small changes in the slopes, shown by distances between markers on the flanks and summits of active volcanoes, are another precursor of volcanic activity. Techniques for detecting such changes include conventional leveling, reflected light beams, and instruments called tiltmeters capable of measuring changes in slope of less than one part in one million. One type of tiltmeter uses brass pots filled with water and fastened to support posts connected by hoses. The posts are placed a few meters apart. Micrometers that measure fractional water-level change are placed inside the pots.

With the rise of magma toward the surface, the earth's magnetic field is disturbed by molten lava, which loses its magnetic properties at high temperatures. Such changes may be detected by a magnetometer carried aerially over the volcano. Moving magma within the volcanic chambers may cause changes in the electrical currents within the ground. Sensitive resistivity meters placed under the surface are used to measure electrical conductivity. An increase in temperature of hot springs, fumaroles, and groundwater may signal an eruption; several months before the 1965 eruption of Taal in the Philippines, the temperature of the water inside the crater increased by 12 degrees Celsius. The chemical composition of waters and volcanic emissions may also change during eruptions. An increase in sulfur dioxide or hydrochloric acid in groundwater or surface emissions can precede an eruption.

Direct visual observations are helpful in monitoring volcanoes. Ice and snow on a volcano may melt from the intense heat. A volcano may bulge from movement of magma along one of its flanks, as was observed prior to the 1980 eruption of Mount St. Helens. Active volcanoes are observed from nearby observatories or sentry outposts. Valuable contributions are made by volcanologists in these facilities, but close observation can be very dangerous.

Significance

The majority of the world's six hundred active volcanoes lie within the Pacific Ocean Basin, with about one-half in the western Pacific region. They form an almost continuous pattern known as the Ring of Fire along the edges of the Pacific. Stratovolcanoes that form along the ocean margins are termed marine stratovolcanoes. Their initial structure resembles basaltic seamounts, with pillow basalts building up on the sea floor, followed by pyroclastics and less effusive eruptions.

Deep marine volcanoes are less explosive than volcanoes on land, because the pressure of the seawater retards the expansion of steam within the magma. Subduction-zone and island-arc volcanoes, in contrast, contain a high concentration of gases in the upper parts of their magma chambers, making them more explosive than are volcanoes in other locations. The lava tends to have a greater amount of silica, so it is more viscous and resistant to flow. The lava that solidifies within the vent, or throat, forms a hardened plug that is blasted into pieces, or pyroclastics, by the pressure of the gas trapped below. Lava flows that are rough and blocky are termed aa, while those with a smooth, ropy texture are called pahoehoe.

Island-arc stratovolcanoes have had an explosive and dangerous history. Volcanoes such as Krakatau and Tambora in Indonesia as well as Rabaul in Papua New Guinea are examples. It is believed that these volcanoes are economically important for the deposition of metallic sulfides. The magma chambers beneath may be regions where copper, molybdenum, and gold are concentrated during the final consolidation.

It is important to study volcanoes to minimize volcano damage, including loss of lives. Various methods have been devised to reduce volcano damage; these include the digging of channels and construction of levees to divert lava flows. Lava streams have been doused with water in attempts to slow down and solidify the lava. In Java, dams have been constructed to direct volcanic mudflows away from cities and agricultural lands. In Hawaii, the U.S. Air Force has bombed lava flows emerging from Mauna Loa, with limited success.

Some volcanoes eject considerable amounts of ash and dust into the atmosphere. Certain massive volcanic eruptions had a significant effect on global temperatures for months afterward. The 1963 eruption of Mount Agung, a stratovolcano in Bali, Indonesia, ejected clouds of gas and dust 10 kilometers into the stratosphere. The average worldwide temperature dropped 0.5 degree Celsius for three years after the eruption. Whether volcanic eruptions do effect significant changes in the weather and climate cannot be satisfactorily answered until scientists can obtain accurate measurements of the volcanic dust and debris entering the stratosphere and the heating and cooling of the earth's atmosphere that result from such dust.

Michael Broyles

Further Reading

Bullard, F. M. *Volcanoes of Earth.* Austin: University of Texas Press, 1984. A definitive study of volcanoes that includes sections on mythology and early speculation, how volcanoes are classified, the types of volcanic eruptions, and the distribution, current activity, theory, and environmental effects of volcanoes. Many diagrams, photographs, and tables make this a valuable resource for the geology student.

Chester, D. K., et al. *Mount Etna: The Anatomy of a Volcano.* Stanford, Calif.: Stanford University Press, 1985. A comprehensive study of the major volcano on the island of Sicily. An interesting historical chronology of both flank and summit eruptions is included. Another table presents the effects of eruptions and earthquakes on settlements in the region. Includes many black-and-white photographs. Some chapters are technical and are best suited to the geology student, while other sections, especially the historical activity, would be of interest to the general reader.

Decker, Robert, and Barbara Decker. *Volcanoes.* 4th ed. New York: W. H. Freeman, 2005. This readable and well-illustrated text is geared toward the general public. The authors explain geologic phenomena in a clear and simple manner. Includes a section on stratovolcanoes. The appendices list the world's 101 "most notorious" volcanoes, as well as Internet sources for volcano information.

Erickson, Jon. *Volcanoes and Earthquakes.* Blue Ridge Summit, Pa.: TAB Books, 1987. An excellent resource for the general reader. A chronology is given for the major eruptions and their toll in human lives; more detailed descriptions of fourteen different volcanic eruptions follow this table. Chapters have been added on the formation of the solar system and the evidence for plate tectonics and continental drift. Volcanoes that have been detected on the other planets and moons are also described.

Francis, Peter, and Clive Oppenheimer. *Volcanoes.* 2d ed. New York: Oxford University Press, 2004. This text covers volcanic activity ranging from lava flows to pyroclastic currents. Provides many examples from Hawaii, Italy, Ethiopia, Japan, Mount St. Helens, and Krakatau. Written for the layperson, but some knowledge of geology is helpful.

Lockwood, J. P., and Richard W. Hazlett. *Volcanoes: Global Perspectives.* Chichester: Wiley-Blackwell, 2010. An up-to-date and well-illustrated undergraduate-level text that is also accessible to the interested reader.

Sigurdsson, Haraldur, ed. *Encyclopedia of Volcanoes.* San Diego, Calif.: Academic Press, 2000. This book contains a complete summary of the scientific knowledge of volcanoes. It includes eighty-two well-illustrated overview articles, each of which is accompanied by a glossary of key terms. Although this is a college-level text, it is written in a clear and comprehensive style that makes it generally accessible. Cross-references and index.

See also: Calderas; Eruptions; Forecasting Eruptions; Island Arcs; Krakatau; Mount Pelée; Mount Pinatubo; Mount St. Helens; Mount Vesuvius; Popocatépetl; Recent Eruptions; Shield Volcanoes; Spreading Centers; Volcanic Hazards.

STROMATOLITES

Stromatolites are the most common megascopic fossils contained within ancient rocks, dating to 3.5 billion years in age. In both the living and fossil form, they are created by the trapping and binding of sediment particles and the precipitation of calcium carbonate to the sticky surface of matlike filaments grown on a daily cycle by blue-green algae. Modern stromatolites are found throughout the world and are of particular use in the creation of hydrocarbon reservoirs, in geologic mapping, and as indicators of paleoenvironments.

PRINCIPAL TERMS

- **algae:** primitive, one-celled, chiefly aquatic plant-like organisms lacking stems, roots, and leaves
- **blue-green algae:** any of the algae classified within the division Cyanophyta
- **cyanobacteria:** algaelike bacteria commonly known as "blue-green algae"
- **fossil:** the remains, trace, or imprint of any plant or animal that lived during the geologic past
- **oncolite:** an organosedimentary rock structure, concentrically laminated, formed by blue-green algae and smaller in size than a stromatolite
- **organosedimentary structure:** a sedimentary rock feature developed by the life processes of blue-green algae
- **prokaryote:** an organism characterized by simple protoplasmic structure and lacking a nucleus
- **stromatolite:** an organosedimentary structure formed by the trapping, binding, or precipitation of sediment upon the laminae surface of blue-green algae
- **thrombolite:** a stromatolite-like fossil characterized by an obscurely clotted, unlaminated internal structure

Stromatolite-Producing Algae

Stromatolites ("layered rock" in Greek) are organosedimentary structures associated with certain types of the sedimentary class of rock; they develop through the metabolic processes of plant microorganisms. Stromatolites can also be considered to be fossils. In this sense, they do not represent the remains of actual organisms but rather the remains of material deposited and collected by a living organism in a manner such that the original size, shape, and morphology of the organism are preserved.

The organisms principally responsible for the development of stromatolites are photosynthetic cyanobacteria, commonly referred to as "blue-green algae." Structurally, blue-green algae are among the most primitive of this class of life, lacking true stems, roots, and leaves. They grow on a daily cycle, in response to the rising and setting of the sun. In response to this cycle, a sticky, filamentous, organic surface is produced in the form of wavy, matlike laminae. This mucus-coated mat traps and binds fine- and coarse-grained sediment to its surface during daylight hours, increasing the thickness of the laminae. During evening hours and nighttime, new algae growth penetrates the sediment-coated layer, producing a new gelatinous filament that in turn will trap and bind more sediment during the following daylight period. As this process continues, a layered, or laminated, structure is built.

Blue-green algae are prokaryotic (prenuclear) microorganisms, as they possess neither a cell nucleus nor a specialized cellular organelle (that portion of a cell that functions as an organ). Blue-green algae are also asexual and thus restricted in their degree of variability, resulting in many living species that are almost indistinguishable from species that lived more than 1 billion years ago. Because of this relative lack of evolution over geologic time, much of what is known regarding ancient stromatolites has been gained by the study of living stromatolites produced by blue-green algae. The mineral matter associated with extant stromatolites is primarily composed of calcium carbonate, although a few species are associated with siliceous material, as evidenced by stromatolites studied in Yellowstone National Park, Wyoming. In the fossil form, the original calcium carbonate content is sometimes altered through replacement by magnesium, iron, and silica.

Modern blue-green algae have a wide geographic distribution and grow in diverse aquatic environments, including salt water, fresh water, and even moist soils. Some blue-green species thrive in hot springs and geysers with temperatures only a few degrees below the boiling point, while others have been collected in Arctic and Antarctic regions. Regardless of their aquatic environment, all modern blue-green algae depend upon sunlight for growth and survival. Because of this need for light, it was formerly thought that marine varieties were restricted to shallow waters near shore. However, it is now known that the photic zone—that region of the ocean penetrated by sunlight—extends to depths approaching 150 meters, allowing growth of deep-water varieties.

Stromatolite-producing algae are found in aquatic environments that are generally hostile to grazing and burrowing invertebrates. Shark Bay in western Australia, a classic site for the study of modern (living) stromatolites, is conducive to blue-green algal growth because the hypersaline environment limits the activity of grazing snails. In Yellowstone National Park, alkaline waters up to 59 degrees Celsius in temperature present an environment supporting the development of unique siliceous stromatolites produced by bacteria rather than by algae. In the shallow marine environment of the Bahama Banks in the Caribbean, blue-green algae are abundant in supertidal channels with current flow too strong for the effective colonization of stromatolitic-grazing invertebrates such as gastropods and ostracods.

Distribution of Stromatolites

In the geologic record, stromatolites are the most abundant of fossils found in rocks dating to the Precambrian era, that period of time from the origin of Earth (approximately 4.6 billion years ago) up to 544 million years ago. The oldest fossil stromatolites are contained in the 3.3-billion- to 3.5-billion-year-old Warrawoona group of rocks in Australia. Close in age are the 3.4-billion-year-old stromatolites of the Swaziland group of South Africa; somewhat less removed are those associated with the 2.5-billion- to 2.8-billion-year-old Bulawayan Limestone, also in Africa. All these examples originated in the Archean eon, the first recorded period of geologic history, extending by definition from approximately 4 billion to 2.5 billion years ago. During the Proterozoic eon, immediately following the Archean eon and extending to 544 million years ago, stromatolites became prolific. This Proterozoic expansion is probably reflective of the initial development of continental landmasses and associated warm, photic continental shelf regions, as plate tectonics became a controlling process in the early development of earth's crust.

Throughout the Archean and Proterozoic eons, blue-green algae underwent a steady and progressive state of biologic evolution, recognized today as the singular, common megascopic fossil of the Precambrian time period (the Archean and the Proterozoic eons). For this reason, the Precambrian is often referred to as the "age of algae," or, more specifically, the "age of blue-green algae." Throughout the Phanerozoic eon, defined as 544 million years ago to the present, blue-green algae underwent minimal evolution, probably because their evolutionary state had become adapted to a variety of environments, reducing the need for further diversification.

Stromatolitic-building algae maintained their dominance of the aquatic world during the Early Phanerozoic eon (544 million to about 460 million years ago). With the rather abrupt appearance of shelled, grazing, and cropping invertebrates in the early Phanerozoic eon, however, blue-green algae began to decline in significance. Today, as compared to their Precambrian domination, they have, on a relative scale, become endangered.

Geographically, fossil stromatolites are ubiquitous on every continent, especially within sedimentary carbonate rock sequences older than 460

A stromatolite formation from Grand Canyon National Park, Coconino County, Arizona. (U.S. Geological Survey)

million years. On southeastern Newfoundland, stromatolites built by blue-green algae of the genus *Girvanella* are found in conglomerate and limestone strata of the Bonavista Formation (approximately 550 million years in age). In the Transvaal region of South Africa, delicately banded stromatolitic structures compose one of the most widespread of early Proterozoic shallow-water carbonate deposits in the world, extending over an area exceeding 100,000 square kilometers. In nearby Zambia, algal stromatolites are closely associated with rock sequences containing economic levels of copper and cobalt. Upper Permian (250 million years ago) stromatolite horizons can be traced over an area of northern Poland exceeding 15,000 square kilometers. Miocene age (15 million years old) algae of the species *Halimeda* compose the limestone-forming rocks of the island of Saipan in the Mariana Islands of the Pacific Ocean. In North America, fossil algal-bearing rocks include the 2-billion-year-old Gunflint (Iron) Formation of Ontario, Canada, and the well-developed stromatolitic horizons of Early Paleozoic era age (450 million years ago) composing the Ellenberger Formation of Oklahoma and Texas.

Classification and Identification

The classification and identification of stromatolites are often concluded on the basis of overall morphology, particularly the size, shape, and internal construction of the specimen. The relevant literature makes use of a variety of morphological terms, including the adjectives "frondose" (leaflike), "encrusting," "massive," "undulatory" (wavelike), "columnar," "laminar," "domed," "elliptical," and "digitigrade" (divided into fingerlike parts). Through the study of modern blue-green algae, it is suggested that three environmental criteria are of importance in stromatolite geometry development. These are direction and intensity of sunlight, direction and magnitude of water current, and direction of sediment transport. As an example, the extant elliptical stromatolites of Shark Bay, western Australia, are oriented at right angles to the shoreline as the result of strong current-driven wave and scour action. Under certain environmental conditions, cyanobacteria growth surfaces are not preserved, producing fossil algal structures characterized by a lack of laminae. These structures are termed "thrombolites," in contrast to laminar-constructed stromatolites.

While stromatolites are generally described as megascopic in size, discussion of specific dimensions relates both to laminae thickness and to overall size. Stromatolite laminae of the Precambrian-aged Pethei Formation, an outcropping along the shores of Great Slave Lake in the Northwest Territories of Canada, are both fine and coarse in dimension. The coarse-grained layers, formed of lime-mud pellets and calcium and magnesium carbonate rhombs, are principally less than 5.0 millimeters in thickness. The fine-grained laminae, composed of calcium carbonate clay and silt-sized particles, are, on average, only 0.5 millimeter thick.

In size, individual stromatolites can range from centimeters up to several meters. Fossils of the common Precambrian genus *Conophyton* occur in a range of sizes, from pencil-sized shapes to columns up to 10 meters in diameter. Subspherical varieties of stromatolite-like structures, formed by the accretion of successive gelatinous mats of blue-green algae and generally less than 10 centimeters in diameter, are termed "oncolites." Stromatolitic complexes in the Great Slave Lake district measure 80 meters long by 45 meters wide by 20 meters in thickness and can be continuously traced for distances exceeding 160 kilometers.

Albert B. Dickas

Further Reading

Dott, R. H., Jr., and D. R. Prothero. *Evolution of the Earth.* 8th ed. New York: McGraw-Hill, 2009. A well-illustrated textbook on the subject of the geologic history of the earth. Contains several sections on stromatolites, their significance to Precambrian history, and their role in the understanding of very early forms of life.

Fedonkin, Mikhail A., et al. *The Rise of Animals: Evolution and Diversification of the Kingdom Animalia.* Baltimore: Johns Hopkins University Press, 2007. An up-to-date account of all the known Ediacaran sites that is extensively and beautifully illustrated. Accessible to all interested readers.

Gebelein, C. D. "Biologic Control of Stromatolite Microstructure: Implications for Precambrian Time Stratigraphy." *American Journal of Science* 274 (1974): 575-598. This lengthy paper discusses the various environmental factors, including water and wind currents, that affect the detailed structure of modern stromatolitic organisms. Such knowledge is useful in worldwide mapping of stromatolite horizons. Written for the undergraduate college student.

Levin, H. L. *The Earth Through Time.* 9th ed. Hoboken, N.J.: Wiley, 2009. This college-level textbook discusses fossil stromatolites of the Proterozoic eon and their importance as heliotrophic rock structures. The author places these fossils within the context of the evolution of prokaryotic organisms. An easy-to-read source complete with numerous photographs.

McMenamin, Mark. *Discovering the First Complex Life: The Garden of the Ediacara.* New York: Columbia University Press, 1998. This entertaining study of the earliest complex life-forms on the planet details the author's work on these organisms. Written for the interested student but understandable by the general reader.

Schopf, J. William, ed. *Major Events in the History of Life.* Boston: Jones and Bartlett, 1992. An excellent overview of the origin of life, the oldest fossils, and the early development of plants and animals. Written by specialists in each field but at a level that is suitable for high school students and undergraduates. Although technical language is used, most terms are defined in the glossary.

Stewart, W. N. *Paleobotany and the Evolution of Plants.* London: Cambridge University Press, 2010. The first four chapters discuss the preservation, preparation, and age determination of fossil plants, the geologic fossil record, and the Precambrian eon, when stromatolites were common throughout the aquatic world. Suitable for college-level readers.

Tewari, Vinod, and Joseph Seckbach, eds. *Stromatolites: Interaction of Microbes with Sediments.* New York: Springer, 2011. This text provides information on the formation, natural history, evolution, and present existence of stromatolites. Chapters discuss research techniques, new data, geochemistry, and ecology, among other topics. Accessible to undergraduate and graduate students, researchers, and scientists of geology and related fields.

Walter, M. R., ed. *Stromatolites.* New York: Elsevier, 1976. Contains papers on zonation, paleoenvironments, environmental diversity, morphologies, origin, geologic significance, and distribution of fossil stromatolites. Several papers discuss modern algal stromatolites and the organisms that build them. A valuable reference for the serious student.

__________. "Understanding Stromatolites." *American Scientist* (September-October 1977): 563-571. An in-depth paper on stromatolites, thrombolites, and oncolites. Discusses means of development of differing morphologies, modern analogues, and uses in stratigraphic mapping of stromatolitic-bearing rocks. Contains a complete references section. Written for the high school science reader.

Young, Emma. "Earth's Earliest Life so Far, and Lots of It." *New Scientist* 190 (2006): 10. This short article discusses the biodiversity of stromatolites and the hypothesis that ancient microbial mats included ecologically diverse organisms. Provides an example of the direction of stromatolite research.

See also: Archaebacteria; Cambrian Diversification of Life; Carbonate Rocks; Catastrophism; Clastic Rocks; Colonization of the Land; Dinosaurs; Ediacaran Fossils; Eukaryotes; Evolution of Birds; Evolution of Flight; Evolution of Humans; Evolution of Life; Fossilization and Taphonomy; Fossil Plants; Fossil Record; Fossils of the Earliest Life-Forms; Fossils of the Earliest Vertebrates; Geoarchaeology; Life's Origins; Mammals; Mass Extinctions; Microfossils; Neanderthals; Paleobiogeography; Paleobotany; Petrified Wood; Prokaryotes; Proterozoic Eon; Systematics and Classification.

SUPERCONTINENT CYCLES

Supercontinent cycles, which recur over periods of 400 million to 440 million years, are helpful in understanding the distribution of certain natural resources, fluctuations throughout geologic time of sea level and climates, the process of mountain building, and the evolution of life.

PRINCIPAL TERMS

- **continental drift:** the theory that continental fragmentation and displacement caused the creation of new ocean basins and the formation of mountain ranges
- **convergence:** the process that occurs during the second half of a supercontinent cycle, whereby crustal plates collide and intervening oceans disappear as a result of plate subduction
- **divergence:** the process of fracturing and dissecting a supercontinent, thereby creating new oceanic rock; divergence represents the initial half of the supercontinent cycle
- **ophiolite suite:** a unique vertical sequence of peridotite rock overlain by gabbro, basalt, and oceanic sediments representative of ancient seafloor material
- **plate tectonics:** a theory describing the earth's surface as composed of rigid plates continually in motion over the interior, causing earthquakes, mountain building, and volcanism
- **seafloor spreading:** the continual creation of new seafloor bedrock along mid-ocean ridges through the process of ascending thermal currents
- **supercontinent:** a single vast continent formed by the collision and amalgamation of crustal plates
- **Wilson cycle:** the creation and destruction of an ocean basin through the process of seafloor spreading and subduction of existing ocean basins

CONTINENTAL DRIFT THEORY

One of the most persistent questions relating to the historical development of the earth concerns the processes whereby extensive mountain chains, such as the Himalaya, Appalachian, and Ural ranges, have been formed. During the nineteenth century, the planet-contraction hypothesis, supported by the doctrine of permanence of continents and ocean basins, suggested that such linear features of the earth's crust were comparable to the wrinkles within the skin of a dried apple. After the discovery of radioactivity in 1896, studies suggested heat formed by radioactive decay of rock minerals approximately equaled heat lost to the atmosphere by the gradual cooling of the earth from its supposed original liquid stage. This balance of heat gain and loss did not support a shrinking-earth concept. With the additional knowledge that linear mountain ranges had formed during different stages of geologic history, rather than simultaneously, the contraction hypothesis gradually lost favor.

Between the publication of his 1915 book *Die Entstehung der Kontinente und Ozeane* (*The Origin of Continents and Oceans,* 1924) and his death during an expedition to the Greenland icecap in 1930, German meteorologist and geologist Alfred L. Wegener gained international repute as the father and chief advocate of the theory of continental drift. Wegener postulated that mountain ranges were created by the collision of large blocks of continental crust moving through oceanic crust, following fragmentation and dispersion of the vast supercontinent he called Pangaea (Greek for "all lands"), surrounded approximately 200 million years ago by the universal ocean Panthalassa (from *thalassa,* Greek for sea). An intriguing set of evidence supports this theory, including the unusual degree of geometric fit of present-day continents (especially South America with Africa) and, through the reassembly of Pangaea, the reconstruction of truncated salt, fossil-reef, glacial-deposit, and mountain-range trends. While many scientists at the time became "pro-drift," many others questioned what possible mechanism could displace solid continental crust through equally solid oceanic crust. Various earth forces, ranging from centrifugal to tidal to rotational axis wobble, were investigated and rejected as inadequate by physicists and mathematicians. By the 1930's, continental drift as a theory was no longer considered viable and was increasingly mentioned only within the context of the history of science.

EXPLORING THE OCEAN FLOOR

Following World War II, a new era of earth investigation began, focused primarily on the ocean basins. Employing newly developed military technology, including methods of water-depth sounding (bathymetry) and magnetic-body detection, surplus military aircraft and surface-vessel equipment began to collect a wide array of information. By the International Geophysical Year of 1957-1958, ocean-floor depth-profile analyses confirmed the existence of a previously unknown, 65,000-kilometer-long global submarine mountain range system traversing the Atlantic, Pacific, Arctic, and Indian Oceans. This feature defied immediate explanation.

During the same period, ocean-evaluation programs sponsored by both private and U.S. government interests began the task of measuring the magnetism of the ocean floor. By the mid-1950's, sufficient data had been collected off the West Coast of the United States to reveal a repetitive north-south pattern representing alternate zones of above-average and below-average magnetism. Soon similar patterns were shown to exist in the Atlantic and Indian Oceans.

The discovery of these unexplainable phenomena prompted the collection of any form of additional data that would help explain the existence of ocean floors dissected by a universal mountain range and masked by symmetrical magnetic patterns. New oceanographic programs gathered information on bedrock temperature, radiometric age, and ocean-sediment thickness and studied the worldwide distribution of earthquakes and volcanic eruptions.

Plate Tectonics Theory

By the early 1960's, broad-based analyses of these various forms of oceanographic data were being conducted in Canada, the United Kingdom, and the United States. Gradually, consensus began to form that perhaps at least some of the ideas of Alfred Wegener were worthy of reconsideration. Paramount among these resurrected ideas was that of the assembly of the supercontinent Pangaea. Rather than postulating continental crust as floating through oceanic crust, the revised continental drift theory, termed "plate tectonics," envisioned the outer layer of the earth as divided into a series of major plates, each composed of both continental and oceanic bedrock.

As examples, the North American plate is made up of the continent of North America (including Greenland), the western half of the north Atlantic Ocean, and eastern Siberia, while the Indian-Australian plate is composed of the continent of Australia, the country of India, and portions of the Pacific and Indian Oceans. Major plate displacement was considered possible because of the movement of global thermal-convection cells, which form within the mantle of the earth and rise toward the surface until they are blocked by the presence of a supercontinent. The blockage of further transmission of heat, caused by the insulating nature of continental rock, divides the convection current into lateral, horizontally directed segments, which gradually dome by thermal expansion and then dissect the supercontinent.

As the new subcontinent begins to diverge, the separating void fills with high-density gabbro and basalt-type rock, which forms the floor of a newly developed ocean. Because the earth is neither expanding nor contracting in size, continuing divergence cannot proceed indefinitely without experiencing resistance caused by the convergence of antipodal plates. The effect of plate convergence depends on whether such plate margins are oceanic (basaltic) or continental (granitic) in nature. Where margins are oceanic, the more dense margin will subduct, or plunge under, the less dense margin. Where margins are both continental, subduction is unlikely, and the result is massive folding and faulting (earthquaking). Finally, where a continental margin converges with an oceanic margin, the latter, being more dense, will subduct beneath the former. In all three possible convergence cases, crust shortening is accomplished. Rock volume harmony results, as the formation of new oceanic crust through plate divergence is matched by the destruction of older crust through the process of subduction.

Processes of divergence and convergence are believed to have been ongoing throughout a large portion of geologic time and to continue today. Divergence, accompanied by new-ocean development in region A, continues simultaneously with convergence in region B, resulting in the gradual destruction of region A ocean by way of subduction. Eventually, region A ocean will cease to exist, and the cycle of ocean birth and death will be complete. This sequence of events, during which it is estimated 2.6 square kilometers of ocean floor rock is created and destroyed each year, is termed a "supercontinent cycle" (also known as a Wilson cycle, after the Canadian geologist J. Tuzo Wilson, an early advocate of the plate tectonics theory). Conversely, the dispersion and amalgamation of continental crustal masses, as opposed to oceanic masses, constitute the principal phases of what has been termed the Pangaean cycle. The operation of plate tectonics continuously creates and recycles ocean basins, while continental regions increase in age geologically even as they are agglomerated and dissected by ongoing seafloor spreading.

Future Supercontinent Cycles

A maxim of geology states that the validity of any hypothesis or theory is determined by the degree to which that concept can be examined through the analyses of extant geology. Where, then, might there be modern-day examples of a supercontinent cycle in its various stages of tectonic development?

The Great Basin of the western United States has been portrayed as a model of very early continental rifting that may, in the future, separate the North American plate from a newly constituted Pacific plate. The approximately one hundred block-faulted mountains composing this geographic terrane are caused by the same extensional forces responsible for the early continental-rift stage of dissection of a supercontinent.

Similar forces have formed the more structurally advanced rift systems of East Africa, the most illustrative of which are those broad-basin and steeply dipping escarpment topographies of Tanzania, Kenya, and Ethiopia, which continue to yield a record of mammalian and hominid evolution. The fresh waters that partially cover these rift valleys, such as Lake Tanganyika, are evidence of a late stage of continental rifting. To the north, the central valley or rift of the Red Sea is filled with salt water characteristic of an incipient ocean developing during an early stage of oceanic rifting separating the Arabian from the African plate.

Finally, the Atlantic Ocean is a mature example of oceanic rifting representative of the midpoint of a supercontinent cycle. According to the concepts of plate tectonics, the Atlantic Ocean has been created through some 200 million years of seafloor spreading driven by extensional rifting of Pangaea. This stage, representing the first half of a supercontinent cycle, will in theory be followed by assembly of the world's continents over the next 200 million years into a new supercontinent. This assembly may have already begun, as India, an island subcontinent up to approximately 35 million years ago, has since that time been colliding with the Eurasian plate, resulting in the formation of the Himalaya Mountains.

Evidence of Past Supercontinent Cycles

Following the general acceptance of the plate tectonics theory in the 1960's, many ideas were advanced regarding the existence and nature of pre-Pangaea supercontinents. Since the supercontinent cycle not only creates but also destroys oceans, pre-Pangaea supercontinents must be reconstructed from continental geologic data that become, with increasing geologic age, more difficult to interpret. Certain criteria have been developed in the attempt to discern a pattern of supercontinent cycling since the earliest periods of geologic time.

The convergence of continents will largely eradicate any intervening ocean. Such loss by subduction is seldom complete, as attested by the presence of remnants of preexisting ocean floor rock contained within the deformed (suture) zone caused by plate collision. Basalt, gabbro, and olivine-rich rocks, termed an "ophiolite suite," are scraped off, or "obducted" from, the subducting ocean floor and thus preserved in the developing mountain belt. Obducted ophiolitic rock from the former Tethys Sea, which lapped onto the eastern shores of Pangaea, is present in the Alps and the Himalayas, while ophiolites within the Ural Mountains of central Asia are evidence of an ocean that was destroyed by the collision of Baltica and Siberia, continental masses that preceded the formation of Pangaea. The presence, location, and age-dating of ophiolite suites are helpful in the identification of former supercontinents.

Paleontological and high-pressure rock evidence collected from the Appalachian Mountains—created by the convergence of proto-Africa (the earliest form of Africa) with proto-North America—suggests the existence approximately 500 million years ago of a proto-Atlantic Ocean. The existence of this body of salt water, the Iapetus Ocean, attests the existence of an associated supercontinent.

The presence of exotic terranes forming the collisional edge of former crustal plates in Japan, New Zealand, and the Apennines of northern Italy is further evidence of former supercontinents and their cycles. These terranes, packages of rock possessing similar mineral and fossil character, are accreted to enlarging supercontinents by the same obduction process as ophiolites.

Two Schools

There are two major contemporary schools of thought regarding supercontinent cycles. The oldest earth specimen found to date is the 4.28-billion-year-old metamorphic continental rock forming a portion of the Northwest Territories of Canada. This and slightly younger rock terranes from Greenland, Antarctica, and Australia may have combined to form the first amalgamated continental masses. John Rogers, a professor of geology at the University of North Carolina, proposed the existence 3 billion years ago of an early subcontinent he calls "Ur" (from the German for "original"). Approximately 500 million years later, a second subcontinent, Arctica (predecessor to Canada, Greenland, and eastern Russia) formed, followed another 500 million years later by Baltica (proto-Western Europe) and Atlantica (eastern South America and western Africa). Then, 1.5 billion years ago, plate tectonic forces formed the sub-supercontinent Nena (from the Russian for "motherland") through the merging of Baltica and Arctica. This lengthy chain of events culminated 1 billion years ago in the formation of the first supercontinent, Rodinia (also known as proto-Pangaea), as a result of the joining of Ur, Atlantica, and Nena. After a period of stability lasting some 300 million years, Rodinia subdivided, forming numerous proto-continents and Iapetus, the proto-Atlantic. Finally, the reassembly of these proto-continents and subsequent subduction of Iapetus brought about the creation of the second supercontinent, Pangaea, about 250 million to 300 million years ago.

A second school of thought differs principally in the suggestion that plate tectonic processes did not begin until some 2.5 billion years ago, on the occasion of the development of the first distinct oceanic and continental crust. Prior to that time, the very high temperature of the earth and the relative thinness of primordial crust forestalled the onset of supercontinent-cycle processes such as divergence, convergence, and subduction. While differing in detail, both schools of thought generally recognize Rodinia and Pangaea as supercontinents.

The length of a typical supercontinent cycle is variously estimated at from 400 million to 440 million years. Such a cycle would constitute three phases. Once formed, a supercontinent would exist for 100 million to 120 million years before the accumulation of thermal convection heat initiated crustal dissection. During the second phase, lasting from 150 million to 160 million years, maximum dispersal of subcontinents would take place, with resultant development of new oceans. Finally, subduction would gradually destroy the intervening oceans over a period of 150 million to 160 million years, creating a new supercontinent and terminating the cycle.

Albert B. Dickas

Further Reading

Dalziel, Ian W. D. "Earth Before Pangaea." *Scientific American* 272 (January 1995): 58-63. An easy-to-read account of the nomadic wanderings of the North American plate prior to the assembly of Pangaea.

Davidson, J. P., W. E. Reed, and P. M. Davis. *Exploring Earth: An Introduction to Physical Geology.* 2d ed. Upper Saddle River, N.J.: Prentice-Hall, 2001. Chapters 6 through 11 are an excellent undergraduate-level introduction to the natural consequences of plate tectonics and their role in the supercontinent cycle.

Nance, R. Damian, T. R. Worsley, and J. B. Moody. "The Supercontinent Cycle." *Scientific American* 259 (July 1988): 72-79. Combining specialties of tectonics, oceanography, and geochemistry, the authors discuss in a clear manner several supercontinent cycles and their effects on climate, evolution, and geologic changes.

Nicolas, A. *The Mid-Ocean Ridges.* New York: Springer-Verlag, 1995. A review presenting the European view of the construction and destruction of ocean basins. Written at the knowledgeable adult level.

Nield, Ted. *Supercontinent: Ten Billion Years in the Life of Our Planet.* Cambridge, Mass.: Harvard University Press, 2007. This book covers the supercontinent cycle. Discusses continents of the past and future. Written in a manner accessible to scientists in general.

Rogers, John J. W., and M. Santosh. *Continents and Supercontinents.* New York: Oxford University Press, 2004. Discusses many aspects of the creation, transformation, and destruction of continents. Includes a summary of plate tectonics. Discusses the supercontinents of Pangaea, Gondwana, and others; includes information on the effects of continents on evolution and climate. Accessible to undergraduate students.

Sullivan, W. *Continents in Motion.* 2d ed. New York: American Institute of Physics, 1993. The author, a science editor for *The New York Times,* presents a recommended compilation of the history and dynamics of plate tectonics. Suitable for the educated layperson.

Wicander, Reed, and James S. Monroe. *Historical Geology*. Belmont Calif.: Brooks/Cole, Cengage Learning, 2010. This textbook is accessible to university students with or without a background in geology. It covers Earth's geological systems and plate tectonics, physical and biological history, geological time, and evolution. There are many diagrams, images, and charts along with a CD-ROM to supplement the text.

See also: African Rift Valley System; Continental Crust; Continental Drift; Continental Growth; Continental Rift Zones; Continental Shelf and Slope; Continental Structures; Continents and Subcontinents; Gondwanaland and Laurasia; Ocean Ridge System; Ophiolites.

SYSTEMATICS AND CLASSIFICATION

The modern classification of both fossil and living organisms attempts to organize life-forms in a logical and uniform manner and to illustrate the evolutionary relationships between different groups of animal and plant life. Only 250,000 fossil species have been described, while it has been estimated that some 500 million species have existed over the past 3.4 billion years. Thus, there are enormous numbers of fossil species still awaiting discovery and classification.

PRINCIPAL TERMS

- **binomial system:** the current system of classifying organisms, which gives each organism a dual name consisting of the genus and the species
- **genus:** a group of closely related species that share a common ancestry and similar morphological characteristics
- **kingdom:** one of the five large subdivisions of life, differentiated on the basis of gross body plan (single-celled versus multicellular) and method of obtaining food or nutrients (produces own food or obtains it from other organisms)
- **nomenclature:** the names and terms used in a classification system
- **phylogeny:** the study of the evolutionary relationships between organisms
- **phylum:** a major grouping of organisms, distinguished on the basis of basic body plan, grade of anatomical complexity, and pattern of growth or development
- **species:** a group of actually or potentially interbreeding organisms; the basic taxonomic unit
- **taxon:** one of several systematic groups or categories into which a particular organism may be placed
- **type specimens:** the original fossil specimens upon which a species was first erected

BINOMIAL SYSTEM

The system of classification utilized by biologists and paleontologists to classify both living and fossil organisms is based on the premise that a similarity in morphology or structure provides evidence of a biological kinship between organisms. The biological classification employed by biologists and paleontologists not only expresses the similarities and differences between groups but also summarizes or conveys a vast amount of information concerning these groups, especially the general phylogenetic or evolutionary relationships between groups.

"Systematics" is the study of taxonomy, which is concerned with the identification and classification of fossil organisms into morphologically and evolutionarily related groups. These groups or taxonomic categories are referred to as taxa. The system of classification in use worldwide is the binomial system of nomenclature, proposed by the Swedish naturalist Carolus Linnaeus in the 1700's. The fundamental unit, or taxon, in this classification is the species. A species can be defined as a group of morphologically similar individuals whose ecological demands and physiologic functions are the same and whose members interbreed to produce fertile offspring. Groups of closely related species that share a common ancestry and possess a set of similar morphological characteristics are termed "genera" (singular, genus). Under the binomial system, each plant or animal is given two names, the genus name first, the species second. The human species, for example, is referred to as *Homo* (genus) *sapiens* (species). The names of both the genus and the species are printed in italics. Species names are given in Latin, as originally this was the universal language of science in the Western world, and naturalists were looking for a standardized nomenclature recognizable to scientists from all countries. The Latinized names of species have remained universal names, recognizable to biologists and paleontologists worldwide.

There are also taxonomic categories above the genus level. Similar genera are grouped into a single family; groups of related families are placed into an order; similar orders form a class; and closely related classes form a phylum. The more than thirty-five phyla are placed into one of five large groups termed kingdoms. These include the Monera (single-celled organisms with no cell nucleus), the Protista (single-celled organisms with a cell nucleus), the multicellular Plantae (organisms that produce their own food via photosynthesis), the multicellular Animalia (organisms that gain nutrients from the ingestion of other organisms), and the Fungi (multicellular organisms that absorb nutrients from their environment).

Whereas the species represents a real, natural taxonomic unit, all taxonomic categories above the species level are artificial and subjective. The classification of the carnivorous dinosaur *Tyrannosaurus rex* would be as follows: kingdom, Animalia; phylum, Chordata; class, Reptilia; order, Saurischia; family, Tyrannosauridae; genus, *Tyrannosaurus;* species, *Tyrannosaurus rex.* The basic categories of higher taxa are sometimes supplemented by additional categories with the use of a sub-prefix or super-prefix.

Hierarchical Classification

This type of classification is referred to as a hierarchical system. In such a system, each fossil specimen belongs to one "lower" taxonomic category or species and one of each of the "higher" or superspecific categories. From the highest rank (kingdom) to the lowest rank (species), these categories become progressively narrower in scope. For example, 1.5 million or more species form the kingdom Animalia, but only one species constitutes the species *Homo sapiens.* The criteria used to differentiate taxonomic groups also become narrower and more trivial as one goes from higher to lower taxa.

The binomial system attempts not only to divide organisms into structurally differentiated groups but also to illustrate the phylogenetic, or evolutionary, relationships between these groups. This classification scheme serves as a phylogenetic outline for the evolution of life on Earth and as an indicator of the diversity of basic body plans available to organisms through time. This is often diagrammatically depicted in the rough form of a branching tree, with the more primitive, structurally simpler organisms at the base and the more advanced, structurally more complex forms in the crown. In such figures, one can see how animals evolved from simple, single-celled protozoans through increasingly more complex multicellular organisms with successively more sophisticated grades of development. These evolutionary pathways are determined from studies of the comparative anatomy of living organisms and by information provided by the fossil record.

The identification of fossils and the integration of fossil organisms into the system of biological classification allows for reconstructing the history of life on earth. The present records just one instant in the long history of life on this planet—a "snapshot" of life as it exists today. The fossil record, however, extending back some 3.4 billion years, provides the whole epic motion picture. Although some frames are missing or blurred by the effects of time, fossils provide the only record of life on Earth and therefore offer the sole source for the factual documentation of the evolution and extinction of organisms through time. The systematic study of fossil organisms provides four types of evolutionary information: It allows scientists to reconstruct the phylogenetic relationships that exist between various taxonomic groups; it reveals the times of major evolutionary events, such as major evolutionary radiations and so-called mass extinction events; it gives some idea as to the rates of evolutionary change; and it reveals patterns of evolutionary change.

Fossil Removal and Description

Determining the identity of a particular fossil provides an outline of its biological affinities, evolutionary pedigree, and anatomical complexity, and also provides some idea of how the organism lived in its environment. The fossil then becomes more than a curiously shaped stone; it becomes the remains of a once-living organism that, in itself, is a time capsule for information about that interval of geologic time during which the organism lived and eventually died.

The basic tools of the systematic paleontologist traditionally have been the various implements used to remove the fossils from the rock matrix and prepare them for description, calipers to measure morphological dimensions, and an adequate reference library to aid in determining the identity and relationships of the fossils under study and compare them with those already described. Today, computers can store and compare the dimensions of large suites of specimens and statistically analyze large sets of data in order to determine the possible relationships between fossil taxa. Studies of shell microstructure and geochemistry have also provided new approaches to fossil identification and classification, beyond the study of the external features of a fossil specimen.

The basic approach to fossil identification, however, remains essentially the same as it has been for the past two centuries. A fossil is discovered and collected, with the details of its occurrence—that is, the rock layer or stratum and geographic locality where it was found—being determined and recorded with the fossil. Ideally, associated fossil species should also be noted, as well as the orientation of the fossil specimens in the rock and the nature of their preservation. This may help the paleontologist to infer something about the life habits and habitats of the fossils. A single specimen of a particular fossil taxon is not enough; the larger the number of specimens of the fossil available for collection from the site, the better for the purposes of statistical analysis of the species.

The fossil specimens are removed from the enclosing rock matrix, and rock matrix adhering to the specimens is removed using preparatory equipment ranging from hammers and chisels to air-abrasive equipment and vibra-tools. This preparation is necessary to expose the main morphological features of the shell or bone, features critical in determining the fossil's identity. Paleontologists occasionally use sonic cleaners, weak acids, or strong detergents to remove more stubborn matrix adhering to the fossils. Fortunately, many fossils weather cleanly from soft sedimentary rocks like shales, marls, and poorly cemented sandstones and require little cleaning and preparation.

Following this, a detailed description of the specimen is formulated, including the dimensions of the specimen (measurements of length, width, height, thickness, and so on), the general arrangement of the parts that compose the fossil, and a description of the external and internal morphological features that distinguish the specimen. For certain fossil groups, especially corals, bryozoans, and nautiloid cephalopods, it is necessary to cut open specimens with rock saws, both longitudinally and tangentially, to expose critical internal structures and the microstructure of the shell if it is preserved. If more than one specimen was collected of the fossil type under study, each specimen is described and then compared with its associates in order to determine any morphological variation present within the fossil population or to determine whether one or more distinct species are present in the collection.

Fossil Classification

To classify the fossil specimen, the fossil at hand is compared with fossil forms already described in the scientific literature. The usual process for identifying a fossil specimen is to determine its phylum first and then proceed from the general to the specific. Information to aid in the identification of fossils can be found in a number of reference books, textbooks, and popular field guides on fossils. Invertebrate fossils (remains of animals that lack backbones) constitute the majority of fossils found. These primarily marine animals are also the easiest to identify, largely as a consequence of the possession of hard parts consisting of mineralized external or internal shells that usually are composed of a minimal number of parts that are preserved intact.

The fossil remains of vertebrate animals and plants are much less common and more difficult to identify and classify. This is attributable, in part, to the fact that these organisms are most common in terrestrial environments where the rapid and complete burial necessary for preservation as a fossil is not a common process. The internal skeletons of vertebrates consist of a multitude of individual bones held together by tendons, cartilage, and fleshy tissue. Upon the death of the organism, these soft parts decay, and the skeleton readily disaggregates into a pile of loose bones that, upon transport by stream action or scavengers, are spread over a broad area and are difficult to reassemble into the original skeleton. Plant remains lack mineralized hard parts and consist of multiple parts (roots, stems, leaves, seeds) that also are easily separated from one another in the process of burial.

After comparing the fossil specimens with species that have already been described, one of two conclusions can be made. Either the fossil specimens can be identified as those of an existing species, or they constitute a species new to science. If the specimens are identified as a preexisting species, they should be closely compared with the type specimens of the known species to ascertain if any variation exists between the two suites of specimens. The "type specimens" are the original specimens for which the fossil species was first erected. They are usually preserved for posterity in collections held at major natural history museums. The age of the rock unit and the geographic locality of the collecting site should be compared with that of the existing species to determine whether these new specimens extend the range of the species in time or space. If it is determined that the fossil specimens are sufficiently distinct from existing species, then a new species is created for these specimens.

Description of New Species

The description of a new species must follow a specific format that has been specified in the Code of Zoological Nomenclature established by the International

Commission on Zoological Nomenclature in the early 1900's to bring a measure of uniformity to animal taxonomy. A comparable set of procedures has been established for plants. This code specifies the choice of the species name, publication of the name, description of the new species, and the designation of one or more type specimens.

The name given to a new species must be binomial, consisting of the genus name and the species name. The name used for the species cannot already be in use with the genus name with which it is to be associated. The names of the species and the genus must be Latin words or words that have been Latinized. Species names can be Latinized place-names, the names of people, or descriptive words. A species cannot be named anonymously, and the author's name is part of the official species name. For example, the official name for the living species of the chambered nautilus from the southwest Pacific would be *Nautilus pompilius* Linnaeus, Carolus Linnaeus being the original author of the species.

In order for a new species to be officially recognized, the name and a description of the distinctive features that identify it must be published in an approved medium. That is, it must be in print, published in quantity, and circulated to libraries to ensure that the announcement of the new name is readily available to scientists around the world. The description of a new species should include a diagnosis listing the characteristics distinguishing the new species from similar species as well as a fuller description of the various morphological aspects of the species, along with its geographic locality, geologic occurrence, and a reference to the museum where the designated type specimens for the species are stored. It is also strongly recommended that an illustration of the new species be included in this description.

The International Code also specifies that each new species description must be accompanied by the designation of a type specimen or set of type specimens. These are the actual original fossil specimens upon which the species is based. The code requires that type specimens be clearly labeled as such and that measures be taken to ensure their preservation and accessibility to interested scientists. This usually means that specimens are deposited in a major natural history museum, where such facilities are available. If the new species is based on a single specimen, this specimen is designated the type specimen, or "holotype." If several specimens serve this purpose, they are designated the "syntypes" for the species.

The naming of new species is best left to a professional paleontologist who is familiar with the rules of zoological nomenclature and the accepted format for publication. Nevertheless, amateur fossil collectors, who typically find the bulk of new species in the field, are often recognized for their contributions by being named a coauthor of the new species or having the new species named for them.

Cladistics

Reconstruction of the phylogenies, or evolutionary family trees, of fossil organisms operates in a manner opposite that of fossil identification and classification. A paleontologist starts at the species level, reconstructing evolutionary relationships between species, and then works up the taxonomic hierarchy, from the narrow to the broad. Phylogenies are reconstructed using a system called "cladistics," in which organisms are grouped according to the presence of shared characteristics that are particularly informative. This system assumes that as evolution progresses, new, heritable traits emerge and are passed on; two groups of animals sharing such new traits will be more closely related to each other than to groups that only share the original traits. Analyses of these characters are presented as treelike diagrams called "cladograms," which show the order in which the new characters, and thus the new species, evolved. Each branching point in the diagram reflects the emergence of an ancestor that founded a group that exhibited advanced characters not present in the groups that had developed earlier. Through detailed studies, or "monographs," of groups of related species and genera, the phylogenies of most of the major animal groups have been mapped out. These reconstructions, however, are constantly undergoing change, primarily as the result of the description of newly discovered fossil taxa and the restudy of older, established species.

Robert C. Frey

Further Reading

Argue, D., et al. "*Homo floresiensis*: A Cladistic Analysis." *Journal of Human Evolution* 57 (2009): 623-639. This article discusses the discovery of *Homo floresiensis*, and the impact on human systematics. Multiple hypotheses are addressed in this article regarding the classification of this species. Cladistic analysis is utilized by this study to determine the species' placement in the tree of life.

Benton, Michael, and David Harper. *Introduction to Paleobiology and the Fossil Record.* Chichester: Wiley-Blackwell, 2009. Comprehensive coverage of invertebrate and vertebrate organisms and the basic principle of paleontology. Written for undergraduate students but suitable for the interested reader also.

Briggs, Derek E. G., and Peter R. Crowther, eds. *Paleobiology: A Synthesis II.* Oxford: Blackwell Scientific Publications, 2003. This multiauthored text deals with all aspects of paleontology in a series of short articles by experts in each field. Intended for university students but also useful to high school students and interested readers.

Briggs, D. E., and R. A. Fortey. "Wonderful Strife: Systematics, Stem Groups, and the Phylogenetic Signal of the Cambrian Radiation." *Paleobiology* 31 (2005): 94-112. The authors discuss the unresolved issues of systematics in reference to the morphological disparity and divergence times of the Cambrian period. The article provides specific examples to support the broad concepts. Best suited for graduate students and professionals interested in theories of evolutionary biology.

Clarkson, E. N. K. *Invertebrate Paleontology and Evolution.* 4th ed. Hoboken, N.J.: Wiley-Blackwell, 1998. A college invertebrate textbook up-to-date with good description of the major invertebrate phyla, classes, and orders. Good illustrations of the major morphological features that distinguish each group. A brief general description of taxonomy and fossil systematics.

Cleal, C. J., and B. A. Thomas. "Botanical Nomenclature and Plant Fossils." *Taxon* 59 (2010): 261-268. This article addresses the changes made to the Vienna Code for naming fossilized plant taxa. Discusses the concept of morphotaxa.

Foote, Michael, and Arnold Miller. *Principles of Paleontology.* 3rd ed. San Francisco: W. H. Freeman, 2007. An excellent college-level text emphasizing paleontological concepts, including thorough sections on fossil classification, the description of new species, and reconstruction of fossil phylogenies. Suitable for college-level readers with some background in biology or paleontology.

Kitching, Ian J., Peter L. Forey, Christopher J. Humphries, and David M. Williams. *Cladistics: The Theory and Practice of Parsimony Analysis.* 2d ed. Oxford: Oxford University Press, 1998. An introduction to the study of systematics using the method of phylogenetic analysis. Intended for the college student.

Lemey, Philippe, Marco Salami, and Anne-Mieke Vandamme, eds. *The Phylogenetic Handbook: A Practical Approach to Phylogenetic Analysis and Hypothesis Testing.* 2d ed. New York: Cambridge University Press, 2009. This text covers theory and application of phylogenetic analysis. The book focuses on hypothesis testing, which makes this text a useful tool for researchers, graduate students, and advances undergraduates.

Margulis, Lynn, and Michael J. Chapman. *Kingdoms and Domains: An Illustrated Guide to the Phyla of Life on Earth.* 4th ed. Boston: Elsevier, 2009. Provides a comprehensive overview (current at the time of publication) of life's classification and phylogeny. An improvement on Margulis's previous work, *Five Kingdoms,* this text includes more recent discoveries and systematics. The classification is a modified version of Linnaean taxonomy.

Moore, R. C., ed. *Treatise on Invertebrate Paleontology.* 40 vols. Lawrence: University of Kansas Press, 1953-2010. This series of volumes includes descriptions and illustrations of all fossil invertebrate groups, down to the genus level. Each volume concentrates on a particular phylum or class and includes general chapters on evolution, ecology, and classification. The standard reference for the identification and classification of fossil invertebrates. Some basic knowledge of biology or paleontology required.

Prothero, Donald R. *Bringing Fossils to Life.* 2d ed. Boston: McGraw-Hill, 2004. This well-illustrated and entertaining text covers a broad range of paleontological topics, including systematics and classification. Glossary, bibliography, and index.

Radinsky, Leonard B. *The Evolution of Vertebrate Design.* Chicago: University of Chicago Press, 1987. A very good introductory text to the vertebrates. Well written and illustrated, with a unique approach to vertebrate classification, relating important morphological features in each group to their function. Highly readable and understandable to the college-level reader.

Smith, Andrew B. *Systematics and the Fossil Record.* Oxford: Blackwell Press, 1996. An overview of approaches to systematics and biostratigraphy, with particular emphasis on their application to fossil organisms. For college students.

See also: Archaebacteria; Biostratigraphy; Cambrian Diversification of Life; Catastrophism; Colonization of the Land; Dinosaurs; Ediacaran Fossils; Eukaryotes; Evolution of Birds; Evolution of Flight; Evolution of Humans; Evolution of Life; Fossilization and Taphonomy; Fossil Plants; Fossil Record; Fossils of the Earliest Life-Forms; Fossils of the Earliest Vertebrates; Geoarchaeology; Life's Origins; Mammals; Mass Extinctions; Microfossils; Neanderthals; Paleobiogeography; Paleobotany; Petrified Wood; Prokaryotes; Stromatolites.

T

THRUST BELTS

Thrust belts are long, narrow zones composed of many thrust faults that record significant horizontal displacements of the continental crust. Thrust faults are surfaces that emplace older rock above younger rock. Exploration of thrust belts has led to the discovery of vast reserves of oil and natural gas.

PRINCIPAL TERMS

- **continental crust:** the upper 50 kilometers of the earth below continents; it is composed primarily of rock that is lighter than that which makes up oceanic crust
- **continental margin:** the edge of a continent next to an ocean basin; it is both exposed on land and submerged below water
- **faulting:** the process of fracturing the earth such that rocks on opposite sides of the fracture move relative to each other; faults are the structures produced during the process
- **folding:** the process of bending horizontal layers of rock so that they dip; folds include anticlines, which are arches, and synclines, which are shaped like the letter U or V
- **geosyncline:** a major depression in the surface of the earth where sediments accumulate; geosynclines lie parallel to the edges of continents and are long and narrow
- **orogenic belt:** a mountain belt composed of a core of metamorphic and plutonic rocks and an adjacent thrust belt

Discovery of Thrust Belts

Thrust belts are arcuate zones in which the surface of the earth has been shortened in its horizontal dimension, primarily by faulting. The zones are expressed as mountain ranges hundreds of kilometers wide and thousands of kilometers long. They are composed of faulted sedimentary rock and are integral parts of the linear orogenic belts that mark the edge of every continent. Active thrust belts are areas of intense deformation, rugged topography, and frequent earthquakes. Thrust faults are the principal features of thrust belts and are gently inclined surfaces along which older rocks, originally near the bottom of a pile of sediments, are pushed up and over younger rocks at the top. The direction in which the rocks move is perpendicular to the length of the thrust belt. Thrust belts frequently merge with fold belts. Fold belts are similar to thrust belts except that the principal structures formed are folds several kilometers wide. Fold and thrust belts form along continental margins, which are also convergent plate boundaries, and record significant horizontal movements within the continental crust.

Thrust faults were first recognized in the mid- to late nineteenth century in the Scottish Highlands. Geologists working there prior to the middle of the nineteenth century believed that sedimentary strata were interbedded with metamorphic rocks, an interpretation that arose from a lack of knowledge about the process of metamorphism. It is now known that during metamorphism, heat transforms sedimentary rocks into metamorphic rocks. The transformation occurs gradually both in space and in time.

Rocks adjacent to each other should therefore exhibit only slight differences in metamorphism, an implication that led to the conclusion that interbedding of the metamorphic rocks and the sedimentary strata in Scotland was unlikely. At about the same time, another important discovery was made: The metamorphic rocks high in the sediment pile contained older fossils than did the sedimentary strata below. Because the sedimentary sequence was not believed to be upside down, the idea of great gliding of the older, metamorphic rocks up and over the younger, sedimentary strata was proposed. This idea of thrust faulting was quickly accepted and applied to other areas, including the United States, where, in the early twentieth century, Bailey Willis documented that rocks greater than 500 million years old were thrust above strata fewer than 100 million years old in Glacier National Park.

In Glacier National Park, rocks more than 500 million years old have been thrust above strata less than 100 million years old. (PhotoDisc)

Characteristics of Thrust Belts

Thrust belts are elongate areas of faulted and relatively unmetamorphosed sedimentary strata that border linear zones of metamorphic and plutonic rock (a pluton is a body of rock that crystallizes from a liquid below the surface). Together, the thrust belt and the zone of metamorphic and plutonic rocks define two parts of an orogenic belt: the core of the orogenic belt, where temperatures were high enough to transform sedimentary strata into metamorphic rocks, and the thrust belt, where temperatures were too low for metamorphism. Deformation in the core of the orogenic belt is complex. In contrast, the deformation in the neighboring thrust belt, although still severe, is fairly simple to unravel. Thrust belts taper from thicknesses on the order of 10,000 to 20,000 meters next to the metamorphic core to less than 5,000 meters at the opposite side, defining a wedge of faulted sedimentary strata that is inclined gently toward the core. Sediments closest to the core typically have been thrust farther than 100 kilometers.

In most thrust belts, deep-water oceanic rocks are thrust above shallow-water marine strata. The thrusts originate where the deep-water rocks were deposited and move toward the region of shallow marine sedimentation. The area toward which the thrusts move is termed the foreland and is away from the core of the orogenic belt, which is called the hinterland. The thrust faults are not planar surfaces of uniform orientation but rather are composed of a series of flat faults that parallel layering in the sedimentary rocks and ramp faults that cut obliquely across layering to join flat faults at different depths. The rocks above a ramp fault move upward, whereas rocks above a flat fault move horizontally, relative to those below. The depth of the flat segments of a single thrust fault decreases toward the foreland, imparting a staircase shape to the thrust fault and forcing deep-water rocks to climb to the surface, away from the core of the orogenic belt. The sliding of rocks along a ramp produces folds of a characteristic shape, broad anticlines and narrow synclines, as the rocks bend to accommodate the motion.

After one thrust moves older rocks up and over adjacent younger sediments, a new thrust parallel to the old one forms in the younger strata on the foreland side of the old thrust. Parallelism of the two thrusts allows the thrust slice (the rock between the two thrusts) to be tens of kilometers long. The horizontal distance between the old and new thrusts, usually on the order of hundreds of meters, defines the width of the thrust slice and depends primarily on the thickness of the pile of rocks being faulted. The new thrust then carries the younger strata and the old thrust toward the foreland and above still younger sediments. This process, called piggy-backing, continues such that the thrust slices in the thrust belt are stacked like a series of cards. Because piggy-backing is relatively straightforward, deformation in thrust belts is well understood. The cumulative horizontal displacement resulting from piggy-backing can be as large as 200 kilometers.

Décollement

The many thrusts in a thrust belt root into one flat fault at a depth of 10 to 20 kilometers, called the detachment surface or the décollement. Each separate thrust begins as a ramp fault that joins the décollement to a flat fault closer to the surface. The style of deformation above the décollement differs from that below. Thrusting occurs only above the décollement; rocks below, however, resist faulting and deform by changing their shape. The depth of the décollement commonly is controlled by a discontinuity that marks a change from brittle material above to ductile (changes shape without breaking when subjected to stress) below.

An example of a material that can be either ductile or brittle, depending on its temperature, is candle wax or paraffin. When warm paraffin is squeezed, it simply changes shape without breaking. Cold paraffin, however, fractures under compression. The warm paraffin is ductile; the cold paraffin is brittle. Rocks behave similarly. Such a transition from brittle to ductile material exists at depth in the continental crust where sedimentary strata lie above metamorphic or plutonic rocks. The sedimentary strata are brittle and are termed "cover." Conversely, the metamorphic and plutonic rocks are ductile and are called "basement." Because faulting in thrust belts is restricted to the brittle "cover" rocks, the deformation is known as thin-skinned. Thin-skinned thrusting significantly shortens the horizontal dimensions and thickens the vertical dimensions of the upper 10 to 20 kilometers of the continental crust. The effects of thin-skinned thrusting on basement rocks are largely unknown.

Active Thrust Belts

Active thrust belts are localized along the edges of continents, which are also convergent plate boundaries. The shortening of the upper 10 kilometers of continental crust by thin-skinned thrusting provides a mechanism by which the horizontal compressive forces generated during the convergence of two plates can be accommodated. Essential to the formation of thrust belts are thick sequences of layered rock such as those that accumulate in geosynclines. If the rock being deformed is not thick enough, it will not transmit the compressive forces necessary for the development of thrust faults. Examples of active thrust belts include the foothills to the Himalaya in India and Nepal where the Asian and Indian plates are colliding, the Andes of South America where the Pacific and South American plates are converging, and the Transverse Ranges of California where the Pacific plate and North American plates are converging slightly.

Large basins filled with sediment border all these continental margins. The Appalachian Mountains of Tennessee and North Carolina are an ancient thrust belt. There, slices of sediment that were deposited in geosynclines next to North America were thrust westward 225 million years ago as the continents of Africa and North America collided. The Appalachian Mountains are part of an orogenic belt that is greater than 3,000 kilometers long and almost 500 kilometers wide. In contrast to the mountains in Tennessee and North Carolina, the northern part of the Appalachian Mountains in Virginia and Pennsylvania is a fold belt. The folds are several kilometers wide and are responsible for the valley and ridge topography characteristic of the northern Appalachian Mountains. Another ancient thrust belt is the Rocky Mountains of Idaho and Wyoming. It was there that piggy-backing was first documented.

Pamela Jansma

Further Reading

Cubas, N., et al. "Prediction of Thrusting Sequence Based on Maximum Rock Strength and Sandbox Validation." *Trabajos de Geologia* 29 (2009): 189-195. Development of procedures by which thrust faulting events can be identified, including experimentation with sandbox models and field research. Lots of diagrams are provided; the article is written in a manner that requires some scientific background.

Finkl, Charles W., ed. *The Encyclopedia of Field and General Geology*. New York: Van Nostrand Reinhold, 1988. This reference contains many useful articles about field and geophysical investigations and oil and natural gas exploration, including petroleum geology, petroleum exploration geochemistry, geological survey and mapping, and geophysical methods. The article on geophysical methods is much less technical than are the other two sources cited here. Well illustrated. Suitable for college-level readers.

Hatcher, Robert D., Jr. *Structural Geology: Principles, Concepts, and Problems*. 2d ed. Englewood Cliffs, N.J.: Prentice-Hall, 1995. This undergraduate textbook covers folds in three chapters. Intended for the more advanced reader.

Lacombe, Olivier, et al., eds. *Thrust Belts and Foreland Basins*. Berlin: Springer, 2007. A compilation of articles discussing geological processes, methods of study, modeling, and tectonic forces of thrust belts. Recommended for graduate students and professional geologists.

Nemcok, M., S. Schamel., and R. Gayer. *Thrust Belts; Structural Architecture, Thermal Regimes and Petroleum Systems.* New York: Cambridge University Press, 2005. Extremely detailed explanation of thrust belts, covering all aspects of mechanics, thermal regimes, and petroleum systems. The authors go into great depth when covering analysis techniques.

Plummer, Charles C., and Diane H. Carlson. *Physical Geology.* 13th ed. Boston: McGraw-Hill, 2009. This is a straightforward introduction to geology intended for those with little or no science background. Discusses the development of thrust belts. Includes many excellent illustrations.

Poblet, J., and R. J. Lisle. *Kinematic Evolution and Structural Styles of Fold-and-Thrust Belts,* Special Publication 349. London: Geological Society of London, 2011. Provides an overview of current research of fold-and-thrust structures and processes. It covers theoretical modeling, methodology, and experimentation.

Sharma, R. S. *Cratons and Fold Belts of India.* New York: Springer, 2009. This text begins with an overview of plate tectonics and mountain building. It discusses cratons, as well as young and old fold belts, and focuses the second half of the book on the fold belts of India. Written for the graduate student.

Short, N. M., and R. W. Blair. *Geomorphology from Space: A Global Overview of Regional Landforms.* NASA SP-486. Washington, D.C.: National Aeronautics and Space Administration, 1986. This book contains beautiful pictures taken by various satellites that orbit the earth. Text accompanies each picture and explains the tectonic setting. Many thrust belts are shown. Although the text is fairly technical, the photographs are worth examining by anyone. Recommended for college-level students.

Skinner, Brian F., et al. *The Dynamic Earth: An Introduction to Physical Geology.* 5th ed. New York: John Wiley & Sons, 2006. This book contains excellent descriptions of thrusts and thrust belts. Beautiful illustrations abound. The text is similar to *The Earth* by Press and Siever but has an approach that is geared more to the practical aspects of geology, such as mineral and petroleum exploration. Suitable for senior high school or college-level students.

See also: African Rift Valley System; Continental Crust; Displaced Terranes; Faults; Fold Belts; Folds; Geoclines; Joints; Mountain Belts; Ophiolites.

TRACE FOSSILS

Trace fossils or ichnofossils are the remains of the activity of an organism, recorded in the fossil record as disturbances in the sediment. Trace fossil formation generally requires that the movement of an organism occurs in an area containing fine-grained sediment in a semiaquatic medium.

PRINCIPAL TERMS

- **ichnofauna:** the collective remains of trace fossil records of animal activity
- **ichnofossil:** a trace fossil created by the disturbances or activity caused by living animals that leave impressions in the sediment of the earth
- **ichnology:** a branch of paleontology concerned with tracks, burrows, and other fossilized remains representing the activity of living animals
- **igneous rock:** a rock formed by the solidification of volcanic magma
- **paleoecology:** the study and use of fossils and other geological evidence to reconstruct ecosystems from the past
- **replacement:** a process in which the organic materials of an organism are replaced by mineral deposits after the animal's death, leading to the formation of a body fossil
- **sedimentary rock:** a rock formed from the consolidation of clay minerals
- **sedimentology:** a branch of science dealing with the formation, classification, and study of sedimentary rock
- **stratigraphy:** the study of geological strata, including the age, deposition, and composition of sedimentary rock
- **taphonomy:** the study of the ways in which organisms and traces of organisms become fossilized

Types of Fossils

Paleontologists define a "fossil" as the remains or impressions of an organism preserved for a minimum of ten thousand years. Fossils generally form in sedimentary rock, which is composed of hardened layers of clay and silt, deposited through millions of years. Fossils do not form in igneous rock, which results when magma from the earth's core cools at the surface or underwater, forming solid structures. Because magma is superheated at the time of its deposition, it usually destroys any potential fossils before cooling sufficiently to allow for fossil formation.

Paleontologists use the term "body fossil" to refer to the fossilized remains of all or part of an organism's body. In a process called replacement, the decay of the organic material making up an organism's body is coupled with the slow accumulation of minerals from surrounding sediment, forming a cast that can preserve the three-dimensional structure of the organism within. Most body fossils form from decay-resistant tissues, including bones, teeth, shells, and other "hard" parts of an organism. In some cases, body fossils can preserve the outlines of soft internal organs and other parts of the organism's anatomy.

Another class of fossils called trace fossils or ichnofossils comprises the fossilized remains of an organism's activity within its environment. Trace fossils include fossilized footprints, burrows, and other impressions left by the movement of living organisms in soft sediment. These traces of activity become preserved when the surrounding sediment is buried and then hardens over time, producing a stone copy of the original impression.

Fossilization is rare, and the vast majority of organisms will not leave any lasting fossilized trace of their existence. Despite more than 165 million years of dinosaurs roaming the earth, approximately 50 percent of dinosaur species are known from only a single specimen, while more than 80 percent are known from fewer than five specimens. A variety of other factors influence the likelihood that an organism will be fossilized, including its habitat, general lifestyle, position in the food chain, and whether or not the organism possessed hard, resistant tissues. In 3.4 billion years of life on Earth, hundreds of thousands of fossils have accumulated in the sediment, leaving paleontologists with a detailed record of life on the planet.

Formation of Trace Fossils

Fossilization is much more likely to occur in marine environments, in part because minerals mix with water in aqueous environments; this liquid solution is better able to infiltrate a potential fossil site. Many trace fossils result from animals that can burrow into solid or partially solid materials. An organism that burrows through rock is far more likely to leave a trace fossil than is an animal that lives in soft substrate, because softer substrate is more likely to deform before hardening to preserve a trace fossil. For this reason, the burrows of marine animals form a large portion of the existing trace fossil record.

Trace fossils form best in fine-grained sediment, such as sandstone and limestone. Sandstone is composed of both crystallized mineral fragments and tiny fragments of rock eroded from larger stones. Before it hardens, sandstone begins as fine silt that gathers in moist or aquatic environments. Semisolid sandstone is the best material for the preservation of trace fossils from terrestrial animals. The material is both soft enough that a moving organism can leave a distinct trace and firm enough to maintain the shape of the impression after the organism has moved away.

After the depression is formed, heat from the sun may begin to harden the silt in the fossil site. If left exposed to the elements, the fossil erodes and deforms, so it is important that the entire fossil site be covered by water and additional silt before erosion damages the depression. In many cases, trace fossils result from situations in which the initial "mold" is later filled by a different type of silt (such as mudstone or limestone), which, when solidified, will have different qualities from the stone making up the mold.

Through thousands of years, the entire fossil site is buried under layers of additional sediment. Later, as the environment changes, the terrain surrounding the fossil site erodes, exposing the hardened sediment containing the trace fossil. The stone filling the fossil depression erodes at a different rate from that containing the depression, leaving the fossil exposed.

Trace fossils are often preserved in environments where body fossils are rare because the chemical and physical processes that destroy organic remains do not have the same effect on trace fossils. For instance, potential body fossils are often damaged by a process called diagenesis, which involves chemical and physical changes within a sample of sediment after deposition. Diagenesis does not damage trace fossils and may actually enhance trace fossil formation by increasing differences between the sediment of the mold and that of the material filling the mold. Therefore, areas like deep marine shales and shallow aquatic environments, which resist the formation of body fossils, often contain a variety of trace fossils. In some cases, trace fossils are the only record of life in chemically or physically active environments where body fossil formation is hindered.

Classification of Trace Fossils

In most cases, it is difficult or impossible for paleontologists to determine the species responsible for a specific trace fossil. Trace fossils left by burrowing marine mollusks and worms, for instance, are often indistinguishable from one another and may have been created either by the same species or by two species of similar size, shape, and habits. To organize trace fossils, paleontologists have created a set of ethological classes, based on the type of activity that formed the fossil. There are five or six major groups of trace fossils, which are further divided into ethological subcategories.

Cubichinia. Cubichinia (resting traces) usually take the form of shallow depressions and are thought to result from animals temporarily resting in the substrate. In exceptionally well-preserved samples, cubichinia may also indicate the morphology of the resting organism, making it possible, occasionally, for paleontologists to determine the species responsible for the resting trace fossil.

Pascichnia and Fodinichnia. Pascichnia (grazing traces) and fodinichnia (feeding traces) are left by organisms feeding on smaller organisms within the substrate of the marine environment. Pascichnia take the form of shallow grooves and corresponding burrows formed by an animal that feeds by rubbing its jaws across the surface of the sediment, in a characteristic pattern, and grazing off small organisms in the soil. Fodinichnia appear as cylindrical depressions in the sediment, left by organisms known as deposit feeders that feed on the detritus that accumulates at the bottom of the ocean.

Fugichnia. Fugichnia (escape traces) generally appear as burrows in the sediment, occasionally with angles indicating rapid lateral movement. Paleontologists believe that fugichnia result when an organism has been rapidly covered by sediment and thereafter moves rapidly upward in an effort to escape the sediment. Fugichnia generally involve upward movement, whereas most trace fossils result from organisms burrowing deeper into the sediment.

Domichnia. Domichnia (dwelling traces) show up as burrows or small chambers excavated by ancient marine animals seeking temporary or permanent shelter. In some cases, domichnia may be surrounded by nodules resulting from repeated digging against the sides of the chamber. In other cases, domichnia may include a branching system of interconnected tunnels created when the domicile created by one animal is used again and again by other animals in repeated excavations.

Repichnia. Repichnia (crawling traces) are traces left by the movement of an animal across the surface of the sediment. These include the most famous type of trace fossil, the dinosaur track-ways, or imprints of dinosaur footprints, found in Mesozoic excavation sites around the world. Other types of repichnia include depressions made by slithering or crawling animals observed in marine environments. Repichnia are especially interesting in that they provide information about the locomotive patterns of the animals making the tracks.

Another method for classifying trace fossils is to group them into assemblages called ichnofacies, which contain multiple fossil varieties from a single area, representing both different species and different ethological categories. Ichnofacies provide paleoecologists with a snapshot of ancient marine communities, showing how different species divided a given territory according to their varying needs.

Ichnofacies are divided according to whether the environment was terrestrial, marine, or freshwater and according to the type of sediment represented in the sample. For example, terrestrial and freshwater habitats with firm sediment are part of the Scoyenia ichnofacies, within which paleontologists find tracks and trails of dinosaurs and other terrestrial or semi-aquatic animals that lived in ancient environments. Scoyenia ichnofacies also contain remnants of plant life from ancient environments, sometimes in the form of fossilized seeds and pollen grains, which can be used to further specify the type of climate that existed when the ichnofacie began to form.

Marine ecosystems are represented by a variety of ichnofacies divided according to the type of sediment and the depth of the environment in which they were formed. The Nereites ichnofacies, for instance, contain feeding burrows, locomotor pathways, and a variety of other trace fossils from soft-sediment, deep-water marine environments. The Cruziana ichnofacies, by contrast, represent assemblages of organisms living in mid-depth marine environments where the soil bears evidence of occasional environmental disturbances. The Skolithos ichnofacies are characterized by a variety of feeding and burrowing structures left by organisms living in intertidal marine environments marked by frequent, cyclical changes in water depth.

Importance to Paleoecology

Paleoecology is the study of the interrelationships between extinct animals and their environments. Trace fossils are important to the study of paleoecology because they provide information not given by other types of fossils. Trace fossils provide clues to the types of organisms that lived in environments where body fossils are rare. In addition, trace fossils can be created by organisms lacking the stiff anatomical structures most conducive to body fossil formation. Trace fossils may therefore provide the only evidence of species that would otherwise be absent from the fossil record.

In addition, body fossils do not always provide evidence about the types of organisms that lived in specific environments because the bones and other body parts that become fossils can move from one location to another after the death of the animal and before the fossilization process begins. By contrast, trace fossils are necessarily tied to the environment in which they were created. For this reason, when paleontologists find a specific kind of trace fossil in a certain environment, they can reason that the organism responsible for the fossil appeared in that exact area at the time of deposition.

Micah L. Issitt

Further Reading

Briggs, Derek E. G., and Peter R. Crowther, eds. *Paleobiology II.* New York: Wiley-Blackwell, 2001. Introduction to the study of paleobiology aimed at students of related disciplines. Contains a discussion of trace fossils, body fossils, and the fossilization process and of methods used to evaluate biological activity from fossil remains.

Bromville, Richard Granville. *Trace Fossils: Biology, Taphonomy, and Applications.* New York: Psychology Press, 1996. Detailed overview of the processes used to find, examine, and evaluate trace fossils. Includes a description of methods used to identify species and evaluate behavior from fossil remnants and a thorough discussion of physical processes involved in fossil deposition.

Foote, Michael Joseph, et al., eds. *Principles of Paleontology.* New York: Macmillan, 2007. Revised edition of a classic text on paleontology giving a basic introduction to field and laboratory methods used in the study of fossils and prehistoric life. Contains a discussion of trace fossils and body fossils and of the methods used to find and classify fossil remains.

Fortey, Richard. *Fossils: The History of Life.* New York: Sterling, 2009. Introduction to fossil research including modern examples of paleontology and geological research in both laboratory and field settings. Contains a discussion of trace fossils and their significance to paleontology.

Martin, Anthony J. *Introduction to the Study of Dinosaurs.* New York: Wiley-Blackwell, 2006. Covers many aspects of paleontological research. Includes discussions of fossilization processes and of important trace fossil discoveries as related to theories of dinosaur behavior.

Miller, William Charles. *Trace Fossils: Concepts, Problems, Prospects.* St. Louis, Mo.: Elsevier, 2007. Updated introduction to the history and modern methods used in trace fossil research. Provides a detailed discussion of the problems encountered in using trace fossils to investigate biodiversity and behavior.

Nichols, Gary. *Sedimentology and Stratigraphy.* New York: John Wiley & Sons, 2009. Introductory text covering various aspects of earth science research, theory, and practice. Chapter 11 contains information on the study of trace fossils in the context of modern geological research.

See also: Biogeography; Carbonate Rocks; Clastic Rocks; Dinosaurs; Fish-Amphibian Transition; Fluvial Processes; Fossil Fish; Fossil Reptiles and Amphibians; Geomicrobiology; Life's Origins; Minerals; Molecular Paleontology; Paleoecology; Pangaea; Permo-Triassic Boundary; Radiometric Dating.

TRANSGRESSION AND REGRESSION

Transgression and regression are two of the most common phenomena of the geologic record. The changes in sea level that they represent have a major impact on the interpretation of earth history and in the reconstruction of ancient environments. Understanding these phenomena in the geologic past is essential if present sea-level changes are to be dealt with effectively.

PRINCIPAL TERMS

- **continental shelf:** the margin of the continents, usually covered by shallow seas
- **eustacy:** any change in global sea level resulting from a change in the absolute volume of available sea water
- **ocean basins:** the large worldwide depressions that form the ultimate reservoir for the earth's water supply
- **regression:** the retreat or withdrawal of the sea from land areas and the evidence of such a withdrawal
- **transgression:** the extension of the sea over land areas and the evidence of such an advance

GLOBAL PROCESSES

Transgression and regression are two phenomena by which changes in sea level are recognized. The two are not processes as such but rather are the effects of changes in sea level. Transgression is any rise of the sea over a land area (including the continental shelf) or any change in the physical marine conditions involving a progression into a deeper water environment. Regression is a withdrawal of the sea from any land area or any change resulting in a progression from deep water to shallow water or nonmarine environments. The concepts of transgression and regression are themselves independent of a causal mechanism. They are the results of the processes that can cause sea-level change.

Transgression and regression are very diverse phenomena and are the result of many different processes. Some of these processes involve an actual change in global sea level, or eustacy. Because transgression and regression are effects, however, they can be produced by a local process rather than a global one. It is important to understand the various processes involved in transgression and regression by examining first the global processes, then the local ones.

The global, or eustatic, processes resulting in transgression and regression are those that cause a change in the volume of seawater on the continents. Such global changes are accomplished by two major processes: first, an actual change in the volume of the global seawater supply and, second, a change in the volume or holding capacity of the ocean basins. A change in the amount of seawater would not seem an easy task to accomplish. The earth's environment is essentially a closed system; although small amounts of water are added from volcanism and by human activity, it is not enough to account for wide fluctuations in global sea level. Even if enough new water could be added to the system to account for a sea-level rise, the problem of getting rid of large amounts of water to account for a sea-level fall remains. Rather than add or subtract new water, a redistribution of existing water at the earth's surface seems a more likely way to change the volume of seawater. One way that may be accomplished is by the growth and melting of the earth's continental glaciers.

Glacial ice is different from sea ice. Sea ice forms in high latitudes over open ocean. As such, there is a constant connection with the oceans and a seasonal interchange as the ice melts in the summer and freezes in the winter. Glacial ice, however, forms from accumulated snow on the continents. Once a glacier grows, the snow that forms glacial ice is no longer available to the oceans. It is locked on the continents for the duration of the glacial episode. As glaciers grow, more water is locked within them. As glaciers melt, the meltwater is available to the oceans. The effect on sea level can be dramatic. During a glacial advance (an "ice age"), sea level will fall. During a glacial melt, sea level will rise. This sea-level mechanism has characterized the geologic record of sea-level change over the past 20 million years and perhaps longer. It has also operated during various other times in earth history as long as there was continental-scale glaciation.

Another process by which transgression and regression can occur is by a change in the geometry of the ocean basins. The ocean basins are the ultimate reservoir of standing water on earth. The depths in these basins are great, and it would seem that a change in the volume would not have a great effect on sea level. Yet if the volume of the ocean basins decreases, for example, the excess water must go somewhere. When that happens, a flooding of the continents will occur. Once the volume of the ocean basins increases, the water will withdraw from the continents. The actual mechanism in the ocean basins is related to heat flow from the earth's interior at the mid-ocean ridges. As the ridges become more volcanically active, an increase in heat flow occurs; the high heat flow causes an increase in the elevation of the mid-ocean ridge. This increase in elevation causes a decrease in the volume of the ocean basins, resulting in a transgression. As the volcanic activity decreases, heat flow decreases, and the elevation of the mid-ocean ridge drops so that the volume of the ocean basins is increased. This results in a regression. Sea-level changes brought about by this process produce large-scale fluctuations resulting in the flooding not only of the continental shelves and the coastal plains but also of the interiors of the continents. Much of the geologic record contains rocks deposited in seas in continental interiors. This process may result in transgressions where ancient sea level could be as much as 300 meters above present sea level. This process is not the dominant process at the present. The last sea-level transgression apparently related to this process was during the early Paleocene epoch (about 66 to 58 million years ago). The sea reached from the Gulf of Mexico up the Mississippi Valley into present-day southern Illinois and reached as far west as the Dakotas.

LOCAL PROCESSES

While the global changes in sea level would seemingly result in uniform transgression or regression throughout the world, that is not apparent when the sea-level record at any given place is studied. The reason for that is the effect of local processes on sea level. In fact, many transgressions and regressions are local phenomena and may not be recognizable in other regions.

Two local processes that control sea level are sedimentation and tectonics. By operating in coastal regions, these two processes can mask the global trend of sea level. Sedimentation is the process whereby sediment is deposited by water or wind. Sedimentation can affect local sea level if the accumulation rate is greater than is the global sea-level rise. As the sediment builds up in the coastal regions, it displaces the ocean, thus "filling in" the former sea floor. As an area that was once a part of the sea is now land, a regression has occurred. This process is presently operating in the modern environment. A very good example of this phenomenon is the modern Mississippi Delta in southern Louisiana. In this area, sedimentation rates are very high. Although global sea level is rising as the remaining continental glaciers melt, new land is being created around the Mississippi Delta. At this location, sediment is accumulating faster than sea level is rising. Thus, a regression at the delta is taking place. Away from the delta, sea level is still rising, and more land each year in southern Louisiana is being flooded by marine waters. Therefore, even though the Mississippi Delta is a site of high sedimentation rates and locally affects sea-level change, that is not enough to alter the global process.

Tectonism is another local process by which the pattern of sea-level change can be controlled. Although tectonics is a global process, the effects of this process can be very localized, and some include phenomena not related to global tectonics at all. For sea-level change, the processes of interest are uplift and subsidence. Uplift and subsidence in a coastal region can be related to many causes, but it is the result rather than the process that is important to transgression and regression. If the process of uplift is more rapid than is global sea-level rise, a regression will be apparent. Similarly, if subsidence is greater than is global sea-level fall, a transgression will be the interpretation. These two examples illustrate the extremes of the tectonic effects, but many variations exist. In the modern environment, global sea level is rising, but the apparent change in any specific area is controlled by the tectonic setting. An example is the Baltic Sea coast in Sweden, where uplift resulting from glacial rebound is not reversing the trend of global sea level but is slowing the apparent rate of sea-level transgression. Tectonic factors are one of the most difficult effects to take into consideration when examining transgression and regression because of the geographic variability and the nature of the tectonic processes themselves.

Transgression and regression are seemingly simple phenomena: Sea level is either rising or falling. Yet the contribution to the observed pattern by local processes is significant and, in some cases, is the dominant effect observed. In order to gain meaningful insight into the global pattern of sea-level change from local data, a thorough understanding of the geologic history of the geographic region being studied is essential.

Transgression and regression appear to be very dramatic and perhaps even catastrophic phenomena when their effects are considered. Nevertheless, both must be kept in the proper time context. Transgression and regression are not unique phenomena in the geologic record. Sea level has been in constant fluctuation throughout most of geologic time, and there is very little evidence that there were times of global sea-level stasis. Transgressions and regressions generally occur over long periods of time and, while some sea-level changes such as glacially induced ones occur over a relatively short span of geologic time (several hundred thousand years), many occur over millions of years. Thus, the observed effect of sea-level change may appear minor unless viewed over long intervals of geologic time.

Richard H. Fluegeman, Jr.

Further Reading

Brookfield, Michael E. *Principles of Stratigraphy.* Hoboken, N.J.: Wiley-Blackwell, 2004. Written for undergraduate students, this text provides an overview of the principles and applications of stratigraphy. Includes stratigraphic techniques and case studies. It is organized into three sections, beginning with foundational material, followed by data collection and research topics, and completed with interpretations and analysis.

Coe, Angela L., ed. *The Sedimentary Record of Sea-Level Change.* New York: Cambridge University Press, 2003. The text discusses changes in the sea-level throughout time. Discusses the factors influencing the sea-level including ice ages and sedimentation. Includes multiple case studies, chapter summaries, references, and an index. This text is suited for undergraduates and graduate students.

Cooper, John D., Richard H. Miller, and Jacqueline Patterson. *A Trip Through Time: Principles of Historical Geology.* 2d ed. New York: McGraw-Hill, 1990. This book is designed for first-year college students but is written in such a way that senior high school students should be able to read it with little difficulty. Contains a specific section on sea-level change with good illustrations of concepts. Summaries of sea-level change through time are also throughout the book. A glossary is included.

Hallam, A. *Phanerozoic Sea-Level Changes.* New York: Columbia University Press, 1992. Deals with the succession of alternating transgressions and regressions of the sea over the continents through time. Discusses the causes of these events and their relationship to other events, such as mass extinctions. Bibliography and index.

Lemon, Roy R. *Principles of Stratigraphy.* Columbus, Ohio: Merrill, 1990. This college-level textbook deals with all aspects of stratigraphy, including sea-level changes and transgressions and regressions. Glossary and index.

Levin, H. L. *The Earth Through Time.* 9th ed. Hoboken, N.J.: Wiley, 2009. This book is an excellent reference for the general reader. Although designed as a first-year college text, it has a writing style such that readers from high school on will find it a valuable reference. A well-illustrated section on sea-level change as well as summaries of transgression and regression through time are presented.

Prothero, Donald R. and Robert H. Dott. *Evolution of the Earth.* 8th ed. New York: McGraw-Hill, 2009. This basic textbook on historical geology is aimed at students of geology. However, it is very readable by anyone with a background in science. Presents an up-to-date account of the earth's history from the viewpoint of plate tectonics. Includes a glossary

Stanley, Steven M. *Earth System History.* 2d ed. New York: W. H. Freeman, 2004. A comprehensive overview of the development of the earth. Explains the mechanism of transgression and regression and its effect on the depositional history of the earth. Illustrations, glossary, bibliography, and CD-ROM. Readable by high school students.

Yanko-Hombach, Valentina, et al., eds. *The Black Sea Flood Question: Changes in Coastline, Climate, and Human Settlement.* Dordrecht: Springer, 2007. This text is a compilation of articles discussing aspects on both sides of the Noah's Flood hypothesis. The editors' intention is to provide information without bias to allow readers to draw their own conclusions. These articles were previously only available in Cyrillic.

See also: Biostratigraphy; Cenozoic Era; Cretaceous-Paleogene Boundary; Fossil Record; Geoclines; Geologic Time Scale; Ice Ages; Land Bridges; Mesozoic Era; Paleoclimatology; Paleogeography and Paleocurrents; Paleoseismology; Paleosols; Paleozoic Era; Proterozoic Eon; Stratigraphic Correlation; Unconformities; Uniformitarianism.

TRANSVERSE RANGES

The Transverse Ranges are a prominent east-west-trending mountain chain that forms a spectacular backdrop behind the most populated region in California. The range is unique in several ways: The east-west orientation departs markedly from the usual north-south trend of other geologic structural blocks in California. There is a greater variety of unusual rocks and structures in the range than elsewhere in the state and the range has important mineral and energy resources.

PRINCIPAL TERMS

- **basement:** an underlying complex generally of igneous and metamorphic rocks
- **reverse fault:** a steeply to moderately inclined fault in which the overlying block of rock has moved upward over the underlying block
- **right-slip:** sideways motion along a steep fault in which the block across the fault appears displaced to the right; left-slip faults are the opposite
- **tectonics:** a branch of geology that deals with the study of regional large-scale structural or deformational features and their origins, mutual relations, and evolution
- **thrust fault:** a variety of a reverse fault in which the inclination of the fault plane is at a low angle to horizontal

LOCATION AND CHARACTERISTICS

The Transverse Ranges are a narrow mountainous belt in California that extends from Point Arguello, 80 kilometers west of Santa Barbara, eastward for more than 450 kilometers to the Eagle Mountains in the Mojave Desert; they are from 16 to 80 kilometers wide. The Transverse Ranges are a distinctive, if not unique, geologic province of California because of an east-west orientation (which is rare in North America and unique in California), a greater range of rock types and geologic structures than any other region in California, the oldest plutonic and metamorphic basement rocks in California, and the largest oil fields in the state outside the San Joaquin Valley and Los Angeles County.

Initial observations of the overall alignment of the Transverse Ranges might suggest that the province constitutes a single, apparently homogeneous east-west-trending mountainous belt. That is not the case, however, as closer examination reveals that the Transverse Ranges comprise three unrelated segments, each with a widely differing geologic history and origin, that have been brought into alignment coincidentally by the large-scale tectonic regime that is peculiar to Southern California. On the basis of geologic differences, the three segments of the Transverse Ranges may be subdivided into the eastern segment, which includes the San Bernardino, Little San Bernardino, Hexie, Pinto, and Eagle Mountains; the central segment, the major portion of which is the San Gabriel Mountains but which also includes the lesser Verdugo, Sierra Pelona, Liebre, and Sawmill Mountains; and the western segment, which consists of the Santa Susana, Santa Monica, Topatopa, and Santa Ynez Mountains and the northern four Channel Islands.

The mountains of the eastern segment of the Transverse Ranges are bounded largely by faults, of which the San Andreas fault forms the southern margin. The eastern segment is about 250 kilometers long and includes the highest elevation in Southern California at San Gorgonio Mountain, 3,505 meters above sea level, in the San Bernardino Mountains. Rocks underlying the eastern segment are predominantly Mesozoic granitic rocks, gneiss and schist of probable Precambrian age, and lesser amounts of fossil-bearing Late Paleozoic metamorphosed marine sedimentary rocks. Most of the summit region of the San Bernardino Mountains is a plateaulike old erosion surface of moderate relief that developed prior to uplift. This surface now stands at elevations of about 1,950 to 2,290 meters. The preservation of the old erosion surface is explained by the comparatively recent and rapid uplift of the range, which is thought to have occurred within only the last 2 million years. A second elevated plateaulike old surface exists in the Little San Bernardino Mountains. This surface has been eroded into the picturesque, bold rock formations for which Joshua Tree National Monument is famous.

The central segment of the Transverse Ranges is dominated by the San Gabriel Mountains, a high, rugged upland about 100 kilometers long and as much as 38 kilometers wide. The range is bordered on the north and east by the San Andreas fault zone and by reverse faults on the south and southwest. The range consists mainly of a crystalline basement complex of intrusive igneous rocks and a variety of metamorphic rocks. Within the rock complex are two generations of early Precambrian gneisses (1,045 million and 1,700 million years old, respectively), which are the oldest rocks in California. Summit elevations of the peaks in the range vary from about 1,950 to 2,450 meters, the highest being Mount San Antonio (Mount Baldy) at 3,072 meters.

There are many faults within the central segment of the Transverse Ranges, some of which have right-slip movement of several tens of kilometers. The range has been uplifted essentially as a unit by broad arching across its northern margins and central part and by reverse faulting along its southern margin. Most of the uplift has occurred within the last 15 to 20 million years. Because uplifting of the ranges of the central segment began long before that of the eastern segment, erosion has had greater time to wear away any uplifted old erosion surfaces. Consequently, no erosion surfaces exist in the San Gabriel Mountains, in contrast to those present in the more recently elevated ranges of the eastern segment. A thick sequence of Cenozoic fossiliferous nonmarine sedimentary rocks and some volcanic rocks occurs around the base of the central segment. These rocks have been eroded into the scenic landforms at Devil's Punchbowl County Park near Pearblossom, Vasquez Rocks County Park on State Highway 14, and at Cajon Pass.

Unlike the segments of the Transverse Ranges to the east, the mountains of the western segment are composed almost entirely of marine sedimentary rocks of late Mesozoic and Cenozoic age. Locally within the sedimentary rocks are extensive volcanic rocks, many of which were produced by submarine eruptions. The ranges of the western segment have a combined length of about 190 kilometers. Summit elevations of the mountains are modest, reaching their highest elevation of 2,050 meters at Hines Peak in the Santa Ynez Range.

Part of the western Transverse Ranges is underlain by a large syncline about 190 kilometers long, named the Soledad Basin in the east and the Ventura Basin in the west, where it is contiguous with the Santa Barbara Channel farther to the west. The Ventura Basin, filled with primarily marine sedimentary rocks, merges eastward with the nonmarine sediments of Soledad Basin. The Santa Susana Mountains, South Mountain, Oak Ridge, Ventura Anticline, and Sulphur Mountain are prominent anticlinal hills that have formed in the last 200,000 to 300,000 years within the Ventura Basin. The Ventura Basin is geologically famous for its immense combined thickness of late Mesozoic and Cenozoic marine sedimentary rocks totaling more than 17,600 meters, for the recency of its deformation, and for its great petroleum production.

Study of Transverse Ranges

Geologists use their full armamentarium of devices and techniques to study the geology of the Transverse Ranges. The most important technique has been field mapping, the process of on-site research on foot, by car, or in an airplane in order to locate and plot the various rock units, faults, and folds on aerial photographs and topographic base maps. Rock and fossil samples are collected and labeled in the field. These samples are taken into the laboratory for detailed examination and identification and compared with samples from other areas of the same or similar geology that have been studied previously.

The San Gabriel Mountains, in Southern California, constitute the central segment of the three geologically distinct sections of the Transverse Ranges. (AP Photo)

Paleontologists, experts in the study of fossils, may be consulted to assist in fossil identification. Correlation (the determination of age relationships between rock units or geologic events in separate areas) of rocks and fossils in exposures on opposite sides of faults allows the determination of the amount and direction of movement along the faults, assuming that rocks with identical properties and fossils were once a single body.

Added to the work of field geologists is subsurface information obtained from oil wells: drilling logs, electrical resistivity and self-potential logs, gamma ray-neutron logs, cores, and side-wall samples. This subsurface information is immensely useful to geoscientists, enabling them to "see" the geology underground. Other techniques to discover what is underground are seismic reflection and seismic refraction profiling. Seismic reflection profiling is based on the travel time of sound waves excited by precisely timed "thumping" of the ground that are reflected off buried rock layers or fault surfaces. Seismic refraction profiling is based on the travel time of sound waves that are generated by a small controlled explosion. Data generated by these techniques are plotted on graphs, from which the depths and orientations of the various layers and structures in the subsurface geology may be interpreted.

Supplementary subsurface information important in the study of the Transverse Ranges is obtained by a geophysical technique called gravimetry. By using gravimeters, sensitive instruments that can precisely measure very slight differences in the earth's gravitational field, it is possible to deduce much about the physical characteristics of the rocks underlying the surface. For example, gravimetry reveals that the San Gabriel Mountains lack "roots," lower-density rocks such as granite, which usually exist beneath mountain ranges, where they extend into the mantle for tens of kilometers. Such underlying roots provide the buoyant lifting force that supports mountain ranges as elevated masses above the average elevation of the earth's surface. Thus, most of the world's mountains literally float on the mantle buoyed by their roots much as the hidden keel of an iceberg supports its visible portion. An explanation of how the high-standing San Gabriel Mountains can exist without such roots comes from another geoscientific discipline, geodetic surveying.

Geodetic surveying is a branch of civil engineering that requires consideration of the earth's curvature. Instrumental techniques used in geodetic surveying incorporate laser optical devices, pulsed infrared lightwave and microwave electronic surveying equipment. These instrumental techniques produce extremely accurate measurements of distances and angles to a high degree of precision. Precision surveys conducted since the 1970's reveal strains in the earth's crust across the San Gabriel Mountains that cause the range to be compressed in a north-south direction. Thus, the San Gabriels are supported essentially by being squeezed upward by crustal compression, a process that is much different from that supporting most of the world's mountain ranges.

Radioactive age dating, which determines absolute ages of rocks, is another technique used in the study of the Transverse Ranges. Knowing the rate of decay of radioactive elements allows geochronologists (geoscientists who specialize in radioisotopic age dating of rocks) to determine the ages of rocks in years. The technique is an involved laboratory procedure requiring the use of a mass spectrometer and similar specialized equipment. Radioactive age dating is especially useful in determining the ages of igneous and metamorphic rocks.

Environmental Impacts and Resources

It is important that scientists and others understand the relation between rock types, slope angles, and watershed management in the Transverse Ranges in order to protect the lowland communities from flooding during the episodic winters of high precipitation. A second concern with the Transverse Ranges is their role as a watershed that provides runoff water used by the residents of the foothill communities that border those ranges.

The mineral wealth of the eastern and central segments of the Transverse Ranges is of economic importance, with numerous mines and mining prospects scattered throughout the province. Mineral deposits in the central segment of the Transverse Ranges have been exploited for almost 150 years. The first gold discovery in California was made in the San Gabriel Mountains, at Placerita Canyon, in 1842. Exploration during the fifty years following the discovery produced important gold discoveries in several areas of both the San Gabriel and San

Bernardino Mountains and in some of the lesser ranges of the eastern segment. Iron ore was mined for many years in the Eagle Mountains. Silver, copper, lead, tungsten, and numerous nonmetallic minerals also have been sought and mined from time to time in the central and eastern segments. Limestone deposits in Lower Cushenbury Canyon on the north side of the San Gabriel Mountains are the basis for the current mining activity in the Lucerne Valley limestone district and the existence of the largest cement plant in the world.

Probably even more significant is the great petroleum production from the western segment of the Transverse Ranges. The marine sedimentary rocks making up the western segment in concert with the faulting, folding, and rapid uplift of the rocks during the last 1 million years not only have created the impressive modern landscape but also have formed the geologic structures that are responsible for the numerous oil fields of the region. Of historic importance is that the first petroleum production in California, in 1857, came from oil seeps near Ventura and that the first oil well drilled in California and put into serious commercial production was completed near Ojai in 1866. This early well is located in what is now known as the Silverthread area of the Ojai field. Even as late as the 1980's, the production from the Silverthread area was still economically feasible, yielding upwards of 1 million barrels of oil per year.

D. D. Trent

Further Reading

Ehlig, P. L. "Origin and Tectonic History of the Basement Terrane of the San Gabriel Mountains, Central Transverse Ranges." In *The Geotectonic Development of California, Rubey Vol. 1*, edited by W. G. Ernst. Englewood Cliffs, N.J.: Prentice-Hall, 1981. This article is probably the most authoritative source of information on the basement rocks of the San Gabriel Mountains. The text is suitable for college-level readers who are not intimidated by technical language.

Fife, D. L., and J. A. Minch, eds. *Geology and Mineral Wealth of the California Transverse Ranges.* Santa Ana, Calif.: South Coast Geological Society, 1982. Probably the most all-inclusive work on the geology, mineral, and energy resources of the Transverse Ranges. Especially relevant are eight articles by Thomas W. Dibblee, Jr., on the regional geology of virtually the entire Transverse Ranges province. Although written for professional geologists, the maps and diagrams in Dibblee's articles are essential to an understanding of the geology of the province.

Gamache, Tom, and Matthew Jaffe. *The Santa Monica Mountains: Range on the Edge.* Los Angeles: Angel City Press, 2006. Although this publication is considered a "coffee-table book," there are nuggets of information and excellent photographs contained within its pages. As one of the many Transverse Mountain Ranges, the Santa Monica Mountains are unique due to their location within a major city.

Hatcher, Robert D., Jr. *Structural Geology: Principles, Concepts, and Problems.* 2d ed. Englewood Cliffs, N.J.: Prentice-Hall, 1995. This undergraduate textbook provides an overview of the structural development of the San Andreas fault system and the Transverse Ranges. Intended for the more advanced reader.

Grotzinger, John, and Tom Jordan. *Understanding Earth.* 6th ed. New York: W. H. Freeman, 2009. This comprehensive physical geology text covers the formation and development of the earth. Contains extensive sections on the deformation of the continental crust and the development of the Transverse Ranges in relation to the San Andreas fault system. Readable by high school students, as well as by general readers. Includes an index and a glossary of terms.

Plummer, Charles C., Diane H. Carlson, and Lisa Hammersley. *Physical Geology.* 13th ed. New York: McGraw-Hill Science, 2009. A first-semester college-level introductory geology textbook that is clearly written and wonderfully illustrated. An excellent sourcebook of basic information on geologic terminology and the fundamentals of geologic processes. The chapters on sedimentary rocks, geologic time, structural geology, and plate tectonics are especially relevant to an understanding of geologic processes involved in the formation of the Transverse Ranges.

Salcedo, Chris. *San Bernardino Mountain Trails.* Berkeley: Wilderness Press, 2003. A series of maps developed for the hiker, but featuring locations of human-made and natural structures of the San Bernardino and San Jacinto area.

San Gabriel Mountains. 5th ed. Berkeley, Calif.: Wilderness Press, 2005. A series of maps developed for the hiker. Features locations of human-made and natural structures of the Angeles National Forest, Cucamonga, San Gabriel, and Sheep Mountain Wildernesses.

Stanley, Stephen M. *Earth System History*. 2d ed. New York: W. H. Freeman, 2004. A comprehensive overview of the development of the earth. A chapter is devoted to the development of the Transverse Ranges. Illustrations, glossary, bibliography, and CD-ROM. Readable by high school students.

Trent, D. D. "Geology of Joshua Tree National Monument, Riverside and San Bernardino Counties." *California Geology* 37 (April 1984): 75-86. An article intended for the interested layperson. One of the few "popular" sources of information on the origin of the unusual scenery for which the monument is famous.

__________. "Mount Baldy Mining Area, Los Angeles and San Bernardino Counties, California." *California Geology* 39 (August 1986): 179-184. A review of the mining history and the geology of the higher-elevation regions on Mount San Antonio, the highest peak in the San Gabriel Mountains. Discusses the role of the Vincent thrust fault in localizing deposits of gold in the central part of the range. Intended for the general reader.

See also: Alps; Andes; Appalachians; Basin and Range Province; Carbonate Rocks; Cascades; Himalaya; Minerals; Mountain Belts; Rocky Mountains; Sierra Nevada.

U

UNCONFORMITIES

Unconformities are surfaces of erosion, nondeposition, or both that have, through renewed sedimentation, been covered by younger rock. Where they occur, they represent significant portions of geologic time not recorded by rock and, as such, serve to divide the geologic history of a region into chapters of more or less continuous deposition separated by depositional breaks.

PRINCIPAL TERMS

- **angular unconformity:** an unconformity in which the beds below are at an angle to the beds above
- **diastem:** a short depositional break of geologically unmeasurable duration caused by a limited period of nondeposition
- **disconformity:** an unconformity separating parallel strata and exhibiting physical evidence of erosion
- **intrusive contact:** contact resulting from injection of molten igneous rock into previously formed rock
- **nonconformity:** an unconformity in which rock below is homogeneous igneous or metamorphic rock and rock above is layered
- **paraconformity:** an unconformity for which the only evidence is a gap in the fossil record
- **regression:** a shift of sedimentary facies away from land, commonly caused by a drop in sea level
- **sedimentary facies:** different types of sediment that collected simultaneously in adjacent areas of the seafloor
- **transgression:** a shift of sedimentary facies toward land, commonly caused by a rise in sea level

Rock Record Boundaries

The units that compose the rock record of a region are separated by two fundamental types of boundary: conformable and unconformable. Rock layers are said to be conformable if deposition of the layer above followed the one below without an appreciable lapse of time. For example, periodic delivery, as during storms, of similar sediment to a depositional site will produce a sequence of homogeneous layers separated from one another by simple bedding planes that represent the short, geologically unmeasurable periods between storms. Such layers are considered conformable. Rocks of contrasting types can also exhibit a conformable relationship. In present oceans, an appreciable area of the sea bottom exhibits different types of sediment collecting simultaneously at adjacent sites. The type of sediment that accumulates at a particular site depends on the type of grains being supplied and the wave and current energy operating there. Adjacent environments of relatively quiet water, such as a lagoon, and agitated water, such as a beach, will collect fine and coarse sediment, respectively. The fine and coarse sediments are said to represent sedimentary facies (*facies* is the Latin word for the general appearance or aspect of something). Should the agitated conditions spread so as to come to operate in the region formerly occupied by quiet-water conditions,

A small mesa near San Lorenzo Arroyo in Arizona shows an unconformity overlying upthrust rock. (U.S. Geological Survey)

then the coarser grains will be delivered to the site in which previously only finer grains accumulated. At the same time, because of the increase in energy, the finer grains would no longer be allowed to settle where they formerly did but would be moved on to another site. This change in energy within the area would result in deposition of a layer of coarse material over fine material with essentially no break in sediment accumulation between the two layers.

Most conformable contacts between unlike rock layers in the sedimentary record are interpreted as resulting from such shifts in sedimentary facies through time. These shifts often accompany a rise or fall in sea level. A shift of sedimentary facies toward land during a rise in sea level is termed transgression, and one in which facies are displaced away from land during a lowering of sea level is termed regression. A conformable sequence of strata will generally exhibit an interlayering of grain types or a gradation of grain types between layers.

The rock record exhibits many examples of widespread marine regression that, for a period of time, exposed previously formed rock to subaerial weathering and erosion. In these examples, subsequent transgression of the sea initiates renewed sedimentation, covering the erosional surface with younger layers. Because the time involved allows for broad changes in the distribution of geographic elements and life-forms, these new sediments characteristically differ in grain type and fossil content from those below. The rocks of the younger series above the erosional surface are said to exhibit an unconformable relationship to those below, and the buried erosional surface separating the two series is termed an unconformity. The erosional surface, which is to say the unconformity, records an interval of geologic time not represented by the rock succession of the region.

Unconformities divide the history of a region into conformable groups of rock that are separated from one another by depositional breaks. They are the divisions between the major chapters of geologic history. The conformable units between a pair of unconformities compose a related group of sedimentary facies that have shifted through time and become intimately interbedded with one another. Because earth history is understood and narrated in terms of the broad divisions outlined by unconformities, the identification and analysis of depositional breaks constitute one of the most important pursuits undertaken by geologists who specialize in the study of the stratigraphic record.

Development of Unconformity Studies

Although unconformable relationships were observed and sketched by naturalists in the sixteenth and early seventeenth centuries, the credit for understanding their significance to earth history is generally given to James Hutton, the Scottish geologist whose basic observations are often considered as the beginning of modern geological investigations. As early as 1786, Hutton observed stratal relations in which the overlying series of beds rested upon the tilted, eroded edges of the underlying series. In the succeeding two years, Hutton found other examples of angular relations between strata in northern Scotland. He interpreted the sequence of events necessary to develop such a relationship as beginning with marine deposition of the underlying beds, followed by their lithification, deformation (tilting), uplift, and exposure to subaerial erosion, and concluding with submergence by the oceans and deposition of the overlying series of horizontal beds. Hutton was particularly impressed by the time he perceived necessary to produce the angular relationship. In fact, such observations strengthened his belief that the earth was of great antiquity, a discovery generally accredited to Hutton by science historians.

Hutton did not apply the term "unconformity" to the stratal relationship he elucidated. S. I. Tomkeieff credits the origin of the term to a translation, in the first decade of the nineteenth century, of the German phrase *abweichende Lagerung* ("deviating bedding") into the English word "unconformity," which carried the same meaning as the ecclesiastical term "nonconformity" (dissent from the established church). Throughout the nineteenth century, the term "unconformity" was applied exclusively to the angular relation described by Hutton, although some, particularly Charles Darwin (in order to explain gaps in the record of the evolution of life), argued that the earth's strata were replete with subtler, yet equally significant, depositional breaks that lacked discordance.

The first half of the twentieth century was a period of describing criteria for the recognition of unconformities, identifying the types of stratal relations represented by unconformities, and proposing terminology to be applied to the different types of relations. Building on a century of study, Eliot Blackwelder in

1909 clearly outlined the three basic types of relations found at erosional surfaces: angular discordance between beds above and below the erosional surface (the "unconformity" of Hutton), parallel strata above and below the erosional surface, and layered rocks resting on an erosional surface developed on homogeneous metamorphic or plutonic igneous rock, such as granite, which lacks bedding.

Following Blackwelder, a variety of terms were suggested to describe the different types of unconformable relations, with the term "unconformity" itself most commonly retained to designate Hutton's original angular relationship. For American geologists, the question of terminology essentially was settled by Carl Dunbar and John Rodgers in their popular advanced geology textbook *Principles of Stratigraphy* (1957). In a review of the subject, these authors accepted the term "unconformity" in a generic sense to refer to all types of significant depositional breaks in the geologic record. They identified four types of structural relations that previous geologists had associated with these breaks. For three of these, they recommended retaining previously proposed terms: angular unconformity, for the discordant relation of Hutton; nonconformity, for situations where layered rock rests on eroded, unlayered igneous or metamorphic rock; and disconformity, for an erosional surface of "appreciable relief" separating parallel strata. They introduced the term "paraconformity" for the fourth structural type, which had been considered by earlier geologists as a special kind of disconformity.

Unconformities

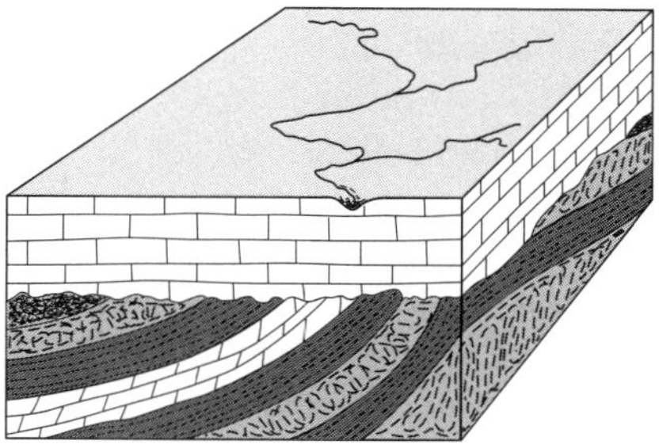

angular unconformity

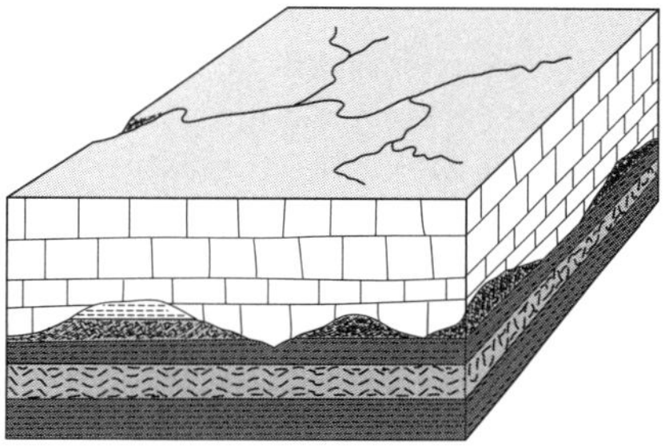

disconformity

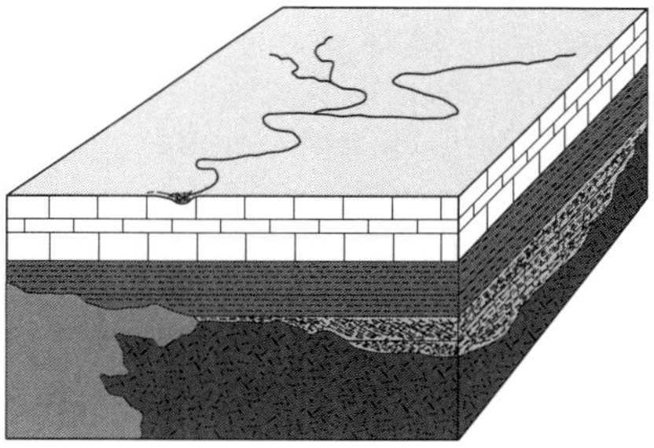

nonconformity

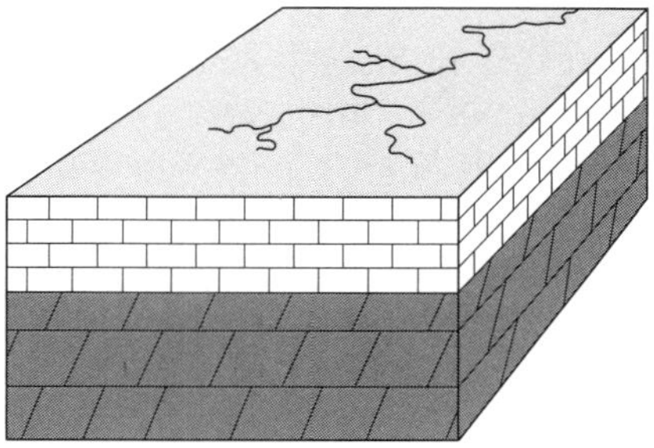

paraconformity

IDENTIFYING UNCONFORMITIES

Disconformities and paraconformities are similar inasmuch as both are interpreted as erosional surfaces separating parallel strata. They differ in that a disconformity exhibits physical evidence of erosion, preferably an undulatory surface that truncates beds and features in the rock below, whereas a paraconformity is manifested as a simple bedding plane essentially lacking relief. The paraconformity's interpretation as a surface that represents a significant portion of geologic time not recorded by rock is based on an abrupt change in fossil types, commonly accompanied by a sharp change in rock composition, across the boundary. Because abrupt changes in fossil content and rock composition can be explained by natural processes not involving loss of record, paraconformities are the most subjective of unconformities.

It is for this reason that Dunbar and Rodgers recommended that these apparent breaks in sequence, which otherwise exhibit no evidence of erosion, be separately identified from disconformities.

Angular unconformities are the most vivid and unequivocal of depositional breaks. The special significance of angular unconformities is that they record crustal movement prior to the development of the unconformity. Care must be taken not to confuse angular beds resulting from crustal movement with beds naturally deposited at an angle, the so-termed cross beds characteristic of subaqueous and subaerial sand dunes.

Nonconformities are also reasonably simple to identify, but they need to be distinguished from intrusive contacts in which igneous rock has been emplaced as a molten body (magma) beneath layered rock. Although both nonconformable and intrusive contacts produce similar results—that is, layered rock above unlayered plutonic rock—the two differ fundamentally in their history. The former qualifies as an unconformity because the older igneous rock was exposed to weathering and erosion at the earth's surface for some time prior to the deposition of the layered rock. In the intrusive relationship, which is not an unconformity, the younger igneous rock was intruded into previously formed layered rock. Geologists separate the two by examining the evidence at the contact. Indications of baking of the sedimentary rock would suggest an intrusive relation, whereas a weathered rind along the margin of the igneous rock would point toward a nonconformable contact.

Evidence for unconformities within parallel strata is more subjective than that for angular unconformities and nonconformities. The criteria developed fall into three broad categories: sedimentological, structural, and paleontologic. Sedimentological criteria revolve around phenomena that develop when a land surface that has undergone weathering and erosion for a period of time is transgressed by the seas. As an example, portions of the soil horizon developed on the layers below the unconformity can be preserved, generally significantly modified in appearance. If the underlying rock contains quartz sand or chert nodules, both relatively insoluble by the weak organic acids found in nature, these grains will, in the weathering process, become scattered through the soil. The energy of the advancing sea will remove the finer materials of the soil, and the coarser insoluble grains will become concentrated in the base of the unit overlying the erosional surface, generally in sharp contact with the underlying unit.

Structural criteria involve demonstration of truncation (erosion) of features in the rock underlying the postulated erosional surface. Commonly, that involves, in the base of the overlying unit, channels cutting through whole beds in the underlying sequence, demonstrating the presence of topography—in some examples with relief of several hundred feet—on the erosional surface. On a smaller scale, fossil shells and other relatively coarse grains and structures can be cut off at the erosional surface.

Paleontologic evidence for an unconformity is constituted by a sharp change in fossil species, commonly associated with missing forms known to be present in the established sequence of fossils in nearby regions. The thought is that if sedimentation had been continuous, a complete paleontologic sequence would be present. Paleontologic evidence is almost always the only indication of the time value of an unconformity. If it is the only evidence for an unconformity, then the postulated erosional surface is most properly called a paraconformity.

Evaluation of Disconformities and Paraconformities

Because most evidence for unconformities in parallel strata can be explained also by natural processes not involving loss of record, some postulated disconformities and many paraconformities are the subject of debate among geologists. For example, the presence of a sandy layer with a sharp basal contact could be the result of a rapid shift in depositional facies rather than grains concentrated through weathering of the rocks below. A stream can break out of its channel and cut a new channel in the adjacent floodplain. Truncation of the bedding of the floodplain sediments would be pronounced. Yet the contact between the channel deposits and those of the floodplain into which the channel scoured would not be considered unconformable because the age difference between the two would at most be only a few years, a geologically insignificant value.

Paraconformable surfaces are the most difficult to evaluate. Those involving large gaps in the fossil record, expressible in terms of units of the geologic time scale, generally exhibit some sedimentological or structural evidence when examined closely over a large geographic extent. They are nevertheless enigmatic inasmuch as the indication is that large areas lay exposed to subaerial erosion for considerable lengths of time without the development of distinct criteria for weathering and removal of rock. As the fossil gap represented at paraconformities decreases, interpretation becomes more contentious. The problem is that fossils were once life-forms, all of which are controlled, at least to some extent, by environmental conditions. Absence of a few fossil forms at a boundary could result from the absence of the environments to which the forms, when living, were adapted, rather than from loss of record through erosion. Norman Newell suggests that some paraconformities might develop during a prolonged standstill of sea level on a shallow shelf adjacent to a low-lying land area. Under such conditions, sedimentation on the shelf would eventually raise the sediment surface into water so shallow that the available wave energy would not allow more sediment to accumulate but instead would move it across the shelf to deeper water. If such conditions of nonaccumulation persisted long enough, ultimately surfaces that represented a significant extent of geologic time, but which exhibited no physical evidence of erosion, could develop. These surfaces would qualify as paraconformities.

It is apparent that no one criterion serves as unequivocal proof of an unconformity in parallel strata. In attempting to establish the position of an unconformity in undisturbed beds, geologists pursue several lines of sedimentological, structural, and paleontologic evidence. Geologists have become more and more aware that the stratigraphic record contains a spectrum of sedimentological breaks ranging from a few days of nondeposition to several hundred million years involving uplift, subaerial exposure, and erosion. In general, they agree that the term "unconformity" should apply only to those surfaces that involve a "significant" amount of time, although just how much time qualifies as significant is undefined. It is apparent that if an area is undergoing erosion, the sediment produced must be deposited somewhere. Therefore, all unconformities diminish laterally and eventually pass into conformable surfaces. In practice, geologists apply the term "unconformity" to those surfaces where the missing record can be expressed in terms of the regional stratigraphic column or in units of the international geologic time scale. Sedimentological breaks too small to be expressed in local or international terms are commonly referred to as diastems.

William W. Craig

Further Reading

Bennison, George M., Paul A. Olver, and A. Keith Moseley. *An Introduction to Geological Structures and Maps.* 8th ed. London: Hodder Education, 2011. This text covers stratigraphy, sedimentation, folding, and faulting. Chapter 5 discusses unconformities. Written in a manner accessible to undergraduate students with little geology background.

Blackwelder, Eliot. "The Valuation of Unconformities." *Journal of Geology* 17 (1909): 289. The earliest comprehensive discussion of unconformities, well written and still a valuable introduction to the subject. Suitable for college and advanced high school students. The journal should be available in most university and comprehensive technical libraries.

Brookfield, Michael E. *Principles of Stratigraphy.* Hoboken, N.J.: Wiley-Blackwell, 2004. Written for undergraduate students, this text provides an overview of the principles and applications of stratigraphy. Includes stratigraphic techniques and case studies. It is organized into three sections, beginning with foundational material, followed by data collection and research topics, and completed with interpretations and analysis.

Lemon, Roy R. *Principles of Stratigraphy.* Columbus, Ohio: Merrill, 1990. This college-level textbook deals with all aspects of stratigraphy, including unconformities. Glossary and index.

Lisle, Richard J. *Geological Structures and Maps.* 3rd ed. Burlington, Mass.: Butterworth-Heineman, 2003. Provides a concise introduction to geological structures. Chapters cover folding, faulting, unconformity, and geological maps.

Ogg, James G., Gabi Ogg, and Felix M. Gradstein. *The Concise Geologic Time Scale.* New York: Cambridge University Press, 2008. This book is a complete overview of the geological time scale, including stratigraphy topics such as chronostratigraphy and magnetic stratigraphy. The book is organized by geological time periods. Also contains a reference appendix, the geological time scale table, and indexing.

Prothero, Donald R. *Bringing Fossils to Life.* 2d ed. Boston: McGraw-Hill, 2004. This well-illustrated and entertaining text covers a broad range of paleontological topics, including the development of sedimentary rocks and their use in stratigraphy. Glossary, bibliography, and index.

Stanley, Steven M. *Earth System History.* 2d ed. New York: W. H. Freeman, 2004. A comprehensive overview of the development of the earth. Explains the development of unconformities and their use in stratigraphy. Illustrations, glossary, bibliography, and CD-ROM. Readable by high school students.

Tomkeieff, S. I. "Unconformity: An Historical Study." *Proceedings of the Geologists' Association* 73 (1962): 383. The most comprehensive examination available on the development of ideas about unconformities. Well written and illustrated. Easy to follow. Suitable for advanced high school students. Periodical available in most comprehensive university libraries.

Wicander, Reed, and J. S. Monroe. *Historical Geology.* 6th ed. Belmont, Calif.: Brooks/Cole, Cengage Learning, 2010. A beginning college-level textbook with a clear, succinct discussion of unconformities and facies (pp. 47-51). Well illustrated. Suitable for advanced high school students.

See also: Biostratigraphy; Carbonate Rocks; Cenozoic Era; Cretaceous-Paleogene Boundary; Fluvial Processes; Geologic Time Scale; Ice Ages; Land Bridges; Mesozoic Era; Paleoclimatology; Paleogeography and Paleocurrents; Paleoseismology; Paleosols; Paleozoic Era; Proterozoic Eon; Stratigraphic Correlation; Transgression and Regression; Uniformitarianism.

UNIFORMITARIANISM

Uniformitarianism, in its most basic form, describes a methodology used in the study of earth history. It assumes that the earth can be interpreted in terms of natural processes that are operational today. In its broader context, uniformitarianism also embraces certain conclusions about earth history: that the rates at which these processes operate are constant through time and that the earth has experienced a cyclic history.

PRINCIPAL TERMS

- **catastrophists:** adherents to the belief that the present-day slow rates of natural processes are not sufficient to explain many features of the rock record
- **diluvial theory:** the belief that the Mosaic flood was responsible for shaping many of the earth's surface features
- **doctrine of final causes:** the belief that a purpose or design is revealed in the organization of nature
- **methodological uniformitarianism:** that aspect of uniformitarianism that describes a procedural approach to the study of the earth—that is, that ancient phenomena can be interpreted in terms of present-day natural processes
- **Neptunists:** adherents to Abraham Werner's belief that granite and basalt formed by chemical precipitation from seawater
- **Plutonists:** adherents to James Hutton's belief that granite and basalt formed by crystallization from molten mineral matter
- **rock cycle:** the path of earth materials as erosional products of rock are deposited and reformed again into rock, which then can be eroded again
- **substantive uniformitarianism:** that aspect of uniformitarianism that draws conclusions about the history of the earth—that is, that natural processes have been taking place at constant rates through time and that the earth has had a cyclic history
- **surficial gravels:** alluvium found at different places on the earth's surface, originally interpreted as the product of the Mosaic flood
- **uniformity of nature:** Charles Lyell's formulation describing a nature in which processes are both consistent and constant through space and time; a mixture of methodological and substantive uniformitarianism

Origin of Uniformitarianism

Uniformitarianism owes its origin and development to three British geologists: James Hutton, who laid the groundwork for the concept; John Playfair, who elucidated and elaborated on Hutton's ideas; and Charles Lyell, who championed the uniformitarian cause in the mid-nineteenth century.

At the end of the eighteenth century, there was fairly close agreement between theories of the earth based on direct, though limited, observations of earth material and structure and the account of earth history revealed in the biblical book of Genesis. The earth was thought to be of no great antiquity, considered by most to be not much older than six thousand years. The surface features of the earth, its mountains with their tilted rock, its river valleys, and its ubiquitous surficial gravels, were attributed to a great debacle, assumed to be the biblical flood of Noah. These features existed in equilibrium with the elements, possibly waiting for the next great debacle. Some erosion was happening, but not enough to alter the topography greatly in the short amount of time available. It was against this intellectual backdrop that Scottish scientist James Hutton presented to the Royal Society of Edinburgh a paper entitled *System of the Earth* (1785), which contained the seeds of a controversy that continued for fifty years. The paper was published in the first volume of the Society's Transactions under the title *Theory of the Earth: Or, An Investigation of the Composition, Dissolution, and Restoration of the Land Upon the Globe* (1788).

Hutton's work essentially outlined the rock cycle. He recognized, as others had before him, that the fossiliferous strata of the continents had formed on the bottom of the sea as unconsolidated sediment. In order to be added onto the continent, in places at angles of inclination too steep to represent the original repose of water-charged sediment, a mechanism was needed to fuse the sediment into rock and uplift the sea bottom. In Hutton's scheme, the energy required to do that was derived from the internal heat of the earth. Once exposed to the elements, the rocks were eroded to form sediment again. In Hutton's view, the present land was derived from a former continent, and the disintegration and erosion of rock observed at its surface were providing material out of which a future continent would be constructed. Hutton emphasized that these processes that shaped the earth were natural and operated very slowly, hardly causing noticeable change during the history of humankind. He eschewed calling upon preternatural causes in explaining natural phenomena.

Hutton's 1788 paper contains the major elements of what was later termed uniformitarianism. The earth was to be interpreted in terms of present-day natural processes, excluding supernatural explanations. The processes operated very slowly and probably at fairly constant rates, although Hutton was not clear on this latter point. The major emphasis on constant rates was added by Charles Lyell almost fifty years later. Lastly, the earth had gone through a cyclic development, or, as Hutton phrased it, a "succession of worlds," with the destruction of one world providing the material for the next.

Hutton's view of cyclic "worlds" destroyed and reconstructed by slow processes led him to conclude that the earth is very old, in his phraseology, an earth with "no vestige of a beginning" and, in reference to the cyclicity, "no prospect of an end." Hutton's "discovery" of an earth of inscrutable age is considered his major contribution to geology. Hutton is also credited as the first to recognize the true origin of those rocks now known as igneous. His belief that these bodies were injected as molten masses into stratified rock molded his attitude toward the efficacy of heat in driving his cycles.

Neptunism and Plutonism

The prevailing view of these rocks at the time was that of the German mineralogist and natural philosopher Abraham Gottlob Werner, whose science of geognosy maintained that basalt and granite had been chemically precipitated from an ocean that from time to time had flooded the earth. Hutton's adherents became known as Plutonists and Werner's as Neptunists. A few years after publication of Hutton's 1788 paper, his ideas were acrimoniously attacked by the Neptunists. Although the argument ostensibly was over the origin of basalt, a major undercurrent centered on Hutton's timeless earth, a story that was not in line with that of the Scriptures. An ailing Hutton attempted to answer these attacks in 1795 with a wordy expansion of his ideas that is notorious for its unreadability. He died in 1797.

Fortunately, Hutton's good friend and sometime field companion John Playfair engaged the battle with the Neptunists. Playfair's prose was as facile as Hutton's was cumbersome, and it is from Playfair's *Illustrations of the Huttonian System of the Earth* (1802) that comes most of posterity's knowledge of Hutton's thoughts. In the period following Playfair's elucidation, most geologists came to accept the igneous origin of granite and basalt and to adjust their thinking in line with the unavoidable conclusion that the earth is certainly much older than the six thousand years afforded it by a literal reading of Genesis.

The most important discovery in the first two decades of the nineteenth century was that the earth's strata record a progression of life-forms, with primitive types in lower layers and more advanced types above. Foremost among researchers of the fossil record was the French comparative anatomist Georges Cuvier, who in 1812 published a monograph that was to serve as a major impetus for renewed controversy surrounding uniformitarianism. Cuvier, one of the first naturalists to reconstruct fossil vertebrates, was recognized as the premier scientist of his day. His early work had been on fossil vertebrates, the mammoth among many others, found in the surficial gravels. Because he felt that the extinction of this fauna could not be explained by processes presently acting on the surface of the earth, he concluded that their demise resulted from a sudden but prolonged and localized inundation of the land by the sea brought about by unknown natural causes. He referred to the event as a revolution. Later research in the rocks around Paris caused him to expand his idea of earth history to include several revolutions.

With his theory of periodic revolutions, Cuvier traditionally has been considered in English-speaking countries as a proponent of an earth shaped through divinely instituted catastrophes, which in turn were followed by special creations to replace the extinguished life. (Martin J. S. Rudwick, in an excellent account entitled *The Meaning of Fossils*, points out the injustice of this characterization. Cuvier was a strict empiricist who believed that science and religion should be pursued separately.) Cuvier's research was immediately seized by William Buckland, England's most prestigious geologist, who took the latest of the revolutions and transformed it from its original form, a localized inundation of long duration, into a catastrophic, short-lived universal deluge—in short, the flood of Noah. Buckland's flood theory, the so-called diluvial theory, gained much popularity in England in the 1920's but was essentially abandoned toward the end of the decade as it became clear that the surficial gravels, the supposed deposits of the deluge, did not exhibit the distribution or character expected of a worldwide catastrophic event.

Catastrophism

Still, a contingent of prominent British Diluvialists believed that the observations on which Cuvier's revolutions were based could not be explained by any of the processes in operation on the earth. In response to the threat posed to the uniformity of nature theory by the position of these individuals, who became known as catastrophists, Charles Lyell published his landmark book *Principles of Geology* (1830-1833), a work built upon the basic tenets of Hutton's theory of the earth. The ensuing debate between the catastrophists and Uniformitarians, as Lyell and his followers became known, was a conflict in viewpoint among scientists, not a struggle between science and the church, as is commonly assumed. In many important respects, the two sides were in agreement. The catastrophists willingly accepted the idea of an ancient earth and were devoted in most situations to the methodology of uniformitarianism. They were firmly opposed, however, to the interpretive aspects that the doctrine imposed on earth history: namely, that the slow rates of present-day processes are representative of all time (steady rate) and that earth history is cyclic, with one cycle looking much like another (steady state).

On the surface, the catastrophists' argument against steady-rate uniformitarianism seemed to be a simple request for accelerated rates to explain phenomena such as those on which Cuvier's revolutions were based. It is clear from their writings, however, that the catastrophists were anxious to interpret these episodes as times during which natural law was suspended by the regulatory hand of God. Their view of God was as both creator and referee, willing to intervene in the course of his creation if it veered off track. To them, any movement to remove the punitive hand of God from the course of things would lead to moral decay and the disintegration of British society. Lyell, in contrast, was unwilling to concede any portion of earth history, no matter how small, to a supernatural cause that fell outside the pale of scientific inquiry. As succinctly analyzed by Charles Coulston Gillispie in his revealing account of the times entitled *Genesis and Geology* (1951), the catastrophists were apprehensive that without any cataclysms there would be no God, whereas Lyell was concerned that without the uniformity of nature there would be no science. Lyell's opinion eventually prevailed, and supernatural explanations, for at least the physical world, were expunged from the methodology of historical interpretation. As a companion to this development, the efficacy of gradual processes in explaining all historical phenomena became firmly established in geologic dogma, but not without some unfortunate side effects.

To the catastrophists, Lyell's stubborn support of steady-state uniformitarianism not only seemed palpably false in the light of a fossil record that showed progression but also was contrary to physical laws that required the earth to evolve as its energy dissipated. In the light of today's knowledge, Lyell's arguments against the evidence of the fossil record seem foolish. He went so far as to predict that ichthyosaurs—marine reptiles seemingly extinct for more than 65 million years—would return to populate the seas when their "world" once again rolled around. Stephen Jay Gould, in his short, provocative book, *Time's Arrow, Time's Cycle* (1987), argues that differing views of history occupied a central position in the conflict between the two sides. Lyell's repetitious earth (time's cycle) was essentially ahistorical because it developed no unique configurations through time. Opposed to this was the catastrophists' vision of an ever-changing earth (time's arrow), unique at each stage and thus amenable to sophisticated historical analysis. Lyell abandoned his stand on steady-state uniformitarianism shortly after Charles Darwin's convincing documentation that life on earth had evolved.

William W. Craig

Further Reading

Ager, Derek V. *The Nature of the Stratigraphical Record.* 3rd ed. Hoboken, N.J.: John Wiley & Sons, 1993. A short, readable book emphasizing the incompleteness of the stratigraphic record (record of layered rocks) and the prominent role of natural cataclysms in earth history. Well illustrated and suitable for college-level students.

Bowler, Peter J. *Evolution: The History of an Idea.* 3rd ed. Berkeley: University of California Press, 2003. Bowler covers many topics in his survey of evolution's history, including Catastrophism, Darwinism, Lamarckism, Natural Selection, and Uniformitarianism. The writing provides clear descriptions of topics and puts them in historical context. The text was developed for the undergraduate student.

Doyle, Peter, and Matthew R. Bennett. *The Key to Earth History: An Introduction to Stratigraphy.* New York: John Wiley & Sons, 2001. This text begins with an explanation of the basic principles of stratigraphy followed by the explanation of uniformitarianism and actualism, geological time, and interpretation of the stratigraphic record.

Geikie, Archibald. *The Founders of Geology.* Reprint. Mineola, N.Y.: Dover, 1962. A long book, providing complete coverage on ideas about the earth and its development extending back to the Greeks. Good discussions of Hutton, Playfair, and Lyell, and debates between Uniformitarians and opposing groups, are provided by a geologist not far removed from the times. Well written and understandable at the high school level.

Gould, Stephen J. *Time's Arrow, Time's Cycle: Myth and Metaphor in the Discovery of Geological Time.* Cambridge, Mass.: Harvard University Press, 1987. Treating what he calls the "Hutton myth," one of the country's leading geological scholars reassesses the contributions to geology of Hutton and Lyell and refocuses the debate between the catastrophists and Uniformitarians. This short, cleverly written, provocative book, which contrasts sharply with most previously written analyses, should be read only after an introduction to the subject from other sources. It is suitable for college-level students.

Hutton, James. "Theory of the Earth: Or, An Investigation of the Laws Observable in the Composition, Dissolution, and Resolution of the Land Upon the Globe." Vol. 1 in *Royal Society of Edinburgh Transactions.* Reprint, with a foreword by George W. White. Darien, Conn.: Hafner, 1970. This article began the debate over what became known as uniformitarianism. This short, readable account of Hutton on his own ideas is understandable at the high school level.

Lemon, Roy R. *Principles of Stratigraphy.* Columbus, Ohio: Merrill, 1990. This college-level textbook deals with all aspects of stratigraphy. Chapter 2 discusses "the new uniformitarianism" and its relation to modern views on catastrophism. Glossary and index.

Lyell, Charles. *Principles of Geology, Being an Attempt to Explain the Former Changes of the Earth's Surface by Reference to Causes Now in Operation.* 3 vols. London: John Murray, 1830-1833. A well-written argument in support of uniformitarianism, directed toward the catastrophists. Understandable at all levels and an excellent example of nineteenth-century scientific prose, the book went through several editions, one of which should be available in most university libraries.

Playfair, John. *Illustrations of the Huttonian Theory of the Earth.* Reprint, with an introduction by George W. White. Urbana: University of Illinois Press, 1956. Beautiful prose describes Hutton's theory in detail. The book provides an excellent insight into the knowledge and ideas of geology of the time and is understandable at all levels.

Prothero, Donald R., and Fred Schwab. *Sedimentary Geology.* 2d ed. New York: W. H. Freeman, 2003. This college-level textbook discusses the development of sedimentary rocks and their use in stratigraphy. Includes a chapter on uniformity and its application in lithostratigraphy. Bibliography and index.

Rudwick, Martin J. *The Meaning of Fossils: Episodes in the History of Paleontology.* 2d ed. Chicago: University of Chicago Press, 1985. A fine, well-written account by a leading scholar of science history of the development of the study of paleontology, this book is of moderate length and is easy reading. Contains a well-balanced analysis of the meaning of uniformitarianism and is suitable for advanced high school and college levels.

Stanley, Steven M. *Earth System History.* 2d ed. New York: W. H. Freeman, 2004. A comprehensive overview of the development of the earth. The first chapter deals with the principle of uniformity, and Chapter 6 discusses its importance in stratigraphy. Illustrations, glossary, bibliography, and CD-ROM. Readable by high school students.

Wicander, Reed, and James S. Monroe. *Historical Geology.* Belmont Calif.: Brooks/Cole, Cengage Learning, 2010. This textbook is accessible to university students with or without a background in geology. It covers Earth's geological systems and plate tectonics, physical and biological history, geological time, and evolution. There are many diagrams, images, and charts along with a CD-ROM to supplement the text.

See also: Biostratigraphy; Cenozoic Era; Cretaceous-Paleogene Boundary; Evolution of Life; Fossil Record; Geologic Time Scale; Ice Ages; Land Bridges; Mesozoic Era; Paleoclimatology; Paleogeography and Paleocurrents; Paleoseismology; Paleosols; Paleozoic Era; Proterozoic Eon; Stratigraphic Correlation; Transgression and Regression; Unconformities.

V

VOLCANIC HAZARDS

Hazards associated with volcanic activity include explosions that eject liquid magma and rock fragments ranging from dust-sized particles to blocks of solid rock, lava flows, glowing avalanches, ash flows, volcanic mudflows, and gases. Active volcanoes typically lie within relatively narrow zones in close proximity to tectonic plate margins, coinciding with principal earthquake zones. The type of volcanism, with associated hazards, differs from one kind of plate boundary to another.

PRINCIPAL TERMS

- **ash flow:** a density current composed of a highly heated mixture of volcanic gases and ash, which travels down the flanks of a volcano or along the ground surface
- **composite volcano:** a volcano built of alternating layers of lava and pyroclastic deposits, along with abundant dikes and sills
- **lahar:** a mudflow composed chiefly of volcanic debris on the flanks of a volcano
- **magma:** a body of molten rock, including any dissolved gases and suspended crystals
- **mudflow:** a general term for a flowing mass of predominantly fine-grained earth material that possesses a high degree of fluidity during movement
- **pyroclastic:** composed of clastic (rock-derived) material formed by volcanic explosion or aerial expulsion from a volcanic vent
- **shield volcano:** a volcano in the shape of a flattened, broad, and low dome built by flows of very fluid lava or ash flows
- **tephra:** a general term for all pyroclastic material and shreds of liquid magma formed by volcanic explosion or aerial expulsion from a volcanic vent

Magmas and Gases

Volcanoes are essentially vents in the earth's surface through which magma—a combination of molten rock, dissolved gases, and suspended crystals—and associated gases and ash erupt or are ejected from the interior onto the surface of the earth. The nature of the eruption is largely dependent on the proportion of gas to molten rock and how readily the gas escapes. Eruptions can occur only once or frequently, with the proportion of gas to molten rock varying widely. Hazards associated with volcanic activity commonly occur in the form of lava flows, glowing avalanches, ash flows, volcanic mudflows, tephra falls, glacier bursts, and volcanic gases.

Gases

Volcanic gases originate in the magmas that rise out of the earth to activate the volcano. Gases in magma are soluble at the high temperatures found deep in the earth, but as magma rises, the pressure exerted at depth decreases, and some of the gas is no longer in solution. The gas, which first appears as small and widely spaced bubbles, increases in proportion and begins to coalesce as the magma continues to rise. As the gas pressures then increase, an explosion is prevented by the confining pressure of the magma and, partially, by the viscosity of the magma.

When the gas pressures are high enough, however, the enclosing liquid bursts, gas escapes, and an explosion occurs. If the confining pressures are low and the enclosing magma is relatively fluid, gas bubbles escape readily, with rather mild eruption of magma at the surface. If the confining pressures are high, however, and continue to increase, with gas bubbles escaping with much difficulty, eruptions are very violent and of an explosive nature. If the magma comes in contact with water, in the form of seawater, surface water, or groundwater, violent steam explosions are possible, as documented at the Kilauea volcano on the island of Hawaii. Magma evolution, however, eventually causes an increase in viscosity and gas content, which leads to progressively shorter and thicker lava flows, formation of domes composed of many individual flows formed by the extrusion of highly fluid lava, and increasingly explosive eruptions.

Volcanoes can be broadly divided into two groups based on magma composition: basaltic volcanism and acidic volcanism. Basaltic volcanism is associated with the oceanic crust, commonly producing large quantities of lava with minimal explosive activity. Those active volcanoes situated on the Hawaiian Islands and Iceland's Kirkefell are examples of this type of volcanism. Acidic volcanism is associated with destructive plate margins and tends to be violent, with an unpredictable pattern of behavior and long periods between eruptions. The relatively high viscosity of the magma can induce a buildup of gas pressures, resulting in massive explosions such as those of Krakatau and, more recently, Mount St. Helens. This type of volcanism can produce flowing gas clouds or surges of pyroclastic material without any warning during periods of prolonged and less turbulent activity. Hot gases charged with varying proportions of red-hot ash can descend a mountainside at speeds greater than 100 kilometers per hour, bringing instant destruction as in the case of Pompeii by the eruption of Vesuvius in 79 C.E.

Location of Volcanoes

The most explosive volcanoes exist in zones where subduction of oceanic plates occurs. The volcano develops on the overlying plate within a few tens of kilometers from its edge. The eruptions of this type of volcano are characterized by tephra constituting from 45 to 99 percent of the volcanic product. Mount St. Helens and the other volcanoes that form the Cascade chain are examples of this type of volcanism. Associated hazards include heavy ashfalls, glowing avalanches, ash flows, mudflows, and lava flows.

Ejecta and Lava Flows

Explosive-type volcanoes result in the ejection of liquid magma or fragments, or a combination of both, into the air. These fragments, referred to as pyroclastic ejecta, or simply tephra, fall back to the ground or are carried away by the wind hundreds to thousands of kilometers distant. Sand-sized and smaller fragments are referred to as ash or dust; fragments measuring 2 to 60 millimeters in diameter are called lapilli; and fragments having diameters greater than 60 millimeters are called bombs or blocks, depending on whether the material was ejected in a liquid or in angular chunks of solid rock, respectively.

Glowing avalanches, nuées ardentes, are rapidly expanding clouds of dust that are typically black in daylight but a glowing dull red at night. The principal part of this phenomenon, however, is not the dust cloud itself but rather the avalanche of incandescent lava blocks, sand, and dust beneath it. Nuées ardentes can travel at great speeds, with documented cases exceeding 150 kilometers per hour. As a result of such great speed, when opposing hill slopes or the bend of a valley wall are encountered, the avalanche is capable of climbing vertically as much as several hundred meters. Ash flows closely resemble glowing avalanches in mobility and cause; however, the flow is composed primarily of an emulsion of bits of very hot glass, referred to as ash, and occasional lumps of pumice. Debris avalanches and landslides result from the sudden downslope movement of loose materials on the flank of a volcano.

Mudflows

Mudflows, which are also known as lahars, are slurries of solid fragments in cool to hot water, associated with explosive-type volcanoes where loose fragmented material is abundant. Flowing downhill under the influence of gravity, the flows are largely controlled by topography and can attain speeds of 90 kilometers per hour, with some documented at 160 kilometers per hour. Mudflows have the notoriety of having destroyed more property than any other single volcanic process. Also attributed to them is the loss of thousands of lives.

Glacier Bursts

Glacier bursts occur by volcanic eruptions generated beneath glaciers, as in Iceland. For periods of a few hours, these bursts may achieve volumes greater than that of the world's largest river, with documented cases exceeding velocities of 92,000 cubic meters per second and volumes of liquid exceeding 6 cubic kilometers.

Tsunamis

As an effect of volcanic activity, huge waves known as tsunamis may be generated by abrupt displacement of the seabed. Great volumes of water are involved, and the travel speeds of the waves can reach 800 kilometers per hour; a tsunami may arrive at a shore, after being slowed by the shallower water and increasing in amplitude, as a breaker up to 20 meters high. In 1896, a tsunami was responsible for the loss of 27,000 lives in Japan. The tsunami resulting from the explosion of Krakatau in 1883 caused nearly all the deaths from that disaster. Fortunately, tsunamis

Aerial photograph of houses and buildings in Plymouth, the former capital city of Montserrat Island in the Caribbean Sea. The city was destroyed by pyroclastic flows from the Soufriere Hills volcano in August 1997, and then partially buried under ash and mud. The city, with an estimated population of 3500, had already been evacuated. (Bernhard Edmaier/Photo Researchers, Inc.)

are relatively infrequent, occurring less than once every ten years; nearly all of them originate in the Pacific Ocean.

Evaluating Volcanic Risks

Evaluation of volcanic risks and hazards on or near any volcano is heavily dependent on an understanding of its historic behavior; the assumption is that a volcano is most likely to behave like it has behaved in the past. This evaluation is based on the recorded history of eruptions and geological studies of the composition and structure of the cone.

Attention to volcanic eruptions has been increasingly directed toward the assessment of risks to life and property. A problem arises, however, in defining when a volcano is actually dead or extinct and thus presents no further risk. For example, volcanoes can remain quiet or dormant for thousands of years and then, abruptly, erupt. In fact, in some of the most destructive volcanic events, no documentation of previous eruptions exists and the volcanoes were thus considered to be inactive; the eruption of Vesuvius in 79 C.E. is an example.

As far back as the beginning of the twentieth century, maps have been prepared identifying areas around known dangerous or active volcanoes, based on the historical evidence. Noted have been the eruptive characteristics of a particular volcano, especially the effect of topography on pathways of flows. Certain areas are typically coded to determine the types of risks to be expected given differing types of activity. For example, the primary risks associated with the Hawaiian volcanoes (Kilauea and Mauna Loa) are almost wholly from lava flows, whereas with the Cascade and Alaskan volcanoes, the primary risks are from ashfalls and mudflows. Thus, with the former, topography is considered principally, whereas with the latter, meteorological conditions are monitored.

Forecasting Eruptions

Earth scientists contribute to the reduction of volcanic hazards through an understanding of the nature and geographic distribution of these hazards. Attention presently is directed toward the assessment of risks to life and property. The solution is to minimize these risks for those people who live in close proximity to volcanoes. Thus, the practical goals of volcanology are to be in a position to provide ample warning of the timing, type, and location of future eruptions; to minimize the effects of eruptions and associated phenomena; and to rehabilitate devastated areas in an efficient and effective manner. Great loss of life was averted during the 1991 eruption of Mount Pinatubo in the Philippines by the action of volcanologists, whose warnings led to the evacuation of sixty thousand people prior to the main eruption.

With sufficient warning, people and property can be evacuated. Prediction of the timing of an eruption uses a combination of indicators. Unfortunately, no volcanic district as a whole is monitored sufficiently in order to avoid the loss of life and property completely. Monitoring the temperature of fumaroles, hot springs, and crater lakes may serve to warn of an eruption because, in some cases, a rise in temperature has precluded an eruption. Changes in gas discharge and composition also may forewarn of an imminent eruption. Additionally, an alteration of the strength or orientation of the earth's magnetic field may precede an eruption. This indicator is based on the hypothesis that increased temperature of a volcano should reflect a decrease in the overall strength of its local magnetic attraction.

Innovative methods of attempting to predict volcanic eruptions, which appear to show much promise, have included monitoring of deformation of the ground surface and earthquake behavior. Eruptions are commonly preceded by local earthquakes, which result from the opening of fissures at depth through which the magma rises toward the surface. Earthquakes are also produced by the movement of blocks in and around the volcano as it swells or, in the case of Kilauea, as an indication that the underlying magma reservoir is being filled and an eruption is possible. Volcanic tremors (rhythmic vibrations of the ground surface, reflecting the movement of magma) often occur just before or during eruptions, although tremors may also occur without an eruption. Changes in the earth's electrical currents within the volcano—where rapid changes in earth currents have occasionally been observed prior to eruptions—are monitored and inferred to precede a potential eruption. Finally, the disturbed and excited behavior of animals on or near the volcano previous to its eruption has been observed, suggesting that certain animals may sense earthquakes that are too small in magnitude to be felt by humans.

Volcanic Benefits, Too

Volcanoes are probably the most impressive as well as the most devastating of phenomena. Overall, however, volcanoes have done more good than harm by creating thousands of square kilometers of very fertile land surface. In addition, the use of volcanic heat in the form of natural steam and hot water continues to be developed to produce electricity and heat at a low cost and with minimum pollution to the environment. In fact, most episodes of volcanic unrest typically end without an eruption; eruptions are thus the exception rather than the rule.

Nevertheless, millions of people continue to live in close proximity to active volcanoes, risking the potential destruction all too often unpredictably associated with them. As developable land becomes scarcer, urban development progressively encroaches upon these areas. During the last five hundred years, more than 200,000 people have lost their lives as a result of, at minimum, five hundred active volcanoes—and that figure does not include the casualties resulting from tsunamis. Lives were lost not only directly, by the volcanic activity itself, but also by the destruction of food crops and livestock, which led to deaths from starvation.

Stephen M. Testa

Further Reading

Abbott, Patrick L. *Natural Disasters.* 7th ed. New York: McGraw-Hill, 2008. Focuses on how the normal processes of the earth create disasters at the human level. Two chapters deal with volcanic hazards. Written for undergraduates but accessible to interested readers.

Best, David M., and David B. Hacker. *Earth's Natural Hazards.* Dubuque, Iowa: Kendall Hunt, 2010. Written at the high school and undergraduate level but accessible to the interested reader. Chapter 3 covers volcanic hazards in detail. Extensive bibliography and very well illustrated.

Chester, Roy. *Furnace of Creation, Cradle of Destruction.* New York: AMACOM Books, 2008. This text discusses the turbulent processes of the planet. It covers earthquakes, volcanoes, and tsunamis in reference to plate tectonics, natural disasters, and prediction and mitigation of these events. Multiple chapters explore seafloor spreading. The author also discusses hydrothermal activity. This text takes on an immense range of content, but still explains concepts clearly and with detail.

de Boer, Jella Zeilinga, and Donald Theodore Sanders. *Volcanoes in Human History: The Far-Reaching Effects of Major Eruptions.* Princeton, N.J.: Princeton University Press, 2002. Provides an overview of volcanism followed by specific volcanoes and eruptions. Includes the interaction of human history and volcanic activity. Covers Hawaiian Islands, Crete, Vesuvius, Iceland, Tambora, Krakatau, Pelee, Tristan da Cunha, and Mount St. Helens.

Decker, Robert, and Barbara Decker. *Volcanoes.* 4th ed. New York: W. H. Freeman, 2005. This readable and well-illustrated text is geared toward the general public. The authors explain geologic phenomena in a clear and simple manner. Includes chapters on forecasting eruptions and reducing volcanic risk. The appendices list the world's 101 "most notorious" volcanoes, as well as Internet sources for volcano information.

Dobran, Flavio. *Vesuvius: Education, Security and Prosperity.* Oxford: Elsevier, 2006. Provides an all-encompassing account of Vesuvius, both historically and presently. This document represents a project to understand the volcanism of Vesuvius for the purpose of protection against and prevention of future disasters. Addresses geology, economics, historical progression of volcano knowledge, geophysics and ballistics, and computational analysis.

Francis, Peter, and Clive Oppenheimer. *Volcanoes.* 2d ed. New York: Oxford University Press, 2004. This text covers the volcanic activity from lava flows to pyroclastic currents. Provides many examples from Hawaii, Italy, Ethiopia, Japan, Mount St. Helens, and Krakatau. Written for the layperson, but some knowledge of geology is helpful.

Grattan, John, and Robin Torrence, eds. *Living Under the Shadow: Cultural Impacts of Volcanic Eruptions.* Walnut Creek, Calif.: Left Coast Press Inc., 2007. A collaborative work from geologists, anthropologists, biologists, and other scientists aimed at presenting the resilience of humans to volcanic activity. Discusses historical events and the implications of future events.

Gribbin, John R. *This Shaking Earth: Earthquakes, Volcanoes, and Their Impact on Our World.* New York: G. P. Putnam's Sons, 1978. This book provides a popularized, well-illustrated discussion of current scientific theory, volcano types and mechanisms, and their impact on human society. Suitable for high school readers.

Lipman, Peter W., and D. R. Mullineaux. *The 1980 Eruptions of Mount St. Helens, Washington.* U.S. Geological Survey Professional Paper 1250. Washington, D.C.: Government Printing Office, 1981. This source presents the early but extensive and comprehensive results of studies of the eruption and associated volcanic events that occurred in 1980, including the geophysical monitoring and studies of volcanic deposits, effects, and potential hazards. Suitable for college-level readers and those who are more technically inclined.

Marti, Joan, and Gerald G. J. Ernst, eds. *Volcanoes and the Environment.* New York: Cambridge University Press, 2005. This text provides information on volcanoes from the physical mechanisms and behaviors to hazards and effects on humans and nature. Written for students and geologists studying environmental science and geology.

Sigurdsson, Haraldur, ed. *Encyclopedia of Volcanoes.* San Diego, Calif.: Academic Press, 2000. This book contains a complete summary of the scientific knowledge of volcanoes. It contains eighty-two well-illustrated overview articles (including a sequence of twelve articles on volcanic hazards), each of which is accompanied by a glossary of key terms. Although this is a college-level text, it is written in a clear and comprehensive style that makes it generally accessible. Cross-references and index.

Smith, David G., ed. *The Cambridge Encyclopedia of Earth Sciences.* New York: Crown Publishers, 1981. Chapter 26, "Geologic Hazards," presents a general discussion of volcanic hazards in relation to the two major magma compositional groups. A well-illustrated and carefully indexed reference volume suitable for the general and college-level reader.

Trombley, R. B. *The Forecasting of Volcanic Eruptions.* Lincoln, Neb.: iUniverse, 2006. The author has a PhD in volcanism. The book examines probability and distribution models used to predict volcanic activity. Data analyses using current software are discussed. The final chapter addresses the concept of supervolcanoes. Each chapter contains algorithms and references.

See also: Calderas; Eruptions; Forecasting Eruptions; Hawaiian Islands; Krakatau; Mount Pelée; Mount Pinatubo; Mount St. Helens; Mount Vesuvius; Popocatépetl; Recent Eruptions; Shield Volcanoes; Spreading Centers; Stratovolcanoes.

W

WEATHERING

Weathering is a chemical and physical process in which rocks fragment or break into smaller sedimentary rocks. Weathering plays a significant role in the creation and deposition of minerals in the soil. It also is involved in the cycling of elements such as oxygen, carbon, and hydrogen and of vital compounds, including water and carbon dioxide.

PRINCIPAL TERMS

- **chemical weathering:** a form of weathering in which rocks are broken down through chemical interactions between surface minerals and water and acidic compounds
- **clastic rock:** a type of rock formed by the cementing of sedimentary rocks
- **erosion:** a process in which weathered rock particles are transported by water, wind, and other natural means
- **igneous rock:** a type of rock created from cooling magma beneath the earth's surface
- **physical weathering:** a form of weathering in which rocks are broken down by environmental conditions, crystal growth, or interaction with animals or vegetation
- **rock cycle:** a process in which all three basic types of rock are formed, brought to the earth's surface, weathered, and returned to the earth's interior in a cyclical pattern
- **sedimentary rock:** a type of rock that breaks from a larger igneous, metamorphic, or clastic rock
- **stratigraphy:** the study of layered clastic rocks, revealing the type of sedimentation and weathering that has occurred in time
- **uniformitarianism:** a theoretical concept that sedimentation occurs because of events transpiring through Earth's history instead of through a single, biblical flood
- **weathering:** a geological process in which igneous, metamorphic, or sedimentary rock is broken into smaller particles

Basic Principles

Weathering is a geological process involving the breaking down of rocks into smaller pieces or minerals by chemical or physical means. In some cases, weathering leads to the formation of new minerals that compose the earth's soil. In other cases, the process creates even smaller particles, such as sand and dust.

The rock particles produced out of this breakdown are known as sedimentary rocks. Once those particles are formed, they may be transported to another location by water, wind, or other natural process through erosion. When those particles cement together to form new rocks, the new stones are known as clastic rocks.

Two other basic types of rock are subject to weathering. The first of these is igneous rock, which is produced as molten rock deep beneath the earth's crust. The magma flows outward, and in some cases it simply pools in small pockets; in other cases, magma is ejected more violently through volcanic eruptions. In both situations, the magma cools to form igneous rocks.

The second basic type of rock is metamorphic. This rock originally was igneous or sedimentary, but under heat or extreme pressure (such as that which occurs when the earth's tectonic plates contact each other) this rock "transforms" in composition to reestablish equilibrium, or balance.

Weathering leads to the formation of many different types of soil at the surface of the earth. A number of factors contribute to these variations, including climate, the physical and chemical environment, the type of basic rock undergoing the weathering process, and the amount of time that transpires. Weathering is one of the most significant contributors to the formation of Earth's many landscapes, shaping mountains and forming plains and plateaus.

Background and History

Before the eighteenth century, prevailing scientific thought held that the earth's surface features, its mountains, valleys, and plains, were formed by the biblical flood. Most scientists believed that even the rocks and soil had been carried and then deposited by such a cataclysmic event. However, in the late eighteenth century, Scottish scientist James Hutton, considered the founder of modern geology, developed an alternative point of view.

In observing the remnants of flooding that occurred in the rivers near his home, Hutton noticed a large volume of sand, pebbles, and other rocks. He then theorized that instead of by means of a great biblical flood, the rocks and soil that appeared on the earth's crust (and on the mountains and hills) had been deposited through millions of years of flooding, mountain erosion, and other natural processes. The sediment that covers the earth is deposited in an ongoing, lengthy process of recycling and erosion. Hutton's concept was later dubbed uniformitarianism.

Weathering plays an important role in another of Hutton's concepts, in which each type of rock (metamorphic, igneous, and clastic) goes through a cycle whereby it is formed, moved to the earth's surface, broken down, and returned to the earth's interior; here, Hutton added, the process restarts. Weathering is the primary mechanism by which the minerals and compounds contained in the soil

Joshua tree and juniper growing from a crack in a rock at Joshua Tree National Park, California. (Bruce M. Herman/ Photo Researchers, Inc.)

and rocks are prepared for this recycling process. Hutton's concepts of the rock cycle and uniformitarianism led to the notion that Earth's history is far more expansive than Hutton's predecessors believed. With the earth's geologic time dating to 4.6 billion years ago (a figure now accepted), a great deal more can be learned about how the planet's geological and geodynamic processes have evolved.

Types of Weathering

Weathering occurs in two general forms. The first is physical weathering, in which a rock, for example, which has long been situated beneath the great weight of other rocks or under glaciers and bodies of water, expands when the pressure of those elements is removed. If preexisting fractures (or joints) exist in these rocks, the relief of this pressure can cause the rock to split along those cracks. Physical weathering also occurs when ice and other crystals within the rock push outward, when plants grow into fractures in the rock, when animals burrow through them, and when a rock is subjected to extreme heat.

The second type of weathering is chemical. In this manifestation of weathering, compounds and elements surrounding the rock react with the minerals contained within the rock, breaking down the stone. For example, carbon dioxide that is found in the atmosphere may be captured in falling rain. When that rainwater reaches a rock, it causes the less stable minerals on the rock's surface to react to the carbon, resulting in the corrosive substance carbonic acid. In other situations, acidic compounds (such as sulfuric acid), resulting from the combination of oxygen and certain minerals found in the soil or in vegetation, are formed and transported through rivers and streams. These compounds strip and break down rocks.

Weathering can have a number of effects on a rock's appearance. In some rocks, weathering is evident on an external layer, but inner layers remain unaltered but exposed. In other examples, a rock is broken down by sloughing away small fragments from its exterior. Furthermore, some weathering occurs from within, in which water, ice, and minerals entering the rock's fractures break apart the rock in cube-like pieces.

Environmental and Structural Factors

Weathering occurs in a variety of forms because of many contributing factors. One of the key elements in weathering is the rock's structure. Chemical weathering typically occurs when unstable minerals on the rock's surface contact the corrosive element. Some minerals, however, are less susceptible to weathering than others. Quartz, for example, is highly resilient to chemical and physical weathering. Limestone, however, comprises calcite, a mineral that is more easily weathered. Studies show that limestone can be broken down even by some types of algae and seaweed.

Also playing an important role in the degrees to which weathering occurs is the rock's immediate environment. For example, some minerals are more susceptible to weathering when they are in a warm environment. Sodium, according to one study, shows a higher rate of weathering in warmer climates than in cooler temperatures. Warmer climates also foster the growth of vegetation, which plays an important role in weathering. Furthermore, certain compounds and elements in an environment are factors in weathering. Iron, for example, typically reacts to the presence of free oxygen when exposed, causing a weathering process called oxidation. Meanwhile, when calcite is submerged in water, it quickly dissolves.

The Role of Water in Weathering

Water plays a significant role in the weathering process and in the process of erosion. In terms of physical weathering, water that seeps into fractures and cracks in a given rock may freeze and therefore expand, breaking the rock into smaller pieces. High water pressure also can contribute to the physical weathering of rocks, pushing apart fractured rocks and separating them from the main body. In chemical weathering, compounds and minerals flowing in the water can bond with the minerals on the surface of the rock, drawing them away for erosion.

Two important hydrological factors can either foster or inhibit the weathering process: the rate at which water flows and the length of time the rock is submerged under water. Weathering is likely to occur at greater rates when the water flow is slow, because the minerals contained in the water have more time to bond with and remove unstable surface minerals. In a study of granite material, scientists concluded that, as fluid residence (the period in which the rock is submerged in the water) continues for ten or more years, the amount of chemical weathering that will occur is significant when compared with rocks that have shorter periods of fluid residence.

Computer Modeling

In the study of the geological phenomenon of weathering, scientists look for common characteristics and trends by employing computer models to analyze patterns within different environmental conditions. For example, in one study, statistical data were collated from a number of sedimentary rock formations in shallow-water environments. Using these data, scientists created a computer model that simulated sedimentary rock formation and deformation according to high and low sea-level conditions. Because of the complexity of studying clastic rock formation and composition, computer models are proving to be an increasingly reliable tool for scientists.

Stratigraphy

When clastic rocks are carried to their ultimate destination, they frequently cement together in basins to form large, solid slabs that, in time, become layered. The study of these layers, a scientific approach known as stratigraphy, helps scientists find clues about the myriad geological changes that have taken place in and since the earth's prehistory. Stratigraphic analysis of different clastic basins, for example, can yield information about the geodynamic activity that occurred during a particular era or epoch.

One stratigraphic study of sites in Nevada and Estonia revealed a number of radioactive minerals (namely, strontium and a carbon isotope) that were deposited in those two regions more than 400 million years ago. An analysis of the curves within those layers suggested that the minerals came from weathered source rocks that were pushed outward during a continental collision nearly 50 million years earlier. In this case, a stratigraphic study of these regions revealed that weathering plays a major role in the recycling and transport of minerals from their basic sources.

Lithology

One of the more useful ways in which weathering is studied is by analyzing the sedimentary rocks that result from weathering. A microscopic and chemical analysis of the minerals found in soil, sand, and clay can reveal much about the geochemical characteristics (lithology) of the rocks from which the particles are separated. Scientists studying lithology can therefore assess how weathering occurs with regard to specific minerals and certain environments. Lithology also can reveal the basic rocks from which weathered particles were formed and composed. For example, a hydrochemical analysis of soil and sediment samples from hundreds of sites in Japan revealed large volumes of silicate (which is known to undergo weathering at a high rate). Scientists concluded that silicate, coupled with water runoff, plays a dominant role in the geological processes in this region. This study of Japan's overall lithology reveals much about the overall process of weathering.

Relevant Organizations and Institutions

The study of weathering has a wide range of implications, for persons interested in studying the earth's past as well as those who seek prime locations for exploratory drilling and mining. Groups and institutions having an interest in the study of weathering include the following.

The U.S. government plays a significant role in the study of weathering. The U.S. Geological Survey (USGS), for example, studies weathering in a wide range of environments. Scientists within the USGS have authored a number of scholarly papers on chemical and mechanical weathering and its effects (including monitoring rates of weathering in certain regions). Meanwhile, the National Science Foundation provides funding for scientists working in the field of geological weathering.

Universities and their faculties and research staffs play key roles in studying weathering processes. The University of California system, Tulane University, and the University of Colorado are among the myriad institutions whose geology and earth science departments offer courses in weathering and sedimentary geology. It is at many of these universities that the leading theories and research in this field are developed and investigated.

Energy companies are always looking for new oil, coal, and gas deposits. Such findings are usually unearthed in or near clastic rock basins. For this reason, the petroleum industry often calls upon full-time geophysicists, mineralogists, and geodynamics consultants to help them find such deposits.

Implications and Future Prospects

The study of weathering has a number of implications for the overall study of Earth's geodynamic and geological processes. Sedimentary rocks produced by weathering provide a number of clues to geodynamic processes, including volcanism and tectonics (the geodynamic processes surrounding the earth's outer crust). Scientists who seek to learn about climate change also may look to the strata of clastic rock dating for an understanding of the conditions that existed during prehistoric periods that featured significant climate shifts. These samples might not have been available had the weathering process not broken down the basic rocks that contained such evidence.

The study of weathering has been greatly aided by the technologies available in the twenty-first century. Computers capable of collating large volumes of data and creating two- and three-dimensional models based on this data are helping scientists gain a better understanding of how weathering occurs under a wide range of environmental conditions. Furthermore, the Internet allows scientists from around the world to quickly share data on samples found in various basins and to collectively analyze such data.

Michael P. Auerbach

Further Reading

Atkinson, David. *Weathering, Slopes, and Landforms.* London: Hodder Education, 2004. In this book, the author describes how major rocks and landscapes are formed. He then discusses how those masses are subjected to different types of weathering (and the various rates at which it occurs), citing examples from different climates around the world.

Beaulieu, E., et al. "Modeling of Water-Rock Interaction in the MacKenzie Basin: Competition Between Sulfuric and Carbonic Acids." *Chemical Geology* 289, nos. 1/2 (2011): 114-123. In this article, the authors review the various factors contributing to chemical weathering, such as climate, soil content, plate tectonics, and the presence of acids. The authors discuss the presence of two types of acid, carbonic and sulfuric, found in a Canadian river basin and the rates at which these two acids foster weathering in that area.

Bland, Will, and David Rolls. *Weathering: An Introduction to the Scientific Principles.* London: Hodder Education, 1998. This book reviews the scientific principles involved in physical and chemical weathering processes. The authors discuss the rates and intensities by which certain manifestations of weathering occur and the effects weathering has on the earth's landscapes.

Calcaterra, D., and M. Parise. *Weathering as a Predisposing Factor to Slope Movements.* Engineering Geology Special Publication 23. London: Geological Society of London, 2010. In this book, the authors provide an outline of how weathering occurs along rock masses on slopes. Discusses the relationship between weathering and slope movements, using examples from sites all over the world.

Dagdelenler, Gulseren, Ebru A. Sezer, and Candan Gokceoglu. "Some Non-linear Models to Predict the Weathering Degrees of a Granitic Rock from Physical and Mechanical Parameters." *Expert Systems with Applications* 38, no. 6 (2011): 7476-7485. Discusses why weathered granite cannot be used in construction and why it remains necessary to study the degrees to which granite may be subjected to weathering. Reviews analytical approaches to this issue.

Horbe, Adriana M. C., and R. R. Anand. "Bauxite on Igneous Rocks from Amazonia and Southwestern of Australia: Implication for Weathering Process." *Journal for Geochemical Exploration* 111, nos. 1/2 (2011): 1-12. Discusses how severe weathering on certain types of bedrock can lead to the formation of a thin layer of hardened soil called duricrust. Cites examples from the Amazon River watershed in Brazil and from Boddington, Australia, wherein duricrust consisting largely of the mineral bauxite has been formed after weathering.

Maher, K. "The Dependence of Chemical Weathering Rates on Fluid Residence Time." *Earth and Planetary Science Letters* 294, nos. 1/2 (2010): 101-110. The author discusses the process of chemical weathering as it relates to fluid residence. Analyzes the chemical weathering process that has been occurring on granitic rock in a period of a few days through ten years and beyond.

See also: Biogeography; Carbonate Rocks; Clastic Rocks; Coastal Processes; Deserts and Dunes; Fluvial Processes; Geomicrobiology; Groundwater; Mass Wasting; Minerals; Paleoclimatology; Paleoecology; Snowball Earth; Trace Fossils.

Y

YELLOWSTONE NATIONAL PARK

Yellowstone National Park is known throughout the world for its volcanic and thermal features. Over the past 2 million years, Yellowstone has been the site of three major volcanic eruptions, reflecting its position over a "hot spot" in the earth's crust and upper mantle.

PRINCIPAL TERMS

- **ash-flow tuff:** a sheetlike pyroclastic deposit that is laid down as a hot mixture of gas, crystals, pumice, and volcanic ash
- **caldera:** a large, flat-floored volcanic depression that is formed on top of a large, shallow magma chamber during the eruption or withdrawal of magma
- **hot spot or mantle plume:** a zone of hot, upwelling rock that is rooted in the earth's upper mantle; as plates of the earth's crust and lithosphere glide over a mantle plume, a trail of hot spot volcanoes is formed and the earth's surface bulges upward
- **obsidian:** a dense variety of silica-rich volcanic glass; the same material as pumice but does not contain gas bubbles
- **Plinian eruption:** a violent, explosive type of volcanic eruption named either for Pliny the Elder, a Roman naturalist who died while observing the eruption of Mount Vesuvius in 79 C.E., or for Pliny the Younger, his nephew, who chronicled the eruption
- **pumice:** frothy, silica-rich volcanic glass that is commonly produced during explosive eruptions of rhyolite magma
- **resurgent dome:** a broad, oval area of uplift within a volcanic caldera that is marked by the upwarping and fracturing of caldera-filling deposits
- **rhyolite:** a viscous, gas-rich type of magma that contains greater than about 70 percent silica; rhyolite erupts nonexplosively as thick, slow-moving lava flows and explosively as widespread sheets of frothy pumice

YELLOWSTONE AND ITS ENVIRONS

The first recorded observations of the Yellowstone country were made by fur trappers in the early nineteenth century, and later government-sponsored surveys confirmed the geological wonders of the area. As a result, much of what is now northwestern Wyoming and adjacent portions of Idaho and Montana were set aside in 1872 as the Yellowstone National Park. The largest thermal area in the world, Yellowstone is best known for its outstanding assemblage of hot springs, geysers, and mud pots.

Yellowstone National Park is an 899,152-hectare, wooded mountain area in the northern Rocky Mountains. The rugged and geologically youthful terrain is mostly of volcanic and glacial origin, with many waterfalls and lakes. Its eastern border is within the Absaroka Range, and the Teton Range and Jackson Hole lie to the south. The northeastern sector of the park is cut by the Grand Canyon of the Yellowstone River. The park occupies the central part of the Yellowstone Plateau, a 6,500-square-kilometer, high plateau that is three times larger than the park and averages 2,500 meters in elevation. Immediately west of Yellowstone, Island Park, Idaho, is a shallow basin that is partly filled with basalt lava flows and sediments. To the southwest, the Yellowstone Plateau drops gradually to the eastern Snake River plain of Idaho, a downwarped region (1,000 to 1,500 meters elevation) of low topographic relief that is largely covered by basalt lava flows.

A strong geological relationship exists between Yellowstone and the Snake River plain, an 800-kilometer-long, low-lying volcanic belt that crosses southern Idaho. Near the Idaho-Oregon border, the oldest volcanism on the western Snake River plain began about 17 million years ago. Since that time, volcanic activity has slowly progressed northeastward toward Yellowstone at an average rate of about 4 centimeters per year. The youngest basalt eruptions on the eastern Snake River plain occurred about 2,100 years ago, and the youngest rhyolite eruptions at Yellowstone occurred around 50,000 years ago.

The most popular explanation for the progression of volcanism toward Yellowstone is that a stationary plume of upwelling, perhaps partially molten rock, is rooted in the earth's upper mantle beneath the northwestern United States. Although the plume seems to be fixed in place, the tectonic plate of North America has been moving westward for at least 17 million years. Like a candle flame held beneath a moving piece of paper, the volcanoes form an eastward-bound trail, first with explosive rhyolite eruptions and then with basalt lava outpourings. Thus, the old rhyolites of the Snake River plain in southern Idaho are covered almost completely by basalt lava flows, but Yellowstone is the youngest rhyolite volcanic center, and very little basalt has erupted there. The mantle plume, or hot spot, is now considered to lie beneath Yellowstone and is the ultimate source of magma and geothermal heat under that area.

Yellowstone is known for its geysers. The Africa Geyser, its spray frozen during winter, melts from its base, where the ice meets the warm water. (U.S. Geological Survey)

VOLCANIC ACTIVITY

For the past 2 million years, Yellowstone has been the site of huge but infrequent eruptions of rhyolite lava and pumice; the youngest eruptions of rhyolite lava occurred about 50,000 years ago. The products of those eruptions comprise the Yellowstone volcanic field, which formed during three cycles of volcanism. Each cycle began and ended with the eruption of lava flows of rhyolite obsidian, a dense, black, silica-rich volcanic glass. The lava flows issued mostly from curved fractures of the Yellowstone caldera and now cover much of the caldera floor. All three volcanic cycles culminated with explosive eruptions of voluminous rhyolite ash and pumice. The ash and pumice deposits from the three climactic eruptions of the Yellowstone-Island Park area are known as the 2.1-million-year-old Huckleberry Ridge tuff, with an estimated volume of 2,500 cubic kilometers; the 1.3-million-year-old Mesa Falls tuff, with an estimated volume of 280 cubic kilometers; and the 0.6-million-year-old Lava Creek tuff, with an estimated volume of 1,000 cubic kilometers. For comparison, the total volume of material emitted from Mount St. Helens on May 18, 1980, was only 1 to 2 cubic kilometers.

Frothy rhyolite pumice and ash were explosively discharged from curved fractures around the rim of the Yellowstone caldera. Plinian eruptions ejected large quantities of rhyolite to heights of tens of kilometers into the atmosphere. Some of the ejecta then cascaded back to the earth like a water fountain or "boiled over" as hot, ground-hugging flows of ash and pumice, forming three major ash-flow deposits that fill the

Yellowstone caldera and blanket the surrounding terrain. The ash flows were sufficiently hot when they came to rest (about 550 degrees Celsius or more) that the particles of pumice and ash were compacted into a dense mass, forming welded ash-flow tuffs. Each of the catastrophic eruptions probably lasted only hours or days. The rapid eruption of voluminous rhyolite pumice from shallow (perhaps 5 to 10 kilometers deep) magma chambers led to the collapse of the roof rocks overlying the magma chambers, forming large rhyolite calderas ringed by 50-kilometer-wide sheets of welded ash-flow tuff. Finer particles of ash and pumice were ejected to great heights and carried away by high-altitude winds. The finest particles probably took months or years to sift out of the atmosphere. Ash-fall deposits from the three major Yellowstone eruptions are found as far away as Louisiana, California, and southern Canada, where they are important marker beds for the dating of Ice Age fossils and glacial deposits.

The Yellowstone caldera—a broad, oval-shaped depression about 40 by 60 kilometers in size—was formed during the climax of the last cycle of volcanism 600,000 years ago, when the Lava Creek tuff was emplaced. Since that time, voluminous lava flows of rhyolite obsidian—the Plateau Rhyolites of Yellowstone—have obscured the caldera rim and now largely cover the caldera floor. The most voluminous Plateau Rhyolite lava flows compose the Madison and Pitchstone Plateaus in the western and southwestern sectors of the park.

During the Ice Age, large masses of glacial ice formed in northern Yellowstone National Park. Two main periods of ice formation have been identified. An older ice cap formed during the Bull Lake glaciation, about 150,000 years ago. Most of the Bull Lake glaciers flowed to the west, and the West Yellowstone Basin to the west of Yellowstone Park is filled with Bull Lake outwash gravels and other deposits. A later ice cap formed during the Pinedale glaciation, between about 30,000 and 15,000 years ago. Most of the Pinedale ice flowed to the north because rhyolite lava flows of the Madison Plateau blocked westward passage of the ice. Both ice caps were probably 1 kilometer or more thick, and much of Yellowstone Park is covered by a thin mantle of glacial deposits. Glacial deposits play an important role in the thermal activity of Yellowstone, since permeable sands and gravels become capped with mineral deposits but remain permeable to hot fluids at depth. The formation of impermeable caprock over permeable strata is important in the development of the shallow plumbing systems beneath hot springs and geysers.

Seismic Activity

Yellowstone is the most seismically active region of North America. During the fifteen-year period between 1973 and 1988, about fifteen thousand earthquakes greater than Richter magnitude 2 were recorded. In 1959, the Hebgen Lake earthquake, centered about 10 kilometers west of Yellowstone and estimated at between 7.3 and 7.5 on the Richter scale, was the largest tremor ever recorded in the Rockies; it caused changes in the behavior of hundreds of geysers and hot springs. The frequent earthquakes of Yellowstone are probably important in maintaining the thermal activity of the region, since permeable fault zones that channel hot fluids to the surface would quickly be sealed with mineral matter if not for regular disturbances from earthquakes. Beneath the Yellowstone caldera, the velocities of earthquake waves traveling through the earth's crust are measurably decreased, suggesting the presence of a plumelike body of partly molten rock extending downward from a depth of about 8 kilometers. The amount of heat flowing from the ground beneath Yellowstone is estimated as thirty to forty times the heat flow elsewhere in North America and accounts for about 5 percent of the total heat flow from all North America.

Resurgent Domes

Since the first precise topographic surveys were done during the 1920's, resurveying has shown that the central caldera floor is slowly warping up at an average rate of about 14 millimeters per year. Two broad areas of uplift, known as resurgent domes, have been identified in the southwestern and northeastern Yellowstone caldera. Both resurgent domes are about 15 kilometers across and 200 meters high. Resurgence of the caldera floor, together with the high heat flow and seismicity, suggests the presence of rhyolite magma in the upper crust, either left over from the last cycle of volcanism or in the form of a new magma batch that may appear at some time in the future as a fourth volcanic cycle. The hot rock, or magma, beneath Yellowstone is the source of heat for nearly ten thousand hot springs and several hundred active geysers. Rain and snowmelt soak into the ground and percolate through fractured volcanic rocks to depths of several kilometers, where the fluids become heated. The hot water then buoyantly rises to the surface along north-trending faults of the northern Rocky Mountains, the ringlike fractures of the Yellowstone caldera, and the fractured rocks around the edges of the two resurgent domes. Hence, the thermal features of Yellowstone are not randomly distributed throughout the park but tend to occur where fluid pathways are available, especially along the rim of the Yellowstone caldera and near the two resurgent domes.

Hydrothermal Activity

The Norris Geyser Basin is on the northern edge of the Yellowstone caldera, where circular fractures of the caldera rim are intersected by a major, north-trending fault that extends northward from the caldera. The intersection of two major fault zones enhances the upward migration of deep, hot fluids, and Norris is considered to be the hottest and most active geyser basin in Yellowstone. The basin contains a great variety of thermal features, including geysers, clear boiling springs, mud pots, and fumaroles. Porcelain Basin is the world's largest sinter-covered area (sinter is a variety of hydrated silica that forms around certain hot springs). The geothermal waters at Norris are neutral-chloride to slightly acid-sulfate in composition. The content of dissolved sulfates may be a result of the hot waters traveling through underground sedimentary rock formations that are intersected by the faults at depth.

Old Faithful is a famous geyser in the Upper Geyser Basin, one of several major thermal areas along the Firehole River in the western part of Yellowstone Park. The Upper Geyser Basin is situated on the edge of a resurgent dome within the Yellowstone caldera, where fractured volcanic rocks allow the upward passage of hot fluids. The basin occurs in a topographically low-lying area between three lobes of rhyolite lava, and the valley is underlain by permeable sand and gravel deposits from glaciers and rivers; those deposits allow the hot fluids to disperse over a wide area. As a result of such ideal conditions, the Upper Geyser Basin contains the largest concentration of geysers in the world.

The Mud Volcano area is to the north of Yellowstone Lake, on the southwestern edge of a resurgent dome within the Yellowstone caldera. The water table is boiling underground, and fumaroles are a major feature of this vapor-dominated thermal area. The muddy pools and mud volcanoes are composed of clay, formed by the chemical alteration of near-surface rocks and glacial deposits by acid vapors. The source of the surface water is rainfall and steam condensation. Mammoth Hot Springs is one of the few major thermal areas that occur outside the Yellowstone caldera. Hot, underground water flows northward along the same fault system as Norris, cooling to subboiling temperatures and becoming alkaline in composition during its passage through sedimentary rocks. Calcium carbonate is dissolved from the sedimentary rocks and is later deposited on the surface as mounds and rimmed terraces of travertine, a variety of hot-spring limestone. Travertine grows at much more rapid rates than sinter, and substantial changes in the configuration of Mammoth Hot Springs are observed from year to year.

William R. Hackett

Further Reading

Bryan, T. Scott. *Geysers: What They Are and How They Work.* 2d ed. Missoula, Mont.: Mountain Press Publishing Company, 2005. Provides a succinct, nontechnical explanation of geysers and hot springs. Focuses mostly on Yellowstone, but also discusses Iceland, Chile, New Zealand, and elsewhere.

Decker, Robert, and Barbara Decker. *Volcanoes.* 4th ed. New York: W. H. Freeman, 2005. This readable and well-illustrated text is geared toward the general public. The authors explain geologic phenomena in a clear and simple manner. Includes sections on Yellowstone and its hydrothermal activity. The appendices list the world's 101 "most notorious" volcanoes, as well as Internet sources for volcano information.

Fournier, Robert O. "Geochemistry and Dynamics of the Yellowstone National Park Hydrothermal System." In *Annual Review of Earth and Planetary Sciences,* edited by G. W. Wetherill, A. L. Albee, and F. G. Stehli. Vol. 17. Palo Alto, Calif.: Annual Reviews, 1989. This technical and comprehensive review article emphasizes the geochemistry of hot spring waters. It is readable by people with a knowledge of college chemistry. A number of graphs and a summary table of fluid compositions are given.

The references list is extensive and current, but the article has no index or glossary.

Francis, Peter, and Clive Oppenheimer. *Volcanoes.* 2d ed. New York: Oxford University Press, 2004. This text covers volcanic activity ranging from lava flows to pyroclastic currents. Provides many examples from Hawaii, Italy, Ethiopia, Japan, Mount St. Helens, and Krakatau. Written for the layperson, but some knowledge of geology is helpful.

Fritz, William J., and Robert C. Thomas. *Roadside Geology of the Yellowstone Country.* 2d ed. Missoula, Mont.: Mountain Press, 2011. This paperback is written for the general public. It contains numerous maps and black-and-white photographs of the geological and thermal features of Yellowstone and is meant to be used in self-guided tours of the roadside geology. The introductory chapters are concise, and the book contains a glossary and extensive reference list.

Harris, Stephen L. *Fire Mountains of the West: The Cascade and Mono Lake Volcanoes.* 3rd ed. Missoula, Mont.: Mountain Press Publishing Company, 2005. Although this book focuses mostly on the Cascades, it includes information on Yellowstone National Park, Mount St. Helens, Mammoth Mountain, and Crater Lake. Chapter 2 covers the Ring of Fire and hot spots. Written in a manner accessible to anyone with some interest in volcanoes.

Lockwood, J. P., and Richard W. Hazlett. *Volcanoes: Global Perspectives.* Chichester: Wiley-Blackwell, 2010. An up-to-date and well-illustrated undergraduate-level text that is also accessible to the interested reader.

Pierce, Kenneth L. *History and Dynamics of Glaciation in the Northern Yellowstone National Park Area.* U.S. Geological Survey Professional Paper 729-F. Washington, D.C.: Government Printing Office, 1979. This 90-page paperbound publication with four plates is geared to professional geologists who are interested in glacial features; it should, nevertheless, be useful to people with a good background in general science. Outlines the glacial history of northern Yellowstone. Well illustrated with maps and diagrams. Contains an extensive list of technical references plus an index.

Sigurdsson, Haraldur, ed. *Encyclopedia of Volcanoes.* San Diego: Academic Press, 2000. This book contains a complete summary of the scientific knowledge of volcanoes. It contains eighty-two well-illustrated overview articles (including several relating to geothermal systems in Yellowstone), each of which is accompanied by a glossary of key terms. Although this is a college-level text, it is written in a clear and comprehensive style that makes it generally accessible. Cross-references and index.

Smith, Robert B., and Robert L. Christiansen. "Yellowstone Park as a Window on the Earth's Interior." *Scientific American* 242 (February 1980): 104-117. An excellent overview of the geological and geophysical significance of Yellowstone, this article is readable by persons with a sound high school science background. Well illustrated but contains no glossary or references list.

Smith, Robert B., and Lee J. Siegel. *Windows into the Earth.* New York: Oxford University Press, 2000. This book describes the geology of Yellowstone National Park and Grand Teton National Park. Discusses hot spots and other geological forces, while maintaining a casual writing style. An easy transition into geology for the layperson.

Steingisser, Alethea, and W. Andrew Marcus. "Human Impacts on Geyser Basins." *Yellowstone Science* 17 (2009): 7-18. This article discusses the effects of tourism and geothermal energy development on geysers' activity. Focuses specifically on the geysers within Yellowstone Park. Contains maps, tabulated data, and many images.

U.S. Geological Survey. "Geologic Map of Yellowstone National Park." Miscellaneous Investigations Map I-711. Denver: Author, 1972. A 1:125,000-scale, color geological map that shows the areal distribution of rock units exposed at the earth's surface. Understandable by anyone who has taken an introductory geology course. Would be useful for persons wanting to know the general distribution of rock units within the park.

Vink, Gregory E., W. J. Morgan, and Peter R. Vogt. "The Earth's Hot Spots." *Scientific American* 242 (April 1985): 50-57. This article reviews the locations and criteria for recognition of the world's hot spots, including Yellowstone. Written from a global perspective, it emphasizes the significance of hot spots in the breakup of continents and in reconstructions of the earth's plate motions. Readable by persons with a sound high school science background, it is well illustrated but contains no glossary or references list.

See also: Calderas; Eruptions; Flood Basalts; Forecasting Eruptions; Geysers and Hot Springs; Hawaiian Islands; Hot Spots, Island Chains, and Intraplate Volcanism; Hydrothermal Vent Faunas; Island Arcs; Lava Flow; Ring of Fire; Spreading Centers; Volcanic Hazards.

APPENDIXES

GLOSSARY

aa flow: a Hawaiian term (pronounced "ah-ah") that has been adopted for basaltic lava flows with rough, clinkery surfaces characterized by large angular blocks.

abiogenesis: a process involving the generation of living systems from nonliving components.

ablation: the loss of material from a glacier through melting and evaporation; the opposite of "accumulation."

absolute date: a date that gives an actual, though approximate, age of an artifact determined by direct means from the present.

accretion: the process of growth of a larger crustal unit, such as a continent, by collision with smaller tectonic terranes, such as volcanic arcs or microcontinents; the gradual accumulation of matter in one location, typically under the influence of gravity.

Acheulian: a cultural tradition distinguished by stone tools characterized by large (about 15-centimeter), crudely fashioned, leaf-shaped hand axes.

acoustic echo sounding: a method of determining the depth of the ocean floor that measures the time of a reflected sound wave and relates that to distance.

actuopaleontology: the method of paleontological study in which organismic features found on modern living animals are transposed onto extinct organisms.

aerobes: prokaryotes (usually bacteria) that live in the presence of elemental oxygen.

agglutination: the clumping of particles to form a shell around a particular organism or object.

aggradation: the process in which deposition of sediment from flowing water gradually raises the level of the terrain through which it flows.

agnathan: any member of the class of jawless fish characterized by the lack of any mobile jaw structure around the mouth.

albedo: the amount of solar energy reflected back into space by Earth's surface; an increase in albedo is believed to lead to lower temperatures and stimulates the expansion of glaciers.

algae: primitive, one-celled, chiefly aquatic plantlike organisms lacking stems, roots, and leaves.

alien: a species of plant or animal that is not native to, but is found to have successfully colonized a particular place; generally referring to one that has been introduced as a result of human activity.

allochem: large, coarse grains about the size of sand or gravel, found in carbonate rocks.

allopatric: of animals or plants, occupying separate, non-overlapping distribution areas; of speciation, occurring as a result of such geographic separation.

alluvial: referring to sediment deposited by flowing water.

alluvium: all sediments deposited by flowing water.

alpine glacier: a small, elongate, usually tongue-shaped glacier commonly occupying a preexisting valley in a mountain range.

amber: hardened, fossilized resin from prehistoric trees; some is known to contain intact, preserved organisms.

amino acid: an organic chemical compound characterized by the structure in which an amino group (—NH2) and a carboxylic acid group (—COOH) are bonded to a common carbon atom in a molecule; the basic structural unit of protein molecules.

amnion: the sac-like structure that lines the inside of a shelled egg and contains the embryo, the yolk sac, and the amniotic fluid medium.

amniote: descriptive of eggs consisting of a tough, mineralized, air-permeable outer shell enclosing a fluid-filled sac called the amnion.

amphibian: meaning "of double lives" and referring to creatures whose life cycle includes both aquatic and terrestrial existences.

anaerobes: prokaryotes that can live only in an atmosphere that is free of elemental oxygen.

anapsid: animals characterized by the lack of skull apertures.

ancestral Rockies: the mountain ranges that preceded the Rockies by many millions of years and no longer have surface topographic expressions of their former existence.

andesite: a light-colored volcanic rock with intermediate silica content that contains the minerals plagioclase, biotite, and hornblende; the predominant lava expelled by stratovolcanoes such as those in the Cascades.

angular unconformity: an unconformity in stratigraphic beds in which lower beds are at an angle to the beds above them.

anion: an atom or stable molecular fragment that has one or more unbalanced electrons, making it a negatively charged ion.

anoxia: the condition resulting from an absence of oxygen.

anoxic: a physical state in which oxygen is low.

anticline: a structure of stratigraphic folds created when rock strata arch upward; the limbs of the fold dip in opposite directions and the oldest rocks typically are exposed in the middle of the fold.

apes: mammals, including gibbons, orangutans, chimpanzees, and gorillas, that are similar to monkeys but lacking a tail and having five low cusps on their molars instead of four sharp ones.

aquifer: the water-bearing stratum of porous, permeable rock, sediment, or soil.

***Archaeopteryx*:** the oldest known bird, fossils of which were first discovered in the Upper Jurassic Solnhofen Limestone of Bavaria in 1860.

Archean eon: the early Precambrian period, from about 2.5 to 4 billion years ago, roughly corresponding to the earliest known fossils.

arête: an extremely narrow mountain ridge created between adjacent U-shaped valleys.

Archosauria: the diapsid reptile group of the Permian and early Triassic periods that eventually gave rise to the crocodiles, dinosaurs, pteranodons, and marine reptiles of the Mesozoic Era, as well as the earliest mammalian ancestors.

aridisol: a type of soil that contains little or no organic matter, is typically alkaline and consists mostly of salts such as chlorides, sulfates, and carbonates; the largest order in soil taxonomy.

artesian aquifer: an aquifer that is bounded above and below by impermeable strata called aquicludes and that experiences hydraulic pressure.

artesian pressure: pressure exerted on groundwater in an artesian aquifer; it is significantly greater than atmospheric pressure and will naturally expel water from the aquifer via any suitable conduit.

arthropods: invertebrates with rigid proteinaceous external skeletons that are jointed at the points of movement and that must be shed regularly.

aseismic: without seismic activity or active fault structures conducive of earthquake activity.

ash: rocky, unconsolidated volcanic ejecta of sand-grain size (4 millimeters or less).

ash flow: a moving mass of hot ash and gases that flows from a volcano; the hot material is deposited as it spreads out on flatter areas.

ash-flow tuff: a sheetlike pyroclastic deposit that is laid down as a hot mixture of gas, crystals, pumice, and volcanic ash.

asteroid: one of the numerous small rocky bodies bigger than about 10 meters in size orbiting the sun, mostly between Mars and Jupiter; some, however, cross Earth's orbit.

asthenosphere: a thin, relatively liquid boundary layer between Earth's solid crust material and the molten magma of Earth's mantle.

atmospheric circulation: the movement of air mass as a result of regional pressure differences.

australopithecines: erect-walking early human ancestors whose cranial capacity and body size were within the range of modern apes rather than of humans.

autoclastic breccia: the clinkery or blocky rubble that forms on some lava flows.

avalanche: a type of high-velocity granular flow of large volumes of materials (e.g., snow, ice, rocks) resulting from the failure of underlying supporting material and driven by gravity.

axial plane: a surface connecting all hinges of a fold that may or may not be planar; that is, the axial plane of a fold may vary from a flat plane to a complexly folded plane.

axis: a line parallel to the hinges of a fold; also called a fold axis or hinge line.

azimuth: degrees of arc measured clockwise from the north.

barrier: any sort of physical, chemical, or biological obstacle that prevents the migration of a species into or out of a given geographic area.

barrier island: a long, low sand island parallel to the coast and separated from the mainland by a salt marsh and lagoon.

basal slip: flow movement of the basal material of glacier, somewhat slower than that of the overlying material and dependent upon the amount of liquid water present for lubrication.

basalt: a dark-colored igneous rock containing minerals such as feldspar and pyroxene, high in iron and magnesium, low in silica, relatively free-flowing and tends to be the basement rock underneath sediments in the ocean basins.

base level: the theoretical vertical limit below which a stream cannot cut or erode its bed.

basement: an underlying complex structure, generally of igneous and metamorphic rocks.

basin: a regionally depressed topographical structure available for the collection of sediments.

batholith: body of rock crystallized from magma, or molten rock, and having an exposed surface greater than 100 km^2 in extent; compositions are typically granitic.

beach nourishment: the restoration of a beach by the mechanical placement of sand on the beach for purposes of recreational and shore protection.

bedrock: the solid rock underlying soil or any other unconsolidated superficial cover.

Benioff zone: the seismically active zone dipping below a continental shelf or mass on the landward side of a subduction trench, named after Hugo Benioff, the seismologist who first defined it.

benthic: that which pertains to the bottom of a lake or sea, or the ecosystem at the bottom of a lake or sea.

beta (β) particle: a subnuclear particle emitted during nuclear fission as an electron (e^-) from the decomposition of a neutron.

binomial system: the current system of classifying organisms, which gives each organism a two-part latinized name consisting of the genus and the species.

bioabrasion: physical and chemical erosion or removal of rock as a result of the activities of marine organisms.

biogenic sediments: the sediment particles formed from skeletons or shells of microscopic plants and animals living in seawater.

biogeographic province or realm: a geographic region distinguished by a unique set of environmental conditions and endemic organisms that live in the region.

bioremediation: a geomicrobiological process whereby bacteria, microbes, and other microorganisms consume pollutants and deposit water, nutrients, and other by-products.

biostratigraphy: the identification and organization of strata based on their fossil content and the use of fossils in stratigraphic correlation.

block lava: lava flows whose surfaces are composed of large, angular blocks that are generally larger than those formed in aa flows and that have smooth, rather than jagged, faces.

blue-green algae: bacteria-like, primitive, single-celled prokaryote organisms that are among the oldest known living species, having existed on Earth for more than 3.5 billion years.

body fossil: petrified remains exhibiting the body structure of a plant or animal.

bolide: a meteorite or comet that explodes upon striking Earth.

borax: a hydrated sodium borate mineral that occurs as surface crust or large crystals in the muds of alkaline lakes; used in glass, ceramics, cleansing agents, water softeners, and other industrial applications.

brachiopod: a member of the Brachiopoda phylum of marine animals having two shells, superficially similar but unrelated to such mollusk species as clams and oysters.

breccia: a general term for any deposit formed by the amalgamation mainly of coarse volcanic rock fragments.

bryozoan: a colonial marine animal very much like modern sponges.

calcite: a common, rock-forming mineral form of calcium carbonate that is the main component of limestone and marble; it is soluble in dilute acids.

caldera: the large, basin-shaped crater atop a volcano, created by the collapse of the underlying magma chamber.

caliche: a type of soil or paleosol that contains a high proportion of calcium carbonate, calcium sulfate, or both.

Cambrian period: the period from about 542 to 488.3 million years ago, marked by the appearance of hard-shelled organisms.

carapace: the bony plate that forms the dorsal side or back of a turtle or tortoise; also, a generic term for the corresponding segment of an exoskeleton.

carbonatite: an igneous rock containing a relatively high percentage of carbon in its composition.

carbonic acid: H_2CO_3, a weak acid formed by mixing water and carbon dioxide; it plays an important role in the dissolving and erosion of the most common karst rock, limestone.

Carboniferous period: the fifth of the six periods in the Paleozoic era; lasting from 359.2 to 299.0 million years ago, it preceded the Permian period.

carbonization: the driving of carbon dioxide and water out of tissues, leaving only free carbon.

cartography: the science of making maps or charts of terrestrial land areas and the ocean floor, blending aspects of geography and several related fields.

cast: a fossil that displays the form of the original organism in true relief.

catastrophists: adherents to the belief that the present-day geology of the planet was formed through unusual catastrophic events rather than slow natural processes.

cation: an atom or stable molecular fragment that has a deficiency of one or more electrons, making it a positively charged ion.

catkin: a dense cluster of scale-like flowers that are typically unisexual.

cave: an enclosed cavity in underground rock formations, typically formed as permeable or chemically susceptible limestone is dissolved away, leaving surrounding material intact.

Cenozoic era: the youngest of the three Phanerozoic eras, from 66 million years ago to the present; it encompasses two geologic periods, the Paleogene (older) and the Quaternary.

ceratopsian: any of a large group of herbivorous dinosaurs characterized by the evolution of cranial appendages including horns and protective shields.

charophyte: a kind of freshwater green alga with filamentous branches arranged in whorls around a threadlike central stem, characterized by egg cells that are large compared with those of other algae.

chemical weathering: a form of weathering in which rocks are broken down through chemical interactions between susceptible mineral structures, water and acidic compounds.

chemosynthesis: the synthesis of organic compounds by living organisms using energy derived from chemical reactions rather than energy from light.

chert: a mineral sediment that often contains the pulverized remains of diatoms, shelled microbes that inhabit marine and aquatic ecosystems.

chloroplast: the organelle in eukaryotes in which photosynthesis is carried out.

Chondrichthyan: any member of the class of fish having cartilaginous skeletons.

cinder cone: a cone-shaped mound made of volcanic granules as pyroclasts and scoria, generally not more than 450 m (1,500 ft) in height.

cirque: a steep-sided, gentle-floored, semicircular hollow produced by glacial erosion of bedrock high on mountain peaks, normally at the head of an alpine glacier valley.

cladistics: a method of assigning biological relationships according to the extent to which characteristics exhibited by the organisms are shared.

class: a group of related biological orders.

clast: pieces or particles of other rocks that, when consolidated together, form clastic rocks.

clastic rock: types of rock formed by the cementing together or consolidation of rock particles in a matrix of sedimentary rock.

clay: a very fine-grained sedimentary deposit consisting primarily of clay mineral particles less than 0.002 mm in size.

clay mineral: any type of hydrous silicate material that form most of the fine-grained particles in clay, formed by erosion, weathering, or similar alteration of other silicates.

cleavage: the tendency of a rock or mineral to break apart in a certain direction, or along parallel planes defined by its internal crystal structure.

climate: the accumulative effects of weather over a particular region and period of time; its basic elements include radiation, temperature, atmospheric moisture and precipitation, evaporation, and wind.

coelomate: an organism possessing an internal cavity termed the coelom separating the body wall from the gut and internal organs.

collagen: a protein having a tough, rubbery consistency that is a principal component of connective tissue.

collage tectonics: a complex patchwork of different types of terranes thought to represent a region in which accretion has joined together suspect terranes.

colonization: the long-term establishment of a successfully reproducing population of a particular species of plant or animal, in a habitat where it was not previously found.

columnar jointing: the formation of columns, often with hexagonal cross sections, as joints grow inward from the outer surfaces of cooling igneous rock bodies.

compaction: the process by which loose grains of sedimentary material are brought closer together into a higher state of consolidation.

composite volcano: a steep-sided volcano built of alternating layers of lava and pyroclastic deposits, along with abundant dikes and sills.

cone: the hill or mountain, more or less conical, surrounding a volcanic vent and created by its ejecta, normally surmounted by a crater.

conglomerate: sedimentary rock composed of gravel in a sandy matrix.

conjugate shear sets: two sets of joints that make angles with each other of close to 60 degrees and 120 degrees.

continental crust: near-surface rocks primarily composed of low-density, light-colored minerals and constituting the bulk of continental landmasses.

continental drift: the horizontal movement of continental masses relative to one another, as a result of tectonic plate displacement, driven by convection currents in the magma of the mantle.

continental ice sheet: an unconfined glacier at least 50,000 square kilometers in extent and of considerable thickness that covers a large part of a continent, obscuring the relief of the underlying surface.

continental margin: the region of a continental landmass both exposed on land and submerged below water, defining the transition from continental landmass to ocean basin; said to be active if subduction occurs along them.

continental shelf: the submerged offshore portion of a continent, ending where water depth increases rapidly from a few hundred to thousands of meters at the ocean floor.

continental shield: the oldest exposed Precambrian rocks that form the nuclei of the continents.

convection cell: a region within a fluid medium in which heat transfer takes place in a cyclic manner as hotter material from a lower depth rises because of its lesser density while cooler, denser material sinks back to lower depths.

convection: the transfer of heat by mass movement in a fluid medium.

convection currents: the cyclic transfer of material through a fluid medium caused by differences in density, usually brought about by heating.

convergence: the process that occurs during the second half of a supercontinent cycle, whereby crustal plates unite and intervening oceans disappear as a result of plate subduction.

convergent organisms: evolutionarily unrelated organisms that come to resemble one another through adaptation to similar habitats and life conditions.

converging plates: a tectonic plate boundary where two plates are approaching each other rather than being separated from each other.

cordillera: a major mountain chain or system of mountain chains, especially as a dominant feature of a continent-sized landmass.

coring devices: devices that drill into ocean-floor and other sediments to extract a continuous stratigraphic sample, providing scientists with information on the composition of the seabed or the underlying structure.

coronae: ring-shaped geological structures on Venus consisting of alternating concentric ridges and valleys, higher in elevation than the surrounding terrain; possibly volcanic in origin.

correlation: the determination of identity of rocks observed in different areas according to age, fossil content, or physical continuity.

correlation spectrometer (COSPEC): an instrument that measures the output of sulfur dioxide from volcanic gas plumes.

cosmopolitan: of animals or plants, occupying distribution areas that span the entire or almost the entire globe.

cranial capacity: the internal volume (in cubic centimeters) of the braincase of a skull.

crater: an abrupt circular depression formed by volcanic eruption or by the impact of an asteroid or meteorite.

craton: a region of unaltered original magmatic crustal material, typically as continental shield structures that form the nucleus of a continental landmass.

creep: the slow downslope movement of soil and soil materials on a grade, caused by the force of gravity.

crestal plane: a plane or surface that goes through the highest points of all beds in a fold; it is coincident with the axial plane when the axial plane is vertical.

Cretaceous period: a period of time that lasted from about 146 to 65 million years ago, the end of which was marked by the extinction of the dinosaurs.

crust: the outermost layer of the earth composed of granitic continental crust material 30 to 35 kilometers thick and basaltic oceanic crust material 5 to 10 kilometers thick.

crustal discontinuity: a boundary within the crust that results from the different densities of crustal layers and that is detected by a change in the velocity of seismic waves passing through.

crystal: a solid form of a substance having a regular geometry and symmetrical faces resulting from the orderly arrangement of its component atoms or molecules.

Curie point, or Curie temperature: the temperature at which materials containing iron oxides lose their magnetic pattern and align with Earth's magnetic field.

cuticle: a thin, noncellular outer cover to the bodies of land plants and arthropods that is waterproof and waxy or proteinaceous.

cyanobacteria: an algaelike bacteria previously called "blue-green algae"; these oxygen-producing, photosynthetic microbes often live in matlike clusters.

décollement: the detachment surface beneath a fold belt, usually located in a weak layer of rock such as shale; the maximum depth to which the rocks are folded and faulted.

dacite: a fine-grained, extrusive volcanic rock of fairly silica-rich composition with plagioclase, alkali feldspar (a potassium, sodium, and aluminum-silicate mineral), and quartz (a silica mineral), similar to andesite and typically gray in color.

debris avalanche: a large mass of soil and rock that falls and then slides as a unit very rapidly downhill on a cushion of air.

debris flow: the movement of mobile slurry in which more than half the solid material is greater than sand size, often resulting from heavy rains and having velocities between 1 meter per year and 100 meters per hour.

deflation: the removal of unconsolidated particles of clay, dust, sand, and rock from dry soil by strong winds.

deoxyribonucleic acid (DNA): the stable nucleic acid molecule made up of two strands of nucleotides arranged in a double helix, in which the genetic composition of the corresponding living system is chemically encoded.

deposition: accumulation of weathered, wasted, and eroded sediments that have been transported by air, water, or ice.

***Dermopteran*:** a gliding squirrel-sized mammal probably most closely related to primates; the so-called flying lemur of Southeast Asia.

desert pavement: a ground surface covered by a thin layer of closely packed small, round, stones; the residual deposits left behind after the process of deflation.

desert (rock) varnish: a smooth, dark, hard, shiny or glazed iridescent (red, brown, black) coating, usually composed of manganese or iron oxides, that forms on exposed rock in desert environments by weathering of dust films and by microbial action.

detachment fault: a horizontal or gently dipping, regionally extensive fault; the hanging wall usually contains numerous smaller, steeper normal faults that end at the detachment fault.

diagenesis: the physical and chemical alterations of a sediment by which it is transformed into sedimentary rock.

diapsid: animals characterized by the presence of two apertures in the skull.

diastem: a short depositional break of geologically unmeasurable duration caused by a limited period of nondeposition.

differential weathering: physical and chemical weathering that occurs at irregular or different rates, caused by variations in composition and resistance of a rock or by differences in intensity of weathering.

diluvial theory: antiquated belief that the Mosaic flood was responsible for shaping many of Earth's surface features.

dip: the vertical angle between a fault plane and the horizontal plane, assessed perpendicular to the strike of the plane.

disconformity: an unconformity separating parallel strata and exhibiting physical evidence of erosion.

disjunct: of animals or plants, having a fragmented distribution area that includes two or more geographically separate regions.

dispersal: for any given species, the movement out of its distribution area of the minimum number of individuals it will take to colonize a new area.

distribution area: the geographic range of a particular group of organisms at any level of classification, such as species, genus, family, or class.

divergence: the movement of tectonic plates, or parts thereof, away from each other, as at a mid-ocean ridge or a rift.

divergent plates: tectonic plates that are moving apart as at a mid-ocean ridge or rift.

diversity: the variety of individual types present in an identifiable realm, usually described in terms of the number of species present, whether biological, geological, or something else.

doctrine of final causes: the belief that an overall purpose or design is revealed in the organization of nature.

dolomite: a mineral consisting of calcium magnesium carbonate compounds that often forms from precipitation from seawater.

dome: a small, steep-sided mass of volcanic rock formed from nonexplosive, viscous lava that solidifies in or above a vent.

drag: the component of force that slows the speed of an object moving through a fluid medium as a result of friction and turbulence.

drainage system: a network of stream branches bounded by topographical divides with a common outlet.

drift: an all-inclusive term for any kind of material that is deposited by glaciers and their meltwater.

drumlin: a smooth, elongated, oval-shaped hill or ridge, up to 60 m in height and 1 km in length, formed and molded from rocky debris by a moving glacier.

dryopithecines: extinct kinds of apes that walked and climbed like monkeys but that lacked the four-cusped molars of monkeys, the tree-swinging specializations of gibbons and orangutans, and the knuckle-walking specializations of chimpanzees and gorillas.

ductile shear zone: a planar zone of rock that accommodates relative movement like a fault, though movement has occurred by processes of solid-state flow rather than by fracture.

dyke: a sheet-like structure of impermeable igneous rock that has intruded into and across the planes of the layers of an overlying structure of preexisting rock strata.

Early Paleozoic: that part of geologic history from 542.0 to 359.2 million years before the present.

earthquake: a relatively violent shaking or trembling of locations on Earth's surface caused by a break or rupture that releases accumulated stress in the rocks at a fault in Earth's outer crust.

earthquake focus: the point within Earth's crust at which the accumulated stress at a fault is released and from which seismic waves of an earthquake are generated.

earthquake swarm: a number of generally minor earthquakes that occur closely spaced in time and focal location, often associated with a volcanic eruption.

ecogeographic rule: a principle describing correlations between the morphology of birds and mammals and the climatic or latitudinal gradients at which they are found.

ecology: the study of the relations between organisms and their environments.

ectosymbiont: a symbiotic organism living on the outer surface of another organism.

Ediacaran fossils: fossils of marine animals with moderately complex body structures (metazoans) that lived some 570 to 670 million years ago; the term "Ediacaran" is derived from the Ediacara Hills locality in south Australia.

ejecta: material thrown out by a volcano, such as rock fragments, lava and ash.

elastic rebound: in the process of the buildup of geologic stress between two plates moving in opposite directions at a fault, the release of the strain in the rocks that allows the two opposing points to achieve their normal relative locations and results in earthquakes.

element: a substance whose component atoms consist of one and only one type of atom, each having the identical number of protons in its respective nucleus.

elemental cycling: the process by which elements move in various forms through the planet's subterranean, ground-level, and above-ground environments.

Enantiornithes: a diverse group of Cretaceous birds that filled many of the avian niches now inhabited by modern bird groups.

endemic species: species occurring naturally in a particular area and environment.

endosymbiont: a symbiotic organism living within the body of another organism.

endosymbiotic theory: the concept that eukaryotes arose from prokaryotes by incorporating free living microbes into symbiotic relationships.

environment: the interaction of conditions affecting the existence, growth, and overall status of an organism.

enzyme: a proteinaceous substance that facilitates and thereby increases the rate of a specific biochemical reaction or type of reaction.

Eocene epoch: part of the Cenozoic era of geologic time, dating from 55.8 to 33.9 million years ago.

eolian (also aeolian): erosion, transportation and deposition of materials by the influence of wind.

epeiric sea: a shallow sea that temporarily (in geologic terms) covers a portion of a craton; also termed an epicontinental sea.

epeirogeny: uplift or subsidence of the crust within a region, without the internal disturbances characteristic of orogeny.

epicontinental sea: a sea covering part of a continental block, as opposed to a sea underlain by oceanic crust; generally less than 200 meters deep and the depositional site of most exposed sedimentary rocks.

epifauna: organisms that live on the sea floor.

epigenetics: the effect of environmental influences on inheritable characteristics with their eventual incorporation into the structure of the affected deoxyribonucleic acid (DNA).

epoch: a relative time unit designated as a subdivision of a period in the geologic time scale.

equilibrium line: the line or zone that divides a glacier into the upper zone of accumulation and the lower zone of wastage.

era: one of the major divisions of geologic time, including one or more periods.

erg or erg desert: any vast area of desert covered in sand or where excessive sand forms a "sand sea" applied especially to the Sahara Desert.

erosion: a process in which weathered rock particles are transported by water, wind, and other natural means.

erratic: a rock fragment that has been transported by a glacier and left resting on bedrock unlike that from which the fragment was derived.

eruption: volcanic activity of such force as to expel significant amounts of magmatic materials from a volcanic vent.

esker: a sinuous ridge of stratified glacial deposits believed to form as meltwater flows in a tunnel under stagnant ice, typically 3-30 m in height and up to 100 km in length.

eubacteria: minute, prokaryotic organisms that inhabit a range of habitats considerably greater in diversity than those occupied by other organisms; exclusive of archaebacteria, they constitute the majority of monerans.

eukaryotic cell: the cell type present in all animals, plants, fungi, and protists; each such cell has a distinct nucleus and mitochondria, chloroplasts, and other subcellular structures absent in prokaryotic cells.

eustasy: changes in global sea level resulting from a change in the absolute volume of available sea water and not due to uplifting or sinking of land masses.

evaporation: the transformation of a liquid to a gas that occurs normally at the surface of a liquid at ambient temperatures.

evolutionary trends: statistical directions of evolutionary changes; major trends have been toward increased body size, brain size, and complexity.

exfoliation: the splitting off of curving sheets from the outside of a body of rock, typically granite, normally relieving internal stresses; also called sheeting.

exoskeleton: a bony armor covering the outside of an animal; also called a cuticle when on an animal that does not have an internal skeletal structure.

exponential decay: a process of decomposition, particularly nuclear fission, that obeys the time-related function $A = A_o e^{kt}$.

extension: broadening of the expanse of a rock structure through forces acting in opposite directions, as with the forces exerted on crustal rock by the movement of underlying magma, generally accompanied by thinning of the overlying rock strata.

extinction: the disappearance of a species from the earth, as when the last representative of the species dies without leaving any potential offspring.

extremophile: an organism that has evolved to live in extreme environments, which may include extremes of temperature, chemical composition, and other variables of the physical environment.

facies: the characteristic set of features that distinguish a given rock, such as its physical appearance, composition, and method of formation.

family: a grouping of types of organisms having related genera (sing. "genus"), the family name always ending in *-idae* for animals or *-aceae* for plants.

fathom: a unit of length equal to 1.8 meters, used principally in the measurement of marine depths.

fault: a fracture in Earth's crust along which there has been some relative displacement or deformation.

faulting: the process of fracturing the earth such that rocks on opposite sides of the fracture move relative to each other; faults are the structures produced during the process.

fault line: the two-dimensional trace of a fault plane at the surface.

fault offset: the extent of visible misalignment caused by fault movement, apparent in a feature such as a road, creek bed, or tree line that crosses a fault.

fault plane: the relatively planar surface along which the strata, or layers of rock making up Earth's crust, have broken under tectonic forces, observable at the surface as the fault line.

fault scarp: a steep cliff that is formed when tectonic plates or opposite segments of a faulted terrane move vertically against each other along a fault plane.

fault slip: the direction and amount of relative movement between the two blocks of rock separated by a fault.

fault zone: the area around a particular fault plane within which numerous small fractures in Earth's surface have occurred.

faunal succession: the sequence of life-forms, as represented by the fossils within a stratigraphic sequence.

fissible: capable of undergoing nuclear fission.

fission: the splitting of a nucleus through the emission of nuclear particles to form atoms of a different element.

fissure: an extensive crack or fracture, such as a linear volcanic vent, along which there is a distinct separation providing an opening into an otherwise closed system.

flanks: the sides of a fold, mountain or ridge; also called limbs, legs, shanks, branches, or slopes; anticlines share syncline flanks, and synclines share anticline flanks.

flat: that portion of a thrust fault where the fault is subparallel to adjacent layers of rock; also called treads, flats are usually located in weak rocks such as shale or evaporite (salt, anhydrite, gypsum).

flood basalt: basaltic lava that has flowed over, or flooded, an extensive land area during an eruption forming a broad, flat basalt plain.

fluvial: processes related to or geological structures formed by water flowing in rivers and streams

fold: an upward or downward bend in layered rock strata caused by external pressure.

folding: the process of bending horizontal layers of rock so that they dip; folds include upward-arching anticlines, and downward-arching synclines and can be as small as millimeters and as big as kilometers long.

footwall: the upward-shifting side of a normal (or tension) fault or the downward-shifting side of a reverse (or thrust) fault.

foraminiferan: a single-celled aquatic animal with a shell composed of compacted minerals or grains of sand from surrounding sediment.

fossil: naturally preserved evidence of past life or organic activity, usually in sedimentary rock strata, such as the remains, trace, or imprint of a plant or animal that lived during the geologic past.

fossilization: the gradual process by which the remains of a once-living organism are naturally preserved, usually by mineralization in sedimentary rock.

fractography: the study of fracture surfaces to determine the propagation history of the crack.

fracture: the distinctive way in which a mineral form breaks.

fracture zones: large, linear zones of the sea floor characterized by steep cliffs, irregular topography, and faults; such zones commonly cross and displace oceanic ridges by faulting.

free fall: a type of slope failure where individual rocks on a steep slope break loose and fall to the ground; also known as rock fall.

fumarole: a volcanic vent from which only gases are emitted.

fusion: the combination of two atomic nuclei to produce a single nucleus of a higher element; the opposite of fission.

gabbro: a coarser-grained, low-silica, intrusive igneous rock; the equivalent of basalt formed by slow cooling of the parent magma.

gabions: boulders and rocks that are bound into wire mesh cages and placed in front of areas likely to experience heavy to moderate erosion.

gem or gemstone: a mineral that is valued for its color, sparkle, rarity, and hardness when cut and polished.

gene: the basic unit of heredity; a component encoding segments of the deoxyribonucleic acid (DNA) molecule; genes are located in the cell nucleus on chromosomes.

genus (plural: genera): a group of closely related species that share a common ancestry and similar morphological characteristics; for example, *Homo* is the genus of humans, including the species *Homo sapiens* and *Homo erectus*.

geode: a small, hollow rock whose interior cavity is lined with crystal formations or other mineral matter.

geological column: the order of rock layers formed during the course of Earth's history.

geologic map: a map illustrating the age, structure, and distribution of rock units.

geologic record: the history of Earth and its life as recorded in successive layers of rock strata and the fossil specimens they contain.

geologic terrane: a crustal block with a distinct group of rocks and structures resulting from a particular geologic history; assemblages of terranes form the continents.

geomorphic domain: a geologic area of characteristic features that define the appearance of the area.

geomorphology: a branch of geology that studies deformations in Earth's crust with regard to their origin, evolution, form, classification, and spatial distribution.

geosyncline: a major, elongated depression in the crustal rock formations in which great thicknesses of sedimentary and volcanic rocks generally accumulate; geosynclines lie parallel to the edges of continents and are long and narrow.

geothermal gradient: the rate at which temperature increases with depth in Earth's crust.

glacial erratic: see erratic.

glacial till: the unstratified and unsorted drift deposits believed to originate in the movement of ancient glacial ice across underlying layers of sediment.

glaciation: the process of being covered by an ice sheet, glacier, or major formation of land ice; also the period in which this process occurs.

glacier: a mass of snow and ice that persists for two or more years and is capable of viscous flow movement by internal deformation under the force of its own weight; it forms from snow in areas where accumulation of snow exceeds the speed of melting and sublimation; such systems of moving ice can occur at any time in high elevations and periodically will occur on the continents in the high latitudes during periods known popularly as ice ages.

glaciologists: scientists who specialize in the study of glaciers and ice.

gneiss: a coarse-grained metamorphic rock that shows compositional banding and parallel alignment of minerals.

gondwana: a supercontinent that emerged from the breakup of Pangaea in the Jurassic period and contained the continents that now constitute Africa, South America, Australia, and India.

graben: a roughly symmetrical linear crustal depression formed by the lowering of a crustal block between two normal faults that slope toward each other.

gradual evolution: the theory that evolution occurs throughout much of a species' existence, mostly at slow rates.

gradualism: the principal belief that evolution functions through slow, constant change over time.

granite: a coarse-grained, light-colored igneous rock produced by the underground cooling of magma and composed of feldspars, quartz, mica, and variable amounts of darker minerals.

granular flow: a type of sediment flow that is characteristic of sediment mixtures containing between 0 and 20 percent water.

gravity anomalies: differences between observed gravity readings and expected values after accounting for known irregularities.

grazing and cropping eukaryotes: single-celled (protists) or multicelled (metazoans) eukaryotes that appear in the fossil record during the Precambrian eon, about 1 billion years ago.

greenstone: a field term used to describe any altered basic igneous rock that owes its color to the presence of various green minerals.

groundwater: the water that exists in the pore spaces and fractures in rock and sediment (aquifers) beneath Earth's surface.

guyot: a seamount, whose top has been beveled flat by wave erosion, generally formed in shallow waters but now well below sea level due to transport by seafloor spreading.

habit: the general shape or form of a mineral as it occurs naturally, dependent upon the crystal structure of the material and the environmental conditions under which it was formed.

Hadean eon: the period of time before the Archean and lasting from 4.5 to 3.6 billion years ago.

half-graben: an asymmetrical structural depression formed along a normal fault where the downfaulted block has subsided only on one side and so is tilted toward the fault.

half-life: the length of time required for one-half of an amount of material to decompose or be consumed through an exponential decay process.

hanging wall: the downward shifting side of a normal (or tension) fault or the upward shifting side of a reverse (or thrust) fault.

harmonic tremor: a type of earthquake activity in which the ground undergoes slow, continuous shaking in response to subsurface movement of magma.

heterocercal: descriptive of a tail fin whose upper and lower halves have different radii, as is typical of the tails of sharks.

hinge: the line of maximum curvature or bending of a fold, also called the axis of syncline or axis of anticline.

holotype: the definitive example of a specimen, used to compare all others.

hot spot: the crustal region above a large rising plume of magmatic material in the mantle, presumed to be caused by an intense prominence of the outer core; as a tectonic plate moves over a mantle plume, a trail of hot spot volcanoes forms and Earth's surface bulges outward.

humid-midlatitude: land area with average temperature of the coldest month less than 18 degrees Celsius but at least eight months with average monthly temperatures greater than 10 degrees Celsius; this area has no dry season.

hydrologic cycle: a complete cycle by which water passes via phase changes from the oceans, through the atmosphere, to the land, and back to the ocean.

hydrostatic pressure: the pressure imposed by the weight of an overlying column of water.

hydrothermal (or geothermal): related to hot water, particularly involving the production or dissolution of minerals; referring to natural systems of hot fluids that originate underground.

hydrothermal fluid: a natural hot water steam that seeps through the ground.

hydrothermal vent: seafloor outlets for superheated, highly mineralized springs associated with seafloor spreading centers and that are often the site of deep-sea, chemosynthetic biological communities.

ice age: a period of time during which extensive ice sheets and glaciers exist, covering large terrestrial areas.

ice cap: a large, dome-shaped glacier less than 50,000 square kilometers in extent.

ice sheet: glacial covering of a land mass greater than 50,000 square kilometers in extent.

ice shelf: that portion of an ice sheet that extends into the ocean.

ichnofauna: the remains of creatures whose activities left trace fossil records.

ichnofossil: a trace fossil created by the activity of living animals that left impressions in the sediment of the earth.

ichnology: a branch of paleontology concerned with tracks, burrows, and other fossilized remains indicative of the activity of living animals.

igneous rock: types of rock formed when magma or lava cools and forms mineral crystals; it can form on the surface of the planet when volcanoes erupt, or it can form at depth, without reaching Earth's surface.

ignimbrite: an igneous rock deposited from a hot, mobile, ground-hugging cloud of ash and pumice, high in silica.

index (guide) fossil: the remains of an ancient organism that are useful as reference markers of the age of rock strata; index fossils are abundant and have a wide geographic distribution, a narrow stratigraphic range, and a distinctive form.

infauna: organisms that live all or part of their life cycle in underwater sediment, including bivalves, scaphoids, and others.

infiltration: the penetration of water through the ground surface into subsurface soil, or the penetration of water from the soil into sewer or other pipes through defective joints, connections, or utility hole walls.

inorganic compound: a chemical compound whose molecular structure is built around atoms other than carbon.

insectivore: any animal that eats insects, such as shrews, bats, and many birds.

insolation: the incoming solar radiation that strikes the planet; differences in global insolation at various places on Earth's surface create weather and climate patterns.

interglacial period: the interval of milder climate between two major glacial episodes.

internal drainage: the condition in which a river system has no outlet but drains instead into a saline lake or playa (a lake basin that contains water only shortly after a rainstorm).

internal flow: flow that occurs in glaciers as individual component ice crystals are slowly deformed, enabling the glacier to flow over irregular surfaces and around curves.

intrusive contact: contact resulting from injection of molten igneous rock into previously formed rock.

ion: an atom, molecule, or molecular fragment in which the total number of electrons is not equal to the total number of protons, resulting in a net positive or negative charge on the species.

isobath: the contour lines on maps indicating equal water depth below the surface.

isostatic adjustment: the adjustment of Earth's crust to maintain equilibrium by subsiding when loaded and uplifting when unloaded.

isostatic readjustment: rapid tectonic uplift or subsidence of continental areas in response to the addition or removal of the weight of overlying deposits of glacial ice or seawater.

isotope: atoms of an element having the same atomic number but differing atomic weights because different numbers of neutrons are present in their atomic nuclei.

isthmian links: chains of islands between substantial landmasses.

jökulhlaup: a flood of meltwater rapidly formed by the heat from an erupting volcano.

joint: a fracture in a rock across which there has been no substantial slip parallel to the fracture.

Jurassic period: a period of geological time that lasted from 199.6 to 145.5 million years ago, during which birds originated; it is considered the peak of dinosaur evolution.

kame: a conical hill of well-sorted or stratified drift, as sand and gravel, formed by deposition from a stream of glacial meltwater at the front or edge of a glacier.

karst: a geological formation created when permeable limestone dissolves, resulting in characteristic structures such as sinkholes, towers, caves, and ridges.

kettle: a depression created by the eventual melting of a remnant block of glacial ice that has become covered by glacial debris.

kingdom: one of the five large subdivisions of life, differentiated on the basis of gross body plan (single-celled versus multicellular) and method of obtaining food or nutrients (produces own food or obtains it from other organisms); the Animalia, Plantae, Fungi, Protista, and Monera.

knuckle walking: a form of locomotion found in gorillas and chimpanzees whereby the animal walks on all fours, with feet facing forward but hands curled backward so that the knuckles are placed on the ground.

labeling: the incorporation of atoms of a detectable radionuclide into the normal molecular structure of a material.

lacustrine: descriptive of sedimentary deposits formed from lake-bottom residues.

lagerstatte: an assemblage of exceptionally well-preserved fossils.

lahar: a highly mobile mudflow formed when volcanic ash and minor coarse volcanic particles mix with the abundant rainfall and meltwater that often accompanies an eruption.

land bridges: narrow land formations that connect landmasses.

landform: one of the many features that, taken together, make up the surface structure of Earth.

landscape: broad term for the land setting and topography of an area.

landslide: a sudden, rapid downward movement of soil and rock from a cliff or steep slope.

Laramide revolution: an orogeny that occurred between the Late Cretaceous and the Early Tertiary periods of geologic time, instrumental in producing the ranges known as the Rocky Mountains in North America.

Late Precambrian: that part of geologic time from about 1 billion to 542.0 million years before the present.

lateral accretion: the process by which crustal material becomes conjoined with a continental shield by horizontal compression forces at a subduction zone.

latitudinal gradient: of plant or animal species, an increase in overall numbers found when moving from higher to lower latitudes.

Laurasia: a northern supercontinent that broke away from Pangaea in the Jurassic period and contained the landmasses of North America, Greenland, Europe, and Asia.

lava: molten rock of volcanic origin when above the surface and the extrusive igneous rock that it forms upon cooling.

lava tube: an extended cavern-like structure formed by the draining out of liquid lava from a pahoehoe flow, leaving the surrounding solidified crust.

lift: the component of force acting on a wing in an upward direction as it moves through air.

limestone: a common sedimentary rock containing the mineral calcite, a form of calcium carbonate originating from the fossil shells of marine plants and animals or by chemical precipitation from seawater.

limiting factors: the physical, chemical, and biological factors that govern the distribution of a given organism.

lithic: having to do with rock.

lithification: the process by which soft particulate sedimentation is transformed into hard, compact sedimentary rock.

lithology: a term loosely used by many geologists to refer to the study of the mineral composition and texture rock.

lithosphere: the hard outermost layer of the earth, consisting of both the crust, made up of a collection of rigid plates, and the upper mantle above the asthenosphere; differentiated into oceanic lithosphere and continental lithosphere.

lithospheric plate: any of the several crustal segments of various sizes that constitute Earth's outer crust or lithosphere, capable of movement over the semifluid material of the planet's asthenosphere; their borders are outlined by major zones of earthquake activity.

Little Ice Age: a short-term cooling trend that lasted from about 1450 to 1850, during which mountain glaciers all over the world advanced considerably beyond their present limits.

littoral drift: displacement of sediment, sand, gravel, and debris along the shore.

luster: the perception of light reflected from the surface of a mineral.

macrofossil: a fossil that is large enough to study with the unaided eye, as opposed to a microfossil, which requires a microscope for examination.

mafic rock: rock that contains relatively large amounts of magnesium and iron, found mainly in the oceanic crust and upper mantle; from "magnesium" and "ferric."

magma: molten rock material, including any dissolved gases and suspended crystals, formed deep in Earth's interior; known as lava when ejected or extruded from a volcano.

magmatic: having its origin in the melted rock, or magma, of the planet's mantle layer.

magnetic anomalies: linear areas of ocean crust, parallel to the crest of the mid-ocean ridges, in which reversals of the direction of the paleomagnetic field are observed.

mammary glands: the glands that female mammals use to nurse their young; breasts.

manganese nodules: lumps of minerals consisting mostly of iron, manganese, nickel, and copper that form on deeper parts of continental shelves.

mantle: the layer of Earth's interior between the thin crust and the dense core, either semisolid (lower mantle) or plastic and nonbrittle (upper mantle), making up most of the volume of the planet.

maria: Latin plural meaning "sea," used to describe the moon's dark areas; the light-colored areas of the moon are called "highlands."

marsupials: mammals that give birth to a premature embryo that then finishes its development in a pouch in the mother's abdomen.

mass balance: the net gain or loss of ice and snow mass on a glacier in a year.

mass extinction: an event in Earth's history in which numerous species became extinct on a global scale and at a rate much faster than the normal level of extinction for Earth as a whole.

mass wasting: a general term for the downslope movement of soil and rock under the direct influence of gravity.

matrix: fine particles, typically of mud or silt, that occupy the spaces between larger grains of sediment and so enclose them.

mechanical process: the physical action of forces on matter or material systems.

mechanical weathering: the physical breakdown of rocks into progressively smaller particles.

meltwater: water derived from the melting of ice and snow, often in reference to a glacier.

Mesozoic era: the geologic era spanning the period from about 251.0 million years ago to about 65.5 million years ago, consisting of the Triassic, Jurassic, and Cretaceous Periods.

metallogenesis: the process by which metallic ores are formed.

metamorphic rock: sedimentary rock chemically and physically altered by means of intense heat and pressure.

metamorphism: the change in the mineral composition or texture of a rock because of heat, pressure, or the chemical action of fluids in the earth.

metazoan: a grade of organization of living organisms in which the cells are specialized for various functions and cooperate for the good of the whole organism; another name for animals.

meteorite: a fragment of an asteroid that survives passage through the atmosphere and strikes the surface of the earth.

methodological uniformitarianism: that aspect of uniformitarianism that describes a procedural approach to the study of Earth; that is, the notion that ancient phenomena can be interpreted in terms of present-day natural processes.

microbe: any living organism too small to be seen with the naked eye, many of which are capable of causing fermentation or disease, including viruses, bacteria, yeasts, protozoa, and algae.

microbial mat: a rubbery sheet comprising layers of different species of bacteria and microbes.

microbialite: a biogenic sedimentary structure found fossilized in sedimentary rock strata of various ages and that is attributed to the life activities of monerans.

microcontinent: an independent lithospheric plate that is generally much smaller than a continent but that consists of continental-type crust; examples include Cuba and Japan.

microfossil: a fossil or fossil fragment that cannot be seen by the unaided human eye; the characteristic imprint left by a microscopic organism in a geological formation.

microorganism: a very small, usually one-celled, organism such as algae, bacteria, and protozoans.

mid-ocean ridge: a volcanic ridge on the floor of the ocean basins where a divergent boundary between two oceanic plates occurs; volcanic eruptions and extrusion of basaltic magma occur along these ridges due to seafloor spreading.

millipede: a kind of arthropod with an elongated body of many, nearly identical segments, each with a pair of limbs on each side.

miocene: geological epoch of the Tertiary period in the Cenozoic era, from 23.0 to 5.3 million years ago.

mitochondrion (plural: mitochondria): the eukaryotic organelle in which energy is generated by aerobic respiration; mitochondrial DNA is used to estimate the closeness of genetic relationships between groups of organisms.

Mohorovičić (moho) discontinuity: the boundary between the crust and the upper mantle that was first defined by a rapid change in seismic velocities; it separates low-density crust from the denser mantle.

mold: a fossil that displays the form of the original organism in negative relief.

monerans: generally, single-celled organisms that often grow in colonies and that have a prokaryotic cell structure.

monocline: a bend or dip in an otherwise horizontal portion of strata.

monogenetic: pertaining to an eruption in which a single vent is used only once.

monotremes: primitive mammals, including the platypus and echidna, that lay eggs and have other archaic features.

moraine: rocky debris that has been transported by glacial movement and deposited as an arcuate ridge, often deformed, consisting of unsorted and unstratified material deposited at a glacier margin.

morphologic evolution: changes in the body or skeleton shape of organisms through time.

morphology: study of the appearance (shape and form) of an organism, especially in regard to its soft tissue structure.

motile: having the ability to move spontaneously and independently.

mountain glacier: a glacier confined to a sloping valley within a mountain range.

mudflow: a general term for a flowing mass of predominantly fine-grained earth material that possesses a high degree of fluidity during movement and moves at speeds greater than 1 kilometer per hour along valley floors.

mutation: a spontaneous or induced change in a gene; the ultimate source of variation on which natural selection acts.

nappe: a complex, large-scale strata formation in which some beds have been overturned due to extreme folding pressure.

natural bridge: a bridge over an abandoned or active watercourse; in karst topography, it may be a short cave or a remnant of an old, long cave.

natural selection: the main process of biological evolution; the production of the most offspring by individuals with the most adaptive traits.

nematophyte: an extinct kind of nonvascular land plant consisting of flat sheets, stems, or trunks, supported by densely interwoven, microscopic, proteinaceous tubes.

Neptunists: adherents to Abraham Werner's belief that granite and basalt formed by chemical precipitation from seawater.

niche: in an ecological environment, a position particularly suitable for its inhabitants.

nictitation: the movement, such as blinking, of a specialized protective membrane in creatures without eyelids to cover and uncover the surface of the eye; the movement of a nictitating membrane.

nomenclature: the names and terms used in a classification system.

nonconformity: an unconformity in which rock below is homogeneous igneous or metamorphic rock and rock above is stratified.

normal fault: a steep fault that forms during extension, or stretching, of Earth's surface; the rock on one side of the fault (the hanging wall) moves downward relative to the rock on the opposite side of the fault (the foot wall).

notch or nip: an erosional feature found at the base of a sea cliff, the result of undercutting by wave erosion, bioabrasion from marine organisms, and dissolution of rock by groundwater seepage.

nucleic acid: a molecule, such as deoxyribonucleic acid (DNA) or ribonucleic acid (RNA), that regulates the formation of proteins; essential to reproduction in living systems.

nucleotide: a molecule made up of a purine or pyrimidine base linked to a sugar molecule and a phosphate group, capable of carrying genetic information.

nuée ardente: a hot cloud of rock fragments, ash, and gases that suddenly and explosively erupts from some volcanoes and flows rapidly down the volcano's slope during a violent eruption.

obduction: a tectonic collisional process in which heavier oceanic crust is thrust up over lighter continental crust, opposite in effect to subduction.

obsidian: a dense variety of silica-rich glassy volcanic rock formed by rapid cooling of granite at low pressure, preventing the formation of crystals; the same material as pumice but not containing gas bubbles.

ocean basins: the large worldwide depressions between continental masses in which the major component of Earth's liquid water supply has collected.

oceanic crust: the portion of Earth's crust or lithosphere that forms the floors of oceans and is produced from basaltic magma that is extruded along volcanic mid-ocean ridges.

oceanic ridges: long, narrow elevations of the sea floor, some 2 to 3 kilometers higher than the surrounding ocean basins, that are associated with the creation of new seafloor material.

ocean trenches: long, deep (greater than 6,000 meters deep), and narrow depressions in the sea floor with relatively steep sides marking the boundaries between oceanic and continental crust and associated with the subduction of oceanic crust materials.

offshore transport: movement of sediment or water perpendicular to and away from the shore.

Oldowan: a cultural tradition dating from the Lower Pleistocene age at Olduvai Gorge, characterized by stone tools made from rounded pebbles crudely chipped to a sharp edge on one side.

Oligocene: geological epoch 33.9 to 23.0 million years ago, during the Tertiary period of the Cenozoic era.

olivine: a mineral of differing shades of green, composed of magnesium iron silicate.

oncolite: an organosedimentary rock structure, concentrically laminated, formed by blue-green algae and smaller in size than a stromatolite.

onshore transport: movement of sediment or water perpendicular to and toward the shore.

ophiolite suite: a unique vertical sequence of peridotite rock overlain by gabbro, basalt, and oceanic sediments representative of ancient seafloor material.

order: a group of closely related biological families; order names end in *-formes* (birds and fish), *-ales* (fungi and plants), or *-a* (all other animals).

Ordovician: a time period covering the interval from 488.3 to 443.7 million years ago, following the Cambrian period.

ore: a rock structure from which a metal or mineral can be extracted.

organelles: subcellular membrane-bounded units that perform specific functions within the eukaryotic cell.

organic compound: a compound whose molecular structure is built around a central framework of carbon atoms.

organism: any particular living thing, whether a microorganism, plant, or animal.

organosedimentary structure: a sedimentary rock feature developed by the life processes of blue-green algae.

orogenesis (or orogeny): a deformation of a region of Earth's surface due to contact between tectonic plates leading to the formation of mountain ranges.

orogenic belt: a belt of crust that has been severely compressed, deformed, and heated by convergence of crustal plates, at which orogenesis, or orogeny, takes place.

Osteichthyan: any member of the class of fish having skeletons of true bone, which includes most species of fresh and saltwater fish.

oviparous: of creatures that reproduce by means of eggs that may be fertilized externally or internally, but are incubated and hatched outside the body.

ovoviviparous: of creatures who produce offspring by the internal fertilization, incubation, and hatching of eggs, followed by expulsion of live young that have developed to fully functional form.

pahoehoe flow: a Hawaiian term indicating a lava flow having a ropy or billowy surface of basaltic composition.

palagonitization: the consolidation of loose tephra into tuff through complex chemical reactions under the influence of heat and pressure.

paleobiogeography: the study of the geographic distribution of past life-forms.

paleobiology: the study of the most ancient life-forms, typically through the examination of microscopic fossils.

Paleocene period: the earliest geologic period of the Cenozoic era, lasting from 65.5 to 58.8 million years ago.

paleoclimatology: the study of the climate of an extinct ecosystem as determined through geological evidence and the distribution of fossils.

paleoecology: the study and use of fossils and other geological evidence to reconstruct paleoenvironments or ecosystems from the past.

paleoenvironment: the extinct environment of a certain area during a chosen point in the history of the region.

paleomagnetism: the planet's magnetic polarity signature embedded in molten rock as it cools and solidifies, which provides evidence of the history of Earth's magnetic field and the movements of continents.

paleontologist: a scientist who studies ancient life; invertebrate paleontologists study fossil invertebrate animals, vertebrate paleontologists study fossil vertebrates, and micropaleontologists study microfossils.

paleontology: the study of ancient life-forms and their evolution as revealed through the analysis of fossils.

Paleozoic era: the era of geologic time from 542.0 to 251.0 million years ago, consisting of the Cambrian, Ordovician, Silurian, Devonian, Carboniferous, and Permian periods.

Pangaea: the supercontinent containing all continental crust that existed at the beginning of the Mesozoic era.

panspermia: a theory that life can be carried between planets in the form of bacteria integrated into the mineral components of meteorites.

Panthalassa: a single ocean that existed during the Permian and Triassic periods and that surrounded Pangaea.

paraconformity: an unconformity for which the only evidence is a gap in the fossil record.

parsimony: the theory that, given unexplained phenomena, one should assume the simplest possible explanation for the observed data.

parthenogenic: organisms in which unfertilized females produce viable, fertile offspring without copulation with males of the species.

partial pressure: the proportional pressure exerted by a particular gas relative to the total pressure of a gas mixture.

paterae: inverted, saucer-shaped geological features considered to be of volcanic origin.

pedon: the basic unit of a soil sample envisioned as a hexagonal section extending from the surface to the deepest extent of roots or to underlying bedrock.

peridotite: a coarse-grained igneous rock that consists of olivine, pyroxene, and garnet; it is likely the main rock in the upper mantle of the earth.

periglacial: regions bordering glaciers but not actually covered by ice.

period: a unit of geological time forming part of an era and subdivided, in decreasing order, into epochs, ages, and chrons.

permineralization: the filling of pores and cells with minerals without changing the material surrounding the pores or cells.

petrographic: having to do with the branch of science that describes or analyzes rocks by microscopic study.

petrology: study of the origins, mineral compositions, and other characteristics of rock.

Phanerozoic eon: the period of geologic time with an abundant fossil record, extending from about 544 million years ago to the present.

photosynthesis: the biological process in which chlorophyll and sunlight act upon carbon dioxide to form energy-rich glucose and free oxygen.

phreatic: a zone in the ground below the level of complete water saturation.

phylogenetics: a field of study in which the genetic profiles of long-extinct organisms are determined and their respective similarities to and differences from modern species are established.

phylogeny: the study of the evolutionary relationships between organisms.

phylum: a major grouping of organisms of similar class, distinguished on the basis of basic body plan, grade of anatomical complexity, and pattern of growth or development.

physical weathering: a form of weathering in which rocks are broken down by physical and mechanical influences in their environmental conditions, crystal growth, or interaction with animals or vegetation.

pillow basalt or lava: spheroidal, or "pillow shaped," masses of bulbous, glassy-skinned lava that form only when basaltic magma flows erupt under water.

pistil: the female organ of flowers, which receives the pollen.

placentals: mammals that carry the embryo in the mother until it is born in a well-developed state; it is nourished in the womb by a membrane (the placenta).

placer: a deposit of sand or gravel containing eroded particles of valuable minerals.

placoderm: any member of the extinct class of ancient armored fish, characterized by the presence of external skeletal plates enclosing the front half of the body.

planetary wind system: a global atmospheric circulation pattern, as in the belt of prevailing westerlies.

plastic: that which is capable of being permanently and continuously deformed in any direction without fracturing.

plastron: the bony plate that forms the ventral side or underside of a turtle or tortoise.

plate: a large segment of the lithosphere, either continental or oceanic, that is internally ridged and moves independently over the interior, driven by circulating currents of molten rock in the underlying mantle.

plate tectonics theory (or plate tectonics): the concept that Earth's surface is composed of rigid plates continually in motion over the interior; it combines the ideas of subduction and seafloor spreading to create a theory of continental movement and formation.

playa: a flat, dry lake-bed usually characterized by clay or salt deposits, where rainwater accumulates temporarily but eventually evaporates or moves underground; also called a pan or a dry lake.

Pleistocene epoch: a geologic time period from 1.8 million to 11,477 (± 85) years ago, featuring several expansions and contractions of continental ice sheets.

Plinian eruption: a violent, explosive type of volcanic eruption characterized by a large, vertical plume of volcanic ejecta that subsequently collapses; named for Pliny the Elder, a Roman naturalist who died while observing the eruption of Mount Vesuvius in 79 C.E..

Pliocene: the period of geologic time from 5.3 to 1.9 million years ago.

plunge: the inclination and direction of inclination of the fold axis, measured in degrees from the horizontal.

pluton: a deep-seated igneous intrusion that has cooled slowly within the crust.

plutonic: having its origin and focus in the planet's mantle in the asthenosphere below the level of the crust.

plutonism: magma intrusions and seismic activity having deep mantle sources.

plutonists: adherents to James Hutton's belief that granite and basalt formed by crystallization from molten mineral matter.

pluvial period: an episode in geologic time during which rains were abundant.

pollen: the male cells involved in fertilization of plants.

polygenetic: pertaining to volcanism from several physically distinct vents or repeated eruptions from a single vent punctuated by long periods of quiescence.

porosity: the extent to which a given material, such as rock, contains minute holes or gaps through which air or liquid can travel.

porphyritic rock: igneous rock with large mineral crystals embedded in much finer-grained minerals.

Precambrian: the interval of geologic time from the formation of the earth (4.6 billion years ago) to the beginning of the Cambrian period (542.0 million years ago), which includes nearly 90 percent of geologic time.

pressure melting temperature: the temperature at which ice will melt under a specified pressure; under pressure, liquid water can exist even at temperatures below freezing.

primordial soup: a combination of molecules and gases that were present in the seas at the origin of life and may have contributed to the formation of the first organic compounds.

prokaryote: the cell type found in the kingdom Monera, characterized by a number of criteria, including the absence of a cell nucleus and membrane-bound organelles such as mitochondria and chloroplasts.

protein sequence: the order of amino acids forming a polypeptide chain, which in turn forms part or all of a protein molecule.

protocell: a theoretical structure representing a midpoint between nonliving chemical systems and the earliest true cells.

Proterozoic eon: the late Precambrian era, from about 3.5 billion to 570 million years ago, before the proliferation of macroscopic life.

pteridophyte: simple vascular plants, such as club mosses and horsetails, that have leaves, stems, and roots but no seeds or flowers.

pterosaur: formerly called pterodactyl, any of a group of prehistoric flying reptiles that evolved during the Triassic period and were the dominant flying organisms until the end of the Cretaceous period.

pumice: a solidified volcanic froth of a glassy texture commonly produced during explosive eruptions of rhyolite magma.

punctuated equilibria: an alternative model of the functioning of evolution through rapid speciation events that involve major changes in morphology interspersed with periods of stability marked by little change.

pyroclastic flow: a turbulent, intensely hot, dense mixture of volcanic gases, magma fragments, ash, and rock that is ejected from a vent and flows rapidly down the flanks of a volcano.

pyroclastic material: rocks formed from bits of expelled lava and the debris of explosive volcanic eruptions.

pyroclastic rock: fragmented rock produced during a volcanic eruption (the term includes essentially all volcanic products except lava flows).

pyroclastics: fragmentary igneous rocks that are formed by the forceful ejection of volcanic materials into the atmosphere.

QFL: the three main types of minerals found in sedimentary/clastic rocks: quartz, feldspar, and lithic fragments.

quartz: a form of silicon dioxide or silica, one of the most common minerals on Earth's surface; it occurs in many different forms, including agate, jasper, and chert.

radiometric dating: the use of radioactive elements that decay at a known rate to determine the ages of the rocks in which they occur.

radionuclide: a specific radioactive isotope of an element.

rain shadow: a dry region that forms on the side of a mountainous area protected from the prevailing winds and, therefore, also from the moisture they carry.

ramapithecines: extinct kinds of apes in which the males had prominent canines as in other apes but the females had smaller canines as in humans; otherwise, these creatures were like dryopithecines or orangutans.

ramp: that portion of a thrust fault where the fault cuts across a layer of relatively stiff rock at a higher angle than does the rest of the fault.

recurrence period: the range of time between successive earthquakes.

reflection: the action of a wave in bouncing off of a surface rather than entering the material.

refraction: the change in direction of a wave that passes through a surface rather than being reflected away from it.

regelation: the thawing of ice under pressure followed by refreezing at that pressure.

regolith: the layer of mechanically and chemically weathered rock, sediment, mineral, ash, soil and organic matter covering solid, unweathered bedrock.

regression: a shift of sedimentary facies away from land, commonly caused by the retreat or withdrawal of the sea from land areas, and the evidence of such a withdrawal.

relative date: a date that places an artifact as older or younger than another object, without assigning a specific age for it.

remanent magnetism: the magnetic field imposed upon a rock at the time of its formation in accord with the global magnetic field.

remotely operated vehicle (ROV): a submersible or other robotic device operated from a remote location, such as an unmanned minisub operated by tether from a surface ship, the Mars Rovers, etc.

remote sensing: any of a wide variety of techniques, such as aerial photography, used for collecting data about Earth's surface from a distance.

replacement: a process in which the organic materials of an organism are replaced by mineral deposits after the animal's death, leading to the formation of a body fossil.

reptiles: any of a large class of animals that evolved from amphibian ancestors during the Carboniferous period. characterized by the evolution of scaled skin and egg-laying reproduction that does not depend on a body of water.

resurgent dome: a broad, oval area of uplift within a volcanic caldera that is marked by the upwarping and fracturing of caldera-filling deposits.

reverse fault: a steeply to moderately inclined fault in which the overlying block of rock (the hanging wall) has moved upward relative to the underlying block (the foot wall).

rhizome: an organ of vascular plants with the anatomy of a stem rather than of a root but running along or under the ground more like a root than a stem.

rhyniophyte: an extinct kind of vascular land plant with rhizomes and simple spore-bearing organs but lacking true roots or leaves.

rhyolite: a light-colored, fine-grained extrusive volcanic rock like granite, composed of a viscous lava containing about 70 percent silica, it erupts nonexplosively as thick, slow-moving lava flows and explosively as widespread sheets of frothy pumice.

ribonucleic acid (RNA): a complex compound made up of nucleotides that acts as a template, or "messenger," in the replication of DNA.

ribosome: a large multienzyme associated with cell nuclear materials at the endoplasmic reticulum, composed of ribosomal RNA and protein molecules, by which new protein molecules are constructed.

rift: an area of Earth's crust where tensional forces tear the crust and upper mantle apart, producing a series of faults between which are down-dropped blocks in the central portion, eventually forming a large distinctive valley.

rifting: the splitting of continental masses into separate blocks by the force of convective magma currents that act to drive one from the other as tectonic plates.

rift propagation: the lateral movement of a rifting process that leads to the prying open of a section of the lithosphere, accompanied by the formation of igneous rocks.

right-slip: transverse motion across a fault in which the block on the opposite side of the fault appears to have been displaced to the right; left-slip faults are the opposite.

rill erosion: the formation of small canyons and gullies that follow soil contours, caused by the runoff of accumulated rainfall.

riprap: a protective layer of rock facing a riverbank or levee as a countermeasure to prevent erosion.

rock cycle: a process in which all three basic types of rock are formed, brought to Earth's surface, weathered, and returned to Earth's interior in a cyclical pattern.

rockfall: a relatively free-falling movement of rock material from a cliff or steep slope.

saltation: the gradual movement of sand grains as wind or flowing water causes grains to impact against other grains, bumping them forward in the process.

sand spit: low tongue of land or relatively long, narrow shoal of sand extending from the land.

sandstone: a type of clastic rock composed of sand-sized particles.

sauropods: any of a diverse group of quadrupedal, herbivorous dinosaurs characterized by the evolution of large body size and elongated necks.

savanna (tropical wet-dry): land area with monthly temperatures averaging greater than 18 degrees Celsius and with one to six months of average monthly precipitation of less than 6 centimeters.

scabland: a region characterized by rocky, elevated tracts of land with little soil cover and by postglacial dry stream channels.

schist: metamorphic rock with subparallel orientation of micaceous minerals that dominate its composition.

seafloor spreading: the process whereby crustal plates move away from mid-ocean ridges, creating new crustal material as molten rock moves upward through rifts at the ridge crests.

seamount: isolated elevations on the sea floor, usually rising to higher than 1,000 meters, that are commonly cone-shaped, reflecting their volcanic origin; a flat-topped seamount is known as a guyot.

sediment: solid matter, either organic or inorganic in origin, that settles on a surface; it may be transported by wind, water, or glaciers.

sedimentary facies: different types of sediment that collected simultaneously in adjacent areas of the sea floor.

sedimentary rocks: rocks formed by the consolidation of an accumulation of particles of other rocks, organic skeletons, chemical precipitates, or some combination of these.

sediment flow: debris that flows downslope; mixed with air or water.

sedimentology: a branch of science dealing with the formation, classification, and study of sediments and sedimentary rock.

sediment transport: the movement of sediment from one location to another.

seismic activity: disturbances of the crust related to earthquakes or earth movements, often associated with zones of seafloor subduction and ocean-ridge formation.

seismic gap: a fault region known to have experienced earthquakes but at which none have occurred within that area's most recent recurrence period.

seismic reflection: study of the layered sediments in ocean basins by measuring the properties of sound waves sent into the sea floor and reflected back from the different underlying rock layers.

seismic refraction: examination of the deep structure of the ocean crust using powerful sound waves that are refracted by the underlying crustal layers rather than being immediately reflected back to the ocean surface.

series: a time-rock unit representing rock deposition during a geologic epoch, as subdivisions of layers of rock based on their age.

serpentine: a group of minerals, typically dark green in color, formed of hydrous magnesium silicate.

shale: sedimentary rock composed of silt and clay particles.

shelf dams: geologic formations that hold back sediments on the continental shelf.

shield: a continental block of the Earth's crust that has been stable over a long period of time.

shield volcano: a volcano in the shape of a flattened, broad, and low dome built by flows of very fluid basaltic lava or ash flows.

silicate: a substance whose crystal or molecular structure includes tetrahedral units of silicon surrounded by four oxygen atoms.

silicification: a process in which molecules of silica that are dissolved in groundwater replace the organic material in fallen tree trunks and branches.

silt: sediment that is coarser than clay but finer than sand, having particle sizes between 0.002 and 0.02 mm, often constituting the matrix of clastic rock.

sinkhole: a hole or depression in the landscape produced by dissolving bedrock, ranging in size to openings kilometers wide and hundreds of meters deep.

sinking stream: a stream or river that loses part or all of its water to underground waterways that have formed as underlying bedrock has dissolved; also called a disappearing stream.

sinter: a variety of nonprecious opal that forms around chloride hot springs and geysers.

slide: a type of slope failure that occurs when rocks or debris move spontaneously down a preexisting surface under the force of gravity; also known as translational slide.

slip: a measurement that describes both the direction and the magnitude of the movement of the rocks on either side of a fault plane, in relation to each other.

slip face: the sloping side of a dune that faces away from the prevailing wind; also called lee slope.

slope failure: a sudden failure of a slope resulting in the rapid movement of debris downslope.

slump: a type of slope failure involving the downward rotation of rock or regolith along a concave upward curved surface; also known as rotational slide.

slurry flow: a type of sediment flow in which the mobile material is a mixture containing between 20 and 40 percent water.

snout: the leading edge and lowest point of a glacier that is affected by the warmest weather, commonly experiencing rapid melting and fracturing that produces deep crevasses and ice falls.

snow avalanche: the rapid downslope movement of a mass of snow on a cushion of trapped air and driven by gravity.

soil: the dynamic mixture of mineral and organic materials overlying the bedrock of Earth's surface.

soil horizon: the distinct layers identified in a soil sample or pedon.

solifluction: a slow downslope movement of saturated material at temperatures above freezing, measured at a rate of centimeters per year.

sounding: the measurement of depth; a sounding line is a line used for the measurement of depth.

speciation: the process by which new species arise or populations of the same species become reproductively isolated from each other, either by geographical separation or some other mechanism.

species: a group of similar organisms capable of interbreeding and producing fertile offspring.

speleothem: a mineral deposit formed within a cave.

spore: a small reproductive or resting body from which a new offspring develops, typically single-celled with a tough, acid- and desiccation-resistant, external, proteinaceous coat, produced by lower plants, fungi, and some algae and protozoans.

spring: a place where groundwater reappears at the surface, where the water table and the surface coincide; in karst topography, a spring often represents the discharge point of a cave.

stage: a time-rock unit representing rock deposition during a single geologic age.

stamen: the male reproductive organ of a flower, on which pollen is produced.

steinkern: an internal mold that preserves the interior form of an organism in negative relief.

stoma (plural: stomata): a pore in the surface of green vascular land plants that can be opened and closed by the flexing of adjacent guard cells to control gas exchange with the atmosphere.

strandline: the position or elevation of the portion of the shoreline between high and low tide (at sea level); usually synonymous with "beach"and "shoreline."

stratified drift: a sorted, layered accumulation of sediments derived from glacier ice but subsequently reworked and redeposited by meltwater.

stratigraphic sequence: a set of rock units that reflect the geologic history of a region.

stratigraphic time scale: the chronology of the evolution of life on Earth broken down into time periods based on changes in fossil forms in the sequence of rock layers, with the period names reflecting specific localities or characteristics

stratigraphic unit: any rock layer that can be easily recognized by specific characteristics, such as color, composition, or grain size.

stratigraphy: a branch of geology concerned with studying rock layers and the processes involved in sedimentary layering, including the age, deposition, and composition of sedimentary rock.

stratovolcano (or composite volcano): a large, steep-sided volcano consisting of alternating layers of coherent lava and explosively ejected fragmental material.

stratum (plural: strata): a single bed or layer of sedimentary rock.

stream equilibrium: a state in which a stream's erosive energy is balanced by its sediment load such that it is neither eroding nor building up its channel.

striation: a pattern of parallel scratches cut into bedrock by the passage of a glacier.

strike-slip fault: a fault across which the relative movement is either left- or right-lateral.

stromatolite: a fossilized biogenic sedimentary layered structure produced by sediment trapping, binding, or precipitation as a result of the growth and metabolic activity of photosynthetic microorganisms, principally cyanobacteria (blue-green algae).

subcontinent: a subsidiary continental landmass that is less extensive in size and has a smaller variety of terrains than the larger continental landmass with which it has amalgamated.

subduction: the process occurring at colliding plate boundaries when denser oceanic crust material is forced downward and into the upper mantle by an overriding mass of continental crust material.

subduction zone: a lithospheric plate margin where a converging oceanic plate is forced to slide down into the asthenosphere under the opposing converging plate; a subduction zone forms a deep trench on the ocean floor.

sublimate: solid, crystalline material that is deposited directly from the vapor state, such as crystals of native sulfur around the mouth of a fumarole.

sublimation: a chemical process involving transformation from a solid to a gaseous state without passing through an intermediate liquid state.

submarine canyon: a channel cut deep into seafloor sediments by rivers or submarine currents.

subsidence: a downward vertical movement of an area of the ground, commonly caused by land-cover changes of the surface or by removal of supportive subsurface water or groundwater.

subspecies: the third part of a scientific trinomial, assigned to one of two groups that can mate and produce fertile offspring, but that have some different basic characteristics from other members of the same species.

substantive uniformitarianism: that aspect of uniformitarianism that draws conclusions about the history of Earth based on the idea that natural processes have been taking place at constant rates through time and that the earth has had a cyclic history.

subsurface water: the water in the root zone of the soil and groundwater flowing or stored in the rock mantle of the earth.

supercontinent: a continent formed by the amalgamation of two or more of Earth's continental landmasses into a single larger landmass.

surface water: water flowing in lakes, rivers, and stream channels or held in still pools on the terrestrial surface.

surficial gravels: alluvium found at different places on Earth's surface, originally interpreted as the product of the Mosaic flood.

suture zone: a narrowly definable region of Earth's crust thought to represent a place where two lithospheric plates have collided and subsequently been joined together.

symbiosis: a condition in which organisms live together in close association in the same environment, especially when the relationship is mutually beneficial.

synapsids: the mammal-like reptiles possessing an aperture, or synapse, below the squamosal and postorbital bones of the skull, which lived from about 300 to 200 million years ago and evolved into mammals.

syncline: a structure of stratigraphic folds created when rock strata arch downward; the limbs of the fold dip in opposite directions and the newest rocks typically are exposed in the middle of the fold.

system: a time-rock unit representing rock deposition within a geologic period.

talik: a region of unfrozen ground on, in or below discontinuous permafrost, frequently containing water under pressure.

talus: angular pieces of broken rock that typically collect at the base of the cliff from which they originated.

taphonomy: the study of all the processes that take place between the death of an organism and its discovery as a fossil.

taxon: one of several systematic groups or categories into which a particular organism may be placed.

taxonomy: the branch of biology concerned with classifying and naming living and fossil organisms, or the classification and scientific name of a living or fossil group.

tectonic: of or relating to the large-scale geological processes that both shape the structure of Earth's crust and take place within it; from the ancient Greek word meaning "builder."

tectonics: a general term for the study of the movement, form, pattern, and evolution of regional large-scale structural or deformational features of Earth's crust.

telepresence: human exploration of an area remotely in real time via live video.

temperate zone: those intermediate latitude zones of Earth marked by seasonal changes in climatic conditions.

tephra: a collective term for all pyroclastic material and shreds of liquid magma ejected from volcanoes.

terrane: a coastal subsection of a continent, formed by the accretion of a small continental mass or plate to a larger one under the influence of seafloor subduction movement; generally, a fault-ridden rock mass relocated from its point of origin and unrelated to adjacent rock structures.

terrestrial: of reptiles, those adapted to living solely on land without the need to return to a body of water for reproduction.

terrigenous: originating from the weathering and erosion of mountains and other land formations.

Tertiary period: the period of geologic time from 65.5 to 1.8 million years ago in the Cenozoic era.

Tethys sea: a seaway embayed into Pangaea between the southeast corner of Asia, the western end of the Mediterranean, and the southeast end of Pangaea.

thermokarst: landform characterized by hummocks, low hills, bogs, and pits that form in relatively flat areas of degrading permafrost and melting ice

thermophile: an organism evolved to live in environments marked by excessively high temperatures

theropod: any of a diverse group of bipedal predatory dinosaurs, thought to be ancestral to birds, that had large heads on long necks and a deep, compact body and that ran on two strong, birdlike hind limbs.

tholeiitic basalt: a type of basalt with calcium-rich plagioclase, monoclinic pyroxene, orthorhombic pyroxene, and olivine or quartz; it has low potassium concentrations compared to the more alkali-rich basalts.

tholus: an inverted, bowl-shaped feature considered to be of volcanic origin.

thrombolite: a stromatolite-like fossil characterized by an obscurely clotted, unlaminated internal structure.

thrust belt: a linear belt of rock that has been deformed by thrust faults that emplace older rocks above younger and generally move mass uphill.

thrust fault: a fault that is inclined at a low angle (less than 45 degrees), with rocks above the fault having moved up and over younger rocks below the fault; such faults are generally the result of compressive forces.

till: an unsorted and unconsolidated mixture of rocky materials that was let down directly from glacial ice and does not include material deposited by meltwater.

tiltmeter: an instrument that precisely measures tilting of the ground surface.

tombolo: an offshore rock or island connected to the beach by a sand spit.

tomography: the imaging carried out by illuminating a subject with physical waves.

tongue stones: an ancient colloquial term used to describe what are now recognized as fossil sharks' teeth; if viewed from the convex side, a large shark's tooth might resemble a tongue turned to stone.

topography: general configuration of a land surface, including its relief and the position of its human-made features.

trace fossils: indirect evidence of an organism's presence preserved as traces of animal activity, such as burrows or trackways preserved in sediment.

transform margin: a boundary between two or more tectonic plates where one plate slides horizontally past another; also called a transform boundary or simply a fault.

transgression: a shift of sedimentary facies toward land, commonly caused by the extension of the sea over land areas, and the evidence of such an advance.

transitional: having the characteristics of both the preceding and the successive classes of organisms without clearly belonging to either.

transmutation: the conversion of one element into another by natural or artificial fission or fusion processes.

transpiration: the process by which water vapor is lost to the atmosphere by living plants.

transverse valley: a river-cut valley or gorge that runs perpendicular to the main strike direction of a mountain chain.

travertine: a variety of hot-spring limestone that is deposited from alkaline waters.

trench: a long, narrow, deep depression in the ocean floor, usually with steep sides and often adjacent to island arc systems and continental landmasses where oceanic crust is being subducted.

Triassic era: period of geologic time from 251.0 to 199.6 million years ago at the beginning of the Mesozoic era.

trilobite: a many-legged arthropod named for its three symmetrical lobes; the principal index fossil for the Cambrian period of 542.0 to 488.3 million years ago.

tropical rain forest: land area with monthly temperatures averaging greater than 18 degrees Celsius and monthly precipitation averaging more than 6 centimeters.

trough: a line occupying the lowest points of a bed in a syncline; the trough plane connects the lowest points on all beds.

tsunami: a seismic sea wave generated by vertical movement of the ocean floor, caused by an earthquake or volcanic eruption.

tuff: a coherent rock formed by the consolidation of tephra.

tumescence: a local swelling of the ground that commonly occurs when magma rises toward the surface.

tunnel vent: the central tube in a volcanic structure through which material from Earth's interior travels toward the surface.

turbidity currents: fast-moving submarine avalanches or downslope movement of a stream of water laden with sediment not present in the surrounding water through which the current travels.

type specimens: the original fossil specimens based upon which a species was first described.

unconformity: a significant break in a stratigraphic sequence representing a gap in time in which no geologic records remain because of nondeposition or erosion of the missing rock layers.

underfit stream: a stream that is significantly smaller in proportion than the valley through which it flows.

uniformitarianism: the general principle that the Earth's past history can be interpreted in terms of what is known about present natural laws, on the assumption that these processes differ neither in degree nor in kind over time.

uniformity of nature: Charles Lyell's formulation describing a nature in which processes are both consistent and constant through space and time; a mixture of methodological and substantive uniformitarianism.

unit cell: the basic formation of atoms that repeats to produce the crystal structure of a mineral.

uplift: the rising of a block of lithosphere because of a force pushing it up; coastal areas undergoing uplift tend to emerge above sea level.

upper mantle: the outermost portion of the interior zone between the core and the crust of the planet, believed to progress from liquid to a zone of partial melting that may produce melts that then crystallize to basalts.

vadose zone: a zone in the ground below the surface and above the level of complete water saturation.

valley: part of Earth's surface where stream systems are established; includes streams and their adjacent slopes.

varve: a pair of contrasting layers of sediment deposited over one year's time; the summer layer is light, and the winter layer is dark.

vascular: possessing a series of vessels that form a conducting apparatus for food, blood, sap and other liquids.

vascular plants: those plants with elongated, woody, water-conducting, tubular cells (tracheids) in their stems or veins of leaves; now includes ferns, conifers, and flowering plants.

vein: a thin, sheet-like mineral deposit that fills a crack; veins form by precipitation of minerals from fluids, such as an igneous intrusion.

vent: a break or tear on the side of a mountain through which magma and pressure can escape.

vertebrate: possessing a bony, internal-skeletal structure based on a central vertebral column.

viscosity: the resistance of a fluid such as lava to flow, because of its internal friction, which in turn depends upon the chemical composition, temperature, and crystalline nature of the magma.

viviparous: of creatures that reproduce by means of internal fertilization of an ovum, usually through sexual copulation, and delivery of live young that have developed internally without the intermediate, defined structure of an egg.

volcanic: descriptive of seismic activity having its origin within the crust, above the level of the asthenosphere.

volcanic earthquakes: small-magnitude earthquakes that occur at relatively shallow depths beneath active or potentially active volcanoes.

volcanic tremor: a continuous vibration of long duration, detected only at active volcanoes.

volcanism: the dynamic processes by which magma is transferred from Earth's interior to produce lava flows on the surface accompanied by the ejection of gases and ash into the atmosphere.

volcano: a vent at the planetary surface from which gases, rocks, and magma erupt and build a more or less cone-shaped mountain.

volcanology: the study of volcanoes.

wash: a narrow channel or stream bed that is dry through most of the year but that can become partly or totally submerged during flash floods due to sudden heavy rains; also called wadi.

Warrawoona sequence: a geological formation in Western Australia, dated at 3.5 billion years before the present, where "North Pole" stromatolites were discovered.

water cycle: *see* Hydrologic cycle.

water table: the upper surface of the zone of saturation in permeable rocks not confined by impermeable rocks; it rises and falls with the amount of groundwater.

weather: the condition of the atmosphere at a given moment and at a given place.

weathering: destructive processes by which rock is broken down into smaller particles by exposure to the atmosphere at or near Earth's surface, with little or no transport of the loosened or altered material.

Wilson cycle: the creation and destruction of an ocean basin through the process of seafloor spreading and subduction of existing ocean basin material.

xylem: the structure of plant cells through which water and nutrients are transported.

BIBLIOGRAPHY

Abbott, Lon, and Terri Cook. *Hiking the Grand Canyon's Geology.* Seattle: The Mountaineers Books, 2004.

Abbott, Patrick L. N*atural Disasters.* 7th ed. New York: McGraw-Hill, 2008.

Abraham, Athol D., and Anthony J. Parsons, eds. *Geomorphology of Desert Environments.* 2d ed. London: Chapman and Hall, 2009.

Adams, Ansel. *Yosemite and the Range of Light.* Reprint. New York: Little Brown & Company, 1992.

Adams, George F., and Jerome Wyckoff. *Landforms.* New York: Golden Press, 1971.

Adams, John. "Paleoseismology: A Search for Ancient Earthquakes in Puget Sound." *Science* 258 (December 4, 1992): 1592-1593.

Agashe, Shripad N. *Paleobotany: Plants of the Past, Their Evolution, Paleoenvironment, and Application in Exploration of Fossil Fuels.* Enfield, N.J.: Science, 1997.

Ager, D. V., and M. Brooks, eds. *Europe from Core to Crust.* London: John Wiley and Sons, 1977.

__________. *The Nature of the Stratigraphical Record.* 3rd ed. Hoboken, N.J: John Wiley & Sons, 1993.

Ahr, Wayne R. *Geology of Carbonate Reservoirs: The Identification, Description, and Characterization of Hydrocarbon Reservoirs in Carbonate Rocks.* Hoboken, N.J.: John Wiley & Sons, 2008.

Akazawa, Takeru, Kenichi Aoki, and Ofer Bar-Yosef, eds. *Neanderthals and Modern Humans in Western Asia.* New York: Plenum, 1998.

Albritton, Claude C., Jr. *The Abyss of Time: Changing Concepts of the Earth's Antiquity After the 16th Century.* Mineola, N.Y.: Dover Publications, 2002.

Aleshire, Peter. *Deserts.* New York: Chelsea House, 2008.

Alexander, R. McNeill. *Exploring Biomechanics: Animals in Motion.* New York: Scientific American Library, 1992.

Allison, Peter A., and David J. Bottjer, eds. *Taphonomy: Process and Bias Through Time.* 2d ed. New York: Springer, 2010.

Allmon, Warren, and David J. Bottjer. New York: Columbia University Press, 2000.

Alt, David D., and Donald W. Hyndman. *Roadside Geology of Northern and Central California.* Missoula, Mont.: Mountain Press, 2000.

__________. *Roadside Geology of Oregon.* Missoula, Mont.: Mountain Press, 1994.

Alvarez, Luis. "Mass Extinctions Caused by Large Bolide Impacts." *Physics Today* 40 (July 1987): 24-33.

Alvarez, W. *T. Rex and the Crater of Doom.* Princeton, N.J.: Princeton University Press, 2008.

Amos, Salvador, ed. *International Stratigraphic Guide.* 2d ed. Hoboken, N.J.: John Wiley & Sons, 2008.

Anderson, Bjørn G., and Harold W. Borns, Jr. *The Ice Age World.* Oslo: Scandinavian University Press, 1994.

Anderson, J. G. C. *The Structure of Western Europe.* Oxford, England: Pergamon Press, 1978.

Anderson, Mary P., ed. *Groundwater.* Oxford, England: International Association of Hydrological Sciences, 2008.

Anderson, O. R. *Radiolaria.* New York: Springer-Verlag, 1983

Anderson, Roger N. *Marine Geology: A Planet Earth Perspective.* Rev. ed. New York: John Wiley & Sons, 1989.

Andrews, Henry N. *The Fossil Hunters: In Search of Ancient Plants.* Ithaca, N.Y.: Cornell University Press, 1980.

Andrews, J. T. *Glacial Systems: An Approach to Glaciers and Their Environments.* North Scituate, Mass.: Duxbury Press, 1975.

Archana, G. "Geomicrobiology: An Indian Perspective." *Current Science* 100, no. 4 (2011): 451-452

Archibald, J. D. *Dinosaur Extinction and the End of an Era.* New York: Columbia University Press, 1996

__________. *Extinctions and Radiation: How the Fall of Dinosaurs Led to the Rise of Mammals.* Baltimore: Johns Hopkins University Press, 2011

Argue, D., et al. "*Homo floresiensis*: A Cladistic Analysis." *Journal of Human Evolution* 57 (2009): 623-639.

Arkell, W. J. *Jurassic Geology of the World.* Edinburgh: Oliver and Boyd, 1956.

Armstrong, Howard, and Martin Brazier. *Microfossils.* 2d ed. Malden, Mass.: Blackwell Publishing, 2005.

Atkinson, David. *Weathering, Slopes, and Landforms.* London: Hodder Education, 2004

Attenborough, David. *The Living Planet: A Portrait of the Earth.* Boston: Little, Brown, 1984.

Ausich, William I., and N. Gary Lane. *Life of the Past.* 4th ed. Upper Saddle River, N.J.: Prentice Hall, 1998.

Backus, Richard H., and D. W. Bourne, eds. *Georges Bank.* Cambridge, Mass.: MIT Press, 1987.

Badgley, Peter C. *Structural and Tectonic Principles.* New York: Harper & Row, 1965

Bailey, R. H. *Glacier.* Rev. ed. Alexandria, Va.: Time-Life Books, 1985.

Bailey, Roy A., P. R. Beauchemin, F. P. Kapinos, and D. W. Klick. *The Volcano Hazards Program: Objectives and Long-Range Plans.* U.S. Geological Survey Open-File Report 83-400. Washington, D.C.: Government Printing Office, 1983.

Bakker, Robert T. *The Dinosaur Heresies.* New York: William Morrow, 1986

Baldridge, W. Scott. *Geology of the American Southwest: A Journey Through Two Billion Years of Plate Tectonic History.* New York: Cambridge University Press, 2004.

Ballantyne, C. F., and C. Harris. *The Periglaciation of Great Britain.* New York: Cambridge University Press, 1994.

Ballard, Robert B. *The Discovery of the Titanic.* New York: Warner Books, 1995.

_________. *Exploring Our Living Planet.* Rev. ed. Washington, D.C.: National Geographic Society, 1993.

Barazangi, M., and Larry Brown. *Reflection Seismology: The Continental Crust.* Washington, D.C.: American Geophysical Union, 1986.

Bar-Cohen, Yoseph, and Kris Zacny, eds. *Drilling in Extreme Environments: Penetration and Sampling on Earth and Other Planets.* Weinheim, Germany: Wiley-VCH, 2009

Barry, Roger, and Thian Yew Gan. *The Global Cryosphere: Past, Present, and Future.* New York: Cambridge University Press, 2011.

Barry, W. B. N. *Growth of a Prehistoric Time Scale.* Rev. ed. Palo Alto, Calif.: Blackwell Scientific Publications, 1987.

Bartholomew, Mervin J., and Richard, P. Tollo. "Northern Ancestry for the Goochland Terrane as a Displaced Fragment of Laurentia." *Geology* 32 (2004): 669-672

Bear, Jacob. *Dynamics of Fluids in Porous Media.* Mineola, N.Y.: Dover, 1988.

_________. *Hydraulics of Groundwater.* Mineola, N.Y.: Dover, 2007.

Beaulieu, E., et al. "Modeling of Water-Rock Interaction in the MacKenzie Basin: Competition Between Sulfuric and Carbonic Acids." *Chemical Geology* 289, nos. 1/2 (2011): 114-123.

Beck, B. F., and W. L. Wilson, eds. *Karst Hydrogeology: Engineering and Environmental Applications.* Boston: A. A. Balkema, 1987.

Behrensmeyer, Anna K., and Andrew P. Hill. *Fossils in the Making: Vertebrate Taphonomy and Paleoecology.* Chicago: University of Chicago Press, 1980.

Bengtson, Stefan, ed. *Early Life on Earth.* New York: Columbia University Press, 1994.

Bennett, Matthew M., and Neil F. Glassner, eds. *Glacial Geology: Ice Sheets and Landforms.* 2d ed. Hoboken, N.J.: Wiley-Blackwell, 2009.

Bennett, Sean J., and Andrew Simon, eds. *Riparian Vegetation and Fluvial Geomorphology.* Washington, D.C.: American Geophysical Union, 2004.

Bennison, George M., Paul A. Olver, and A. Keith Moseley. *An Introduction to Geological Structures and Maps.* 8th ed. London: Hodder Education, 2011.

Benton, Michael J. *The History of Life: A Very Short Introduction.* New York: Oxford University Press, 2008.

_________. *The Rise of the Mammals.* New York: Crescent, 1991.

_________. *Vertebrate Palaeontology.* 3rd ed. Malden, Mass.: Blackwell Science, 2005.

_________. *When Life Nearly Died.* New York: Thames & Hudson, 2003

Benton, Michael J., and D. A. T. Harper. *Introduction to Paleobiology and the Fossil Record.* New York: Wiley-Blackwell, 2009.

Berner, Robert A. *The Phanerozoic Carbon Cycle: CO_2 and O_2.* New York: Oxford University Press, 2004.

Berry, William B. N. *Growth of a Prehistoric Time Scale.* Rev. ed. Palo Alto, Calif.: Blackwell Scientific Publications, 1986.

Best, David M., and David B. Hacker. *Earth's Natural Hazards.* Dubuque, Iowa: Kendall Hunt, 2010.

Beus, Stanley S., and Michael Morales, eds. *Grand Canyon Geology.* 2d ed. New York: Oxford University Press, 2003.

Billings, Marland P. *Structural Geology.* 3rd ed. Englewood Cliffs, N.J.: Prentice-Hall, 1972

Billingsley, George H., and Stanley S. Beus, eds. *Geology of the Surprise Canyon Formation of the Grand Canyon Arizona.* Flagstaff: Museum of Northern Arizona, 1999

Blackwelder, Eliot. "The Valuation of Unconformities." *Journal of Geology* 17 (1909): 289.

Blair, J. E., and S. H. Blair. "Molecular Phylogeny and Divergence Times of Deuterostome Animals." *Molecular Biology and Evolution* 22 (2005): 2275-2284.

Bland, Will, and David Rolls. *Weathering: An Introduction to the Scientific Principles.* London: Hodder Education, 1998.

Blatt, H., R. J. Tracy, and Brent Owens. *Petrology: Igneous, Sedimentary, and Metamorphic.* 3rd ed. New York: W. H. Freeman, 2005.

Bloom, Arthur G. *Geomorphology: A Systematic Analysis of Late Cenozoic Landforms.* 3rd ed. Long Grove, Ill.: Waveland Press, 2004.

Boggs, Sam, Jr. *Principles of Sedimentology and Stratigraphy.* 5th ed. Upper Saddle River, N.J.: Prentice Hall, 2011.

Bolius, David. *Paleo Climate Reconstructions Based on Ice Cores: Results from the Andes and the Alps.* SVH-Verlag, 2010.

Bolt, Bruce A. *Earthquakes.* 5th ed. New York: W. H. Freeman, 2005.

__________. *Earthquakes and Geological Discovery.* New York: Scientific American Library, 1993.

Borg, James C. "Birth of an Island." *Oceans* (August 1987): 27.

Borgese, E. M. *The Mines of Neptune: Minerals and Metals from the Sea.* New York: Harry N. Abrams, 1985.

Bott, M. H. P. *The Interior of the Earth.* 2d ed. London: Edward Arnold, 1982.

Bowler, Peter J. *Evolution: The History of an Idea.* 3rd ed. Berkeley: University of California Press, 2003.

Boyer, S. E., and D. Elliott. "Thrust Systems." *American Association of Petroleum Geologists Bulletin* 66, no. 9 (1982): 1196-1230.

Brady, N. C., and R. R. Weil. *Nature and Properties of Soils.* 11th ed. Englewood Cliffs, N.J.: Prentice Hall, 1996.

Brantley, Steve, and Lyn Topinka, eds. *Earthquake Information Bulletin* 16 (March/April 1984).

Brenner, Robert L., and Timothy R. McHargue. *Integrative Stratigraphy Concepts and Applications.* Englewood Cliffs, N.J.: Prentice-Hall, 1988.

Briggs, D. E., and R. A. Fortey. "Wonderful Strife: Systematics, Stem Groups, and the Phylogenetic Signal of the Cambrian Radiation." *Paleobiology* 31(2005): 94-112.

Briggs, Derek E. G., and Peter R. Crowther, eds. *Paleobiology II.* New York: Wiley-Blackwell, 2001.

__________, eds. *Palaeobiology: A Synthesis II.* Oxford: Blackwell Scientific Publications, 2003.

__________., eds. *Paleobiology: A Synthesis.* Oxford: Blackwell Scientific Publications, 1990.

Broadhead, T. W., ed. *Fossil Prokaryotes and Protists: Notes for a Short Course.* Knoxville: University of Tennessee, 1987.

Bromville, Richard Granville. *Trace Fossils: Biology, Taphonomy, and Applications.* New York: Psychology Press, 1996.

Brookfield, Michael E. *Principles of Stratigraphy.* Hoboken, N.J.: Wiley-Blackwell, 2004.

Brothwell, Don, and Eric Higgs, eds. *Science in Archaeology.* With a foreword by Grahame Clark. 2d ed. New York: Praeger, 1970.

Brown, David J., and Martin Lee. "From Microscopic Minerals to Global Climate Change?" *Geology Today* 23, no. 5 (2007): 172-177.

Brown, G. C. and A. E. Mussett. *The Inaccessible Earth.* London: Chapman & Hall, 1993.

Brunsden, Denys, and John C. Doornkamp, eds. *The Unquiet Landscape.* Bloomington: Indiana University Press, 1974.

Bruton, David L., ed. *Aspects of the Ordovician System.* New York: Oxford University Press, 1984.

Bryan, T. S. *The Geysers of Yellowstone.* 4th ed. Boulder: Colorado Associated University Press, 2008.

__________. *Geysers: What They Are and How They Work.* 2d ed. Missoula, Mont.: Mountain Press Publishing Company, 2005.

Buck, W. Roger, Luc L. Lavier, and Alexei N. B. Poliakov. "Modes of Faulting at Mid-ocean Ridges." *Nature* 434 (2005): 719-723.

Bull, William B. *Tectonic Geomorphology of Mountains: A New Approach to Paleoseismology.* Malden, Mass.: Wiley-Blackwell, 2007.

Bullard, Fred M. *Volcanoes of the Earth.* 2d ed. Austin: University of Texas Press, 1984.

Burney, David A. *Back to the Future in the Caves of Kaua'i.* New Haven: Yale University Press, 2010.

Butzer, Karl W. *Archaeology as Human Ecology: Method and Theory for a Contextual Approach.* New York: Cambridge University Press, 1982.

Calcaterra, D., and M. Parise. *Weathering as a Predisposing Factor to Slope Movements.* Engineering Geology Special Publication 23. London: Geological Society of London, 2010.

Canfield, Don, Erik Kristensen, and Bo Thandrup. *Aquatic Geomicrobiology.* Atlanta: Academic Press/Elsevier, 2005.

Caputo, Mario V., James A. Peterson, and Karen J. Franczyk, eds. *Mesozoic Systems of the Rocky Mountain Region, USA.* Denver: SEPM, 1994.

Carey, S. Warren. "A Tectonic Approach to Continental Drift." In *Continental Drift: A Symposium. Being a Symposium on the Present Status of the Continental Drift Hypothesis, Held in the Geology Department of the University of Tasmania, in March, 1956.* Hobart: University of Tasmania, 1958; revised 1959.

Carroll, R. L. *Vertebrate Paleontology and Evolution.* New York: W. H. Freeman, 1987.

Cas, Ray A. F., and J. V. Wright. *Volcanic Successions: Modern and Ancient.* Winchester, Mass.: Allen & Unwin, 1987.

Caseldine, Chris, et al. *Iceland: Modern Processes and Past Environments.* San Diego, Calif.: Elsevier, 2005.

Cavin, L., Alison Longbottom, and M. Richter, eds. *Fishes and the Breakup of Pangaea.* London: Geological Society of London, 2008.

Ceuleneer, Georges. "Earth Science: Mantle Mapped in the Desert." *Nature* 432 (2004): 156-157.

Champ, Michael A., William P. Dillon, and David G. Howell. "Non-living EEZ Resources: Minerals, Oil, and Gas." *Oceanus* 27 (Winter 1984/1985): 28-34.

Chapman, Clark, and David Morrison. *Cosmic Catastrophes.* New York: Plenum Press, 1989.

Charlier, R. H., and B. L. Gordon. *Ocean Resources: An Introduction to Economic Oceanography.* Washington, D.C.: University Press of America, 1978.

Chernicoff, Stanley, and Dona Whitney. *Geology: An Introduction to Physical Geology.* Upper Saddle River, N.J.: Prentice Hall, 2006.

Chester, D. K., et al. *Furnace of Creation, Cradle of Destruction: A Journey to the Birthplace of Earthquakes, Volcanoes, and Tsunamis.* New York: Amacom Books, 2008

__________. *Mount Etna: The Anatomy of a Volcano.* Stanford, Calif.: Stanford University Press, 1985.

Chiappe, Luis M. *Glorified Dinosaurs: The Origin and Evolution of Birds.* Hoboken, N.J.: John Wiley and Sons, 2007.

Chorley, Richard J., S. A. Schumm, and D. A. Sugden. *Geomorphology.* New York: Methuen, 1984.

Chronic, Halka. *Roadside Geology of Colorado.* 2d ed. Missoula, Mont.: Mountain Press, 2002.

__________. *Time, Rocks, and the Rockies: A Geologic Guide to Rocks and Trails of Rocky Mountain National Park.* Missoula, Mont.: Mountain Press, 1984.

Ciochon, Russell L., and John G. Fleagle, eds. *The Human Evolution Source Book.* 2d ed. Englewood Cliffs, N.J.: Prentice Hall, 2005.

Clark, G. A. "Neandertal Archaeology: Implications for Our Origins." *American Anthropologist* 104 (2002): 50-67.

Clark, M. J., ed. *Advances in Periglacial Geomorphology.* New York: Wiley InterScience, 1988.

Clark, Sandra H. B. *Birth of the Mountains: The Geologic Story of the Southern Appalachian Mountains.* U.S. Geological Survey, 2008.

Clarkson, E. N. K. *Invertebrate Paleontology and Evolution.* 4th ed. Hoboken, N.J.: Wiley-Blackwell, 1998.

Clary, Jordan. "Exploring Karst in Guilin, China." *Geotimes* 53 (2008): 34-37.

Cleal, C. J., and B. A. Thomas. "Botanical Nomenclature and Plant Fossils." *Taxon* 59 (2010): 261-268.

Cloud, Preston C. *Oasis in Space: Earth History from the Beginning.* New York: W. W. Norton, 1988.

Cloudsley-Thompson, J. L. *Ecology and Behaviour of Mesozoic Reptiles.* Berlin: Springer, 2005.

Clube, Victor, and Bill Napier. *The Cosmic Winter.* Oxford, England: Basil Blackwell, 1990.

Coe, Angela L., ed. *The Sedimentary Record of Sea-Level Change.* New York: Cambridge University Press, 2003.

Colby, William E., ed. *John Muir's Studies in the Sierra.* San Francisco: Sierra Club Books, 1960.

Coleman, R. C. "The Diversity of Ophiolites." *Geologie en Mijnbouw* 63 (1984): 141.

__________. *Ophiolites: Ancient Oceanic Lithosphere?* New York: Springer-Verlag, 1977.

Compton, R. C. *Geology in the Field.* New York: John Wiley & Sons, 1985.

Condie, Kent C. *Earth as an Evolving Planetary System.* 2d ed. Waltham, Mass.: Academic Press, 2011.

__________. *Plate Tectonics and Crustal Evolution.* 4th ed. Oxford: Elsevier, 2003.

Cook, Frederick A., L. D. Brown, and J. E. Oliver. "The Southern Appalachians and the Growth of Continents." *Scientific American* 243 (October 1980): 156-168.

Cooper, John D., Richard H. Miller, and Jacqueline Patterson. *A Trip Through Time: Principles of Historical Geology.* Columbus, Ohio: Merrill, 1986.

__________. *A Trip Through Time: Principles of Historical Geology.* 2d ed. New York: McGraw-Hill, 1990.

Copland, Luke, Martin J. Sharp, and Julian A. Dowdeswell. "The Distribution and Flow Characteristics of Surge-Type Glaciers in the Canadian High Arctic." *Annals of* Glaciology 36 (2003): 73-81.

Cordani, U., et al., eds. *Tectonic Evolution of South America.* Rio de Janeiro: Brazilian Academy of Science, 2000.

Courbon, Paul, et al. *Atlas: Great Caves of the World.* St. Louis, Mo.: Cave Books, 1989.

Courtillot, V., A. Davaillie, J. Besse, and J. Stock. "Three Distinct Types of Hotspots in the Earth's Mantle." *Earth and Planetary Science Letters* 205 (2003): 295-308

Courtillot, Vincent E., and Paul R. Renneb. "On the Ages of Flood Basalt Events." *Comptes Rendus Geoscience* 335(2003): 113-140.

Courtillot, Vincent, and Gregory Vink. "How Continents Break Up." *Scientific American* 249 (July 1983): 42.

Cox, C. B., and P. D. Moore. *Biogeography: An Ecological and Evolutionary Approach.* 8th ed. Hoboken, N.J.: John Wiley & Sons, Inc., 2010.

Crame, J. A., and Owen, A. W., eds. *Palaeobiogeography and Biodiversity Change: The Ordovician and Mesozoic-Cenozoic Radiations.* London: Geological Society, 2002.

Crowley, John. "Types of Mountain Ranges in the Rocky Mountain Region." *American Geographical Society's Focus on Geography* 50 (2007): 24-29.

Cubas, N., et al. "Prediction of Thrusting Sequence Based on Maximum Rock Strength and Sandbox Validation." *Trabajos de Geologia* 29 (2009): 189-195.

Cuffey, K. M., and W. S. B. Paterson. *The Physics of Glaciers.* 4th ed. Burlington, Mass.: Elsevier, 2010.

Culver, David C., and William B. White, eds. *Encyclopedia of Caves.* Academic Press, 2004.

Currie, Phillip J., et al., eds. *Feathered Dragons: Studies on the Transition from Dinosaurs to Birds.* Bloomington: Indiana University Press, 2004.

Cvancara, Alan M. *Field Manual for the Amateur Geologist.* New York: John Wiley, 1995.

__________. *Sleuthing Fossils: The Art of Investigating Past Life.* New York: John Wiley & Sons, 1990.

Dagdelenler, Gulseren, Ebru A. Sezer, and Candan Gokceoglu. "Some Non-linear Models to Predict the Weathering Degrees of a Granitic Rock from Physical and Mechanical Parameters." *Expert Systems with Applications* 38, no. 6 (2011): 7476-7485.

Dalrymple, G. Brent. *Ancient Earth, Ancient Skies: The Age of Earth and Its Cosmic Surroundings.* Stanford, Calif.: Stanford University Press, 2004.

Dalziel, Ian W. D. "Earth Before Pangaea." *Scientific American* 272 (January 1995): 58-63.

Daniels, Frank J., Richard D. Dayvault, and Brooks B. Britt, eds. *Petrified Wood: The World of Fossilized Wood, Cones, Ferns, and Cycads.* Grand Junction, Colo.: Western Colorado, 1998.

Darwin, Charles. *On the Origin of Species: A Facsimile of the First Edition.* Cambridge, Mass.: Harvard University Press, 1975.

Davidson, J. P., W. E. Reed, and P. M. Davis. *Exploring Earth: An Introduction to Physical Geology.* 2d ed. Upper Saddle River, N.J.: Prentice-Hall, 2001.

Davies, Geoffrey F. *Mantle Convection for Geologists.* New York: Cambridge University Press, 2011.

Davies, W. E., and I. M. Morgan. *Geology of Caves.* U.S. Geological Survey, 1991.

Davis, George H., and Evelyn M. VandenDolder, eds. *Geologic Diversity of Arizona and Its Margins: Excursions to Choice Areas.* Tucson: Arizona Bureau of Geology and Mineral Technology, 1987.

Dawkins, Richard. *The Blind Watchmaker: Why the Evidence of Evolution Reveals a Universe Without Design.* New York: W. W. Norton, 1988.

Dawson, J. B., D. A. Carswell, J. Hall, and K. H. Wedepohl. *The Nature of the Lower Continental Crust.* Palo Alto, Calif.: Blackwell Scientific Publications, 1986.

Day, Michael H. *Guide to Fossil Man.* 4th ed. Chicago: University of Chicago Press, 1986.

Deamer, David W., and Jack W. Szostack. *The Origins of Life: A Subject Collection from Cold Spring Harbor Perspectives in Biology.* Cold Spring Harbor, N.Y.: Cold Spring Harbor Laboratory Press, 2010.

De Angelis, Hernan, and Pedro Skvarca. "Glacier Surge After Ice Shelf Collapse." *Science 299* (2003): 1560-1562.

Dean, Robert G., and Robert A. Dalrymple. *Coastal Processes: With Engineering Applications.* New York: Cambridge University Press, 2004.

de Boer, Jella Zeilinga, and Donald Theodore Sanders. *Volcanoes in Human History: The Far-Reaching Effects of Major Eruptions.* Princeton, N.J.: Princeton University Press, 2002.

Decker, Robert W., Thomas L. Wright, and Peter H. Stauffer, eds. *Volcanism in Hawaii.* 2 vols. Denver, Colo.: U.S. Geological Survey, 1988.

Decker, Robert, and Barbara Decker. *Volcanoes.* 4th ed. New York: W. H. Freeman, 2005.

Dennis, J. G. *Structural Geology: An Introduction.* Dubuque, Iowa: Wm. C. Brown, 1987.

De Pater, Imke, and Jack J. Lissauer. *Planetary Sciences.* 2d ed. New York: Cambridge University Press, 2010.

de Roy, Tui. *The Andes: As the Condor Flies.* Buffalo: Firefly Books, 2005.

Detjen, Philipp Thomas. *Hydrothermal Vents.* Saarbrucken, Germany: VDM, 2011.

Dickinson, William R. "The Basin and Range Province as a Composite Extensional Domain." *International Geology Review* 22 (2002): 1-38.

Dietrich, Richard V. *Geology and Virginia.* Charlottesville: University Press of Virginia, 1970.

Dilek, Y., and P. T. Robinson, eds. *Ophiolites in Earth History.* London: Geological Society Special Publication, 2003.

Dineley, D. L. *Aspects of a Stratigraphic System: The Devonian.* New York: John Wiley & Sons, 1984.

Ditrich, Hans. "The Origin of Vertebrates: A Hypothesis Based on Kidney Development." *Zoological Journal of the Linnean Society* 150 (2007): 435-441.

Dobran, Flavio. *Vesuvius: Education, Security and Prosperity.* Oxford: Elsevier, 2006.

Dobrazanska, Halina, Erzsebet Jerem, and Tomasz Kalicki, eds. *The Geoarchaeology of River Valleys.* Budapest: Archaeolingua, 2004.

Dodd, Robert J., and Robert J. Stanton, Jr. *Paleoecology: Concepts and Applications.* 2d ed. New York: Wiley-Interscience, 1990.

Dolgoff, Anatole. *Physical Geology.* Boston: Houghton Mifflin Harcourt, 1999.

Donovan, Stephen K., ed. *The Process of Fossilization.* New York: Columbia University Press, 1991.

Doran, Peter T., W. Berry Lyons, and Diane M. McKnight, eds. *Life in Antarctic Deserts and Other Cold, Dry Environments.* New York: Oxford University Press, 2010.

Dossetto, Anthony, Simon P. Turner, and James A. van Orman, eds. *Timescales of Magmatic Processes from Core to Atmosphere.* Hoboken, N.J.: Wiley-Blackwell, 2011.

Dott, Robert H., Jr., and Donald R. Prothero. *Evolution of the Earth.* 8th ed. New York: McGraw-Hill, 2009.

Doyle, Peter, and Matthew R. Bennett. *The Key to Earth History: An Introduction to Stratigraphy.* New York: John Wiley & Sons, 2001.

Dudley, Robert. *The Biomechanics of Insect Flight: Form, Function, Evolution. Princeton,* N.J.: Princeton University Press, 2002.

Duff, P. Donald. *Holmes' Principles of Geology.* Cheltenham, England: Stanley Thorne, 1998.

Dyke, Gareth. "Winged Victory." *Scientific American* 303, no. 1 (2010): 70-75.

Dyke, Gareth, and Gary Kaiser. *Living Dinosaurs: The Evolutionary History of Modern Birds.* Hoboken, N.J.: John Wiley & Sons, 2011.

Dyson, Freeman, *Origins of Life.* New York: Cambridge University Press, 1999.

Edmond, John M., and Karen Von Damm. "Hot Springs on the Ocean Floor." *Scientific American* 248 (April 1983): 78-84.

Ehlers J., and P. L. Gibbard. *Quaternary Glaciations; Extent and Chronology: Part II.* New York: Elsevier, 2004.

Ehlers, J., P. L. Gibbard, and P. D. Hughes, eds. *Quaternary Glaciations: Extent and Chronology: A Closer Look.* New York: Elsevier, 2011.

Ehlig, P. L. "Origin and Tectonic History of the Basement Terrane of the San Gabriel Mountains, Central Transverse Ranges." In *The Geotectonic Development of California, Rubey Vol. 1,* edited by W. G. Ernst. Englewood Cliffs, N.J.: Prentice-Hall, 1981.

Ehrlic, Henry Lutz, and Dianne K. Newman. *Geomicrobiology.* 5th ed. Boca Raton, Fla.: CRC Press, 2008.

Elder, John. *Geothermal Systems.* New York: Academic Press, 1981.

Elewa, Ashraf M. T. *Mass Extinction.* New York: Springer, 2008.

Elliott, D. K. "A Reassessment of *Astraspis desiderata,* the Oldest North American Vertebrate." *Science* 237 (July 1987): 190.

Embleton, C., and C. A. M. King. *Glacial Geomorphology.* New York: John Wiley & Sons, 1975.

England, Philip C., and Katz, Richard F. "Melting Above the Anhydrous Solidus Controls the Location of Volcanic Arcs." *Nature* 467 (2010): 700-703.

Engle, A. E., H. L. James, and B. F. Leonard, eds. *Petrologic Studies: A Volume in Honor of A. F. Buddington.* New York: Geological Society of America, 1962.

Erickson, Jon. *Craters, Caverns, and Canyons: Delving Beneath the Earth's Surface.* Chicago: Facts on File, 1993.

__________. *Plate Tectonics: Unraveling the Mysteries of the Earth.* Rev. ed. New York: Facts on File, 2001.

__________. *Volcanoes and Earthquakes.* Blue Ridge Summit, Pa.: TAB Books, 1987.

Erwin, Douglas H. *Extinction: How Life on Earth Nearly Ended 250 Million Years Ago.* Princeton, N.J.: Princeton University Press, 2006.

__________. *The Great Paleozoic Crisis.* New York: Columbia University Press, 1993

Etemad-Saeed, N., M. Hosseini-Barzi, and John S. Armstrong-Altrin. "Petrography and Geochemistry of Clastic Sedimentary Rocks as Evidences for Provenance of the Lower Cambrian Lalun Formation." *Journal of African Earth Sciences* 61, no. 2 (2011): 142-159.

Evans, David A. D. "The Palaeomagnetically Viable, Long-lived and All-inclusive Rodinia Supercontinent Reconstruction." *Geological Society* 327 (January, 2009): 371-404.

Exley, Sheck. *Caverns Measureless to Man.* St. Louis, Mo.: Cave Books, 1994.

Exon, Neville. "Scientific Drilling Beneath the Oceans Solves Earthly Problems." *Australian Journal of Maritime and Ocean Affairs* 2 (2010): 37-47.

Eyles, N., ed. *Glacial Geology.* Oxford, England: Pergamon Press, 1983.

Fagan, Brian M., ed. *The Complete Ice Age: How Climate Change Shapes the World.* New York: Thames & Hudson, 2009.

__________. *Cro-Magnon: How the Ice Age Gave Birth to the First Modern Humans.* New York: Bloomsbury Press, 2011.

Fastovsky, David E., and David B. Weishampel. *Dinosaurs: A Concise Natural History.* New York: Cambridge University Press, 2009.

__________. *The Evolution and Extinction of the Dinosaurs.* New York: Cambridge University Press, 2005.

Fedonkin, Mikhail A., et al. *The Rise of Animals: Evolution and Diversification of the Kingdom Animalia.* Baltimore, MD: Johns Hopkins University Press, 2007.

Fedonkin, Mikhail A., James G. Gehling, Kathleen Grey, Guy M. Narbonne, and Patricia Vickers-Rich. *The Rise of Animals: Evolution and Diversification of the Kingdom Animalia.* Baltimore: Johns Hopkins University Press, 2007.

Feduccia, Alan. *The Origin and Evolution of Birds.* New Haven, Conn.: Yale University Press, 1999.

Feldhamer, George A., et al. *Mammalogy.* 3rd ed. Baltimore, Md.: Johns Hopkins University Press, 2007.

Feldmann, Rodney M., Ralph E. Chapman, and Joseph T. Hannibal, eds. *Paleotechniques.* Knoxville, Tenn.: Paleontological Society, 1989.

Ferguson, James. "The Tragedy of St. Pierre." *Geographical* 74 (2002): 14-19.

Fernández-Jalvo, Y., L. Scott, and P. Andrews. "Taphonomy in Palaeoecological Interpretations." *Quaternary Science Reviews* 30 (2011): 1296-1302.

Fife, D. L., and J. A. Minch, eds. *Geology and Mineral Wealth of the California Transverse Ranges.* Santa Ana, Calif.: South Coast Geological Society, 1982.

Findley, Rowe. "St. Helens: Mountain with a Death Wish." *National Geographic* 159 (January 1981): 2-65.

Finkl, Charles W., ed. *The Encyclopedia of Field and General Geology.* New York: Van Nostrand Reinhold, 1988.

Fisher, George W., ed. *Studies of Appalachian Geology: Central and Southern.* New York: John Wiley & Sons, 1970.

Fiske, R. S. "Volcanologists, Journalists, and the Concerned Local Public: A Tale of Two Crises in the Eastern Caribbean." In *Explosive Volcanism: Inception, Evolution, and Hazards.* Washington, D.C.: National Academy Press, 1984.

Fletcher, Charles. *Physical Geology: The Science of Earth.* Hoboken, N.J.: John Wiley and Sons, 2011.

Flügel, Erik. *Microfacies of Carbonate Rocks.* New York: Springer, 2004.

Foote, Michael Joseph, et al., eds. *Principles of Paleontology.* New York: Macmillan, 2007.

Foote, Michael, and Arnold Miller. *Principles of Paleontology.* 3rd ed. San Francisco: W. H. Freeman, 2007.

Forrester, Glenn C. *Niagara Falls and the Glacier.* Hicksville, N.Y.: Exposition Press, 1976.

Fortey, Richard. *Fossils: The History of Life.* New York: Sterling, 2009.

Fossen, Haakon. *Structural Geology.* New York: Cambridge University Press, 2010.

Foulger, G. R. *Plates vs. Plumes: A Geological Controversy.* New York: Wiley-Blackwell, 2010.

Fournier, Robert O. "Geochemistry and Dynamics of the Yellowstone National Park Hydrothermal System." In *Annual Review of Earth and Planetary Sciences,* edited by G. W. Wetherill, A. L. Albee, and F. G. Stehli. Vol. 17. Palo Alto, Calif.: Annual Reviews, 1989.

Fowler, C. M. R. *The Solid Earth: An Introduction to Global Geophysics.* 2d ed. New York: Cambridge University Press, 2004.

Francheteau, Jean. "The Oceanic Crust." *Scientific American* 249 (September 1983): 130.

Francis, Peter, and Clive Oppenheimer. *Volcanoes.* 2d ed. New York: Oxford University Press. 2004.

Freedman, Roger A., Robert M. Geller, and William J. Kaufmann III. *Universe.* 9th ed. New York: W. H. Freeman, 2010.

Friedman, Matt. "Emerging onto a Tangled Bank." *Science* 324 (2009): 341-342.

Frisch, Wolfgang, Martin Meschede, and Ronald C. Blakey. *Plate Tectonics: Continental Drift and Mountain Building.* New York: Springer, 2010.

Frome, Michael. *The National Park Guide.* Skokie, Ill.: Rand McNally, 1987.

Fuentes, Carlos. *The Buried Mirror: Reflections on Spain and the New World.* New York: Houghton-Mifflin, 1992.

Fukuyama, Eiichi, "Fault Structure, Stress, Friction, and Rupture Dynamics of Earthquakes." In *Advances in Earth Science: From Earthquakes to Global Warming,* edited by Peter R. Sammonds and J. M. T. Thompson. London: Imperial College Press, 2007

Futuyma, Douglas J. *Evolutionary Biology.* 3rd ed. Sunderland, Mass.: Sinauer Associates, 1998.

Gagnier, P.-Y. "The Oldest Vertebrate: A 470 Million Year Old Jawless Fish, *Sacabambaspis janvieri,* from the Ordovician of Bolivia." *National Geographic Research* 5 (Spring 1989): 250.

Gamache, Tom, and Matthew Jaffe. *The Santa Monica Mountains: Range on the Edge.* Los Angeles: Angel City Press, 2006.

Gardner, Jim. "*Multibeam Mapping in the Outer Continental Shelf Region of the Northwestern Gulf of Mexico.*" U.S. Department of the Interior, Minerals Management Service, Gulf of Mexico OCS Region, 2003.

Garner, H. F. *The Origin of Landscapes.* New York: Oxford University Press, 1974.

Garrett, Roger A., and Hans-Peter Klenk, eds. *Archaea: Evolution, Physilogy, and Molecular Biology.* Malden, Mass. Wiley-Blackwell, 2007.

Garrison, Erv. *Techniques in Archaeological Geology.* New York: Springer, 2010.

Garrison, Tom S. *Oceanography: An Invitation to Marine Science.* Belmont, Calif. Brooks/Cole, Cengage Learning, 2010.

Gass, Ian G. "Ophiolites." *Scientific American* 247 (August 1982): 122.

Gebelein, C. D. "Biologic Control of Stromatolite Microstructure: Implications for Precambrian Time Stratigraphy." *American Journal of Science* 274 (1974): 575-598.

Geikie, Archibald. *The Founders of Geology.* Reprint. Mineola, N.Y.: Dover, 1962.

Gibson, G. G., S. A. Teeter, and M. A. Fedonkin. "Ediacaran Fossils from the Carolina Slate Belt, Stanly County, North Carolina." *Geology* 12 (July 1984): 387.

Gillieson, David. *Caves: Processes, Development, and Management.* Oxford, England: Blackwell, 1998.

Gilligan, David Scott. *In the Years of the Mountains: Exploring the World's High Ranges in Search of Their Culture, Geology, and Ecology.* New York: Thunder's Mouth Press, 2006.

Glaessner, M. F. *The Dawn of Animal Life: A Biohistorical Study.* New York: Cambridge University Press, 1985.

Glassley, William E. *Geothermal Energy: Renewable Energy and the Environment.* London: Tyler Francis Group, 2010.

Glazner, Allen F., and Greg Stock. *Geology Underfoot in Yosemite National Park.* Missoula, Mont.: Mountain Press Publishing Company, 2010.

Glennon, J. A., and R. M. Pfaff. "The Extraordinary Thermal Activity of El Tatio Geyser Field, Antofagasta Region, Chile." *Geyser Observation and Study Association (GOSA) Transactions* 8 (2003): 31–78.

Goff, Fraser. *Valles Caldera: A Geologic History.* University of New Mexico Press, 2009.

Goldberg, Paul, and Richard I. MacPhail. *Practical and Theoretical Geoarchaeology.* New York: Wiley-Blackwell, 2006.

Golding, Roland. *Fossils in the Field.* New York: John Wiley & Sons, 1991.

Goodrich, Charles, Kathleen Dean Moore, and Frederick J. Swanson, eds. *In the Blast Zone: Catastrophe and Renewal on Mt. St. Helens.* Corvallis: Oregon State University Press, 2008.

Goodsell, David S. *The Machinery of Life.* New York: Springer, 2009.

Gordon, M. S., and E. C. Olson. *Invasions of the Land: The Transition of Organisms from the Aquatic to Terrestrial Life.* New York: Columbia University Press, 1995.

Gottsmann, Joachim, and Joan Marti, eds. *Caldera Volcanism: Analysis, Modelling and Response.* Oxford: Elsevier, 2008.

Goudie, A. S., and H. A. Viles, eds. *The Earth Transformed: An Introduction to Human Impacts on the Environment.* Oxford: Blackwell, 2003.

Gould, Stephen J. *Hen's Teeth and Horse's Toes: Further Reflections in Natural History.* New York: W. W. Norton, 1994.

__________. *Time's Arrow, Time's Cycle: Myth and Metaphor in the Discovery of Geological Time.* Cambridge, Mass.: Harvard University Press, 1987.

__________. *Wonderful Life: The Burgess Shale and the Nature of History.* New York: W. W. Norton, 1989.

Grant, Michael. *Cities of Vesuvius: Pompeii and Herculaneum.* New York: Spring Books, 1974.

Grattan, John, and Robin Torrence, eds. *Living Under the Shadow: Cultural Impacts of Volcanic Eruptions.* Walnut Creek, Calif.: Left Coast Press, 2007.

Green, Richard E., et al. "Analysis of One Million Base Pairs Neanderthal DNA." *Nature* 444 (2006): 330-336.

Gregory, Kenneth J., and Andrew S. Goudie, eds. *The SAGE Handbook of Geomorphology.* Thousand Oaks, Calif.: SAGE Publications, 2011.

Gribbin, John R. *This Shaking Earth: Earthquakes, Volcanoes, and Their Impact on Our World.* New York: G. P. Putnam's Sons, 1978.

Gross, M. Grant. "Deep-Sea Hot Springs and Cold Seeps." *Oceanus* 27 (Fall 1984): 2-6.

Grosse, G., et al. "Why Permafrost Is Thawing, Not Melting." *EOS* 91 (2010): 87.

Grotzinger, John, and Thomas H. Jordan. *Understanding Earth.* New York: W. H. Freeman, 2010.

Gudmundsson, Agust. *The Geology of Iceland in Relation to Global Tectonics.* New York: Springer, 2011.

__________. *Rock Fractures in Geological Processes.* New York: Cambridge University Press, 2011.

Gueydan, Frederic, Christina Morency, and Jean-Pierre Brun. "Continental Rifting as a Function of Lithosphere Mantle Strength." *Tectonophysics* 460 (2008): 83-93.

Guidoboni, Emanuela, and John E. Ebel. *Earthquakes and Tsunamis in the Past: A Guide to Techniques in Historical Seismology.* New York: Cambridge University Press, 2009.

Gunde-Cimerman, Nina, Aharon Oren, and Ana Plemenita. *Adaptation to Life at High Salt Concentrations in Archaea, Bacteria, and Eukarya.* New York: Springer, 2011.

Guo, Jerry. "Volcanology: Fire and Life." *Nature* 454 (2008): 930-932.

Guyton, Bill. *Glaciers of California: Modern Glaciers, Ice Age Glaciers, the Origin of Yosemite Valley, and a Glacier Tour in the Sierra Nevada.* Berkeley: University of California Press, 2000.

Hag, Bilal, and Anne Boersma, eds. *Introduction to Marine Micropaleontology.* 2d ed. New York: Elsevier, 1998.

Hall, Brian K., ed. *Fins into Limbs: Evolution, Development, and Transformation.* Chicago: University of Chicago Press, 2007.

Hallam, A., ed. *Atlas of Palaeobiogeography.* New York: Elsevier, 1973.

__________. *Phanerozoic Sea-Level Changes.* New York: Columbia University Press, 1992.

Hamblin, William Kenneth, and Eric H. Christiansen. *Earth's Dynamic Systems.* 10th ed. Upper Saddle River, N.J.: Prentice Hall, Pearson Education, 2003.

Hamblin, W. K., and J. D. Howard. *Exercises in Physical Geology.* 12th ed. Upper Saddle River, N.J.: Prentice Hall, 2004.

Hambrey, Michael, and Jurg Alean. *Glaciers.* 2d ed. New York: Cambridge University Press, 2004.

Handy, Mark R., Niels Hovius, and Greg Hirth. *Tectonic Faults: Agents of Change on a Dynamic Earth.* Cambridge, Mass.: MIT Press, 2007.

Harden, Deborah R. *California Geology.* 2ded. Upper Saddle River, N.J.: Prentice Hall, 2003.

Harmon, Russell, and Andrew Parker. *Frontiers in Geochemistry: Contribution of Geochemistry to the Study of the Earth.* Hoboken, N.J.: Wiley-Blackwell, 2011.

Harris, Ann G., and Esther Tuttle. *Geology of National Parks.* Dubuque, Iowa: Kendall/Hunt, 2003.

Harris, David V., and Eugene P. Kiver. *The Geologic Story of the National Parks and Monuments.* 3rd ed. New York: John Wiley & Sons, 1985.

Harris, S. A. *The Permafrost Environment.* Totowa, N.J.: Barnes & Noble, 1986.

Harris, Stephen L. *Agents of Chaos.* Missoula, Mont.: Mountain Press, 1990.

__________. *Fire Mountains of the West: The Cascade and Mono Lake Volcanoes.* 3rd ed. Missoula, Mont.: Mountain Press Publishing Company, 2005.

Hartman, H., J. G. Lawless, and P. Morrison, eds. *Search for the Universal Ancestors: The Origins of Life.* Reprint. Palo Alto, Calif.: Blackwell Scientific Publications, 1987.

Hatcher, Robert D., Jr. *Structural Geology: Principles, Concepts, and Problems.* 2d ed. Upper Saddle River, N.J.: Prentice Hall, 1995.

Haymon, Rachel M., et al. "Manifestations of Hydrothermal Discharge from Young Abyssal Hills on the Fast-Spreading East Pacific Rise Flank." *Geology* 33 (2005): 153-156.

Hazlett, Richard. *Roadside Geology of Hawai'i.* Missoula, Mont.: Mountain Press Publishing, 2005.

Healy, Richard, and Bridget Scanlon. *Estimating Groundwater Recharge.* New York: Cambridge University Press, 2010.

Hearty, P. J., and P. Aharon. "Amino Acid Chronostratigraphy of Late Quaternary Coral Reefs: Huon Peninsula, New Guinea, and the Great Barrier Reef, Australia." *Geology* 16 (July 1988): 579-583.

Hedberg, H. D., ed. *International Stratigraphic Guide.* New York: John Wiley & Sons, 1976.

Heilprin, Angelo. *Mount Pelée and the Tragedy of Martinique.* Philadelphia: J. B. Lippincott, 1903.

Heliker, C., J. D. Griggs, T. J. Takahashi, and T. L. Wright. *Earthquakes and Volcanoes* 18 (1986).

Herring, Peter. *The Biology of the Deep Ocean.* New York: Oxford University Press, 2002.

Hickey, Leo J. *The Forest Primeval: The Geologic History of Wood and Petrified Forests.* New Haven: Yale Peabody Museum, 2010.

Hill, Carol A., and Paolo Forti. *Cave Minerals of the World.* 2d ed. Huntsville, Ala.: National Speleological Society, 1997.

Hill, Mary. *Geology of the Sierra Nevada.* Rev. ed. Berkeley: University of California Press, 2006.

Hill, Robert C. "On the Volcanic Disturbance in the West Indies." *National Geographic* 13 (July 1902): 223-266.

Hodges, Kip. "Climate and the Evolution of Mountains." *Scientific American* 295 (2006): 72-79.

Hoffer, William. *Volcano: The Search for Vesuvius.* New York: Summit, 1982.

Hoffman, Paul F. "United Plates of America, the Birth of a Craton: Early Proterozoic Assembly and Growth of Laurentia." *Annual Review of Earth and Planetary Sciences* 16 (1988): 543-603.

Hofmann, H. J., W. H. Fritz, and G. M. Narbonne. "Ediacaran (Precambrian) Fossils from the Wernecke Mountains, Northwestern Canada." *Science* 221 (1983): 45.

Holcomb, Robin T., James G. Moore, Peter W. Lipman, and R. H. Belderson. "Voluminous Submarine Lava Flows from Hawaiian Volcanoes." *Geology* (May 1988): 400.

Holmes, Tim. *The First Vertebrates: Oceans of the Paleozoic Period.* New York: Infobase, 2008.

Hooke, R. L. "On The History of Humans as Geomorphic Agents." *Geology* 28 (2000): 843-846.

Horbe, Adriana M. C., and R. R. Anand. "Bauxite on Igneous Rocks from Amazonia and Southwestern of Australia: Implication for Weathering Process." *Journal for Geochemical Exploration* 111, nos. 1/2 (2011): 1-12.

Horner, John R., Jack Horner, and James Gorman. *How to Build a Dinosaur: Extinction Doesn't Have to Be Forever.* New York: Penguin, 2011.

Hou, Xian-Guang, et al. *The Cambrian Fossils of Chengjiang, China.* Oxford: Blackwell Publishing, 2010.

Howell, David G. "Terranes." *Scientific American* 253 (November 1985): 116-125.

Howland, John L. *The Surprising Archaea: Discovering Another Domain of Life.* New York: Oxford University Press, 2000.

Hsü, Kenneth J. *The Great Dying: A Cosmic Catastrophe Demolishes the Dinosaurs and Rocks the Theory of Evolution.* New York: Harcourt Brace Jovanovich, 1986.

__________, ed. *Mountain-Building Processes.* New York: Academic Press, 1982.

Huddart, David, and Tim Stott. *Earth Environments: Past, Present, and Future.* Hoboken, N.J.: John Wiley & Sons, 2010.

Huggett, Richard J. *Catastrophism: Systems of Earth History.* London: Edward Arnold, 1990.

__________. *Fundamentals of Geomorphology.* New York: Taylor and Francis, 2011.

Hughes, Terence J. *Ice Sheets.* New York: Oxford University Press, 1998.

Hunt, C. B. *Natural Regions of the United States and Canada.* San Francisco: W. H. Freeman, 1974.

Hunt, Roy E. *Characteristics of Geologic Material: A Field Guide for Geotechnical Engineers.* Boca Raton, Fla.: CRC Press, 2007.

Hutton, James. "Theory of the Earth: Or, An Investigation of the Laws Observable in the Composition, Dissolution, and Resolution of the Land Upon the Globe." Vol. 1 in *Royal Society of Edinburgh Transactions.* Reprint, with a foreword by George W. White. Darien, Conn.: Hafner, 1970.

Hynes, Andrew. "Freeboard Revisited: Continental Growth, Crustal Thickness Change and Earth's Thermal Efficiency." *Earth and Planetary Science Letters* 185 (2001): 161-172.

Imbrie, J., and K. P. Imbrie. *Ice Ages: Solving the Mystery.* Cambridge, Mass.: Harvard University Press, 1986. Reprint: Enslow, 2000.

Jagnow, David H., and Rebecca Rohwer Jagnow. *Stories from Stones.* Carlsbad, N.M.: Carlsbad Caverns-Guadalupe Mountains Association, 1992.

James, D. E. "The Evolution of the Andes." *Scientific American* 229 (August 1973): 61-69.

James, Harold Lloyd. *Harry Hammond Hess.* National Academy of Sciences Biographical Memoirs 43. New York: Columbia University Press, 1973.

Jannsen, Raymond E. *Earth Science: A Handbook of the Geology of West Virginia.* Clarksburg, W.Va.: Educational Marketers, 1973.

Janvier, Philippe. *Early Vertebrates.* Oxford: Oxford University Press, 2003.

Jenkins, J. M., ed. *Applied Micropaleontology.* New York: Springer, 2010.

Jennings, J. N. *Karst Geomorphology.* 2d ed. New York: Basil Blackwell, 1987.

Jerram, Dougal, and Nick Petford. *The Field Description of Igneous Rocks.* 2d ed. Hoboken, N.J.: Wiley-Blackwell, 2011.

Jiskoot, H., and D. T. Juhlin. "Surge of a Small East Greenland Glacier, 2001-2007, Suggests Svalbard-Type Surge Mechanism." *Journal of Glaciology* 55 (2009): 567–570.

Johanson, Donald C., and Maitland A. Edey. *Lucy.* New York: Simon & Schuster, 1981.

John, Brian S. *The Ice Age: Past and Present.* London: Collins, 1977.

Johnson, Kirk R., and Richard Stucky. *Prehistoric Journey: A History of Life on Earth.* Denver, Colo.: Fulcrum, 2006.

Joly, John. *The Surface-History of the Earth.* Boston: Adamant Media Corporation, 2005.

Jones, David L., A. Cox, P. Coney, and Myrl Beck. "The Growth of Western North America." *Scientific American* 247 (November 1982): 70-84.

Jones, Nicola. "Cracked Plates." *Current Science* 89 (2003): 8-9.

Julien, Pierre Y. *Erosion and Sedimentation.* 2d ed. New York: Cambridge University Press, 2010.

Kalb, Jon. *Adventures in the Bone Trade.* New York: Copernicus Books, 2001.

Kandler, Otto, and Wolfram Zillig, eds. *Archaebacteria Eighty-five: Proceedings of the EMBO Workshop on Molecular Genetics of Archaebacteria.* Forestburgh, N.Y.: Lubrecht and Cramer, 1987.

Kapur, Selim, and George Stoops, eds. *New Trends in Soil Micromorphology.* New York: Springer, 2010.

Karner, G. D., G. Manatschal, and L. M. Pinheiro. *Imaging, Mapping, and Modelling Continental Lithosphere Extension and Breakup.* London: Geological Society of London, 2007

Kasting, James B., O. B. Toon, and J. B. Pollack. "How Climate Evolved on the Terrestrial Planets." *Scientific American* 258 (February 1988): 90-97.

Kearay, Philip, Keith A. Klepeis, and Frederick J. Vine. *Global Tectonics.* 3rd ed. Hoboken, N.J.: John Wiley & Sons, 2009.

Kelley, Jane H., and Marsha P. Hanen. *Archaeology and the Methodology of Science.* Albuquerque: University of New Mexico Press, 1988.

Kemp, T. S. *The Origin and Evolution of Mammals.* New York: Oxford University Press, 2005.

Kempe, D. R. C., and Anthony P. Harvey, eds. *The Petrology of Archaeological Artifacts.* New York: Clarendon Press, 1983.

Kennan, George. "The Tragedy of Pelée." *Outlook* 71 (June 28-August 16, 1902): 769-777.

Kennett, James P. *Marine Geology.* Englewood Cliffs, N.J.: Prentice-Hall, 1982.

Kennish, Michael J. *Practical Handbook of Marine Science.* 3rd ed. Boca Raton, Fla.: CRC Press, 2001.

Kenrick, Paul, and Paul Davies. *Fossil Plants.* Washington, D.C.: Smithsonian Books, 2004.

Kiel, Steffen. *The Vent and Seep Biota: Aspects from Microbes to Ecosystems.* New York: Springer, 2010

Kimberley, M. M., and D. E. Grandstaff. "Profiles of Elemental Concentrations in Precambrian Paleosols on Basaltic and Granitic Parent Materials." *Precambrian Research* 32 (1986): 133-154.

King, Geoffrey, and Geoff Bailey. "Tectonics and Human Evolution." *Antiquity* 80 (2006): 265-286.

King, Philip B. *The Evolution of North America.* Princeton, N.J.: Princeton University Press, 1977.

Kirklady, J. F. *Fossils in Colour.* Dorset, England: Blandford Press, 1980.

Kitching, Ian J., Peter L. Forey, Christopher J. Humphries, and David M. Williams. *Cladistics: The Theory and Practice of Parsimony Analysis.* 2d ed. Oxford: Oxford University Press, 1998.

Klein, Cornelius, and Barbara Dutrow. *Manual of Mineral Science.* New York: John Wiley & Sons, 2008.

Klempe, Herald. "Identification of Quaternary Subsurface Glacial Deposits Using 3D Databases and

GIS." *Norwegian Journal of Geography* 58 (2004): 90-95.

Kobashi, T., et al. "Watermass Stability Reconstructions from Greenhouse (Eocene) to Ice House (Oligocene) for the Northern Gulf Coast Continental Shelf (USA)." *Paleoceanography* 19 (2004): PA1022.

Koblick, Ian G., and James W. Miller. *Living and Working in the Sea.* 2d ed. Flagstaff, Ariz.: Best Publishing, 1995.

Koenigshof, P. *Devonian Change: Case Studies in Palaegeogrpahy and Palaeoecology.* London: Geological Society of London, 2009.

Kokelaar, B. P., and I. D. Moore. *Glencoe Caldera Volcano, Scotland.* Nottinghamshire: British Geological Survey, 2006.

Konhauser, Kurt. *Introduction to Geomicrobiology.* Hoboken, N.J.: Wiley-Blackwell, 2006.

Krebs, Robert E. *The Basics of Earth Science.* Westport, Conn.: Greenwood Press, 2003.

Kruckeberg, Arthur R. *Geology and Plant Life: The Effects of Landforms and Rock Types on Plants.* Seattle, Wash.: University of Washington Press, 2004.

Kulkami, Anil V., et al. "Glacial Retreat in Himalaya Using Indian Remote Sensing Satellite Data." *Current Science* 92 (2007): 69-74.

Kurtén, Bjorn. *Dance of the Tiger: A Novel of the Ice Age.* Berkeley: University of California Press, 1995.

Kusky, T. M., ed. *Precambrian Ophiolies and Related Rocks.* San Diego: Elsevier, 2004.

Kusky, T. M., M. G. Zhai, and W. Xiao, eds. *The Evolving Continents: Understanding Processes of Continental Growth.* Geological Society Special Publication, 2010.

Kutter, G. Siegfried. *The Universe and Life: Origins and Evolution.* Boston: Jones & Bartlett, 1987.

Lacombe, Olivier, et al., eds. *Thrust Belts and Foreland Basins.* Berlin: Springer, 2007.

Lageson, David R., and D. R. Spearing. *Roadside Geology of Wyoming.* Missoula, Mont.: Mountain Press, 1988.

Laity, Julie. *Deserts and Desert Environments.* Hoboken, N.J.: Wiley-Blackwell, 2008.

Lall, J. S. *The Himalaya: Aspects of Change.* New Delhi: Oxford University Press, 1981.

Lambert, David. *The Field Guide to Geology.* 2d ed. New York: Checkmark Books, 2006.

Lancaster, N. *Geomorphology of Desert Dunes.* London: Routledge, 1995.

Landa, Edward R., and Christian Feller, eds. *Soil and Culture.* New York: Springer, 2010.

Laporte, Leo F., ed. *The Fossil Record and Evolution.* San Francisco: W. H. Freeman, 1982.

Lasca, N. P., and J. Donahue, eds. *Archaeological Geology of North America.* Boulder, Colo.: Geological Society of America, 1990.

Latrubesse, E. M., J. C. Stevaux, and R. Sinha. "Tropical Rivers." *Geomorphology* 70 (2005): 187-206.

Laurin, Michael. *How Vertebrates Left the Water.* Berkeley: University of California Press, 2010.

Lay, Thorne, and Terry C. Wallace. *Modern Global Seismology.* New York: Academic Press, 1995.

Leakey, Richard E. *The Making of Mankind.* New York: E. P. Dutton, 1981.

Leddra, Michael. *Time Matters: Geology's Legacy to Scientific Thought.* Hoboken, N.J.: Wiley-Blackwell, 2010.

Leet, L. D., Sheldon Judson, and M. E. Kaufman. *Physical Geology.* 7th ed. Englewood Cliffs, N.J.: Prentice-Hall, 1987.

LeGrand, H. E. *Drifting Continents and Shifting Theories.* New York: Cambridge University Press, 1988.

Lemey, Philippe, Marco Salami, and Anne-Mieke Vandamme, eds. *The Phylogenetic Handbook: A Practical Approach to Phylogenetic Analysis and Hypothesis Testing.* 2d ed. New York: Cambridge University Press, 2009.

Lemon, Roy R. *Principles of Stratigraphy.* Columbus, Ohio: Merrill, 1990.

Leppmann, Wolfgang. *Pompeii in Fact and Fiction.* London: Elek, 1966.

Lessing, Peter, ed. *Appalachian Structures: Origin, Evolution, and Possible Potential for New Exploration Frontiers.* Morgantown: West Virginia University, 1972.

Levin, Harold L. *The Earth Through Time.* 9th ed. Hoboken, N.J.: Wiley, 2009.

Levy, Matthys, and Mario Salvadori. *Why the Earth Quakes.* New York: W. W. Norton, 1995.

Lewis, John S. *Rain of Iron and Ice.* New York: Addison-Wesley, 1996.

Lewis, Roger. *Thread of Life: The Smithsonian Looks at Evolution.* Washington, D.C.: Smithsonian Books, 1982.

Lieberman, Bruce S. *Paleobiogeography: Using Fossils to Study Global Change, Plate Tectonics, and Evolution.* New York: Kluwer Academic, 2000.

Lieberman, B. S., and R. Kaesler. *Prehistoric Life: Evolution and the Fossil Record.* Chichester, England: John Wiley & Sons, 2010.

Lipman, Peter W., and D. R. Mullineaux, eds. *The 1980 Eruptions of Mount St. Helens, Washington*. U.S. Geological Survey Professional Paper 1250. Reston, Va.: Department of the Interior, 1981.

Lisle, Richard J. *Geological Structures and Maps*. 3rd ed. Burlington, Mass.: Butterworth-Heineman, 2003.

Little, C. *The Colonization of the Land*. Cambridge, England: Cambridge University Press, 1983.

Livermore, Beth. "Antarctic Meltdown." *Popular Science* 250 (February 1997): 38-43.

Lobley, J. Logan. *Mount Vesuvius*. London: E. Stanford, 1868.

Locat, Jacques, and J. Mienert. *Submarine Mass Movement and Their Consequences*. Boston: Kluwer Academic, 2003.

Lockwood, J. P., and Richard W. Hazlett. *Volcanoes: Global Perspectives*. Chichester: Wiley-Blackwell, 2010.

Lomolino, Mark V., Dov F. Sax, and James H Brown. *Foundations of Biogeography: Classic Papers with Commentary*. Chicago: University of Chicago Press, 2004.

Long, J. A. *The Rise of Fishes: Five Hundred Million Years of Evolution*. Baltimore: Johns Hopkins University Press, 2010.

Losos, Jonathan B., Robert E. Ricklefs, and Robert H. MacArthur, eds. *The Theory of Island Geography Revisited*. Princeton, N.J. Princeton University Press, 2010.

Lowe, J. J., and M. J. C. Walker. *Reconstructing Quaternary Environments*. 2d ed. Upper Saddle River, N.J.: Prentice-Hall, 2005.

Lowry, W. D., ed. *Tectonics of the Southern Appalachians*. Roanoke: Virginia Polytechnic Department of Geological Sciences, 1964.

Lubick, George M. *Petrified Forest National Park: A Wilderness Bound in Time*. Tucson: University of Arizona Press, 1996.

Lucas, Spencer G. *Dinosaurs: The Textbook*. 5th ed. Boston: McGraw-Hill, 2005.

__________., ed. *The Triassic Timescale*. Geological Society Special Publication 334. Bath, England: Geological Society, 2010.

Luisi, Pier Luigi. *The Emergence of Life: From Chemical Origins to Synthetic Biology*. New York: Cambridge University Press, 2010.

Lunar and Planetary Institute, Houston, Texas. *Basaltic Volcanism on Terrestrial Planets*. Elmsford, N.Y.: Pergamon Press, 1982.

Lutgens, Frederick K., Edward J. Tarbuck, and Dennis Tasa. *Essentials of Geology*. 11th ed. Upper Saddle River, N.J.: Prentice Hall, 2011.

Lyell, Charles. *Principles of Geology, Being an Attempt to Explain the Former Changes of the Earth's Surface by Reference to Causes Now in Operation*. 3 vols. London: John Murray, 1830-1833.

Mabbutt, J. A. *Desert Landforms*. Cambridge, Mass.: MIT Press, 1977.

MacDonald, David W., ed. *The Encyclopedia of Mammals*. 2d ed. New York: Facts on File, 2006.

Macdonald, Francis A., et al. "Calibrating the Cryogenian." *Science* 327 (2010): 1241-1243.

Macdonald, G. A. *Volcanoes*. Englewood Cliffs, N.J.: Plinian Press, 1972.

MacDougall, Doug. *Frozen Earth: The Once and Future Story of Ice Ages*. Berkeley: University of California Press, 2006.

MacFarlane, Allison, Rosoul B. Sorkhabi, and Jay Quade, eds. *Himalaya and Tibet: Mountain Roots to Mountain Tops*. Boulder, Colo.: Geological Society of America, 1999.

Mackenzie, Fred T. *Our Changing Planet*. 4th ed. Upper Saddle River, N.J.: Prentice Hall, 2010.

__________. *Sediments, Diagenesis, and Sedimentary Rocks: Treatise on Geochemistry*. Miamisburg, Ohio: Elsevier Science, 2005.

Maeder, Herbert. *The Mountains of Switzerland*. London: Allen & Unwin, 1968.

Maher, K. "The Dependence of Chemical Weathering Rates on Fluid Residence Time." *Earth and Planetary Science Letters* 294, nos. 1/2 (2010): 101-110.

Main, Ian. "Long Odds on Prediction." *Nature* 385 (January 2, 1997): 19-20.

Maisey, John G. *Discovering Fossil Fishes*. New York: Basic Books, 2000.

Mandl, Georg. *Rock Joints: The Mechanical Genesis*. New York: Springer, 2005.

Margulis, Lynn. *Acquiring Genomes: A Theory of the Origin of Species*. New York: Basic Books, 2003.

Margulis, Lynn, and Michael J. Chapman. *Kingdoms and Domains: An Illustrated Guide to the Phyla of Life on Earth*. 4th ed. Boston: Elsevier, 2009.

Margulis, Lynn, and Michael F. Dolan. *Early Life*. 2d ed. Boston: Jones and Bartlett, 2002.

__________. *Symbiosis in Cell Evolution: Life and Its Environment on the Early Earth*. 2d ed. New York: W. H. Freeman, 1992.

Margulis, Lynn, and Dorion Sagan. *Microcosmos: Four Billion Years of Microbial Evolution.* Berkeley: University of California Press, 1997.

Marsh, William M. *Earthscape: A Physical Geography.* New York: John Wiley & Sons, 1987.

Marshak, Stephen, and Gautam Mitra, eds. *Basic Methods of Structural Geology.* Englewood Cliffs, N.J.: Prentice-Hall, 1988.

Marshal, L. G. "Land Mammals and the Great American Interchange." *American Scientist* 76 (1988): 380-388.

Marti, Joan, and Gerald G. J. Ernst, eds. *Volcanoes and the Environment.* New York: Cambridge University Press, 2005.

Martin, Anthony J. *Introduction to the Study of Dinosaurs.* New York: Wiley-Blackwell, 2006.

Mason, Betsy. "Season of Fire." *Discover* 24 (2003): 32-39.

Masselink, G., and M. G. Hughes. *Introduction to Coastal Processes and Geomorphology.* London: Hodder Arnold Publication, 2003.

Mather, Anne. "Arid Environments." In *Environmental Sedimentology,* edited by Chris Perry and Kevin Taylor. Malden, Mass.: Blackwell, 2007.

Mathews, William H. *The Story of Glaciers and the Ice Age.* Irvington, N.Y.: Harvey House, 1974.

Matin, Abdul, and Malay Mukul. "Phases of Deformation from Cross-cutting Structural Relationships in External Thrust Sheets: Insights from Small-Scale Structures in the Ramgarh Thrust Sheet, Darjiling Himalaya, West Bengal." *Current Science* 99 (2010): 1369-1377.

Matsch, Charles L. *North America and the Great Ice Age.* New York: McGraw-Hill, 1976.

Maun, Anwar M. *The Biology of Coastal Sand Dunes.* New York: Oxford University Press, 2009.

Maury, M. F. *The Physical Geography of the Sea, and Its Meteorology.* Cambridge, Mass.: Harvard University Press, 1963.

Mays, Larry W. *Ground and Surface Water Hydrology.* Hoboken, N.J.: Wiley, 2012.

McAnally, W. H., and A. J. Mehta. *Coastal and Estuarine Fine Sediment Processes.* Burlington, Vt.: Elsevier, 2001.

McCalpin, James P., ed. *Paleoseismology.* 2d ed. New York: Academic Press, 2009.

McCarthy, Kevin T., Thomas Pichler, and Roy E. Price. "Geochemistry of Champagne Hot Springs Shallow Hydrothermal Vent Field and Associated Sediments, Dominica, Lesser Antilles." *Chemical Geology* 224 (2005): 55–68.

McClay, K. R., and N. J. Price, eds. *Thrust and Nappe Tectonics.* Boston: Blackwell Scientific, 1981.

McGowran, Brian. *Biostratigraphy: Microfossils and Geological Time.* New York: Cambridge University Press, 2008.

McIntyre, Alasdair. *Life in the World's Oceans: Diversity, Distribution, and Abundance.* New York: Wiley Blackwell, 2010.

McKee, Jeffrey K., Frank E. Poirier, and W. Scott McGraw. *Understanding Human Evolution.* 5th ed. Englewood Cliffs, N.J.: Prentice-Hall, 2004.

McKenzie, David J., Anthony P. Farrell, and Colin J. Brauner, eds. *Primitive Fishes.* Vol. 26 in *Fish Physiology.* San Diego, Calif.: Academic Press, 2007.

McKinney, Frank. *The Northern Adriatic Ecosystem: Deep Time in a Shallow Sea.* New York: Columbia University Press, 2007.

McKnight, Tom L., and Darrel Hess. *Physical Geography: A Landscape Appreciation.* 9th ed. Englewood Cliffs, N.J.: Prentice-Hall, 2007.

McLaughlin, Eugene, and Levan Breaux. *Chemical Mineralogy, Smelting, and Metallization.* New York: Nova Science, 2009.

McMenamin, Mark. *Discovering the First Complex Life: The Garden of the Ediacara.* New York: Columbia University Press, 1998.

McNamara, K. J. *Shapes of Time: The Evolution of Growth and Development.* Baltimore: Johns Hopkins University Press, 1997.

McPhee, John. *Basin and Range.* New York: Farrar, Straus & Giroux, 1981.

__________. *In Suspect Terrain.* New York: Farrar, Straus & Giroux, 1983.

__________. *Rising from the Plains.* New York: Farrar, Straus & Giroux, 1986.

Medina, Socorro, Ellen Sukovich, and Robert A. Houze Jr. "Vertical Precipitation Structures of Cyclones Crossing the Cascades." *Bulletin of the American Meteorological Society* 88 (2007): 1888-1889.

Meng, Zhaoping, Zincai Zhang, and Suping Peng. "Influence of Sedimentary Environments on Mechanical Properties of Clastic Rocks." *Environmental Geology* 51, no. 1 (2006): 113-120.

Menzies, M., ed. *Modern Glacial Environments: Processes, Dynamics, and Sediments.* Oxford, England: Butterworth-Heinemann, 1996.

Merrill, Ronald T. *Our Magnetic Earth: The Science of Geomagnetism.* Chicago: University of Chicago Press, 2010.

Meyer, Charles. "Ore Deposits as Guides to Geologic History of the Earth." *Annual Review of Earth and Planetary Sciences* 16 (1988): 147.

Miall, Andrew D. *The Geology of Stratigraphic Sequences.* 2d ed. New York: Springer, 2010.

Michard, Andre, et al., eds. *Continental Evolution: The Geology of Morocco.* Berlin: Springer, 2010.

Middleton, Gerard V., ed. *Encyclopedia of Sediments and Sedimentary Rocks.* Boston: Kluwer Academic, 2003.

Middleton, John, and Anthony C. Waltham. *The Underground Atlas: A Gazetteer of the World's Cave Regions.* 2d ed. Darby, Pa.: Diane Publishing, 1999.

Miller, Russell, and the editors of Time-Life Books. *Planet Earth: Continents in Collision.* Alexandria, Va.: Time-Life Books, 1983.

Miller, William Charles. *Trace Fossils: Concepts, Problems, Prospects.* St. Louis, Mo.: Elsevier, 2007.

Milliman, J. B., and W. R. Wright, eds. *Environment of the U.S. Atlantic Continental Slope and Rise.* Boston: Jones & Bartlett, 1987.

Molnar, Peter. "The Structure of Mountain Ranges." *Scientific American* 255 (July 1986): 70-79.

Monroe, James S., Reed Wicander, and Richard Hazlett. *Physical Geology: Exploring the Earth.* 6th ed. Belmont, Calif.: Thomson Higher Education, 2007.

Moore, John E. *Field Hydrogeology: A Guide for Site Investigations and Report Preparation.* Boca Raton, Fla.: CRC Press, 2012.

Moore, R. C., ed. *Treatise on Invertebrate Paleontology.* 40 vols. Lawrence: University of Kansas Press, 1953-2010.

Moore, W. G. *Glaciers.* London: Hutchinson Educational, 1972.

Moore, W. S. "Late Pleistocene Sea-Level History." In *Uranium Series Disequilibrium: Applications to Environmental Problems,* edited by M. Ivanovich and R. S. Hannah. Oxford, England: Clarendon Press, 1982.

Moores, Eldridge, ed. *Shaping the Earth: Tectonics of Continents and Oceans.* New York: W. H. Freeman, 1990.

Morell, Virginia. "The Origin of Birds." *Audubon* (March/April 1997): 36-45.

Morgan, Roy P. C., and Mark A. Nearing, eds. *Handbook of Erosion Modelling.* New York: Wiley-Blackwell, 2011.

Morrissey, John, and James L. Sumich. *Introduction to the Biology of Marine Life.* Burlington, Mass.: Jones and Bartlett, 2011.

Morrone, Juan J. *Evolutionary Biogeography: An Integrative Approach with Case Studies.* New York: Columbia University Press, 2009.

Moseley, Charles. "Slow Progress in Earthquake Prediction." *Editorial Research Reports* 2 (July 15, 1988): 354-363.

Müller, A. H. "Fossilization (Taphonomy)." In *Treatise on Invertebrate Paleontology.* Boulder, Colo.: Geological Society of America, 1979. A

Murck, B. W., and B. J. Skinner. *Geology Today.* New York: John Wiley and Sons, 1999.

Murphy, J. B., and R. D. Nance. "Mountain Belts and the Supercontinent Cycle." *Scientific American* (April 1992).

Murray, Marian. *Hunting for Fossils.* New York: Macmillan, 1967.

Nance, R. Damian, T. R. Worsley, and J. B. Moody. "The Supercontinent Cycle." *Scientific American* 259 (July 1988): 72-79.

Nash, J. Madeleine. *El Niño: Unlocking the Secrets of the Master Weather-maker.* New York: Warner Books, 2003.

National Research Council. *Explosive Volcanism: Inception, Evolution, and Hazards.* Washington, D.C.: National Academy Press, 1984.

Nations, J. Dale, and Edmund Stump. *Geology of Arizona.* 2d ed. Dubuque, Iowa: Kendall/Hunt, 1996.

Nemcok, M., S. Schamel, and R. Gayer. *Thrustbelts; Structural Architecture, Thermal Regimes and Petroleum Systems.* New York: Cambridge University Press, 2005.

Nesse, William. *Introduction to Mineralogy.* New York: Oxford University Press, 2011.

__________. *Introduction to Optical Mineralogy.* New York: Oxford University Press, 2004.

Newton, Ian. *Speciation and Biogeography of Birds.* London: Academic Press, 2003.

Nichols, Douglas J., and Kirk R. Johnson. *Plants and the K-T Boundary.* New York: Cambridge University Press, 2011.

Nichols, Gary. *Sedimentology and Stratigraphy.* New York: John Wiley & Sons, 2009.

Nicholson, Sharon E. *Dryland Climatology.* New York: Cambridge University Press, 2011.

Nicolas, A. *The Mid-Ocean Ridges.* New York: Springer-Verlag, 1995.

Nicolson, Nigel. *The Himalayas.* Amsterdam: Time-Life Books, 1975.

Niedoroda, Alan W. "Shelf Processes " In *Encyclopedia of Coastal Science,* edited by M. Schwartz. Dordrecht: Springer, 2005.

Nield, Ted. *Supercontinent: Ten Billion Years in the Life of Our Planet.* Cambridge, Mass.: Harvard University Press, 2007.

Niklas, K. J. *The Evolutionary Biology of Plants.* Chicago: University of Chicago Press, 1997.

Nilsson, T. *The Pleistocene: Geology and Life in the Quaternary Ice Age.* Stuttgart: Ferdinand Enke Verlag, 1983.

Nisbet, E. G. *Leaving Eden: To Protect and Manage the Earth.* New York: Cambridge University Press, 1991.

__________. *The Young Earth: An Introduction to Archean Geology.* Boston: Unwin Hyman, 1987.

Norell, Mark, Eugene S. Gaffney, and Lowell Dingus. *Discovering Dinosaurs: Evolution, Extinction, and the Lessons of Prehistory.* Berkeley: University of California Press, 2000.

Norell, Mark, Luis Chiappe, and James Clark. "New Limb on the Avian Family Tree." *Natural History* 9 (September 1993): 38-42.

Novák, Pavel. *Hydraulic Modelling: An Introduction: Principles, Methods, and Applications.* London: Spon Press, 2010.

Oakeshott, G. B. *California's Changing Landscapes: A Guide to the Geology of the State.* New York: McGraw-Hill, 1978.

O'Brien, Rachel, and Jennifer A. Roberts. "Teaching the Principles of Geomicrobiology and the Process of Experimental Research to Undergraduate and First-Year Graduate Students." *Journal of College Science Teaching* 38, no. 2 (2008): 30-38.

Ochi, Michel K. *Ocean Waves: The Stochastic Approach.* New York: Cambridge University Press, 2005.

Odell, Lee H. *Treatment Technologies for Groundwater.* Denver: American Water Works Association, 2010.

Officer, C., and J. Page. *The Great Dinosaur Extinction Controversy.* Reading, Mass.: Addison-Wesley, 1996.

Ogawa, Yujiro, Ryo Anma, and Yildirim Dilek. *Accretionary Prisms and Convergent Margin Tectonics in the Northwest Pacific Basin.* New York: Springer Science+Business Media, 2011.

Ogg, James G., Gabi Ogg, and Felix M. Gradstein. *The Concise Geologic Time Scale.* New York: Cambridge University Press. 2008.

Ollier, Cliff, and Colin Pain. *The Origin of Mountains.* London: Cambridge University Press, 2000.

Olschak, B. C., A. Gansser, and E. M. Buhrer. *Himalayas.* New York: Facts on File, 1987.

Olsen, Kenneth H., ed. *Continental Rifts: Evolution, Structure, Tectonics.* Amsterdam: Elsevier, 1995.

Olson, Steve. "Neanderthal Man." *Smithsonian* 37 (2006): 76-82.

O'Neill, Dan. *The Last Giant of Beringia: The Mystery of the Bering Land Bridge.* Cambridge, Mass.: Basic Books, 2005.

Oppenheimer, Clive. *Eruptions That Shook the World.* New York: Cambridge University Press, 2011.

Oreskes, Naomi. *Plate Tectonics: An Insider's Guide to the Modern Theory of the Earth.* Boulder, Colo.:

Orr, Brendon B., et al., eds. *Rocky Mountain Geology.* Laramie, Wyo.: *GeoScience World,* 1962-current.

Padian, Kevin, ed. *The Origin of Birds and the Evolution of Flight.* Memoirs of the California Academy of Sciences 8. San Francisco: California Academy of Sciences, 1986.

Padian, Kevin, and Luis M. Chiappe. "The Origin of Birds and Their Flight." *Scientific American* (February 1998): 28-37

Parker, S. P., ed. *McGraw-Hill Encyclopedia of Geological Sciences.* 2d ed. New York: McGraw-Hill, 1988.

Parkes, P. A. *Current Scientific Techniques in Archaeology.* New York: St. Martin's Press, 1986.

Paul, Frank, et al. "Rapid Disintegration of Alpine Glaciers Observed with Satellite Data." *Geophysical Research Letters.* 31 (2004): L21402.

Paul, Gregory S. *Dinosaurs of the Air: The Evolution of Loss of Flight in Dinosaurs and Birds.* Baltimore: John Hopkins University Press, 2002.

__________. *Predatory Dinosaurs of the World.* New York: Simon & Schuster, 1988.

Pentecost, Allan, B. Jones, and R. W. Renaut. "What Is a Hot Spring?" *Canadian Journal of Earth Science* 40 (2003): 1443–1446.

Perkins, Dexter. *Mineralogy.* New York: Prentice Hall, 2010.

Perkins, Dexter, and Kevin R. Henke. *Minerals in Thin Section.* New York: Prentice Hall, 2003.

Perret, Frank A. *Eruption of Mount Pelée, 1929-1932.* Washington, D.C.: Carnegie Institution of Washington, 1935.

Petersen, James, Dorothy Irene Sack, and Robert Gabler. *Fundamentals of Physical Geography.* Belmont, Calif.: Brooks/Cole, 2011.

Pethic, J. *An Introduction to Coastal Geomorphology.* Baltimore: Edward Arnold, 1984.

Philpotts, Anthony, and Jay Aque. *Principles of Igneous and Metamorphic Petrology.* 2d ed. New York: Cambridge University Press, 2009.

Pierce, Kenneth L. *History and Dynamics of Glaciation in the Northern Yellowstone National Park Area.* U.S. Geological Survey Professional Paper 729-F. Washington, D.C.: Government Printing Office, 1979.

Pierrehumbert, Raymond T. *Principles of Planetary Climate.* New York: Cambridge University Press, 2011.

Pinet, Paul R. *Invitation to Oceanography.* Sudbury, Mass.: Jones and Bartlett, 2003.

__________. Invitation to Oceanography. 5th ed. Sudbury, Mass.: Jones and Bartlett, 2009.

Pipkin, Bernard W. *Geology and the Environment.* St. Paul, Minn.: West, 1994.

__________. *Geology and the Environment.* Belmont, Calif.: Brooks/Cole Cengage Learning, 2011.

Pirazzoli, P. A. "Marine Terrace." In *Encyclopedia of Coastal Science* , edited by M. Schwartz. Dordrecht: Springer, 2005.

Playfair, John. *Illustrations of the Huttonian Theory of the Earth.* Reprint, with an introduction by George W. White. Urbana: University of Illinois Press, 1956.

Plummer, Charles C., Diane H. Carlson, and Lisa Hammersley. *Physical Geology.* 13th ed. New York: McGraw-Hill Science, 2009.

Poblet, J., and R. J. Lisle. *Kinematic Evolution and Structural Styles of Fold-and-Thrust Belts,* Special Publication 349. Geological Society of London, 2011.

Pohl, Water L. *Economic Geology: Principles and Practice.* Hoboken, N.J.: Wiley-Blackwell, 2011.

Pokorný, Vladimír. *Principles of Zoological Micropaleontology.* Translated by K. A. Allen. 2 vols. Elmsford, N.Y.: Pergamon Press, 1965.

Powell, John Wesley. *The Exploration of the Colorado River and Its Canyon.* Mineola, N.Y.: Dover, 1961.

Price, Neville J. *Fault and Joint Development in Brittle and Semi-Brittle Rock.* London: Pergamon Press, 1966.

Price, R. J. *Glacial and Fluvioglacial Landforms.* New York: Hafner Press, 1973.

Prothero, Donald R. *After the Dinosaurs: The Age of Mammals.* Bloomington: Indiana University Press, 2006.

__________. *Bringing Fossils to Life.* 2d ed. Boston: McGraw-Hill, 2004.

__________. *Catastrophes!: Earthquakes, Tsunamis, Tornadoes, and Other Earth-Shattering Disasters.* Baltimore: Johns Hopkins University Press, 2011.

Prothero, Donald R., and Carl Buell. *Evolution: What the Fossils Say and Why It Matters.* New York: Columbia University Press, 2007.

Prothero, Donald R., and Robert H. Dott, Jr. *Evolution of the Earth.* 8th ed. New York: McGraw-Hill, 2009.

Prothero, Donald R., and Fred Schwab. *Sedimentary Geology: An Introduction to Sedimentary Rocks and Stratigraphy.* 2d ed. New York: W. H. Freeman, 2004.

Pye, Kenneth, and Haim Tsoar. *Aeolian Sand and Sand Dunes.* New York: Springer, 2008.

Quennell, A. M. *Rift Valleys: Afro-Arabian.* Stroudsburg, Pa.: Hutchinson Ross, 1982.

Quental, Tiago B., and Charles R. Marshall. "Extinction During Evolutionary Radiations: Reconciling the Fossil Record with Molecular Phylogenies." *Evolution* 63, no. 12 (2009): 3158-3167.

Rabinowitz, Phillip, Sylvia Herrig, and Karen Riedel. "Ocean Drilling Program Altering Our Perception of Earth." *Oceanus* 29 (Fall 1986): 36-40.

Radinsky, Leonard B. *The Evolution of Vertebrate Design.* Chicago: University of Chicago Press, 1987.

Radok, Uwe. "The Antarctic Ice." *Scientific American* 253 (August 1985): 98-105.

Ragland, Paul C., and John J. W. Rogers, eds. *Basalts.* New York: Van Nostrand Reinhold, 1984.

Rajasekara, Murthy, C., P. C. Sinha, and Y. R. Rao. *Modelling and Monitoring of Coastal Marine Processes.* New York: Springer, 2008.

Rampino, Michael. "Darwin's Error? Patrick Matthew and the Catastrophic Nature of the Geological Record." *Historical Biology* 23 (2011): 227-230.

Ramsey, Jonathan M., and Frederick M. Chester. "Hybrid Fracture and the Transition from Extension Fracture to Shear Fracture." *Nature* 428 (2004): 63-66.

Ransom, Jay Ellis. *Fossils in America.* New York: Random House, 1970.

Rao, T. Ramamohana. "Geological Evolution of the Caribbean Plate: Some Critical Aspects in the Two Divergent Models." *Current Science* 95 (2008): 736-742.

Rapp, George, Jr., and Christopher L. Hill. *Geoarchaeology.* 2d ed. New Haven, Conn.: Yale University Press, 2006.

Raup, David. *Extinction: Bad Genes or Bad Luck?* New York: W. W. Norton, 1992.

__________. *The Nemesis Affair: A Story of the Death of Dinosaurs and the Ways of Science.* Rev. ed. New York: W. W. Norton, 1999.

Rea, G. Thomas, ed. *Caving Basics: A Comprehensive Guide for Beginning Cavers.* 3rd ed. Huntsville, Ala.: National Speleological Society, 1992.

Reading, Harold G. *Sedimentary Environments: Processes, Facies, and Stratigraphy.* Hoboken, N.J.: Wiley-Blackwell, 1996.

Redfern, Ron. *The Making of a Contitnent.* New York: Times Books, 1983.

__________. *The Making of a Continent.* New York: Times Books, 1986.

Reinhardt, J., and W. R. Sigleo. *Paleosols and Weathering Through Geologic Time: Principals and Applications.* Special Paper 216. Boulder, Colo.: Geological Society of America, 1988.

Renema, Willem, ed. *Biogeography, Time and Place: Distributions, Barriers and Islands.* New York: Springer, 2010.

Restrepo, Juan D., Sergio A. Lopez, and Juan C. Restrepo. "The Effects of Geomorphic Controls on Sediment Yield in the Andean Rivers of Colombia." *Latin American Journal of Sedimentology and Basin Analysis* 16 (2009): 79-92.

Retallack, Gregory J. *Laboratory Exercises in Paleopedology.* Eugene: University of Oregon Press, 1985.

__________. *Soils of the Past: An Introduction to Paleopedology.* 2ded. New York: Blackwell Science, 2001.

Rhodes, Frank H. T., et al. *A Golden Guide to Fossils.* New York: St. Martin Press, 2001.

Rich, Patricia Vickers, et al. *The Fossil Book.* Mineola, N.Y.: Dover Publications, 1997.

Riddell, Janet, and Maurice Colpron. "Lithostratigraphic and Tectonic Framework of Jurassic and Cretaceous Intermontane Sedimentary Basins of South-Central British Columbia." *Canadian Journal of Earth Sciences* 48, no. 6 (2011): 870-896.

Rinehart, J. S. *Geysers and Geothermal Energy.* New York: Springer-Verlag, 1980.

Ritchie, David. *The Ring of Fire.* New York: Atheneum, 1981.

Ritter, Dale F., R. Craig Kochel, and Jerry R. Miller. *Process Geomorphology.* 4th ed. Long Grove, Ill.: Waveland Press, 2006.

Roberts, D. C., and W. G. Hodson. *A Field Guide to Geology: Eastern North America.* Boston: Houghton Mifflin, 2001.

Robinson, Andrew. "Avalanches, Glaciers, and Ice Ages." In *Earthshock: Climate, Complexity, and the Forces of Nature.* London: Thames and Hudson, 1993.

__________. *Earth Shock, Hurricanes, Volcanoes, Earthquakes, Tornadoes, and Other Forces.* New York: Thames & Hudson, 1993.

Robinson, Edwin S., and Cahit Coruh. *Basic Exploration Geophysics.* New York: John Wiley & Sons, 1988.

Rogers, John. *The Tectonics of the Appalachians.* New York: John Wiley & Sons, 1970.

Rogers, John J. M., and M. Santosh. *Continents and Supercontinents.* New York: Oxford University Press, 2004.

Rogers, Raymond R., and Krause, David W. "Tracking an Ancient Killer." *Scientific American* 296 (2007): 42-51.

Romanovsky, V., et al. "Permafrost Temperature Records: Indicators of Climate Change." *EOS, AGU Transactions* 83 (2002): 589-594.

Rona, Peter A., K. Boström, L. Laubier, and K. L. Smith, Jr., eds. *Hydrothermal Processes at Seafloor Spreading Centers.* New York: Plenum Press, 1983.

Rosbacher, L. A. *Recent Revolutions in Geology.* New York: Franklin Watts, 1986.

Rosendahl, B. R. "Architecture of Continental Rifts with Special Reference to East Africa." *Annual Review of Earth and Planetary Sciences* 15 (1987): 445.

Ross, Charles A., ed. *Paleobiogeography.* Stroudsburg, Pa.: Dowden, Hutchinson, and Ross, 1976.

Ruddiman, William F. *Earth's Climate: Past and Future.* 2d ed. New York: W. H. Freeman, 2008.

Rudwick, Martin J. S. *The Meaning of Fossils: Episodes in the History of Palaeontology.* 2d ed. Chicago: University of Chicago Press, 1985.

Russell, Israel C. "The Recent Volcanic Eruptions in the West Indies." *National Geographic* 13 (July 1902): 267-284.

Russo, R. M., and P. G. Silver. "The Andes' Deep Origins." *Natural History* (February 1995).

Rutten, M. G. *Geology of Western Europe.* New York: Elsevier, 1969.

Rutter, Nat. "International Year of Planet Earth 4. Utilizing Paleosols in Quaternary Climate Change Studies." *Geoscience Canada* 36 (2009): 65-77.

Ryan, William, and Walter Pitman. *Noah's Flood: The New Scientific Discoveries About the Event That Changed History.* New York: Simon & Schuster, 1998.

Salcedo, Chris. *San Bernardino Mountain Trails.*

Berkeley: Wilderness Press, 2003.

Samama, J. C. *Ore Fields and Continental Weathering.* New York: Van Nostrand Reinhold, 1986.

Sasowsky, I. D. *Cumulative Index for the National Speleological Society Bulletin, Volumes 1 Through 45, and Occasional Papers of the N.S.S., Numbers 1 Through 4.* Huntsville, Ala.: National Speleological Society, 1986.

Savage, R. J., and M. R. Long. *Mammal Evolution: An Illustrated Guide.* New York: Facts on File, 1986.

Scarth, Alwyn, and Jean-Claude Tanguy. *Volcanoes of Europe.* New York: Oxford University Press, 2001

Schaetzl, Randall J., and Sharon Anderson. *Soils: Genesis and Geomorphology.* New York: Cambridge University Press, 2005.

Scheckenbach, Frank, et al. "Large-Scale Patterns in Biodiversity of Microbial Eukaryotes from the Abyssal Sea Floor." *Proceedings of the National Academy of Sciences of the United States of America* 107 (2010): 115-120.

Schiffer, Michael, ed. *Advances in Archaeological Method and Theory.* 11 vols. San Diego, Calif.: Academic Press, 1978-1987.

__________. *Formation Processes of the Archaeological Record.* Salt Lake City: University of Utah Press, 2002

Schmidt, Stefan M., and Mark Handy. *The Alps; Their Evolution and Role Within the Mediterranean Collision Zone.* New York: Springer, 2011.

Scholle, Peter A., and Darwin Spearing. *Sandstone Depositional Environments.* Washington, D.C.: American Association of Petroleum Geologists, 1982.

Scholle, Peter A., and Dana S. Ulmer-Scholle. *A Color Guide to the Petrography of Carbonate Rocks: Grains, Textures, Porosity, Diagenesis.* Tulsa, Okla.: American Association of Petroleum Geologists, 2003.

Scholz, Christopher H. *The Mechanics of Earthquakes and Faulting.* New York: Cambridge University Press, 2002.

Schopf, J. William, ed. *Earth's Earliest Biosphere: Its Origin and Evolution.* Princeton, N.J.: Princeton University Press, 1983.

__________. ed. *Major Events in the History of Life.* Boston: Jones and Bartlett, 1992.

Schopf, J. William, et al. "Evidence of Archean Life: Stromatolites and Microfossils." *Precambrian Research.* 158 (2007): 141-155.

Schubert, Gerald, ed. *Journal of Geophysical Research 89,* no. B10 (1984).

Schultz, Gwen. *Glaciers and the Ice Age.* New York: Holt, Rinehart and Winston, 1963.

__________. *Ice Age Lost.* Garden City, N.Y.: Doubleday, 1974.

Schumm, Stanley A. *The Fluvial System.* New York: Wiley-Interscience, 1977.

Schuur, Edward A. G., et al. "The Effect of Permafrost Thaw on Old Carbon Release and Net Carbon Exchange from Tundra." *Nature* 459 (2009): 556-559.

Scientific American 249 (September 1983).

Scientific American. The Dynamic Earth. New York: W. H. Freeman, 1983.

Scientific American editors. *Continents.* San Francisco: W. H. Freeman, 1973.

Segar, Douglas. *An Introduction to Ocean Sciences.* 2d ed. New York: W. W. Norton & Company, 2007.

Seibold, Eugen, and Berger, Wolfgang H. *The Sea Floor: An Introduction to Marine Geology.* Reprinted 3rd ed. New York: Springer-Verlag, 2010.

Sepkoski, David, and Michael Ruse, eds. *The Paleobiological Revolution: Essays on the Growth of Modern Paleontology.* Chicago: University of Chicago Press, 2009.

Seyfert, Carl K., ed. *The Encyclopedia of Structural Geology and Plate Tectonics.* New York: Van Nostrand Reinhold, 1987.

Sharma, R. S. *Cratons and Fold Belts of India.* New York: Springer, 2009.

Sharp, R. P. *Living Ice: Understanding Glaciers and Glaciation.* New York: Cambridge University Press, 1988.

Sharpton, Virgil L., and Peter D. Ward, eds. *Global Catastrophes in Earth History.* Boulder, Colo.: Geological Society of America, 1990.

Shea, James H., ed. *Continental Drift.* New York: Van Nostrand Reinhold, 1985.

Shelley, David R. "Migrating Tremors Illuminate Complex Deformation Beneath the Seismogenic San Andreas Fault." *Nature Letters* 463, no. 4 (February 2010): 648-653.

Shell Oil Company—Exploration Department. *Stratigraphic Atlas, North and Central America.* Houston: The Department, 1975.

Shervais, J. W. "Birth, Death, and Resurrection: The Life Cycle of Suprasubduction Zone Ophiolites." *Geochemistry, Geophysics, Geosystems* 2 (January 2001): 1-45.

Shipman, Pat. *Life History of a Fossil: An Introduction to Taphonomy and Paleoecology.* Cambridge, Mass.: Harvard University Press, 1993.

Short, N. M., and R. W. Blair. *Geomorphology from Space: A Global Overview of Regional Landforms.* NASA SP-486. Washington, D.C.: National Aeronautics and Space Administration, 1986.

Shreeve, James. *The Neanderthal Enigma.* New York: William Morrow, 1995.

Shugart, H. H. *How the Earthquake Bird Got Its Name and Other Tales of an Unbalanced Nature.* New Haven, Conn.: Yale University Press, 2007.

Siebe, C., M. Abrams, J. L. Macias, and J. Obenholzer. "Repeated Volcanic Disasters in Prehispanic Time at Popocatépetl, Central Mexico: Past Key to the Future?" *Geology* 24 (1996): 399-422.

Siebert, Lee, Tom Simkin, and Paul Kimberly. *Volcanoes of the World.* 3rd ed. Berkeley: University of California Press, 2011.

Sigmundsson, Freysteinn. *Iceland Geodynamics: Crustal Deformation and Divergent Plate Tectonics.* Chichester, England: Praxis, 2006.

Sigurdsson, Haraldur, ed. *Encyclopedia of Volcanoes.* San Diego, Calif.: Academic Press, 2000.

Sigurdsson, Haraldur, and Steven Carey. "The Far Reach of Tambora." *Natural History* 97 (June, 1988): 67-73.

Sigurdsson, Haraldur, et al. "The Eruption of Vesuvius in a.d. 79." *National Geographic Research* 1 (1985): 332-387.

Simkin, Tom, Richard S. Fiske, and Sarah Melcher, eds. *Krakatau 1883: The Volcanic Eruption and Its Effects.* Washington, D.C.: Smithsonian Institution Press, 1983.

Simmons, I. G. *Environmental History: A Concise Introduction.* Oxford: Blackwell, 1993.

Simpson, George Gaylord. *Life of the Past: An Introduction to Palaeontology.* New Haven, Conn.: Yale University Press, 1968.

Simpson, Sarah. "Violent Origins of Continents." *Scientific American* 302 (2010): 60-67.

Singh, Satish C., et al. "Discovery of a Magma Chamber and Faults Beneath a Mid-Atlantic Ridge Hydrothermal Field." *Nature.* 442 (2006): 1029-1032.

Sintubin, Manuel, et al., eds. *Ancient Earthquakes.* Boulder, Colo.: Geological Society of America, 2010.

Skinner, B. J., et al. *The Dynamic Earth: An Introduction to Physical Geology.* 5th ed. New York: John Wiley & Sons, 2006.

Skinner, Brian J., and Stephen C. Porter. *Physical Geology.* New York: John Wiley & Sons, 1987.

Smith, A. G., D. G. Smith, and B. M. Funnell. *Atlas of Mesozoic and Cenozoic Coastlines.* New York: Cambridge University Press, 2004.

Smith, Andrew B. *Systematics and the Fossil Record.* Oxford: Blackwell Press, 1996.

Smith, David G., ed. *The Cambridge Encyclopedia of Earth Sciences.* New York: Cambridge University Press, 1981.

Smith, Robert B., and Robert L. Christiansen. "Yellowstone Park as a Window on the Earth's Interior." *Scientific American* 242 (February 1980): 104-117.

__________. "Yellowstone Park as a Window on the Earth's Interior." In *Volcanoes and the Earth's Interior.* San Francisco: W. H. Freeman, 1982.

Smith, Robert B., and Lee J. Siegel. *Windows into the Earth.* New York: Oxford University Press, 2000.

Sparks, John. *Planet Earth.* Garden City, N.Y.: Doubleday, 1977.

Sparrow, Andy. *Complete Caving Manual.* Rev. ed. Wiltshire: Crowood Press, 2009.

Spencer, Edgar W. *Introduction to the Structure of the Earth.* 3rd ed. New York: McGraw-Hill, 1988.

Stanley, S. M. *Earth System History.* 2d ed. New York: W. H. Freeman, 2005.

__________. *Earth System History.* 3rd ed. New York: Freeman, 2008.

__________. *Exploring Earth and Life Through Time.* 2d ed. New York: W. H. Freeman, 1989.

__________. *Extinction.* New York: Scientific American Books, 1987.

__________. *The New Evolutionary Timetable: Fossils, Genes, and the Origin of Species.* New York: Basic Books, 1981.

Starr, Cecie, et al. *Biology: The Unity and Diversity of Life.* 12th ed. Belmont, Calif.: Wadsworth, 2008.

Stearn, C. W., and R. L. Carroll. *Paleontology: The Record of Life.* New York: John Wiley & Sons, 1989.

Stearns, H. T. *Geology of the State of Hawaii.* 2d ed. Palo Alto, Calif.: Pacific Books, 1985.

Stebbins, G. Ledyard. *Darwin to DNA, Molecules to Humanity.* New York: W. H. Freeman, 1982.

Stebbins, G. L., and G. J. C. Hill. "Did Multicellular Plants Invade the Land?" *American Naturalist* 115 (1980): 342-353.

Steel, Duncan. *Rogue Asteroids and Doomsday Comets.* New York: John Wiley & Sons, 1995.

Steingisser, Alethea, and Marcus, W. Andrew. "Human Impacts on Geyser Basins." *Yellowstone Science* 17 (2009): 7-18.

Stern, Robert J., and David W. Scholl. "Yin and Yang of Continental Crust Creation and Destruction by Plate Tectonic Processes." *International Geology Review* 52 (2010): 1-31.

Steward, Iain. *Earth: The Biography.* Washington, D.C.: National Geographic Society, 2007.

Stewart, John H. *Geology of Nevada.* Reno: Nevada Bureau of Mines and Geology, 1980.

Stewart, W. N. *Paleobotany and the Evolution of Plants.* London: Cambridge University Press, 2010.

Stewart, W. N., and G. W. Rothwell. *Paleobotany and the Evolution of Plants.* 2d ed. Cambridge: Cambridge University Press, 2010.

Stimac, J. A., et al. "The Crater Lake and Hydrothermal System of Mount Pinatubo, Philippines: Evolution in the Decade After Eruption." *Bulletin of Volcanology* 66(2005): 149-167.

Strahler, Arthur N. *Plate Tectonics.* Cambridge, Mass.: Geo-Books, 1998.

Strassberg, Gil, Norman L. Jones, and David R Maidment. *Arc Hydro Groundwater: GIS for Hydrogeology.* Redlands, Calif.: ESRI Press, 2011.

Stringer, Christopher, and Robin McKie. *African Exodus.* New York: Henry Holt, 1996.

Stuart, G. E. "Teotihuacan." *National Geographic* 188 (1995): 2-35.

Stutz, Matthew L., and Orrin H. Pilkey. "The Relative Influence of Humans on Barrier Islands: Humans versus Geomorphology." *Reviews in Engineering Geology* 16 (2005): 137-147.

Stuwe, Kurt. *Geodynamics of the Lithosphere: An Introduction.* Berlin: Springer, 2007.

Suess, Edward. *The Face of the Earth.* Oxford, England: Oxford University Press, 1909.

Sullivan, Walter. *Continents and Motion: The New Earth Debate.* New York: McGraw-Hill, 1974.

__________. *Continents in Motion: The New Earth Debate.* 2d ed. Melville, N.Y.: American Institute of Physics, 1993.

Summer, W., E. Klaghofer, and W. Zhang, eds. *Modelling Soil Erosion, Sediment Transport, and Closely Related Hydrological Processes.* Wallingford, England: IAHS Press, 1998.

Suppe, John, ed. *Principles of Structural Geology.* Englewood Cliffs, N.J.: Prentice-Hall, 1985.

Sutcliffe, A. J. *On the Track of Ice Age Mammals.* Cambridge, Mass.: Harvard University Press, 1985.

Taminskas, J, et al. "Karst and Ecosystems." In *Geology and Ecosystems,* edited by Igor S. Zektser. New York: Springer, 2010.

Tarbuck, Edward S., Frederick K. Lutgens, and Dennis Tasa. *Earth: An Introduction to Physical Geology.* 10th ed. Upper Saddle River, N.J.: Prentice Hall, 2011.

Tattersall, Ian. *The Last Neanderthal.* Rev. ed. New York: Basic Books, 1999.

Tattersall, Ian, and Jeffrey H. Schwartz. "Hominids and Hybrids: The Place of Neanderthals in Human Evolution." *Proceedings of the National Academy of Science* 96 (June, 1999): 7.

Taylor, Paul D. *Extinctions in the History of Life.* New York: Cambridge University Press, 2004.

Taylor, S. R., and S. M. McLennan. *The Continental Crust: Its Composition and Evolution.* Oxford, England: Blackwell Scientific, 1985.

__________. "The Evolution of Continental Crust." *Scientific American* (January 1996): 274.

Taylor, Thomas N., Edith L. Taylor, and Michael Krings. *Paleobotany: The Biology and Evolution of Fossil Plants.* 2d ed. Boston: Academic Press, 2008.

Tazieff, H., and J. C. Sabroux, eds. *Forecasting Volcanic Events.* Developments in Volcanology 1. New York: Elsevier, 1983.

Tewari, Vinod, and Joseph Seckbach, eds. *Stromatolites: Interaction of Microbes with Sediments.* New York: Springer, 2011.

Thayer, David. *An Introduction to Grand Canyon Fossils.* Grand Canyon: Grand Canyon Association, 2009.

Thomas, M. F. *Geomorphology in the Tropics: A Study of Weathering and Denudation in Low Latitudes.* New York: John Wiley and Sons, 1994.

Thompson, Ida. *The Audubon Society Field Guide to North American Fossils.* New York: Alfred A. Knopf, 1983.

Thybo, H., and C. A. Nielsen. "Magma-Compensated Crustal Thinning in Continental Rift Zones." *Nature* 457 (2009): 873-876.

Tilden, F. *The National Parks.* 3rd ed. New York: Alfred A. Knopf, 1986.

Tilling, Robert I., Christina Heliker, and Thomas L. Wright. *Eruptions of Hawaiian Volcanoes: Past, Present, and Future.* Denver, Colo.: U.S. Geological Survey, 1987.

Time-Life Books. *Volcano.* Alexandria, Va.: Author, 1982.

Tobias, Philip V. "Major Events in the History of Mankind." In *Major Events in the History of Life*, edited by J. William Schopf. Boston: Jones and Bartlett, 1992.

Tomkeieff, S. I. "Unconformity: An Historical Study." *Proceedings of the Geologists' Association* 73 (1962): 383.

Trenhaile, A. S. "Modeling the Development of Marine Terraces on Tectonically Mobile Rock Coasts." *Marine Geology* 185 (2002): 341-361.

Trent, D. D. "Geology of Joshua Tree National Monument, Riverside and San Bernardino Counties." *California Geology* 37 (April 1984): 75-86.

__________. "Mount Baldy Mining Area, Los Angeles and San Bernardino Counties, California." *California Geology* 39 (August 1986): 179-184

Tributsch, Helmut. *How Life Learned to Live.* Cambridge, Mass.: MIT Press, 1982.

Trinkaus, Erik, and Pat Shipman. *The Neanderthals: Changing the Image of Mankind.* New York: Alfred A. Knopf, 1992.

Trombley, R. B. *The Forecasting of Volcanic Eruptions.* Lincoln, Neb. iUniverse, 2006.

Trumpy, Rudolph. *Geology of Switzerland.* Basel: Wepf, 1980.

Tucker, Maurice E. *Sedimentary Rocks in the Field.* Hoboken, N.J.: John Wiley & Sons, 2011.

U.S. Department of the Interior, Minerals Management Service. *Programmatic Environmental Assessment: Arctic Ocean Outer Continental Shelf Seismic Surveys.* Washington D.C.: U.S. Department of the Interior Minerals Management Service, Alaska OCS Region, 2006.

U.S. Geological Survey. "Geologic Map of Yellowstone National Park." Miscellaneous Investigations Map I-711. Denver: Author, 1972.

U.S. Minerals Management Service. *Environmental Survey of Potential Sand Resource Sites Offshore Delaware and Maryland.* Herndon, Va.: Office of International Activities and Marine Minerals, 2000

Uyeda, Seiya. *The New View of the Earth: Moving Continents and Moving Oceans.* Translated by Masako Ohnuki. San Francisco: W. H. Freeman, 1978.

Van Andel, Tjeerd H. *New Views on an Old Planet: Continental Drift and the History of Earth.* 2d ed. New York: Cambridge University Press, 1994.

Van Beynen, Philip E. *Karst Management.* New York: Springer, 2011.

Van der Voo, Rob. *Paleomagnetism of the Atlantic, Tethys and Iapetus Oceans.* New York: Cambridge University Press, 2004.

Van Dover, Cindy. *The Ecology of Deep-Sea Hydrothermal Vents.* Princeton, N.J.: Princeton University Press, 2000.

Van Schmus, W. R., and W. J. Hinze. "The Midcontinent Rift System." *Annual Review of Earth and Planetary Sciences* 13 (1985): 345-383.

Verbyla, D. L. *Satellite Remote Sensing of Natural Resources.* New York: Lewis, 1995.

Verstraete, Michel M., Robert J. Scholes, and Mark Stafford Smith. "Climate Change and Desertification: Looking at an Old Problem Through New Lenses." *Frontiers in Ecology and the Environment* 7, no. 8 (October 2009): 421-428.

Vickers-Rich, Patricia, and Patricia Komarower, eds. *The Rise and Fall of the Ediacaran Biota.* London: Geological Society, Special Publications, 2007.

Vink, Gregory E., W. J. Morgan, and Peter R. Vogt. "The Earth's Hot Spots." *Scientific American* 242 (April 1985): 50-57.

Vogel, Shawna. *Naked Earth: The New Geophysics.* New York: Dutton, 1995.

Voight, B., ed. *Mechanics of Thrust Faults and Décollement.* Benchmark Papers in Geology 32. Stroudsburg, Pa.: Dowden, Hutchinson and Ross, 1976.

Walker, Bryce S. *Earthquake.* Alexandria, Va.: Time-Life Books, 1982.

Walker, Cyril, and David J. Ward. *Fossils: Smithsonian Handbook.* New York: DK Publishing, 2002.

Walker, Gabrielle. "Great Rivers of Ice." *New Scientist* 162 (April 17, 1999): 54-58.

__________. *Snowball Earth.* New York: Bloomsbury Press, 2004.

Walker, Mike. *Quaternary Dating Methods.* Hoboken, N.J.: John Wiley & Sons, 2005.

Walter, M. R., ed. *Stromatolites.* New York: Elsevier, 1976.

__________. "Understanding Stromatolites." *American Scientist* (September-October 1977): 563-571.

Waltham, Tony. *Great Caves of the World.* Buffalo, N.Y.: Firefly Books, 2008.

Wanless, H. R., and Cynthia R. Wright. *Paleoenvironmental Maps of Pennsylvanian Rocks, Illinois Basin and Northern Midcontinent Region.* Boulder, Colo.: Geological Society of America, 1978.

Ward, Peter. "The Extinction of the Ammonites." *Scientific American* 249 (October 1983): 136-141.

__________. *Out of Thin Air: Dinosaurs, Birds, and the Earth's Ancient Atmosphere*, Washington D.C.: National Academies Press, 2006.

Warner, Terence. *Synthesis, Properties, and Mineralogy of Important Inorganic Materials*. New York: Wiley, 2011.

Waters, Tom. "Return to Krakatau." *Discover* 9 (October 1988): 64-67.

Webb, S. David. "The Great American Biotic Interchange: Patterns and Processes." *Annals of the Missouri Botanical Garden* 93 (2006): 245-257.

Webster, Paul. *The Mighty Sierra: Portrait of a Mountain World*. Palo Alto, Calif.: American West, 1972.

Wegener, Alfred. *The Origin of Continents and Oceans*. Translated from the 4th rev. German ed. by John Biram. Mineola, N.Y.: Dover, 1966.

Weide, David L., and Marianne L. Faber, eds. *This Extended Land: Geological Journeys in the Southern Basin and Range*. Las Vegas: University of Nevada Press, 1988.

Weiner, Jonathan. "Glacier Bubbles Are Telling Us What Was in Ice Age Air." *Smithsonian* 20 (May 1989): 78-87.

__________. *Planet Earth*. Toronto: Bantam Books, 1986.

Weinstock, Jaco, et al. "Evolution, Systematics, and Phylogeography of Pleistocene Horses in the New World: A Molecular Perspective." *PLoS Biology* 3, no. 8 (2005): 1373-1379.

Weishampel, David B., Peter Dodson, and Halszka Osmolska, eds. *The Dinosauria*. 2d ed. Berkeley: University of California Press, 2004.

Wellnhofer, P. *Illustrated Encyclopedia of Pterosaurs*. London: Salamander, 1991.

Wenk, Hans-Rudolf, and Andrei Bulakh. *Minerals: Their Constitution and Origin*. New York: Cambridge University Press, 2004.

Wertenbecker, W. "Land Below, Sea Above." In *Mysteries of the Deep*, edited by Joseph J. Thorndike. New York: American Heritage, 1980.

West, Frederick H., ed. *American Beginnings*. Chicago: University of Chicago Press, 1996.

White, D. E. "Some Principles of Geyser Activity Mainly from Steamboat Springs, Nevada." *American Journal of Science* 265 (1967): 641-684.

White, M. E. *The Greening of Gondwana*. 3rd ed. Kenthurst, N.S.W.: Rosenberg Publishing, 1998.

White, S. E. "Popocatépetl: The Ever-Burning Torch." *Volcano News* 17 (1984): 1-3.

White, William B. *Geomorphology and Hydrology of Karst Terrains*. New York: Oxford University Press, 1988.

Whittaker, Robert J., and Jose Maria Fernandez-Palacios. *Island Biogeography: Ecology, Evolution, and Conservation*. New York: Oxford University Press, 2007.

Whittington, H. B. *The Burgess Shale*. New Haven, Conn.: Yale University Press, 1985.

Wicander, Reed, and James S. Monroe. *The Changing Earth: Exploring Geology and Evolution*. 6th ed. Belmont, Calif.: Brooks/Cole, 2011.

__________. *Historical Geology: Evolution of Earth and Life through Time*. 6th ed. Belmont, Calif.: Brooks/Cole, Cengage Learning, 2010.

Wiens, John J. "Paleontology, Genomics, and Combined-Data Phylogenetics: Can Molecular Data Improve Phylogeny Estimation for Fossil Taxa?" *Systematic Biology* 58, no. 1 (2009): 87-99.

Wilderness Press. *San Gabriel Mountains*. 5th ed. Berkeley, Calif.: Wilderness Press, 2005.

Wiley, E. O., and Bruce S. Lieberman. *Phylogenetics: Theory and Practice of Phylogenetic Systematics*. Hoboken, N.J.: Wiley-Blackwell, 2011.

Wilford, John Noble. *The Riddle of the Dinosaur*. New York: Alfred A. Knopf, 1985.

Willett, Sean D., and Mark T. Brandon. "On Steady States in Mountain Belts." *Geology* 30 (2002): 175-178.

Williams, Howell, and A. R. McBirney. *Volcanology*. San Francisco: Freeman Cooper, 1979.

Williams, P. J., and M. W. Smith. *The Frozen Earth: Fundamentals of Geology*. New York: Cambridge University Press, 1991.

Willis, K. J., and J. C. McElwain. *The Evolution of Plants*. New York: Oxford University Press, 2002.

Wilson, J. Tuzo. "Evidence from Islands on the Spreading of the Ocean Floor." *Nature* 197 (1963): 536-538.

Wilson, Stewart, and Gar Rothwell. *Paleobotany and the Evolution of Plants*. 2d ed. New York: Cambridge University Press, 2009.

Winchester, Simon. *The Map That Changed the World*. London: Penguin, 2001.

Windley, Brian F. *The Evolving Continents*. 3rd ed. New York: Wiley, 1995.

Wise, James M. Mount Whitney to *Yosemite: the Geology of the John Muir Trail*. Scotts Valley, Calif.: CreateSpace, 2008.

Witze, Alexandra. "Death of a Continent, Birth of an Ocean." *Science News* 180 (2011): 22-25.

Woese, Carl R. "Archaebacteria." *Scientific American* 244 (June 1981): 98-122.

Wolfe, Edward W. "The 1991 Eruptions of Mount Pinatubo, Philippines." *Earthquakes and Volcanoes* 23, no. 1 (1992): 5-35.

Wood, Charles A., and Jurgen Kienle, eds. *Volcanoes of North America: U.S.A. and Canada.* New York: Cambridge University Press, 1990.

__________. *Volcanoes of North America: United States and Canada.* New York: Cambridge University Press, 1993.

Woodburne, Michael O., ed. *Late Cretaceous and Cenozoic Mammals of North America: Biostratigraphy and Geochronology.* New York: Columbia University Press, 2004.

Worsnop, Richard L. "Earthquake Research." *CQ Researcher* 4 (December 16, 1994): 1105-1128.

Wright, H. E., and D. G. Fry, eds. *The Quaternary of the United States.* Princeton, N.J.: Princeton University Press, 1965.

Wright, V. P., and Alfred Fischer, eds. *Paleosols: Their Recognition and Interpretation.* Princeton, N.J.: Princeton University Press, 1986.

Wyld, Sandra J., and James E. Wright. 'New Evidence for Cretaceous Strike-Slip Faulting in the United States Cordillera and Implications for Terrane-Displacement, Deformation Patterns, and Plutonism." *American Journal of Science* 301 (2001): 150-181.

Wyllie, Peter J. *The Way the Earth Works: An Introduction to the New Global Geology and Its Revolutionary Development.* New York: Wiley, 1976.

Yaacov, Davidov, and Eduard Jurkevitch. "Predation Between Prokaryotes and the Origin of Eukaryotes." *BioEssays* 31 (2009): 738-757.

Yanko-Hombach, Valentina, et al., eds. *The Black Sea Flood Question: Changes in Coastline, Climate, and Human Settlement.* Dordrecht: Springer, 2007.

Yarus, Michael. *Life from an RNA World: The Ancestor Within.* Cambridge, Mass.: Harvard University Press, 2010.

Yeats, Robert S., Kerry Sieh, and Clarence R. Allen. *The Geology of Earthquakes.* New York: Oxford University Press, 1997.

Yirgu, G., C. J. Ebinger, and P. K. Maguire, eds. The *Afar Volcanic Province within the East African Rift System.* London: Geological Society, Special Publications, 2006.

Yockey, Hubert P. *Information Theory, Evolution, and The Origin of Life.* New York: Cambridge University Press, 2011.

Young, Emma. "Earth's Earliest Life so Far, and Lots of It." *New Scientist* 190 (2006): 10.

Yuen, David A., et al., eds., *Superplumes: Beyond Plate Tectonics.* New York: Springer, 2007.

Zhamaletdinov, A. A. "The New Data on the Structure of the Continental Crust Based on the Results of Electromagnetic Sounding with the Use of Powerful Controlled Sources." *Doklady Earth Sciences* 438 (June 2011): 798-802.

Zimmer, Carl. *At the Water's Edge: Macroevolution and the Transformation of Life.* New York: Free Press, 1998.

EARTHQUAKES IN HISTORY

Year	Region	Magnitude m—Richter Scale units	Plate/Plate Boundaries	Impact
464 B.C.E	Sparta, Greece	Unknown	Eurasian/African	20,000 people were reported dead (although scientists believe this figure to be inflated). Increased tensions between Sparta and Athens, leading to the first Peloponesian War.
62	Pompeii, Roman Empire	Unknown	Eurasian/African	No casualty or damage reports available but half of city destroyed. Seventeen years later, the city was destryoed in the eruption of Mt. Vesuvius.
856	Damghan, Iran	Unknown	Arabian/Eurasian	Approximately 200,000 deaths.
1138	Aleppo, Syria	Unknown	Arabian/African	Approximately 230,000 deaths.
1268	Silicia, Asia Minor	Unknown	Eurasian/Arabian	Approximately 60,000 killed.
1290	Chihli, China	Unknown	Indian/Eurasian	Approximately 100,000 deaths.
1556	Shannxi, China	8m	Eurasian/Yangtze	Estimated 830,000 deaths with massive and widespread damage.
1667	Shemakha, Caucasus	Unknown	Eurasian/Arabian	Approximately 80,000 deaths.
1693	Sicily, Italy	7.5m	Eurasian/African	Approximately 60,000 killed.
1727	Tabriz, Iran	Unknown	Eurasian/Arabian	Approximately 77,000 deaths.
1755	Lisbon, Portugal	8.7m	Eurasian/African	Estimated 70,000 deaths with massive tsunami and widespread fires in Lisbon. Earthquake was felt as far away as the United States. Sparked scientific interest in seismology.
1755	Cape Ann, Massachusetts (USA)	Unknown	North American	No reported deaths. Some believe there is a correlation between this quake and the Lisbon earthquake, which took place seventeen days earlier.
1783	Calabria, Italy	Unknown	Eurasian/African	Estimated 50,000 deaths.
1797	Quito, Ecuador	7.1m	South American/ Nazca	Estimated 41,000 people killed.
1811	New Madrid, Missouri (USA)	7.7m	North American	Minimal number of casualties. Mississippi River course altered in several spots. Quake believed to be caused by friction between the fluid mantle and a previously unknown faultline rather than plate boundary activity.
1857	Fort Tejon, California (USA)	8m	Pacific/North American	One person killed. Severe damage to Army post near San Andreas Fault.
1906	Ecuador	8.8m	South American/ Nazca	At least 500-1500 killed. Tsunami reached as far north as Alaska and as far west as Japan. Produced four tsunamis.

Year	Region	Magnitude m—Richter Scale units	Plate/Plate Boundaries	Impact
1906	San Francisco, California (USA)	7.7-8.3m	Pacific/North American	Approximately 3,000 killed. Massive damage Resulting fire destroyed a large part of San Francisco.
1908	Messina, Italy	7.2m	Eurasian/African	Estimated 72,000 deaths with a tsunami reachng as high as 39 feet. Aftershocks were felt for five years.
1920	Haiyuan, Ningxia, China	7.8m	Eurasian	Estimated 200,000 killed. Widespread surface faults and landslides.
1922	Chile-Argentina	8.5m	South American/ Nazca	Approximately 100 killed. A nearly 7-foot tsunami damaged boats in Hilo, Hawaii.
1923	Kanto, Japan	7.9m	Philippine Sea/ Honshu	Approximately 143,000 deaths. Massive damage, including Great Tokyo Fire.
1923	Kamchatka (USSR)	8.5m	Pacific/Okhotsk	No casualty or damage reports available.
1938	Banda Sea, Indonesia	8.5m	Australian/India/ Banda Sea	No deaths reported. Tsunami occurred, creating massive damage from waves.
1939	Erzincan, Turkey	8m	Eurasian/Anatolian	Approximately 33,000 killed, with small tsunami generated. Massive damage in Erzinkan Plain and Kelkit River Valley reported.
1948	Ashgabat, Turkmenistan (USSR)	7.3m	Eurasian/Arabian	Approximately 110,000 deaths, with whole villages leveled.
1950	Assam, Tibet (China)	8.6	Indian/Eurasian	Estimated 780 people killed, with seventy villages destroyed by landslides.
1952	Kamchatka (USSR)	9m	Pacific/Okhotsk	No reported deaths but tsunami caused major damage in Hawaii.
1957	Andreanof Islands, Alaska (USA)	8.6m	North American/ Pacific	No reported deaths. Multiple damaging tsunamis generated, with some traveling as far as Japan and Chile.
1960	Chile	9.5m	South American/ Nazca	Approximately 1,700 killed. Large tsunami caused significant damage as far away as Japan and the Philippines. Largest earthquake in recorded history.
1963	Kuril Islands (USSR)	8.3m	Pacific/Okhotsk	No casualty or damage reports available.
1964	Prince William Sound, Alaska (USA)	9.2m	North American/ Pacific	128 killed, 115 of whom were killed by a subsequent tsunami. Anchorage sustained major damage.
1965	Rat Islands, Alaska (USA)	8.7m	North American/ Pacific	No deaths reported. Cracks found in roads and buildings. 35-foot tsunami generated.
1970	Chembote, Peru	7.9m	Pacific/Nazca	Estimated 70,000 deaths. 20,000 killed as an avalanche buried an entire town.
1976	Tangshan, China	7.5m	Indian/Eurasian	Betweeen 242,000 and 655,000 deaths.

Year	Region	Magnitude m—Richter Scale units	Plate/Plate Boundaries	Impact
1976	Mindanao, Philippines	7.8m	Eurasian/Philippine	Approximately 8,000 killed.
1985	Mexico City, Mexico	8m	North American/ Cocos	Estimated 9,500 killed. Severe damage to Mexico City and surrounding area. Tsunami and rough waves reported off Mexican coastline after quake.
1989	Loma Prieta, California (USA)	7.1m	Pacific/North American	Sixty-three deaths. $6 billion in property damage (most of which occurred in Oakland and San Francisco). Largest earthquake in that region since the 1906 quake.
1990	Iran	7.4m	Eurasian/Arabian	Estimated 50,000 killed. Widespread damage in villages throughout western Iran.
1990	Rasht, Iran	7.4m	Eurasian/Arabian	Approximately 50,000 killed.
1994	Northridge, California (USA)	6.7m	Pacific/North American	60 people killed. 40,000 buildings damaged in greater Los Angeles.
1999	Izmit, Turkey	6.7m	Eurasian/Anatolian	Over 17,000 killed. Major damage to Istanbul.
2001	Gujarat, India	7.9m	Indian/Eurasian	Over 20,000 killed. Major structural damage throughout northern India and neighboring Pakistan.
2004	Sumatra, Indonesia	9.1m	Sunda/Indian	Third-largest earthquake since 1900. Approximately 228,000 deaths. Devastating tsunami produced.
2005	Pakistan	7.6m	Indian/Eurasian	Approximately 86,000 deaths. Widespread and massive damage. Entire villages leveled.
2005	Sumatra, Indonesia	8.6m	Sunda/Indian	At least 1,000 killed. Nearly a 10-foot tsunami damaged an airport and port area.
2007	Sumatra, Indonesia	8.5m	Australian/Sunda	Estimated 25 killed. Tsunami generated. Major structural damage.
2010	Haiti	7m	Caribbean/North American	Between 100,000 and over 300,000 deaths. Multiple local tsunamis.
2010	Bio-Bio, Chile	8.8m	South American/ Nazca	Approximately 523 deaths reported. Pacific-wide and local tsunamis created. $30 billion in damage caused.
2011	Honshu, Japan	9m	Pacific/North American	At least 15,700 people killed. Massive, Pacific-wide tsunami caused major damage throughout Honshu's eastern coastline.
Source: http://earthquake.usgs.gov/				

GEOLOGICAL TIME SCALE

CENOZOIC

PERIOD		EPOCH	MYA
QUATERNARY	NEOGENE	HOLOCENE	
		PLEISTOCENE	0.01
TERTIARY		PLIOCENE	2.6
		MIOCENE	5.3
	PALEOGENE	OLIGOCENE	23.0
		EOCENE	33.9
		PALEOCENE	55.8
			65.5

MESOZOIC

PERIOD	EPOCH	MYA
CRETACEOUS	LATE	65.5
	EARLY	99.6
JURASSIC	LATE	145.5
	MIDDLE	161
	EARLY	176
TRIASSIC	LATE	201.6
	MIDDLE	235
	EARLY	245
		251

PALEOZOIC

PERIOD	EPOCH	MYA
PERMIAN	LATE	251
	MIDDLE	260
	EARLY	271
CARBONIFEROUS	PENNSYLVANIAN	299
	MISSISSIPPIAN	318
DEVONIAN	LATE	359
	MIDDLE	385
	EARLY	398
SILURIAN	LATE	416
	MIDDLE	423
	EARLY	428
ORDOVICIAN	LATE	444
	MIDDLE	461
	EARLY	472
CAMBRIAN	FURONGIAN	488
	SERIES 3	501
	SERIES 2	510
	TERRENEUVIAN	521
		542

PRE-CAMBRIAN

EON	ERA	MYA
PROTEROZOIC	NEOPROTEROZOIC	542
	MESOPROTEROZOIC	1000
	PALEOPROTEROZOIC	1600
ARCHEAN	NEOARCHEAN	2500
	MESOARCHEAN	2800
	PALEOARCHEAN	3200
	EOARCHEAN	3600
HADEAN		3850

GLACIAL RETREAT

Glacier Name	Location	Present Elevation (feet)	Past Surface Area (Year) (square feet)	Present Surface Area (square feet)
Hubbard	Alaska, USA	4,239.00	1,532.40	1,350.00
Perito Moreno	Argentina	180.00	97.00	97.00
Vatnajokull	Iceland	6,963.00	unknown	5,033.10
Kilimanjaro	Tanzania	19,340.00	4.6 (1912)	0.80
Siachen	India/Pakistan	18,480.00	132 (1909)	131.50
Heard Island	Australia	9,059.00	177.30 (1940)	159.70
Lambert	Antarctica	8,202.00	unknown	15,000.00
Pasterze	Austria	12,457.00	14 (1889)	12.29
Yulong	China	12,000.00	5.21 (1984)	5.00
Franz Joseph	New Zealand	8,858.00	9,900.00	14.29
Athabasca	Canada	9,240.00	16.25 (1870)	6.50
Biafo	Pakistan	18,150.00	unknown	39.10
Pine Island	Antarctica	2,567.40	100,849.00	67,954.00
Hispar	Pakistan	16,824.00	unknown	96.25
Fedchenko	Tajikistan	20,790.00	97.52 (1933)	616.40
Aletsch	Switzerland	10,362.00	53.46	53.90
Purog Kangri	China	21,390.60	unknown	262.20
Mer de Glace	France	8,448.00	21.99 (1650)	20.56
Jostedal	Norway	6,420.60	unknown	188.00
Malaspina	Alaska, USA	1,575.00	1566.00 (1962)	1,500.00
Greenland Ice Sheet	Greenland	12,210.00	852800 (1979)	656,000.00
Langjökull	Iceland	4,290.00	unnown	590.30
Pio XI	Chile	11,154.00	786.00	786.00
Kolka	Russia	11,154.00	unknown	4.47
McGinnis	Alaska, USA	13,912.00	unknown	429.00
Petermann	Greenland	1,485.00	120.9 (1962)	111.90
Rhone	Switzerland	9,702.00	18.38 (1880)	17.38
Fox	New Zealand	6,270.00	21.55	21.55
Columbia	Alaska, USA	3,729.00	unknown	679.50
Strongbreen	Norway	990.00	unknown	150.99

Accurate estimates as of Spring 2012
Sources:
http://pubs.usgs.gov/fs/fs-001-03/
http://nsidc.org/data/glacier_inventory/query.html
http://www.nasa.gov/multimedia/imagegallery/image_feature_553.html
http://nsidc.org/cryosphere/quickfacts/icesheets.html
http://pubs.usgs.gov/pp/p1386k/pdf/10_1386K_alaskarange.pdf

TECTONIC PLATES/CRUST

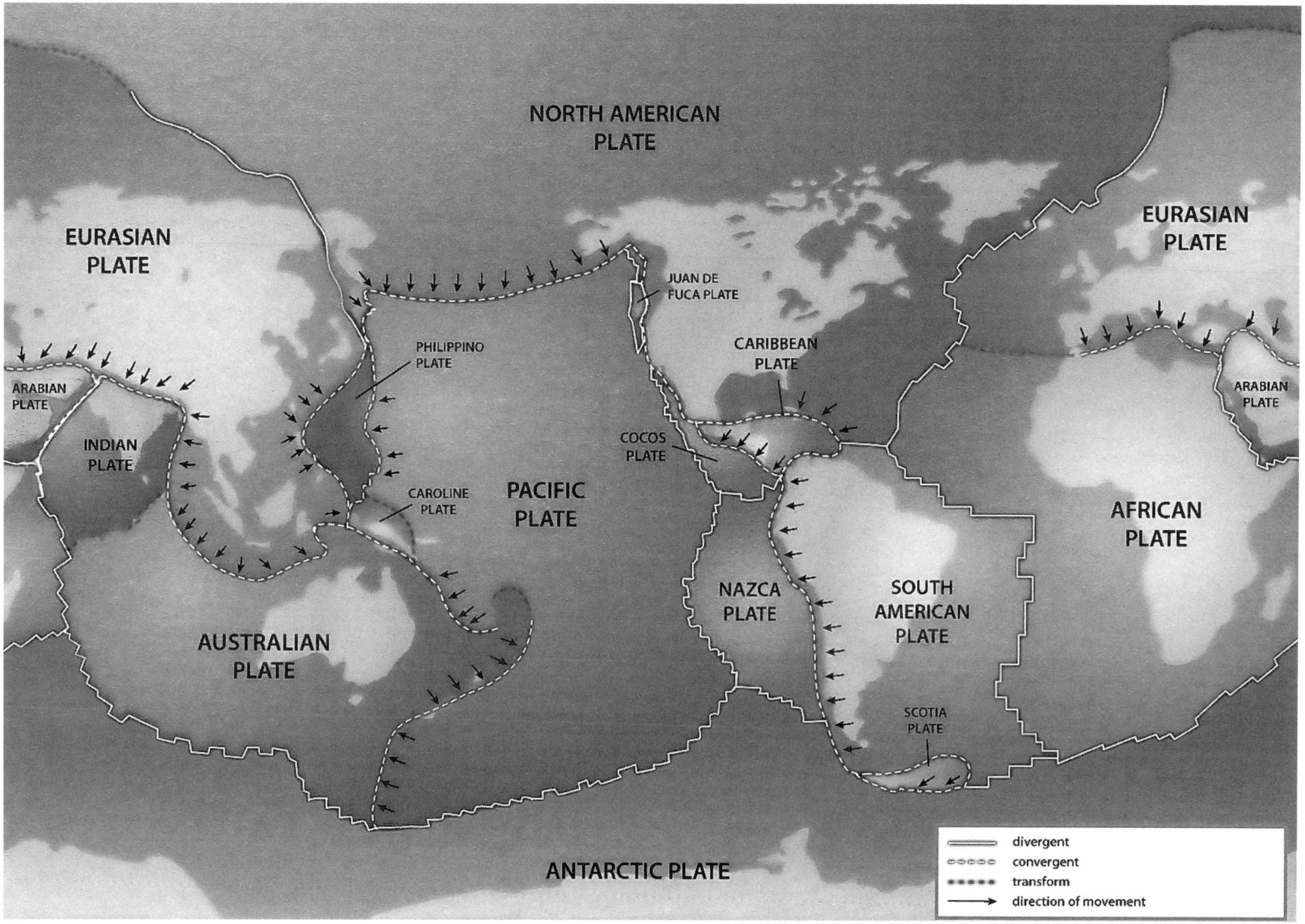

SUBJECT INDEX

Note: Page numbers in bold indicate main discussion.

G

H

I

U